EXPLOSION HAZARDS
AND EVALUATION

FUNDAMENTAL STUDIES IN ENGINEERING

FUNDAMENTAL STUDIES IN ENGINEERING 5

EXPLOSION HAZARDS AND EVALUATION

W.E. BAKER

Department of Energetic Systems, Southwest Research Institute, San Antonio, TX, U.S.A.

P.A. COX

Department of Engineering Mechanics, Southwest Research Institute, San Antonio, TX, U.S.A.

P.S. WESTINE

Department of Energetic Systems, Southwest Research Institute, San Antonio, TX, U.S.A.

J.J. KULESZ

Nuclear Division, Union Carbide Corporation, Oak Ridge, TN, U.S.A.

R.A. STREHLOW

College of Engineering, Department of Aeronautical and Astronautical Engineering, University of Illinois at Urbana-Champaign, Urbana, IL, U.S.A.

ELSEVIER SCIENTIFIC PUBLISHING COMPANY
Amsterdam — Oxford — New York   1983

ELSEVIER SCIENTIFIC PUBLISHING COMPANY
1, Molenwerf
P.O. Box 211, 1000 AE Amsterdam, The Netherlands

Distributors for the United States and Canada:

ELSEVIER SCIENCE PUBLISHING COMPANY INC.
52, Vanderbilt Avenue
New York, N.Y. 10017

Library of Congress Cataloging in Publication Data
Main entry under title:

Explosion hazards and evaluation.

     (Fundamental studies in engineering ; 5)
     Includes bibliographies and index.
     1. Explosions. I. Baker, W. E. II. Series.
QD516.E95 1982   363.3'3   82-8889
ISBN 0-444-42094-0 (U.S.)

ISBN 0-444-42094-0 (Vol. 5)
ISBN 0-444-41756-7 (Series)

Printed in The Netherlands

PREFACE

The precursor to this book was a set of bound course notes which the authors prepared for teaching short courses on explosion hazards evaluation. The notes were first prepared in June 1978, and were extensively revised in April 1980. Because of enthusiastic response to the course, both in the United States and abroad, the authors were encouraged to undertake the effort of converting the notes to the more coherent and consistent format of a reference text. Southwest Research Institute sponsored this conversion with internal research funds, with the proviso that a publisher be found for the revised text.

Knowledge of explosion hazards and ways to control or mitigate these hazards is expanding rather rapidly, with significant research and evaluation studies underway in many parts of the world. To attempt to "freeze" and describe the state of this knowledge at any one time is difficult, but that is what we have tried to do in this book.

The material in this book will hopefully be used by engineers, scientists and plant safety personnel in their own assessments of explosion hazards, investigation of accidental explosions, and design for explosion effects resistance or mitigation. We have tried to be quite comprehensive in our coverage, but still emphasize applications by including a number of example problems which illustrate the use of prediction graphs or equations. (Many of these problems are adapted from class problems worked or suggested by short course students.)

The book is organized into nine chapters, an extensive bibliography, and several appendices. The first two chapters discuss the energy release processes which generate accidental explosions, and the resulting development of pressure and shock waves in a surrounding atmosphere. The manner in which the "free-field" waves are modified in interacting with structures or other objects in their paths is discussed in Chapter 3. Structural response to blast loading and non-penetrating impact is covered in two chapters, with Chapter 4 including simplified analysis methods and Chapter 5 including numerical methods. Chapter 6 includes a rather comprehensive treatment of generation of fragments and missiles in explosions, and the flight and effects of impact of these objects. Large chemical explosions can cause large fireballs, and thermal radiation from the fireballs is the topic of Chapter 7. Explosions may or may not cause damage or casualty, and various damage criteria have been developed for structures, vehicles, and people. These criteria are presented in Chapter 8. Some general procedures have evolved for both the postmortem evaluation of accidental explosions and for design for blast and impact resistance. These procedures are reviewed in Chapter 9.

The literature on explosion effects, hazards, and mitigation is extensive, and this is reflected in our extensive bibliography. A special feature of the bibliography is a short annotated list of those references which the authors felt would be most useful to the reader as supplements to this book.

Various aspects of our topic are technically complex. Where the technical discussion becomes quite detailed, we have tried to summarize the results in the text, and supplement with appendices giving the details.

We hope that this work will prove useful to readers with a variety of backgrounds and interests in explosion hazards.

ACKNOWLEDGEMENTS

The preparation of such a voluminous work is impossible without dedi-
cated help and very hard work by a number of individuals other than the
authors. We cannot possibly acknowledge all of those who have helped in
some way, but instead must single out the organizations and individuals who
have helped the most.

We gratefully acknowledge the financial support of the management of
Southwest Research Institute under an internal research grant to defray the
expense of converting a set of loosely-written course notes into a coherent,
well-illustrated text. Individuals who contributed most significantly were:

- Ms. Jenny Decker for typing and correcting all of the final manu-
  script,
- Mr. Victoriano Hernandez for preparing all line drawings and figures,
- Ms. Deborah Stowitts for editing the entire text,
- Ms. Deborah Skerhut for structural response analyses and computations, and
- Ms. Norma Sandoval for conversion of many figures and example
  calculations to metric units.

TABLE OF CONTENTS

LIST OF ILLUSTRATIONS

LIST OF TABLES

INTRODUCTION

To most individuals, the word "explosion" evokes a destructive image.* One thinks of bombs, warheads, a house blowing up from accumulated fuel gas, a serious chemical plant accident, or other violent and destructive events. Of course, by far, the most common explosions are not destructive at all, but are planned and controlled to do useful work for man. The most ubiquitous example is the internal combustion engine, with untold millions of explosions occurring every minute around the world to provide the power to transport us, to generate electrical power, run many of our power tools, and many other uses. Almost as common is the widespread use of explosions in blasting in quarries, earthmoving, and controlled cratering. Some less common constructive uses of controlled explosions are explosive forming of metals, explosive welding, and demolition of buildings. Explosive devices separate stages of our launch vehicles, and cut cables and bolts, at precisely determined times. We could extend this list for many pages, but instead wish only to illustrate that the vast majority of explosions are controlled and benign.

The events which catch our attention are, however, the accidental explosions, which can, and do, cause unplanned destruction, injury and death. Accidental explosions are the topic of this book.

Accidental explosions have occurred and continue to occur as a result of...

- The storage, transportation, and manufacture of explosives,
- Chemical and petrochemical plant operations,
- The failure of high pressure vessels,

---

*See Chapter 2 for definitions of an explosion.

- Explosions of boilers,

- Molten metal contacting water in foundries,

- Fuel gas leaks in buildings,

- Manufacture, transport and storage of high vapor pressure or cryogenic fuels,

- Cleaning of liquid fuel tanks in tanker vessels, and

- Manufacture, storage and handling of combustible dusts.

Concern about accidental explosions and efforts to evaluate and mitigate their effects have increased in recent years because changes in the economy have dictated an increase in the size of manufacturing and storage facilities and the introduction of larger bulk transport containers as exemplified by supertankers. This increased concern about accidental explosions has accelerated the development of evaluation methods and has resulted in the accumulation of a large body of new data relative to accidental explosion processes. At the present time in this rapidly maturing field, one can estimate with some confidence the properties and effects of such explosions, including:

- Combustion processes which lead to or are the direct source of such explosions,

- Blast properties for free-field and contained explosions,

- Transient loads which blast waves apply to nearby structures,

- Trajectories and kinetic energy of explosively generated fragments or missiles,

- Structural damage caused by blast waves and fragment impact,

- Injury or mortality caused by blast and impact.

In addition, a number of rational design procedures have been developed to enhance the blast resistance of buildings and other structures to either an internal or external blast load. One can allow for partial or complete containment of explosive effects of accidents during the handling or storage of particularly hazardous chemicals.

This book covers the full spectrum of problems encountered in assessing the hazards of such accidental explosions, in designing the proper containment as necessary, as well as developing techniques to reduce incidence of accidents during normal plant and transport operations.

CHAPTER 1

COMBUSTION AND EXPLOSION PHENOMENA

INTRODUCTION

While not all types of accidental explosions are caused by combustion or out-of-control exothermic reactions, most of them are. Because o: this, all exothermic compounds or mixtures should be considered hazardous and special precautions should be taken to reduce the incidence of fire and explosion when such materials are manufactured, transported, stored, or used.

Combustion safety, to be properly implemented, must be based on th known combustion hazards of the compounds and processes involved in any particular industrial setting. Thus, the first purpose of this chapter i to present those fundamentals of combustion science that are required to allow an appreciation of the combustion hazards inherent to industrial en vironments. This introduction to fundamentals will be followed by an ex- tensive discussion of the practical problems involved.

The discussion of practical problems will be divided into the foll ing categories:

(1) A discussion of the dynamics of combustion explosions includir recent experimental and theoretical work.

(2) A discussion of combustion and explosives safety including a ‹ scription and critique of standardized tests that have evolve‹ over the years to characterize the relative danger of handlin different compounds in different environments.

(3) A discussion of risk management and preventive measures inclu ing the use of national codes and various uncodified common sense approaches to protecting personnel, equipment, and builu ings. This section will also reference data banks that contain information which is useful when implementing safety procedures.

FUNDAMENTALS

Thermochemistry of Combustion

Every chemical reaction has a heat of reaction which is defined as follows. First write a balanced chemical equation for the reaction in question. For example, the equation for the complete or stoichiometric oxidation of a general CHONS fuel can be written as follows:

$$C_u H_v O_w N_x S_y + (u + \frac{v}{4} - \frac{w}{2} + y) \, O_2 \rightarrow$$

$$(1-1)$$

$$u CO + \frac{v}{2} H_2 O + \frac{x}{2} N_2 + y \, SO_2$$

In this case, the heat of reaction is called the heat of combustion, ΔH_c, and is defined as the enthalpy that needs to be added to the system when $C_u H_v O_w N_x S_y$ plus $(u + \frac{v}{4} - \frac{w}{2} + y) \, O_2$ at some initial pressure and reference temperature, θ_o, reacts completely to produce carbon dioxide, water, nitrogen and sulphur dioxide at the same initial pressure and reference temperature. The u, v, w, x, y subscripts in Equation (1-1) either represents the empirical formula of a pure fuel, in which case ΔH_c is for one mole (note that throughout this text, the mole is defined as 6.023×10^{23} molecules; in other words, the gram mole is used and the molecular weight of carbon is defined as 0.012 kg) of that fuel, or they are obtained from the ultimate analysis of the fuel (units, percent by weight of the elements in the fuel) and they represent the number of atoms of each element per kg of fuel. In that case, ΔH_c has units of J/kg fuel.

Since the combustion reactions are highly exothermic, the systems must reject heat if the products are to be at θ_o. Therefore, ΔH for such a combustion reaction is a negative number (note that it is reported as a positive number in the handbooks). Also, since water can be assumed to be either a liquid or ideal gas at the end state, there are two values of

ΔH_c that are normally reported; a high value when liquid water is formed and a low value when gaseous water is the end product. The difference between the high and low heats of combustion for any substance is just the heat of vaporization of water. This equals 44.0 kJ/mol at 25°C.

For any general hydrocarbon mixture (an oil, for example), one must know both an ultimate analysis (i.e., the elemental composition) and a heat of combustion. In general, these must be determined experimentally. Ordinarily for solids and liquid fuels, the ultimate is given on a weight percent basis and the heat of combustion is given per unit weight of fuel. For gaseous fuels, the composition is ordinarily given by listing the component gases and their volume percent. Thus, for these fuels, the heat of combustion can be calculated from the analysis.

For any general reaction, the heat of reaction is defined in the same manner. One must first specify a stoichiometric relationship and, once this is done, one can write down a heat of reaction. These heats of reaction are additive just as the stoichiometric equations are.

$$CO + 1/2\ O_2 \rightarrow CO_2\ ;\ \Delta H_1$$

$$+\ [C + 1/2\ O_2 \rightarrow CO\ ;\ \Delta H_2]$$

$$= C + O_2 \rightarrow CO_2\ ;\ \Delta H_3 = \Delta H_1 + \Delta H_2$$

This means that one does not have to tabulate the heats of all possible reactions but can instead tabulate only a limited set from which all others may be derived. The standard set consists of the reactions of formation of one mole of a species from the requisite elements in their idealized standard start at the temperature of interest. For example, gas phase water has a heat of formation at 25°C of

$$H_2(g) + 1/2\ O_2(g) \rightarrow H_2O(g)\ \Delta H_f = -241.99\ kJ/mole$$

and for ethylene at 25°C, we write

$$2C(s) + 2H_2(g) \rightarrow C_2H_4(g)\ \Delta H_f = +52.502\ kJ/mole$$

Note that this compound is endothermic relative to the elements. It is well known that the decomposition of ethylene to form solid carbon and hydrogen gas is an exothermic process.

Thus, the heat of reaction of any general reaction can be written as

$$(\Delta H_r)_j = \sum_{i=1}^{s} \nu_{ij}\ (\Delta H_f)_i$$

where ν_{ij} is the stoichiometric coefficient of the ith species in the jth reaction. Here, if a species appears on the left side of the jth reaction, ν_{ij} is taken to be a negative number and if the species appears on the right side, it is taken to be positive. One example should suffice. Consider the stoichiometric equation for water-gas equilibrium:

$$CO + H_2O \rightarrow CO_2 + H_2\ @\ 1500°K$$

$$\Delta H_r = -1(\Delta H_f)_{CO} - 1(\Delta H_f)_{H_2O} + (\Delta H_f)_{CO_2} + 1(\Delta H_f)_{H_2}$$

$$= -1(115.292) - 1(-250.471) + 1(-395.912) + 1(0)$$

$$= -30.149\ kJ/unit\ of\ reaction\ as\ written$$

8

The last term in the equation is + 1(0) because the element hydrogen is a standard state element and, therefore, its heat of formation is zero.

The numbers for all the above examples have been taken from Stull and Prophet (1972) and converted to SI units. These JANAF thermochemical tables with supplements contain a self-consistent set of thermodynamic properties from 0 to 6000°K for many compounds. In addition, Stull, et al. (1969), contains thermodynamic data consistent with JANAF on many additional compounds over a more limited temperature range.

The heat of explosion of a propellant or explosive compound or mixture is normally determined by exploding or burning the material in a bomb in an inert atmosphere. It is normally less than the heat of combustion because explosives and propellants are usually oxygen deficient. Thus, the final state that is attained during such a measurement is not the fully oxidized state of the substance.

Runaway Exothermic Reactions[+]

1. The Adiabatic Thermal Explosion

Any chemical system that is undergoing a homogeneous exothermic chemical reaction, if physically isolated from the surroundings, will eventually explode. Fundamentally, there are two limit classes of explosions for such an isolated system. The first of these is the purely thermal explosion and the second is the chain branching or purely chemical explosion. While most high temperature gas phase explosions are chain branching in nature, this class of explosions will not be discussed here. This is because the majority of explosions which result in hazardous behavior in industrial environments can be adequately treated using thermal explosion theory. The reader is referred to Strehlow (1968a) for a discussion of all types of adiabatic and vessel explosions.

In all practical systems, there is heat loss during the induc-

---

[+] A good discussion of chemical reactors is given by Denbigh and Turner (1971).

tion period of the explosion development process. However, it is worthwhile to consider the limit case of a pure thermal explosion before discussing the effects of heat loss. Therefore, consider an isolated system which
may be either heterogeneous or homogeneous, that is undergoing an exothermic
chemical reaction. Assume that the overall rate expression for the formation of product by the reaction is

$$\frac{d[P]}{dt} = A[C_1]^n [C_2]^m \exp(-E/R\theta) \tag{1-1}$$

where [] represents concentration, P represents product, C_1 and C_2 represent reactants, n and m are exponents on the composition, A is a pre-expo-
nential factor and E is the Arrhenius constant for the exponential temperature dependence of reaction rate.[+] Further, assume that Q is the heat of
reaction per mole of the product, P, that is produced by the reaction.
Since the system is thermally isolated and assumed to be held at constant
volume, the temperature time history obeys the equation

$$C_v \rho \frac{d\theta}{dt} = -Q \frac{d[P]}{dt} \tag{1-2}$$

where C_v is the heat of capacity at constant volume, ρ is the density, and
the negative sign appears because the quantity Q is a negative number for
an exothermic process. We now must make one further assumption: that Q
is very much larger than $C_v \rho$. This assumption is generally valid for most
exothermic chemical reactions. Because of this, an isolated exothermic reaction will exhibit a very large temperature rise after only a very small
portion of the reactive material has produced products. This means that
one can assume that the concentration of the reactants is constant during

---

[+]The values of n, m, A and E are normally determined by experiment [see
Seery and Bowman (1970) for example].

any temperature excursion. With this assumption, one obtains the equation

$$\frac{d\theta}{dt} = \lambda \exp(-E/R\theta) \qquad (1-3)$$

where λ is a positive constant.

$$\lambda = \frac{-A[C_1]^n [C_2]^m Q}{C_v \rho} \qquad (1-4)$$

Integrating from time $t = 0$, one obtains

$$\lambda t = \int_{\theta_o}^{\theta} \exp(E/R\theta) \, d\theta \qquad (1-5)$$

The integral on the right hand side of Equation (1-5) may be evaluated by a change of variable and then integration by parts to yield the expression

$$\frac{t}{\beta} = 1 - \left(\frac{\theta}{\theta_o}\right)^2 \left[1 + \frac{2R\theta_o}{E}\left(\frac{\theta}{\theta_o} - 1\right) + \dots\right] \exp\left[\frac{E}{R\theta_o}\left(\frac{\theta_o}{\theta} - 1\right)\right] \qquad (1-6)$$

where β is given by the expression

$$\beta = \frac{R\theta_o^2}{E\lambda} \exp(E/R\theta_o) \qquad (1-7)$$

One evaluates the quantity t/β in Equation (1-6) by assuming a series of temperatures θ above the temperature θ_o and plotting the behavior of t/β. Such an evaluation for two values of $E/R\theta_o$ are shown in Figure 1-1 where t/β is plotted against θ/θ_o.

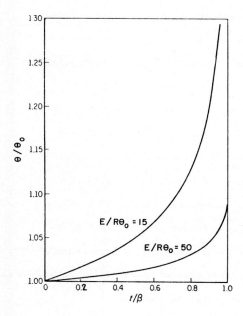

Figure 1-1.   Explosion Behavior of an Isolated System
Undergoing a Pure "Thermal Explosion"

Notice that for small increases in the value of θ over the initial temperature θ_o, the quantity t/β increases rapidly at first and then
increases more slowly as it approaches unity. Since this very simplified
theory is only valid for small extent of reaction, it can only be applied
under conditions where θ/θ_o is near unity. The implication of Figure 1-1
is that when t/β approaches unity, the system explodes. Thus, we can set
the constant β equal to the explosion delay time for an isolated system
undergoing an exothermic reaction

$$t_{ign} = \beta = \frac{R\theta_o}{E\lambda} \exp(E/R\theta_o) \tag{1-8}$$

This theory leads one to the conclusion that any isolated system undergoing an exothermic chemical reaction will always explode with a characteristic time delay given by the constant β in Equation (1-7). Note that the

right-hand side of Equation (1-7) contains the constant λ which is defined in Equation (1-4). Therefore, the delay to explosion is both an exponential function of the temperature and a power function of the concentration of the reactants in the system.

2. Vessel Explosions

When an exothermic reaction is being carried out in a process vessel with external cooling, the system may either "run away" (or "explode") or simply react at a constant rate at a constant temperature depending upon the heat balance in the system. We will assume for simplicity that the system is well stirred and is, therefore, at constant temperature throughout and that heat transfer to surroundings is determined by a heat transfer coefficient to the walls and the actual surface area of the vessel. With the addition of the heat loss term, Equation (1-2) becomes

$$C_v \rho \, \frac{d\theta}{dt} = - Q \, \frac{d[P]}{dt} - \frac{sh}{V} \, (\theta - \theta_o) \qquad (1-9)$$

Here, V is the vessel volume, s is its surface area, h is a heat transfer coefficient, θ is the temperature of the reacting mixture in the vessel and θ_o is the temperature of the wall of the vessel. Before solving Equation (1-9), it is instructive to look at the functional form of the two terms on the right-hand side of this equation. The first term on the right-hand side represents the net rate of heat production due to the chemical reaction in the vessel. We will call this term the chemical gain term. The second term on the right-hand side represents the net loss of heat from the vessel due to conduction through the walls. We will call this the conduction loss term.

Note that the conduction loss term is linearly dependent upon the temperature inside the vessel. The chemical gain term has a functional form which increases as a simple power of initial concentrations of the reactants C_1 and C_2 and also increases rapidly with increasing temperature

due to its exponential temperature dependence. For purposes of argument,
we define the concentration dependent term as $D = [C_1]^n \cdot [C_2]^m$.

A schematic representation of the temperature and pressure sensitivity of these two terms is shown in Figure 1-2. Here we define three
values of D, namely $D_1 > D_{cr} > D_2$. These are chosen relative to the conduction loss curve on Figure 1-2 so that the curves for chemical gain intersect the curve for conduction loss either not at all, once (at a tangency point) or twice as the temperature of the vessel is increased. When
there is no intersection, the curve for chemical gain is always larger than
the curve for conductive loss and the system will always explode because
the temperature will rise without bound. The tangency condition curve,
labelled D_{cr}, represents the highest concentration of reactants at which
conduction loss and chemical gain can just balance each other. All curves
for lower initial concentrations intersect twice and from stability arguments, one can show that the lower intersection is the dynamically stable

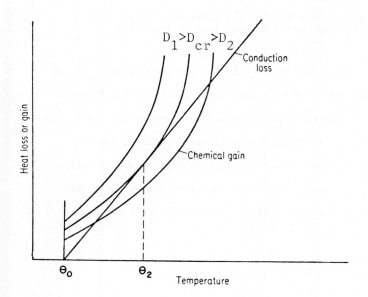

Figure 1-2. Heat Losses and Gains in a Finite Vessel With
Exothermic Reaction. In This Case $D_1 > D_{cr} > D_2$.
Thermal Explosion Only

14

intersection. In other words, if the reactant concentrations are below a certain critical value in this system, the temperature in the vessel will rise somewhat above the wall temperature and will be maintained at that temperature. Under these circumstances, after some initial transient period, the reactor is acting stably and the exothermic chemical reaction is preceeding at some rate which is relatively constant with time. These three behaviors are shown schematically in Figure 1-3, as a function of time after a batch reactor is charged.

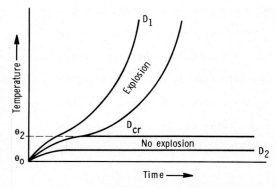

Figure 1-3. The Temperature Time History for a Reactor Containing an Exothermic Reaction and Cooled by Heat Loss to the Walls

As shown in Figure 1-2, we will define the temperature θ_2 to be the temperature at which the conductive loss curve and the chemical gain curve are just tangent. This represents the maximum temperature inside the reactor for a specified set of initial concentrations and specified wall temperature, θ_o, for which stable chemical reaction will occur. In order to find the value of θ_2, we must equate both the heat loss and heat gain term and also equate their slopes. This is shown in the next two equations.

Gain equal to loss

$$-Q \frac{d[P]}{dt} = \frac{sh}{V} (\theta - \theta_o) \Big|_{\theta = \theta_2} \tag{1-10}$$

Tangency requirement

$$\frac{d}{dt}\left\{-Q\frac{d[P]}{dt}\right\} = \frac{d}{dt}\left[\frac{sh}{V}(\theta - \theta_o)\right]\Bigg|_{\theta = \theta_2} \tag{1-11}$$

Substituting the kinetic rate equation [Equation (1-1)] and solving these equations simultaneously yields the equation

$$\frac{R\theta_2^2}{E} = \theta_2 - \theta_o \tag{1-12}$$

or

$$\theta_2 = \frac{1 \pm (1 - 4R\theta_o/E)^{1/2}}{2R/E} \tag{1-13}$$

We choose the minus sign to obtain the lower value of θ_2 since our analysis of the pure thermal explosion showed that the reaction rate for an explosion in an isolated system became very rapid after only a slight temperature rise. Once again, we assume that $E/R\theta_o$ is very much greater than one and, therefore, that the expression for θ_2 may be expanded to yield

$$\theta_2 = \theta_o\left[1 + \frac{R\theta_o}{E}\right] \tag{1-14}$$

Substituting into Equation (1-9) for the condition that $d\theta/dt = 0$ yields the relationship for the explosion limit for a thermal explosion in a vessel

$$\frac{E}{R\theta_2} = \ln\left[\frac{-EQVA\,[C_1]^n\,[C_2]^n}{shR\theta_o^2}\right] \tag{1-15}$$

The quantity on the right-hand side of Equation (1-15) which is in the logarithmic bracket, is actually a positive quantity because the heat of reaction Q is a negative number. Equations (1-14) and (1-15) contain all the
parameters which are important to its heat loss properties. For any particular process vessel, these equations may be used to evaluate the safety
of the design operating conditions for that particular vessel. Specifically, if in the evaluation, the right-hand side of Equation (1-15) is considerably smaller than the left-hand side, the vessel is safe to operate. As
the two sides approach each other, one approaches the incipient explosion
conditions for that particular process vessel. This is sometimes called
the "point of no return" [Townsend (1977)]. If the vessel is a batch process vessel, the initial conditions of concentrations and temperatures
should be used in the evaluation. If it is a continuous flow vessel, the
actual concentration of the reactants in the vessel and the operating temperature of the vessel should be used.

The rate of pressure rise in the vessel during runaway conditions can be evaluated using Equation (1-9) with known kinetic rates for
that particular process. The maximum rate of runaway can be obtained by
assuming that the vessel is actually adiabatic during the runaway process.
In that case, one would use Equation (1-3) to determine the rate of temperature rise in the vessel. The rate of temperature rise as determined
using either of these equations can then be converted into a rate of pressure rise by using the known physical properties of the substance that is
reacting in the vessel. If the vessel contains a liquid system, the rate
of pressure rise will be determined by the vapor pressure of the liquid.
In this case, the rate of energy production in the vessel can be used to
determine the rate of pressure rise and the point at which the pressure
will start to rise catastrophically. If the reactor contains a gaseous
mixture, then either an ideal or some nonideal equation of state for the
gas can be used to calculate pressure rise as a function of the temperature rise. In any case, a knowledge of the rate of the pressure rise dur-

ing an upset or runaway reaction is necessary if one is to determine the proper vent size to be used so that the vessel's integrity will not be lost during the upset. Venting calculations are discussed in Chapter 3. Test techniques to evaluate the possibility of runaway explosions, particularly as applied to propellants or high explosives, will be discussed later in this chapter.

3. Spontaneous Combustion

Another class of runaway exothermic reactions is commonly categorized as spontaneous combustion. Spontaneous combustion occurs when solid organic materials are stored under circumstances where the pile of material has restricted access to fresh air. Under these circumstances, slow oxidation of the organic material by the air produces heating in the interior of the pile, and, if conditions relative to air circulation inside of the pile are correct, can lead to runaway combustion in the central region of the pile. Extensive studies of the spontaneous combustion of coal piles were completed in the mid 1920's. The reader is referred to a review by Davis and Reynolds (1929) for a summary of this work. Additionally, it is well known that cellulosic material such as wood chips, grain dust, etc., are more prone to undergo spontaneous combustion when they are damp.

The process is sufficiently complex that no general analytical guidelines are available to predict the onset of spontaneous combustion. However, external heat sources such as steam lines should not be allowed to contact any pile stored in the open. Also, it is well known that severely restricting the availability of air to the pile can delay and possibly even stop the onset of spontaneous combustion. Furthermore, good housekeeping will always lower the incidence of spontaneous combustion in plant operations that involve the handling of dusty materials such as pulverized coal, grains or pharmaceuticals. An interesting discussion of appropriate precautions is presented in Tuck (1976).

Waves in Premixed Gaseous Systems

1. Laminar Flames

A large collection of mixtures and pure substances which are capable of sustaining a homogeneous exothermic reaction will also support a subsonic "reaction wave" (or flame) which is self-propagating and has definite properties that are dependent only on the initial conditions in the gas. These premixed flames are always rather thick (a few tenths of a mm) and, thus, can be markedly influenced by curvature effects. Furthermore, they can be inherently unstable as one-dimensional waves and, because of their low subsonic velocity, they usually interact very strongly with flow aerodynamics and with buoyancy forces. Two primary properties of such flames are: (a) flame temperature and (b) laminar burning velocity. Since most premixed gaseous flames propagate into fresh unburned gas with relatively low subsonic velocities, there is no pressure rise associated with flame propagation per se. There is, in fact, a slight pressure drop across the flame. This is not to say that a flame propagating in an enclosure will not cause the overall pressure level inside the enclosure to rise. Under these circumstances, the pressure in the vessel will indeed rise. However, if the flame remains a laminar, low speed subsonic flame, the pressure rise throughout the vessel is spatially uniform.

Since the local processes associated with flame propagation are essentially isobaric and since the fluid motions associated with flame propagation are very small, one can calculate flame temperatures assuming that the process is isenthalpic and that the high temperature flame gases are in full chemical equilibrium. A few representative flame temperatures for fuels are tabulated in Table 1-1. Since the mole change for combustion with air is small and pressure rise at constant volume is roughly equal to $\gamma\theta_f/\theta_u$, where γ is the heat capacity ratio, C_p/C_v, these tabulated temperatures show that for these fuel-air mixtures, for combustion under constant volume conditions, the pressure rise will be approximately 6

Table 1-1. Flame Properties for Some Selected Fuel-Air Mixtures

Fuel	Maximum Value		Minimum Value			LEL‡	UEL‡
	θ_f ++ °K	S_u ** m/sec	E min ** millijoules	$d_{||}$ + mm	AIT‡ °K	% Fuel	% Fuel
Hydrogen	2400*	2.70	0.018*	0.55	673	4.0	75.0
Carbon Monoxide (wet)	2370	0.33*	---	---	826++	12.5	74.0
Methane	2230	0.34	0.280	2.50	713	5.0	15.0
Acetylene	2610	1.40	---	0.55	578	2.5	100.0
Ethylene	2395	0.63	---	1.25	763	2.7	36.0
Ethane	2170*	0.44	0.250	2.00*	788	3.0	12.4
Propane	2285	0.39	0.260	2.10	723	2.1	9.5
n-Butane	2170*	0.35	0.260	2.20	678	1.8	8.4

* Lewis and Von Elbe (1961)
+ Potter (1960)
‡ Zabetakis (1965)
++ Alroth, et al. (1976)
** NACA (1959)
++ Calculated using JANAF, Strehlow (1981)

to 8 atmospheres. This is why premixed flame propagation in enclosures
such as buildings, ship compartments, etc., can lead to destruction of
the enclosure.

The determination of the normal burning velocity of a particular combustible system is relatively difficult. Theoretical developments
in this field have shown that a unique burning velocity should exist for
every mixture composition and initial temperature and pressure, and that
this burning velocity is determined by the coupling between diffusion of
heat energy, diffusion of species and the rate of the chemical reaction in
the flame. However, reasonable theoretical predictions of burning velocity have only recently become possible [Smoot, et al. (1976), and Tsatsaronis, G. (1978)]. Experimentally, there are also many difficulties, and

the determination of a burning velocity to an accuracy of $\pm$ 20 percent (except in very unusual situations) is about the best that can be expected at
the present time [Andrews and Bradley (1972)].

Flame temperatures and burning velocities are usually presented
in terms of the equivalence ratio of the mixture. Equivalence ratio is
defined using the following equation

$$\phi = \frac{f/a}{(f/a)_{stoich}} \qquad (1-16)$$

where ϕ is the fuel equivalence ratio, f is the fuel concentration, a is
the air or oxidizer concentration, and subscript stoich represents the
f/a ratio for a stoichiometric fuel-air mixture. For halogen containing
fuels, ϕ is usually defined with the product being the halogenated acid,
HCl for example. Thus, an equivalence ratio of less than one means that
there is excess air in the mixture and the mixture is said to be fuel lean.
An equivalence ratio of greater than one means that there is excess fuel in
the mixture and the mixture is said to be fuel rich. Under these circumstances, combustion cannot be complete. In general, both the flame temperature and the burning velocity increase as one approaches a stoichiometric
mixture. Specifically, most flames show a maximum temperature and burning
velocity at an equivalence ratio of about 1:1. Typical maximum flame temperatures and burning velocities $(S_u)_{max}$, for mixtures of some of the more
common fuels in air are shown in Table 1-1. Other flame properties tabulated in Table 1-1 will be discussed in later sections of this chapter.

Both the initial pressure and the initial temperature of the
premixed gaseous system affect the laminar burning velocity of the flame.
The amount of the effect depends upon the flame temperature. If the flame
temperature is quite high, say above 2100°K, dissociation of the product
gases becomes important. Under these circumstances, if one raises the

initial temperature, one does not appreciably change the final flame temperature because the dissociation process markedly increases the effective
heat capacity of the gas. This means that initial temperature will have a
small effect on the burning velocity of high temperature flames. Just the
opposite effect is true as far as pressure is concerned. In high temperature flames, increasing the pressure greatly suppresses the dissociation
process and raises the flame temperature. This, in turn, causes the burning velocity to increase when the pressure is increased. If the flame temperature is less than approximately 2100°K, just the reverse is true. Under these circumstances, dissociation of the product gases is not important
and raising the initial temperature by a certain amount will essentially
raise the flame temperature by the same amount. This means that the propagation velocity will increase as the initial temperature is increased. However, since there is very little dissociation, changes in the pressure level
will not affect the flame temperature and, therefore, will have a minimal
effect on the burning velocity.

Another property of a flame which is unique is the preheat zone
thickness of the flame. Figure 1-4 is a schematic drawing of the temperature profile in a typical laminar flame. The temperature profile goes

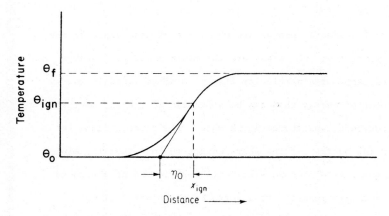

Figure 1-4. Schematic Diagram Showing the Definition of
the Preheat Zone Thickness, η_o

through an inflection point at some intermediate temperature between the
unburned and burned temperature, and if a tangent to the temperature distance curve at this inflection point is extended back to θ_o, the initial
temperature, the distance between the inflection point and the intersection
with θ_o is called the preheat zone thickness, η_o. In this region, the flame
essentially contains no chemistry. The gases are simply being heated by
thermal conduction from the hot reaction zone in the higher temperature regions of the flame. Preheat zone thickness is related to the burning velocity, thermal conductivity, and other flame properties in the following
manner

$$\eta_o = \frac{k_u}{\rho_u S_u (C_p/m)} \left(\frac{\theta_{ign}}{\theta_o}\right) \qquad (1\text{-}17)$$

where k_u is the thermal conductivity of the gas mixture at room temperature, ρ_u is the density of the unburned gas, S_u is the burning velocity,
C_p/m is the heat capacity per unit mass, and θ_{ign}/θ_o is a number whose value has been determined to be around 3 to 4. It is difficult to define exactly θ_{ign} in a flame since it is based on a dynamic concept of the ignition
process. Nevertheless, it does range between approximately 900 and 1200°K
for most flame systems.

Two other fundamental properties which are of importance to the
question of flame propagation and safety are the minimum ignition energy
and the quenching distance of a particular flame. Minimum ignition energy
is the smallest amount of energy that can be deposited in the gas mixture
from a very low inductance capacitance spark which will cause a flame to
propagate away from the spark. If the spark energy is less than the minimum value, a small quantity of the material will burn in the neighborhood
of the spark but the energy imparted to the gas by the spark will not be
sufficient to set up the combustion wave needed for sustained propagation

away from the spark. The quenching distance is the minimum plate separation or gap through which a flame will just propagate. Quenching distances
are measured experimentally using either a parallel plate apparatus or a
tube. Theory shows that these two quenching distances are related in the
following manner:

$$d_{||} = 0.65 \ d_o \qquad (1\text{-}18)$$

Both the minimum ignition energy and the quenching distance when plotted
against equivalence ratio for any particular system show a distinct minimum in the neighborhood of the stoichiometric point. That is, their behaviors are the inverse of the behavior of burning velocity and flame temperature. Experimentally, it has been shown that the quenching distance
is simply related to the pressure (for any one flame, the quantity pd =
constant) and also that it is related to the minimum ignition energy of the
system. It has been found that $E_{min} = Cd^2$ where C is the same constant for
virtually all hydrocarbon-air systems. It has also been shown theoretically that the minimum ignition energy is roughly proportional to the thermal
conductivity divided by the burning velocity multiplied by the flame temperature.

$$E_{min} \approx \frac{k_u}{S_u} (\theta_f - \theta_o) \qquad (1\text{-}19)$$

Table 1-1 also contains values of the minimum ignition energy
and quenching distance for a number of selected fuels in air. Note that
minimum ignition energies are extremely small. As a matter of fact, sparks
with sufficient energy to ignite most hydrocarbons are barely audible and
can be generated by ordinary static electricity. Thus, in ordinary industrial situations, spark ignition of combustible gas or vapor-air mixtures

is an ever present possibility and special precautions must be taken to reduce the possibility of ignition when gases or vapors are present in the
combustible range. It should also be noted that minimum quenching distances
lie in the range of about 0.5 to 2.5 millimeters, and that ordinary electrical equipment is constructed such that flames generated by sparks from motor
bushings or switches can propagate into the room if a combustible mixture is
present.

In addition to the properties mentioned above, combustible systems also exhibit flammability limits. That is, there exists a range of
compositions from some lean to some rich limit between which a flame will
propagate. Outside of this range, the system will not propagate a flame a
long distance from an ignition source. Industrially, the most important of
these limits is the lean flammability limit, LFL. In general, the lean
flammability limit for typical hydrocarbon gas or vapor-air mixtures occurs
at a fuel percentage of about 55 percent of the stoichiometric percentage,
while the rich limit occurs at about 330 percent of stoichiometric [Mullins
and Penner (1959)]. One empirical observation [Bodurtha (1980)] is that for
many hydrocarbon-air mixtures, the lean limit volume percent of fuel
multiplied by its heating value in kJ/mole is approximately 4.35×10^3.
This means, of course, that the flame temperature is approximately constant
at extinction for all hydrocarbon fuels. Typical values for lean and rich
limits for upward propagation in a standard flammability tube are listed
in Table 1-1.

In 1898, Le Chatelier proposed a rule for determining the lean
limit of a mixture of combustible gases, from the known lean limits of the
constituent species in the mixture [Le Chatelier, H. and Boudouard, O.
(1898)]. If we call L_m the volume percent of the fuel mixture at the lean
limit and L_1, L_2, ..., L_n the volume percent lean flammability limits of the
n constituent fuels and c_1, c_2, ..., c_n the volume percent of each constituent
fuel in the fuel mixture, Le Chatelier's rule is given by the formula

$$L_m = \frac{100}{\dfrac{c_1}{L_1} + \dfrac{c_2}{L_2} + \dots + \dfrac{c_n}{L_n}} \qquad (1\text{-}20)$$

This formula works quite well for most hydrocarbons. If organic compounds with widely dissimilar structures are mixed, the rule does not work quite as well.

Unfortunately, there is no good theory for predicting flammability limits. Even more unfortunately when one attempts to measure flammability limits using different types of apparatus, one finds that the measured limits can vary quite markedly from apparatus to apparatus. The practical problem of determining flammability limits will be discussed in more detail in a later section.

When one is dealing with liquid fuels, one must also define a flash point for that fuel. While it is true, as for flammability, that the flash point temperature of the fuel is dependent upon the technique used to measure it, the most reliable technique for pure fuels is the closed cup, equilibrium technique. In this technique, the liquid in question is placed in a cup in a thermostat in contact with air and the system is allowed to come to complete equilibrium at one atmosphere pressure and that temperature. When equilibrium is reached, a portion of the top of the chamber is opened and a small premixed flame is placed in contact with the vapor-air mixture. If the vapor-air mixture burns (or flashes), the thermostat temperature is said to be above the flash point of the liquid. If a flash does not occur, the temperature is said to be below the flash point of the liquid. With repeated tests, one can usually determine the flash point within a few degrees centigrade. A liquid is defined to be "flammable" if its flash point is less than 100°F. It is called "combustible" if its flash point is equal to or greater than 100°F [NFPA (1976a)]. A detailed description of flash point techniques will be presented in a later section.

A rich limit flash point can also be determined because at some temperature the vapor pressure of the liquid will be so large that an equilibrium mixture of fuel with air will be above the upper flammability limit of the fuel. One should be cautious about upper flash points, however, because dilution of such a mixture with air will always cause the mixture to become flammable, even if only momentarily. It is true that if one is lucky, he can extinguish a match by dropping it into a partially filled gasoline can. This is because at room temperature the vapor pressure of gasoline is such that an equilibrium gasoline-air mixture has a composition above the ordinary rich limit. However, if any dilution with air had occurred prior to the introduction of the match, the can would explode. Thus, lean limits are the important limit, from a safety standpoint.

Combustible liquids, vapors, and gas-air mixtures whose composition lie inside the flammable range can also be ignited simply by being heated to a high temperature. This type of ignition is usually studied in a heated vessel by an introduction technique. Alternatively, as described by Strehlow (1968b), a shock tube technique can be used to study this type of ignition for gaseous fuels. Physically, it is observed that when the temperature is very high, the delay time to explosion in any particular configuration may be represented by an equation of the type

$$\log \{\tau[F]^n [O]^m\} = \frac{E}{R\theta} + A \qquad (1-21)$$

where τ is a time delay before ignition in units of seconds, E is an effective activation energy, and the concentration of fuel and oxidizer are represented by [F] and [O]. The exponents n and m and the activation energy E are normally determined by a regression analysis using a number of data points. A good example of such a measurement is the curve for ignition delay time for methane-oxygen mixtures taken from Seery and Bowman (1970), Figure 1-5. At lower temperatures, the ignition delay time starts

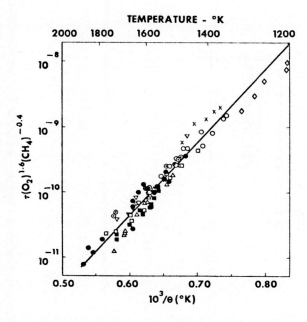

Figure 1-5. Correlation of Measured Induction Times (τ, sec)
With Initial O_2 and CH_4 Concentration (moles/cm^3)
and Reflected Shock Temperature ($^\circ$K)
[Seery and Bowman (1970)]

to increase more rapidly with a decrease in temperature. Eventually,

there is some temperature below which the ignition to explosion cannot oc-

cur even though slow reactions may be occurring in the vessel or chamber.

The temperature at which this happens for any fuel (using a standard test

apparatus) is called the autoignition temperature (AIT) for that fuel in

air. The minimum auto ignition temperature of the fuel gives some indica-

tion of the maximum surface temperatures that should be allowed when that

particular fuel-air mixture is present in an industrial environment, in

order to prevent homogeneous initiation from a hot surface. A typical ex-

ample of a measurement of the delay to ignition to determine an autoigni-

tion temperature (AIT) is given in Figure 1-6 for normal propylnitrate.

In this case, the auto ignition temperature is 170°C. A detailed discus-

sion of the determination of autoignition temperature is deferred to a

later section.

28

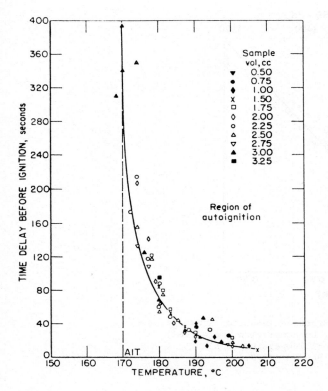

Figure 1-6. Time Delay Before Ignition of NPN in Air at 6900 kPa in
the Temperature Range From 150° to 210°C
[Zabetakis (1965)]

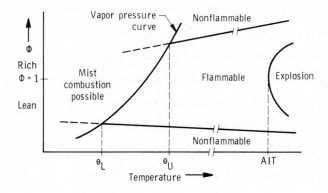

Figure 1-7. Effect of Temperature on Limits of Flammability for
the Vapor of Liquid Fuel in Air. θ_L = Lean (or Lower) Flash
Point; θ_U = Rich (or Upper) Flash Point; AIT = Autoignition
Temperature
[Adapted from Zabetakis (1965)]

The relationship between flash points, flammability limits and the autoignition temperature for a typical liquid hydrocarbon fuel in air is shown schematically in Figure 1-7. Notice that typically the lean flammability limit does not change very much as temperature increases, while the upper flammability limit increases markedly with increased temperature. The intersection of the vapor pressure curve of the fuel with the flammability limit curves defines the equilibrium flash points for lean and rich limits as discussed earlier. However, below this temperature, a mist could be formed which could be flammable. Mist combustion will be discussed in a following section. At some high temperature, auto ignition occurs. The AIT usually has a minimum somewhere near the stoichiometric composition. It has been found that AIT has no unique relationship to either the flammability limits or the flash points.

2. Detonations[†]

Combustible systems can support another type of rather simple conversion wave, namely a detonation wave. A detonation wave is a supersonic conversion wave which passes through the gas and generates a locally high pressure during the combustion process. Typical detonation velocities are tabulated in Table 1-3. These should be contrasted to the flame velocities tabulated in Table 1-1.

A detonation wave is uniquely different from an ordinary flame in many other ways. Around the turn of the century, Chapman (1899) and, independently, Jouguet (1905), formulated a rather simple one-dimensional theory for the propagation of detonation waves. The theory invokes only the inviscid steady equations of motion and the state equation for the substance involved, and it predicts that there is a minimum supersonic propagation velocity if the wave is overall steady. Consider Figure 1-8.

[†]See Strehlow (1968b).

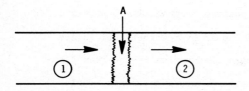

Figure 1-8. Steady and Strictly One-Dimensional Flow From
State 1 to State 2 With a Chemical Conversion
Process Occurring in Region A

The conservation and state equations for a flow of this type are as follows:

$$\rho_1 u_1 = \rho_2 u_2 \quad \text{mass} \tag{1-22}$$

$$P_1 + \rho_1 u_1^2 = P_2 + \rho_2 u_2^2 \quad \text{momentum} \tag{1-23}$$

$$h_1 + (1/2) \ u_1^2 = h_2 + (1/2) \ u_2^2 \quad \text{energy} \tag{1-24}$$

$$P_1 = \rho_1 R_1 \theta_1 \quad P_2 = \rho_2 R_2 \theta_2 \ ; \ R_2 = \text{fn}(\theta_2, P_2) \quad \text{state} \tag{1-25}$$

$$h_1 = \text{fn}(\theta_1) \quad h_2 = \text{fn}(\theta_2, P_2) \quad \text{energy} \tag{1-26}$$

In these equations, the state and energy equations for the incoming flow
(state 1) are for the reactive combustible substance. This substance could
either be a gas phase combustible mixture or a liquid or solid high explosive. In the theory, the equations for the region 2 are assumed to represent full chemical equilibrium in that region at some high temperature.
Thus, the enthalpy of region 2 is a function of both the temperature and
pressure because dissociation processes are dependent on pressure. Note
that we have implicitly assumed that the flow is inviscid and, therefore,
there is no transport by either diffusion or thermal conduction.

In order to examine the nature of these equations, we first solve Equations (1-22) and (1-23) to yield the relationship

$$(\rho_1 u_1)^2 = (\rho_2 u_2)^2 = \frac{P_2 - P_1}{\frac{1}{\rho_1} - \frac{1}{\rho_2}} = \frac{P_2 - P_1}{V_1 - V_2} \qquad (1-27)$$

Equation (1-27) is the Rayleigh equation for steady flow. Volumes V_1 and V_2 in this equation are specific volumes, with units of m^3/kg. Equation (1-27) represents the equation of a straight line on a (P, V) plot, and this line must have a negative slope; otherwise, the mass flow would be imaginary. Thus, in Figure 1-9, the end states for any steady one-dimensional inviscid flow must lie only in the regions that are not marked as excluded.

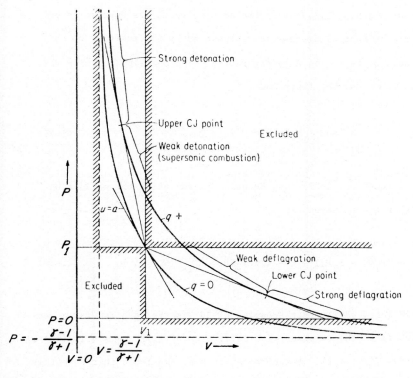

Figure 1-9. Pressure-Volume Plot of End States for a One-Dimensional Steady Process with Heat Addiction Indicating Excluded Regions and Upper and Lower Chapman-Jouguet States

32

The three conservation equations can also be solved to yield the Hugoniot.

$$h_2 - h_1 = (1/2) \, (P_2 - P_1) \, (V_2 + V_1) \tag{1-28}$$

The Hugoniot curve represents the locus of all possible end states of the system on the PV plane. In a general system where full chemical equilibrium exists in the high temperature region in Figure 1-8 (region 2), this equation can be solved only by using iterative techniques. It has been found, however, that very good agreement to such equilibrium Hugoniots can be obtained for gaseous systems initially at one atmosphere pressure by replacing the actual chemical system which involves an equilibrium gas at the high temperature with a heat addition-working fluid model. In this case, the molecular weight and heat capacity of the working fluid are assumed to be constant and the combustion process is represented by the addition of a quantity of heat, Q, to region A in Figure 1-8. The enthalpy function for such a heat addition working fluid model is

$$h_1 = C_p \theta_1 \qquad h_2 = C_p \theta_2 - Q \tag{1-29}$$

Substitution of Equation (1-29) into Equation (1-28) using the definition $\gamma = C_p/C_v$, yields the relationship

$$\frac{\gamma}{\gamma - 1} \, (P_2 V_2 - P_1 V_1) - Q = (1/2) \, (P_2 - P_1) \, (V_1 + V_2) \tag{1-30}$$

It can be shown quite easily that Equation (1-30) is the equation for a rectangular hyperbola in the PV plane with asymptotes of $P_2/P_1 = -(\gamma - 1)/(\gamma + 1)$ and $V_2/V_1 = +(\gamma - 1)/(\gamma + 1)$. Furthermore, it has been found that when this approach is taken, a rectangular hyperbola can be curve fitted to the equilibrium Hugoniot over the pressure range of 1 to 20 atmospheres for stoichiometric fuel-air mixtures and that when this is done, the agreement with the locus of the full equilibrium Hugoniot on a (P, V) plot is excellent, being better than one quarter of one percent in volume for any assumed pressure. Fitting constants, γ and q, are tabulated in Table 1-2 for a number of common fuel-air mixtures. The q in Table 1-2 is defined as $q = Q/C_v\theta_1$, i.e., is nondimensionalized relative to the initial internal energy of the system.

The q = 0 curve in Figure 1-8 represents the Hugoniot for an ordinary shock wave. Only solutions for volumes less than V_1 are available in this case and this curve tangents the Rayleigh line labeled u = a on Figure 1-8. Since the limit of a weak shock wave is a sound wave, all supersonic solutions have a larger negative Rayleigh line slope and yield a high pressure for state 2.

Table 1-2. Hugoniot Curve-Fit Data[*]

P_1 = 101,300 Pa; θ_1 = 25°C

Fuel	ΔH_c Low Value kJ/Mole Fuel	ΔH_c Low Value MJ/kg Fuel	ϕ = 1.0			
			Q_c MJ/kg Fuel	Q MJ/kg Mixture	q	γ
H_2	241.8	120.00	152.34	4.316	7.030	1.192
CH_4	802.3	50.01	64.73	3.549	7.819	1.197
C_2H_2	1256.0	48.22	58.93	4.127	8.779	1.183
C_2H_4	1323.0	47.16	60.53	3.834	8.403	1.188
C_2H_4O	1264.0	28.69	35.34	3.995	8.893	1.183
C_3H_8	2044.0	46.35	60.86	3.649	8.407	1.193

Q is the curve fitted heat addition, Q_c is an effective heat of combustion of the fuel calculated from it, and $q = Q/C_V\theta_1$.

[*] Strehlow (1982).

Note that when q is a positive number, the Hugoniot no longer goes through the point (P_1, V_1), but instead is displaced to the right and upward on the (P, V) plot. Under this circumstance, there is a minimum supersonic velocity and a maximum subsonic velocity for steady flow heat addition. These solutions correspond to the usual thermal choking limits for one-dimensional steady flow and are called Chapman-Jouguet, or CJ, points. The upper CJ point, when calculated assuming complete chemical equilibrium for state 2, yields velocities and pressures for gas phase detonations which agree well with those that are observed experimentally. Values for a number of fuels are tabulated in Table 1-3. Incidentally, a

Table 1-3. Detonation Properties for a Number of
Stoichiometric Fuel-Air Mixtures*

P_1 = 101,325 Pa; T_1 = 25°C

Fuel	% Fuel	Chapman Jouguet		
		Pressure MPa	Temperature °K	Velocity m/s
H_2	29.52	1.584	2951	1968
CH_4	9.48	1.742	2784	1802
C_2H_2	7.73	1.939	3114	1864
C_2H_4	6.53	1.863	2929	1822
C_2H_4O	7.73	1.963	2949	1831
C_3H_8	4.02	1.863	2840	1804

* Strehlow (1982).

35

flame solution on this graph corresponds to a line which has a negative slope but is almost horizontal. Thus, low propagation velocity flames cause a very slight pressure drop as the flame propagates through the mixture.

The curve fit data in Table 1-2 may be used to calculate Chapman-Jouguet conditions using simple algebraic equations. Specifically, if Equations (1-27) and (1-30) are solved to find tangency, one obtains an expression for the Chapman-Jouguet velocity in terms of the curve fit parameters

$$U_{CJ} = M_{CJ} \cdot a$$

$$= \left[\left(\frac{\gamma + 1}{\gamma} q + 1 \right) + \sqrt{\left(\frac{\gamma + 1}{\gamma} q + 1 \right)^2 - 1} \right]^{1/2} \cdot \left[\frac{\gamma R \theta_1}{m_1} \right]^{1/2} \quad (1\text{-}31)$$

where γ and q are taken from Table 1-2, R is the universal gas constant and m_1 is the molecular weight of the initial fuel-air mixture.

Again, solving Equations (1-27) and (1-30) only this time eliminating volume yields the equation

$$P_{CJ} = \frac{1 + \gamma M_{CJ}^2}{\gamma + 1} \quad (1\text{-}32)$$

The end states for constant volume and constant pressure combustion may be determined by solving Equation (1-30) to yield the expressions

$$\frac{P_2}{P_1} = q + 1 \quad (1\text{-}33)$$

for constant volume combustion and

$$\frac{V_2}{V_1} = \frac{q}{\gamma} + 1 \qquad\qquad (1\text{-}34)$$

for constant pressure combustion.

Even though the gross properties of a detonation wave can be predicted using one-dimensional theory, the actual propagation mechanism of the wave itself is much more complex. The wave does have a reaction zone and the thickness of this reaction zone is determined by the rate of the chemistry that converts the reactants to the hot products. This reaction zone contains a number of transverse shock waves propagating across the front as the front propagates in a forward direction. These shock fronts are generated by a combustion instability type phenomenon occurring in the flow behind the lead shock wave and they exhibit a characteristic strength and a characteristic spacing. This characteristic transverse wave spacing is controlled by the rate of the chemical reactions. Using one-dimensional theory, one can calculate a thickness of the detonation wave by assuming that the shock wave is plane and unreactive and that the chemistry occurs at the proper kinetic rate in the flow downstream of the shock wave. If one does this, one obtains an idealized, one-dimensional thickness for the detonation wave. When this thickness is compared with the observed reaction zone thickness, it is found that the observed reaction zone thickness (the hydrodynamic thickness) is approximately ten times the one calculated using one-dimensional theory. Furthermore, it has been observed that the characteristic spacing of the transverse waves in a propagating detonation is about ten times the observed hydrodynamic thickness of the detonation wave. Barthel (1974) has calculated transverse wave spacings for hydrogen-oxygen-Argon detonations using acoustic-acoustic theory and finds excellent agreement over a large range of initial compositions and pressures. It appears that the spacing of the transverse waves on a Chapman-Jouguet detonation is simply related to the

rate of the chemical reactions that occur behind the leading shock wave of
the detonation.

When a detonation propagates down a straight tube of constant
cross section, its transverse wave system tends to couple with standing acoustic modes of the hot column of gas in the region downstream of the front.
As the pressure is lowered or as the composition is changed towards a limit
composition, the preferred spacing of the transverse wave system increases.
This increase is caused by the fact that decreased pressure or dilution with
inerts causes the rate of the chemical reactions to decrease. It has also
been observed that when the transverse wave spacing becomes of the order of
magnitude of the diameter of the tube further dilution or a further drop in
initial pressure causes the detonation to fail. Thus, propagation limits
for detonation are essentially controlled by the size of the apparatus in
which the detonation is propagating. At the present time, there is no theory for the limits of propagation of unconfined detonations even though such
limits should exist. In general, detonation propagation limits are narrower
than flame propagation limits. That is, for most systems, there is a smaller range of composition for detonation propagation than for flame propagation.

Detonations can be initiated by a variety of techniques. The
subject has recently been reviewed by Lee and Moen (1980). In most cases,
before initiation occurs, a shock exists in the system which is strong enough
to cause the exothermic chemistry to become rapid. In other words, a portion
of unreacted gas must be heated and held above its autoignition temperature
(AIT).

Direct initiation from a point can usually be affected by a proper sized high explosive charge, by an exploding wire, by an ordinary capacitance spark or by a laser spark. For all these different types of direct
ignition, there is a minimum energy required, below which only a flame is
generated in the system. Since this type of direct initiation process must
generate a sufficiently strong shock wave which decays sufficiently slowly

such that a reasonable amount of exothermic reaction occurs, the nature of
the deposition process is extremely important. It has been observed, in
fact, that the total amount of energy that one needs in the source region
is directly related to the rate of deposition of energy. This is shown in
Figure 1-10 taken from Bach, et al. (1971). Specifically, for a laser spark
which deposits the energy in approximately 20 nanoseconds, the detonation
has a very low initiation energy. Exploding wires which have a time constant of the order of 10 microseconds require more energy. Ordinary capacitance sparks which deposit the energy in the order of 10 to 60 microseconds require even more energy to produce initiation, and high explosive
charges which do not couple properly to the gas system require even larger
amounts of energy to effect initiation. Bach, et al. (1971) found that
the power density, which is defined as the energy of the source divided by
the volume of the initiation kernel and time of deposition, is approximately a constant, as shown in Figure 1-10. It is true that there is a

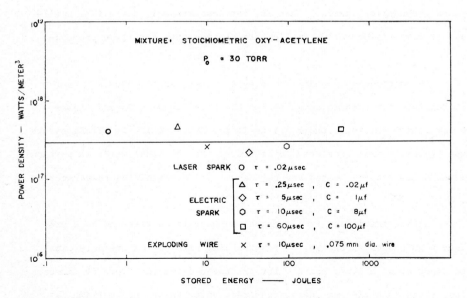

Figure 1-10. The Comparative Correlation of Initiation Data for
Spherical Detonation Waves With Respect to Initiation Energy and
Initiation Power Density at Constant Initial Conditions of the
Explosive Gas [Bach, et al. (1971)]

lower value for initiation energy which is independent of power density. This has been proven experimentally by Knystautas and Lee (1976) and Lee and Ramamurthi (1976), and theoretically by Toong (1981). Also, Lee and Matusi (1978) showed that it is the first excursion of power in a spark that is responsible for detonation initiation and Matusi and Lee (1976) showed that electrode spacing and shape affected initiation energy.

One can also effect initiation in a shock tube either behind a converging incident shock or behind the reflected shock produced when the incident shock reaches the back wall of the tube. In both cases, there is a characteristic delay to explosion just as there is in the case of a homogeneous volumetric explosion. However, in both of these cases, the wave nature of the heating process causes the exothermic reaction zone to be a high subsonic velocity wave system in the gas and this eventually leads to the initiation of detonation [see Strehlow (1968a), Strehlow, et al. (1967), and Lee (1977)].

In all the above cases, initiation occurred because some portion of the explosive mixture was heated to above the autoignition temperature for a sufficient time to cause a local explosion which developed into a detonation. Direct initiation of detonation under conditions where the autoignition temperature is never exceeded has recently been observed. Lee, et al. (1979) irradiated acetylene-oxygen, hydrogen-oxygen and hydrogen-chlorine mixtures in a chamber with ultraviolet light, using quartz windows. The ultraviolet light is absorbed either by the oxygen or chlorine molecule and causes these molecules to dissociate, to form the free radicals O and Cl, respectively. These radicals, if in sufficiently high concentration, will trigger chain reactions and, thereby, cause a local "explosion" in the system. Three different explosion behaviors are observed. If the light intensity is low, the UV light penetrates only a very thin layer near the window and the radicals in this layer ignite a flame or deflagration wave which propagates away from the wall. For very high intensity irradiation, absorption is rather uniform throughout the volume, and the uniform high radical concentration causes a constant volume explosion of the vessel contents. In this case, there are no waves of any consequence in the system during the explosion.

In the case of intermediate levels of radiation, a detonation is observed to form immediately and propagate away from the irradiated window. This behavior can be explained in the following way. The irradiation produces a high concentration of radicals at the wall and, more importantly, a gradient of radical concentration away from the wall. When this gradient has the proper shape and depth, the localized explosion of the layer nearest the wall generates a pressure wave which pressurizes the neighboring layer and shortens the explosion delay time in that layer. This augmentation continues until a fully developed CJ detonation wave is formed. Lee, et al. (1979) called this Shock Wave Amplification by Coherent Energy Release, or the SWACER effect. They make an analogy to the behavior of a LASER because the pressure generated by the simple explosion of one layer would not cause significant reaction in the next layer if it had not been preconditioned to be ready to explode.

In an interesting recent piece of work, Knystautas, et al. (1979) have investigated another SWACER mechanism for the shockless generation of a detonation wave. This effect was first observed by Meyer, et al. (1970) and is very pertinent to the accidental explosion problem. Specifically, they discovered that the proper venting of a jet produced by a combustion process occurring in an enclosure can condition the gas in the external region in such a manner that a detonation will be initiated without the presence of strong shock waves. The schematic of their apparatus is shown in Figure 1-11. An equimolar ignition was effected in a small, round chamber, and at first cold reactant $C_2H_2-O_2$ mixture was forced through the orifice into the large chamber and finally, after some time elapsed, combustion products started to enter the large chamber. They found that after a short time, if

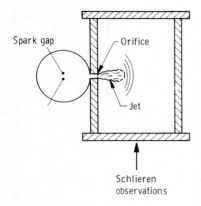

Figure 1-11.   Test Arrangement of Knystautas, et al. (1979)
[The Test Gas was an Equimolar Acetylene-Oxygen Mixture at
20 kPa Initial Pressure]

conditions were correct, a detonation wave propagated directly from the jet

mixing zone. Their description of the phenomena is as follows.

At first, an unreacted jet is formed in the large observation

chamber. Then combustion products start to exit through the slot and fold

into the cold reactive combustible mixture. After some time, the combus-

tion products reach the head of the jet. A plot of the subsequent velocity

at the head of the jet is shown in Figure 1-12. At first, the head of the

jet moves at a velocity of about 800 meters per second and then, after a

short time span, starts to travel at 3.1 km per second. This latter velo-

city is very close to the Chapman-Jouguet detonation velocity for a stoichi-

ometric acetylene-oxygen mixture. The velocity of the edge of the jet be-

fore initiation is so low that the temperature of the cold gas regions must

be well below the autoignition temperature of the mixture. However, they

are either mixing with hot product gases or are being consumed by a flame

which has been folded by the turbulent mixing processes in such a manner as

to have a very large increase in surface area. In both cases the rate of

conversion to products must be so rapid that the pressure in one small re-

gion of the jet rises dramatically. Under these circumstances, neighboring

regions which are ready to react are triggered to reaction by the pressure

42

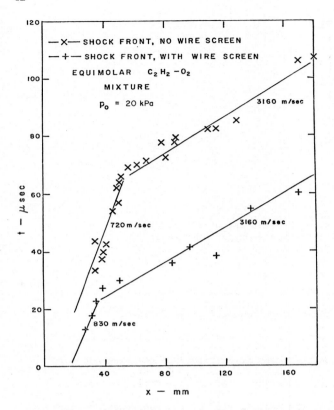

Figure 1-12. Shock Front Positions With Time Frame Experiments
[Knystautas, et al. (1979)]

rise. This enhances the pressure wave while it is still traveling in a mix-

ture of unreactive and active media. Once it leaves the jet region travel-

ing at 3.1 km per second, the reaction wave is a true Chapman-Jouguet deto-

nation wave. Prior to the occurrence of this coupled pressure wave-reaction

wave system, there are no shocks that are strong enough to cause autoigni-

tion. Thus, there is a critical eddy size in the jet which will cause trig-

gering to detonation. This eddy size is undoubtedly related to a critical

volume requirement for initiation such as the shock bubble in direct initia-

tion. This means that in an accidental explosion situation, if the obstacles

in the flame path are a sufficient size, the proper type of eddy folding

during flame propagation can trigger detonation by jet mixing. In this case, there are no strong shocks in the system prior to the appearances of the detonation.

Recently, this type of shockless initiation has been verified by the numerical calculations of Barthel and Strehlow (1979). They used the Oppenheim CLOUD program which is a Lagrangian, constant time step artificial viscosity program to see if enhanced reactivity of a central region could possibly trigger a detonation in the surrounding region without shocks present initially. It is, of course, impossible to trigger detonation in a hydrocarbon-air mixture by bursting a sphere filled with the products of constant volume combustion. However, their preliminary calculations, as exemplified in Figures 1-13 and 1-14, show that there is a threshold above which a proper distribution of enhanced reactivity can directly trigger detonation in a surrounding region that has hydrocarbon-air global kinetics.

The significance of these two mechanisms of direct initiation (radical gradient or hot gas-cold gas mixing, both called SWACER) is that the process is not triggered by the presence of a strong shock wave but, instead offers a chemical mechanism for _generating_ an accelerating shock wave which quickly reaches the CJ detonation velocity.

In addition to the rather pure types of initiation phenomena described above, detonations can also be initiated by flame acceleration processes in enclosures, or in partial enclosures. These types of initiation can be either AIT or radical triggered and they are extremely important during accidental explosions in industrial and transportation environments. The processes are, in general, quite complex and dependent upon the interaction of the flame with the structures present in the enclosure. They will be discussed in some detail in a later section.

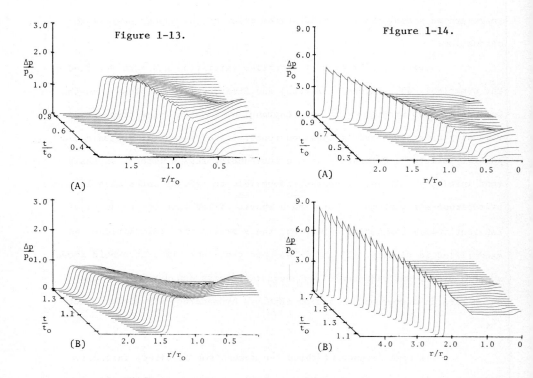

Figure 1-13.

Figure 1-14.

Figures 1-13 and 1-14 present the results of numerical calculations showing how enhanced reactivity of a central core region can directly trigger transition to detonation. Figures show pressure distribution at equally spaced times after reaction is started in central core. In both cases, the central core was very reactive out to $r/r_o = 0.2$ and tapered off in reactivity out to $r/r_o = 1.0$. Outside of that region, normal hydrocarbon-air reactivity was assumed. Figure 1-13: Decay – no detonation (A) early times, (B) later times. At these later times the reaction in the core is complete. Figure 1-14: Transition to detonation. Same conditions as Figure 1-13, only maximum rate of core reactions is twice that of Figure 1-13.

Two Phase and Initially Unmixed Systems

1. Diffusion Flames and Fires

Systems in which the fuel and oxidizer are initially unmixed can burn. Diffusion flames occur under these circumstances. The candle is a very good example of a diffusion flame. In all cases of this type, there is a holding region at which a small premixed flame is established due to the diffusion process [NACA (1959)]. The remainder of the flame consists of a region bounded on one side by pure oxidizer and on the other side by pure fuel. Between these two regions, the fuel and oxidizer diffuse toward each other and the products diffuse outward leading to a

rather thick region where combustion is actively proceeding. There is no
burning velocity, as such, in this case. Also, in general, because large
excesses of fuel exist in regions which are quite hot, considerable pyrol-
ysis and decomposition of the fuel can occur prior to complete oxidization.
For hydrocarbon fuels, this leads to soot formation and the hot soot particles radiate to produce the typical yellow color of the hydrocarbon diffusion flame. These flames can lose a significant amount of their energy by
radiation, and large diffusion flames (i.e., fires) present a hazard to
nearby persons and cellulosic material because radiative transfer can be
sufficient to cause burns and start new fires at a distance. These effects are discussed in Chapter 7. Because of the slowness of the processes that occur inside a diffusion flame, they are all buoyancy dominated.

2. Spray Droplet and Mist Flames

Liquid fuels which are dispersed into a fine spray can support
combustion in one of the two modes. If the droplets are small (less than
25 microns) and if they have high volatility, the droplets will tend to
evaporate in the preheat zone of a laminar flame and the flame will appear
to be a normal premixed flame in which the fuel is a gas. If the droplets
are large and/or they have a relatively low volatility, flame propagation
will involve the individual droplets and each will be surrounded by its
own diffusion flame until the oxidizer or fuel is completely consumed.
The limits of flammability of mists and spray droplet flames is not well
understood at the present time. The upper limits are larger than they are
for a pure gas. This is because combustion of this type will always contain rich and lean regions because the system is not strictly homogeneous
as it is in the vapor phase. Thus, for rich mist-air mixtures, a large
quantity of liquid fuel can pass through the flame unevaporated. The theory of single droplet combustion is quite well developed at the present
time [Williams (1968) and Krier and Foo (1973)]. However, its applicability to the question of spray droplet and mist flames is not well understood.

3. Dust Flames

Palmer (1973), Cybulski (1975) and Bartknecht (1981) have written book length reviews on dust flames and explosions. There are two major types of dust which can burn. These are the organic dusts and the
metal dusts. For metal dusts, the burning process is inherently simpler
than it is for organic dusts. This is because the number of intermediate
steps during the burning process is small and the overall process is,
therefore, simpler. Differences do exist based on whether the metal melts
during the burning process and/or whether the oxide which is formed is
either a solid, liquid or vapor under the combustion conditions.

In the case of organic dusts, a tremendous range of behaviors
is possible. Devolatization and charring can occur prior to actual heterogeneous attack of carbon residuals in the dust. If the mixture is very
rich in dust, it is highly probable that only the volatiles will burn.
In this regard, highly rich dust mixtures do burn. Coal dust, for example, has been burned in a steady flame at equivalence ratios of 10. This
is well above the normal rich equivalence ratio for a gaseous fuel. Under
these circumstances, the fixed carbon in the coal is hardly attacked at
all, and it is quite obvious that only the volatiles are burning.

Dusts are also uniquely different from spray or mist flames because dust particle temperatures can become extremely high during the burning process. In the case of liquid mist flames, the droplet temperature
is limited to the boiling temperature of the liquid under those particular
pressure conditions. This means that spray or mist flames can radiate
significantly only from any soot that is formed in the rich regions of the
flame, while dust flames radiate from the dust particles themselves while
burning progresses. There are strong theoretical indications that radiative transport of energy can be the dominant mechanism when the dust cloud
that is burning becomes very large and opaque. This theoretical work indicates that under these conditions the burning velocity increases by ap-

proximately an order of magnitude from the burning velocity that one would observe on smaller flames where radiation losses are dominant. These results are supported qualitatively by experimental results which show that unshielded small coal dust flames have burning velocities of only about 10 to 30 cm/sec while radiatively shielded flames exhibit burning velocities of about 1 m/sec [Krazinski, et al. (1979)].

4. Mist and Dust Ignition

Very little work has been performed on the spark ignition of mists [see, for example, Law and Chung (1980)]. However, there has been a considerable amount of effort toward understanding the spark ignition of dust clouds [see, for example, Eckhoff (1975), Eckhoff (1976), Eckhoff (1977), and Eckhoff, et al. (1976)]. It has been found that in contrast to the simple minimum ignition energy that exists for spark ignition of gas or vapor-air mixtures, when a dust cloud is involved the nature of the spark itself, the gap length, etc., are very important to the ignition process. Also, it is found that there is a large range of spark energy over which the probability of ignition varies from 0 to 100 percent. Specifically, one usually finds that the spark energy varies by approximately one order of magnitude from the level at which ignition never occurs to the level at which ignition occurs 100 percent of the time. Also, the ignition energy of typical dusts are as much as two orders of magnitude larger than the ignition energy of hydrocarbon gases or vapors.

Dust clouds also exhibit auto ignition at high temperatures. This behavior is similar to the auto ignition temperature exhibited by gases and vapors. In addition, dusts exhibit a layer ignition temperature; that is they suffer from spontaneous ignition or combustion if held at a sufficiently high temperature for a sufficiently long time. This represents an additional hazard when one is handling combustible dusts.

5. Detonative Behavior

There is no question but that spray droplet mist or dust clouds can support detonation waves. Dabora, et al. (1966) and Ragland, et al.

(1968) have shown that droplet sprays of very low vapor pressure hydrocarbon fuels in air can be shock initiated to produce a self-sustaining wave that travels at or near the expected CJ velocity. Nicholls, et al. (1974) and Fry and Nicholls (1974) have also shown that direct initiation of a spray droplet field is possible, using high explosive charges. In addition, Ragland and Nicholls (1969), Sickel, et al. (1971), and Rao, et al. (1972) have shown that a film of low vapor pressure oil on the wall of a tube can be made to support a detonation if triggered by a sufficiently strong shock wave.

In the case of dusts, there is every indication that certain of the more destructive dust explosions in coal mines and grain elevators had propagation velocities which approach that of a CJ detonation. This supposition has not been verified experimentally in the laboratory, however, as it has for mist detonations. The reason for this is that the very large radiative losses of individual dust particles as they burn in small scale apparatus greatly alters the propagation process.

Condensed Phase Systems[+]

1. Characterization of Explosives

All pure substances or mixtures that release heat when they decompose must be considered dangerous. The degree of danger in handling these substances can only be learned by experience. There is really no good way to predict a priori exactly how dangerous a substance or mixture is. In general, exothermic substances or mixtures can be placed into one of four categories based on this accumulated experience. Even though the boundaries between these categories are very diffuse and sometimes depend on the circumstances under which the material is ignited, they do tend to have meaning relative to explosive properties.

The first category contains substances which are extremely dan-

---

[+] A good reference is the U. S. Army Material Command (1972).

gerous. These are substances such as nitrogen trichloride, and certain organic peroxides, which are so unstable that they cannot be handled safely in any quantity. They, therefore, must be considered as laboratory curiosities to be avoided at all cost in an industrial environment. A good example of one of this class of compounds which could be encountered in an industrial environment are the copper acetylides which are formed when acetylene comes in contact with copper or a copper-containing alloy. If acetylene is regularly allowed to contact copper, copper acetylides will deposit in various portions of the apparatus. After a sufficient quantity is deposited, almost any kind of disturbance can cause a local explosion of the deposit and, therefore, lead to possible destruction of the chamber. This is particularly true if the system contains high pressure acetylene because pure acetylene can support a decomposition detonation wave.

The second and somewhat less dangerous class of compounds are called primary explosives. Lead azide is a good example. In general, primary explosives are extremely shock and spark sensitive. Their primary use is in the preparation of the primers that are used to initiate high explosive charges.

The third category contains secondary high explosives. There are many examples of these, ranging all the way from dynamite through TNT, RDX, HMX, pentolite, etc. These are substances which are high explosives, but which require a relatively strong shock wave to initiate the detonation. In general, they can be handled relatively safely and stored for relatively long periods of time without undue hazard.

The fourth and final category of exothermic substances are classified as propellants. These exothermic substances or mixtures are generally so shock insensitive that common experience tells us that they cannot detonate. They are used in rocket motors or guns as propellants. Typical examples are the double base propellants which are mixtures of nitrocellulose and nitroglycerin with other additives and composite propellants which are mixtures of ammonium perchlorate or some other solid oxidizer with an organic binder, such as polyurethane.

The reader should realize that the above classifications have very fuzzy edges. In other words, just because a substance has been classified as a propellant for many years does not mean that it cannot be made to detonate. There have been many cases in history where a very serious accidental explosion occurred simply because it was thought that the substance that was being handled was not detonable. A prime example of this is the disaster that occurred in Oppau, Germany, in 1921. A congealed pile of 4.5 million kg of an ammonium nitrate-sulfate double salt (which had been used as an extender for TNT during the first world war) was being used as a commercial fertilizer after the war and was dynamited to facilitate handling. Instead of simply fracturing, the pile detonated as a unit. The disaster was the largest man-made explosion prior to the advent of the atomic bomb. The explosion killed an estimated 1100 people, left a crater 60 m deep and 120 m in diameter, and caused severe damage to a radius of 6 km. It is now known that ammonium nitrate has a minimum charge diameter for detonation of approximately 300 mm, and the ammonium nitrate-sulfate involved in the incident certainly had a larger minimum charge diameter. However, at the time of this incident, ammonium nitrate was thought to be nondetonable because the standard detonation sensitivity tests were performed using charge diameters of approximately 25 mm. It is best to assume that all exothermic compounds are dangerous and handle them appropriately.

2. Detonation Behavior of High Explosives

A high explosive charge of sufficient diameter exhibits a detonation velocity which can be calculated with resonable accuracy using Chapman-Jouguet theory presented previously for gas phase detonations. The only complication in this case is that because of the very high initial loading density, the pressures during the detonation process reach levels of the order of one million atmospheres. Under these conditions, the gaseous products of the detonation have a density which is close to that of the solid (specific gravity of about 1.4 to 1.6), and the product gas is quite nonideal. Unfortunately, the pressures that are generated under these circumstances are so large that there is no experimental technique available to measure them. These pressures are above the yield strength of any known substance, and they must be determined by using indirect evidence. That is why one can only expect "reasonable" accuracy for CJ calculations on high explosives.

High explosives exhibit a definite charge diameter below which the detonation cannot be made to propagate for long distances. This minimum charge diameter is dependent on the amount of confinement around the charge. A charge that is confined in a heavy steel walled pipe will exhibit a smaller failure diameter than a charge of the same material that is surrounded only by air. The reason for extinguishment in the case of high explosive charges is thought to be due to the strong side relief wave which occurs when a bare charge is fired in air. Under these circumstances, a very strong rarefaction fan propagates towards the center of the charge as the detonation proceeds. If the charge size is too small, this rarefaction fan can interfere with the completion of the chemical reactions at the center line of the charge. Under these circumstances, failure will occur.

It is well known that high explosives can be initiated to detonation only by generating a sufficiently strong shock wave in the high explosive charge. This shock wave can be generated by the impact of a missile, by the detonation of a primary explosive, or locally by frictional sliding. Initiation energies are usually specified in terms of the shock pressure required to produce direct initiation.

High explosives and propellants are all exothermic compounds or mixtures and they, therefore, exhibit the same thermal stability behavior as the reactors discussed earlier. That is, if held under adiabatic conditions, they will all decompose at an ever accelerating rate and eventually explode. Also, if held in a reactor vessel, there will be some vessel

52

size, shape and temperature which will represent the threshold for explosive decomposition. Measured and/or calculated Chapman-Jouguet detonation velocities and pressures for typical condensed phase (high) explosives are tabulated in Chapter 2, Table 2-2.

Thus, in the case of condensed phase exothermic substances and mixtures, the important properties, from a safety standpoint, are the minimum charge diameter that will allow a sustaining detonation to propagate, the shock sensitivity of the explosive to direct initiation and the thermal stability of the explosive. A good review of these properties is given by Macek (1962). Appropriate testing techniques will be described in a later section.

THE DYNAMICS OF COMBUSTION EXPLOSIONS

The Nature of Combustion Explosions

There are a number of exothermic processes which can produce explosions.[†] Two examples, taken from the previous section, are the rupture of a vessel due to a runaway chemical reaction or the detonation of a high explosive charge. There are also many accident situations where an explosion is generated because of combustion following the spill of a flammable liquid or release of a flammable gas or dust, or the combustion of the flammable contents of a container which was not intended to contain such a mixture. Invariably, these gas phase combustion explosions are ignited by a weak energy source and start propagating as an ordinary laminar flame. Thus, under the proper circumstances ignition may lead to a simple fire with no explosion, as such. For example, a small spill in a room, when ignited, would burn and would not generate an explosion. However, a larger spill, when ignited, might easily cause the building to rupture.

[†] A complete discussion of explosions, by types, will be presented in the next chapter, including examples. At this point, we are only attempting to look at combustion mechanisms that can be responsible for explosions.

The same types of considerations apply when one is talking about spills outside of buildings. Generally, two types of explosion behaviors are observed in this case. When a container of high vapor pressure substance is ruptured because of an external fire or runaway chemical reaction, the contents flash evaporate rapidly. This produces very little blast damage but can cause pieces of the container to rocket long distances. This type of explosion is called a BLEVE (Boiling-Liquid-Expanding-Vapor-Explosion) [see Walls (1978)]. Additionally, if the substance is a fuel and is ignited immediately, a very large fireball can be produced.

The other type of combustion explosion which is observed outside enclosures is an unconfined vapor cloud explosion. In this case, a large spill occurs and does not ignite immediately. Because ignition does not occur quickly, the fuel can disperse and mix with air to form a large volume of mixture in the combustible range. Three things can happen in this case. In the first place, the spill may not find an ignition source and be dissipated harmlessly. Secondly, the mixture may be ignited and simply burn. There are many cases of this type on record. Thirdly, the mixture may be ignited, burn and also generate a damaging blast wave. In this case, the accident is called an unconfined vapor cloud explosion.

In the following parts of this section, we will first discuss the fundamentals of flame aerodynamics and then relate how these aerodynamic processes can lead to the development of "explosive" release rates in the source region, whether it be a pipe line, a building, or a nominally unconfined cloud. We will not discuss how the dynamics of the explosion affects the blast wave that is formed. Nor will we discuss specific accidents. These subjects will be left to the next chapter.

Aerodynamics of Flames

1. Flame Propagation in the Open

There is a large body of experimental evidence that points to the fact that flame acceleration processes cannot become severe if the flame is completely unconfined. Dörge and Wagner (1976) and Leyer, et al.

54

(1974) have shown quite conclusively on a small scale that completely unconfined flames do not accelerate. On a larger scale, the work of Lind
and Whitson (1977) allows one to reach the same conclusion. They performed
a number of experiments in five meter radius and 10 meter radius, two mil
thick, polyethylene, hemispherical bags placed on a concrete platform. They
filled the bags with a stoichiometric or slightly rich combustible mixture
of various fuels with air and then ignited the bag centrally at the surface
of the concrete pad. The set-up is shown schematically in Figure 1-15. The
experiments were instrumented with rapid response pressure gauges plus two
high-speed framing cameras looking horizontally at the flame source region
and one overhead camera looking directly down at the source region. A
typical plot of flame radius versus time in both the horizontal and vertical directions is shown in Figure 1-16, and the data for all their runs
are summarized in Table 1-4. Additionally, Figure 1-17 is a photograph of
a typical flame during the later stages of propagation.

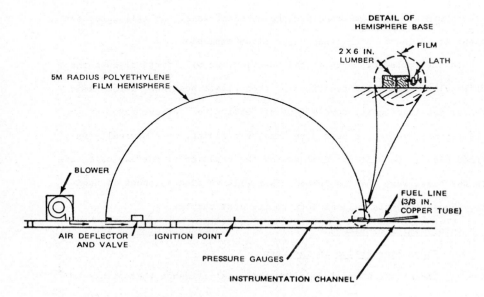

Figure 1.15. Hemispherical Burn Test, Experimental Arrangement
[Lind and Whitson (1977)]

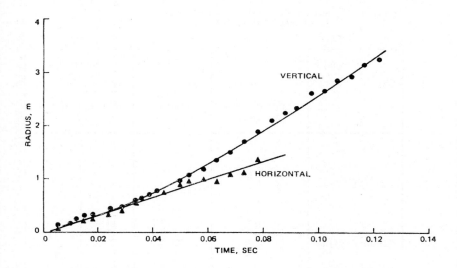

Figure 1.16. Flame Position 7.7 Percent Acetylene Test 18
[Lind and Whitson (1977)]

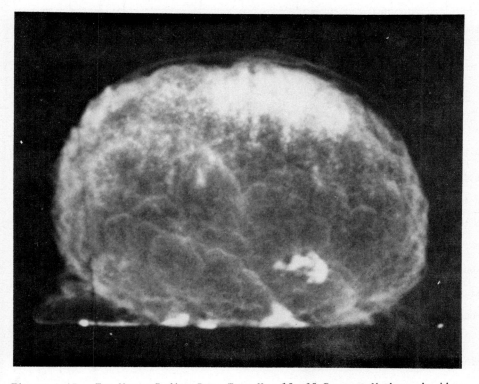

Figure 1-17. Ten Meter Radius Bag, Test No. 13, 10 Percent Methane in Air.
Photograph Taken Near the End of Flame Propagation.
[Lind and Whitson (1977)]

Table 1-4. Summary of Results of Hemisphere Tests
[Lind and Whitson (1977)]

Test No.	Size, meters	Fuel	Concentration, Volume %	Horizontal Velocity, m/s	Vertical Velocity	
					at 3 m, m/s	at 8 m, m/s
5	5	Methane	10.0	5.8	7.3	...
7	5	"	10.0	...[a]	7.3	...
13	10	"	10.0	5.2	6.5	8.9
1	5	Propane	4.0	...[b]	6.3	...
12	10	"	4.0	6.1	7.8	10.6
3	5	"	5.0	...[a]	7.4	...
6	5	"	5.0	6.9	9.5	...
4	5	"	5.0	8.3	10.2	...
11	10	"	5.0	9.6	9.9	12.6
10	10	Ethylene Oxide	7.7	13.4	15.2	22.5
8	10	"	7.7	14.7	16.0	22.4
14	5	Ethylene	6.5	8.8	17.3	...
15	5	Acetylene	3.5	3.6	4.6	...
18	5	"	7.7	23.7	35.4	...
17	5	Butadiene	3.5	3.9	5.5	...

[a] Burning fuel in the instrumentation channel distorted the shape of the flame and no horizontal velocity could be obtained.

[b] Test performed in daylight and flame base was insufficiently visible for horizontal velocity to be obtained.

They were fortunate in these experiments because in all cases, the bags tore at the base as the pressure rose slightly because of the combustion process. Sometimes the tear started at one location on the edge, and if this happened, the bag would gently tilt upward. In other cases, the tear occurred almost uniformly at all points along the base at the same time. In these cases, the very light bag would simply float on the mixture as burning proceeded. This means that the bag's presence did not exert any real influence on the propagation of the flame and the flame was essentially unconfined.

Table 1-5 presents a comparison of the observed horizontal propagation velocities to the normal burning velocity of the fuel as measured under laboratory conditions. Since the observed space velocity, S_s, is related to the effective burning velocity, S', by the equation $S' = S_s \rho_b / \rho_u$, one

Table 1-5. Augmentation of Flame Velocity in Bag Tests

Fuel	Concentration, Volume %	Flame Temperature, °C[a]	Density Ratio Across Flame	Horizontal Flame Velocity, m/s	S´, m/s	S,[a] m/s	ψ
Methane	10.0	1960	0.131	5.8	0.76	0.37	2.1
Ethylene	6.5	2100	0.137	8.8	1.21	0.75	1.6
Acetylene	7.7	2325	0.112	23.7	2.65	1.56	1.7
Propane	4.0	1980	0.128	6.1	0.78	0.43	1.8
Butadiene	3.5	2100	0.117	3.9	0.46	0.60	0.8
Ethylene Oxide	7.7	2140	0.113	14.0	1.58	1.01	1.6

[a] NACA (1959).

can calculate the effective normal burning velocity S' from the bag tests.
The last column of Table 1-5 gives these ratios as ψ = S'/S. Undoubtedly,
butadiene is spurious (probably due to poor mixing). It is interesting to
note that for all other mixtures with the exception of methane, the ratio is
approximately 1.7. The flame that was observed always showed a roughened
structure, and this would explain the somewhat higher effective burning
velocity. Physically, these higher velocities are probably due to the fact
that a large flame with a long lifetime is hydrodynamically unstable as a
plane wave.

In no case did the flame continue to accelerate to a higher velocity. The typical r-t plot for the flame propagation in Figure 1-16 shows
that the effective burning velocity had become constant at the end of flame
travel. Figure 1-16 also shows that buoyancy affected this flame. Typically, the upward propagation velocity always showed a slight acceleration at
first and then leveled out at a value which was somewhat higher than the

horizontal velocity. This is due to the buoyancy of the combustion products behind the flame.

In the five meter bags, vertical acceleration continued until the flame burnout. However, in the 10 meter bags, the flame always reached a constant terminal velocity before burnout.

Another effect which was observed is shown in Figure 1-17. In all cases, the expansion of the hot product gases produced by the flame caused gas motion ahead of the flame away from the center of the source region. At the surface of the pad, this gas motion induced a boundary layer which caused a higher propagation velocity than in the bulk of the gas. This flame traveling along the surface caused an oblique flame sheet to be propagated up into the bulk of the mixture. Thus, in a typical experiment (as shown in Figure 1-17), the flame appears as a spherical flame which is rising above the pad and is sitting on a pedestal (the oblique flame sheet), produced by the higher velocity boundary layer flame propagating along the pad. These boundary layer flames will be discussed later.

Three important conclusions can be drawn from the work of Lind and Whitson (1977). These are: (1) large, completely unconfined flames do not show excessive accelerations, (2) buoyancy of the product gases when the flame is unconfined is not important to the acceleration process, and (3) the turbulent boundary layer produced by motion ahead of the flame along the pad did augment the local burning velocity.

2. Flame Propagation in a Spherical Bomb

The simplest geometry for flame propagation in an enclosure is central ignition by a spark of a premixed combustible mixture in a spherical bomb. Under these circumstances, as shown in Figure 1-18, the flame is always propagating toward the wall and is normal to the wall. Thus, there is no motion along the wall surface and no boundary layers are generated. The pressure rise begins as a cubic with time and then has an apparent rise which is somewhat faster because of an augmentation of the burning velocity with time caused by preheating of the gas ahead of the flame by the adiabatic compression process. With this geometry, the pressure in the sphere is spatially uniform, except for a very small pressure drop at the flame.

Two properties are usually measured. These are P_f/P_o and $(dP/dt)_{max}$ (see Figure 1-18). The maximum pressure rise can be approximated by using a full equilibrium thermodynamic calculation based on $E_f = E_o$, since no external work is performed. This is true even though when the process is over, the temperature of the gas in the center of the bomb is considerably hotter than that near the surface because the entropy of the gas in the central region is considerably higher. This is caused by the fact that central gas is converted to hot products and then compressed at high temperature while the gas near the wall is compressed to a high pressure before conversion to products occurs.

The value of $(dP/dt)_{max}$ is related to the normal burning velo-

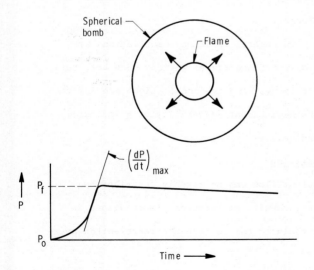

Figure 1-18. Spherical Vessel with Central Ignition
Flame is always propagating normal to and towards the wall. Maximum vessel pressure P_f and rate of pressure rise $(dP/dt)_{max}$ are illustrated in the schematic pressure-time curve. Pressure drops after reaching its maximum value because of heat loss to the walls.

city and vessel size [Bartknecht (1978a) and Bradley and Mitcheson (1978a and 1978b)] by the equation

$$\left(\frac{dP}{dt}\right)_{max} \cdot V^{1/3} = K_g \qquad (1-35)$$

where K_g is a constant for any initial condition in the bomb and V is the bomb volume. This is a useful relationship and has found many combustion safety applications.

It is important to realize that even though there are no gas motions that generate turbulence in this geometry and, therefore, the effective burning velocity is not increased markedly as the flame propagates, nevertheless, there is gas motion in such a bomb during flame propagation. This is because combustion causes a large expansion of the product gases due to temperature rise. Thus, initially, the gases ahead of the flame are pushed away from the center at a high velocity and eventually, when burnout occurs, the product gases near the flame are traveling toward the center of the bomb at a high velocity.

In spherical vessels with central ignition, oscillations are sometimes observed near the end of the flame travel period. However, turbulent propagation is absent in this case and the oscillations are due to an acoustic type interaction and the inherent instability of a laminar flame under the proper circumstances.

3. Turbulent Flame Propagation[+]

The nature of a turbulent flow can be roughly described by characterizing the scale and the intensity of turbulent fluctuations in the flow. Here, the scale of turbulence can be taken to represent an aver-

---

[+]Andrews, et al. (1975) present an excellent recent review of premixed turbulent combustion. See also NACA (1959).

age eddy size at that location in the turbulent flow. To some extent, it
is a function of the geometry of the system that generates the turbulent
flow. Turbulence generating devices at first generate a somewhat ordered
flow (like the Karman vortex streets that form in the wake of a bluff
body). Eventually, these degenerate to form the random fluctuations in
the flow that one characterizes as turbulence. It should be relatively
obvious that the largest scale of turbulence that can be generated in any
device is a function of the size of that device, whether the device be an
obstacle in the flow, a nozzle from which the flow is emitting or the size
of the channel that the flow is in. The intensity of turbulence relates
to the magnitude of the fluctuating velocities relative to a mean velocity
of the flow. It is usually defined as the root mean square of the fluctuating component of the velocity.

Premixed turbulent flames are produced when a laminar flame encounters a region where the flow ahead of the flame is turbulent. The
nature of the premixed turbulent flame that is produced is to a large
measure dictated by the scale and intensity of the turbulence ahead of the
flame relative to the flame thickness and the normal burning velocity of
the flame. In general, low intensity fluctuations whose scale is large
relative to the flame thickness produce highly wrinkled flames whose surface area per unit of frontal area in the propagation direction is somewhat
increased. This has been used to define a turbulent flame to appear as the
mean flow velocity that will cause the turbulent flame to appear time average steady in the observer's frame of reference. This apparent increase in
the effective burning velocity of the flame is not a unique quantity, but
is apparatus dependent.

As the intensity of the turbulence ahead of a premixed flame increases and the scale becomes smaller, the concept of such a turbulent burning velocity becomes more and more difficult to justify. Under these circumstances, there is evidence that the concept of a simple flame front is
no longer valid and that the combustion process extends over a larger re-

gion and becomes more "global" in nature. Thus, high intensity turbulence ahead of the flame can lead to extremely high effective conversion rates. This enhancement of the effective conversion rate by turbulence generated ahead of the flame is extremely important to the development of the explosion process during an accidental explosion. As with laminar flames, however, one can generally assume that if the flame is completely unconfined, pressure will not rise during the combustion process because effective turbulent conversion rates in unconfined situations are still slow relative to the velocity of sound. However, it has been recently discovered that with the proper circumstances, turbulent mixing processes can become so fast in the wake of a sufficiently large object that shockless initiation of detonation can occur. This has been discussed in some detail in a previous section. Also, if the flame is occurring in an enclosure, the background pressure will always rise with time until combustion is complete or the enclosure ruptures. The observed characteristics of specific acceleration processes and pressure time histories are discussed in more detail in the following sections.

In the case of diffusion flames, such as a pool fire, for example, the turbulence that interacts with the flame can be generated by the flame itself and the free convection that is induced by buoyancy. Specifically, one finds that there is a critical size above which the diffusion flame no longer has a laminar character, and turbulent transport processes become important to the rate at which fuel is being consumed. As with laminar diffusion flames, one cannot define anything like an effective burning velocity for a turbulent diffusion flame. Turbulent diffusion flames do not, in general, lead to explosions. Therefore, they will not be discussed further.

4. Taylor Instability

Classical Taylor instability occurs when a contact surface separating a light and heavy fluid is accelerated in the direction of the lighter fluid. In contrast, a contact surface is stable and remains flat in the case when the acceleration is in the direction of the heavy fluid

(for example, a normal water-air interface in the earth's gravity field).

However, when the acceleration vector is reversed and held constant, the

surface develops waves whose amplitude initially increases exponentially

with time.

A premixed gas flame propagating through a fuel-air mixture is

a relatively low velocity wave which produces a large density difference

in the gas. In this case, the burned gases are approximately a factor of

six to eight times less dense than the unburned gas ahead of the flame.

If such a flame and the fluid it is processing is impulsively accelerated

by some external gas dynamic process in such a direction that the cool re-

active gases are caused to push the hot product gases, the Taylor insta-

bility mechanism will cause the flame surface area to increase markedly.

If the acceleration is sufficiently rapid, this rate of increase of flame

area can be very dramatic, and since the overall rate of conversion of

reactants to products is directly proportional to the flame area, the ef-

fective burning velocity, in a global sense, will increase markedly. One

of the earliest observations of this instability in a vessel explosion was

made by Ellis (1928) and is shown in Figure 1-19. In the four cases shown

in this figure, the flame was ignited near the upper end of closed chambers

which had various length to diameter ratios. Note that for the vessel with

a small value of L/D shown on the right-hand side as Figure 1-19(d), flame

propagation occurred smoothly throughout the entire vessel in a somewhat

similar manner to that which was described for a spherical bomb. However,

as the L/D ratio becomes larger and larger, the flame distorts with time

producing an inverted shape near the center of the bomb. This is due to

the fact that the flame area at first increases very rapidly and then, when

it touches the side walls, more slowly. This causes the flame front to de-

celerate and this deceleration generates a Taylor instability wave on the

flame front, which causes a deep trough in the front, and, therefore, causes

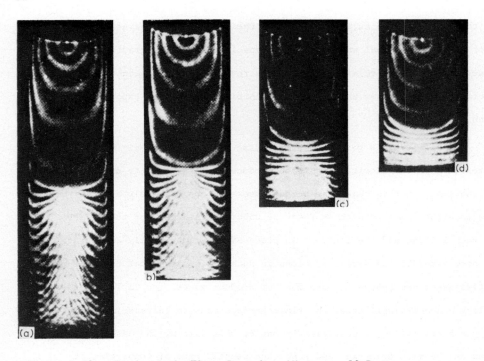

Figure 1-19. Stroboscopic Flame Records. Mixture: 10 Parts
CO + 1 Part O_2, Saturated With Water Vapor at 15°C. Tube Closed
at Both Ends; Diameter 50 mm; Length: (a) 195 mm; (b) 170 mm;
(c) 120 mm; (d) 95 mm. Ignition at Top Center of Each Vessel.
[Ellis, O. C. de C. (1928)]

the area of the flame to increase markedly at later time. This means, of
course, that the rate of pressure rise in a long vessel will be different
than that for a vessel of the same volume with a smaller L/D. In fact, the
vessel with a large L/D ratio would be expected to show a more rapid late
rise in pressure and a very much slower early rise.

Markstein (1958) produced the same type of Taylor instability
in a flame bubble by impacting it from the unburned side with a weak shock
wave. In this case, the fluid is accelerated in the direction from the
heavier to the lighter gas and the flame front area and, therefore, effective conversion rate increases markedly with time. This is shown in Figure 1-20.

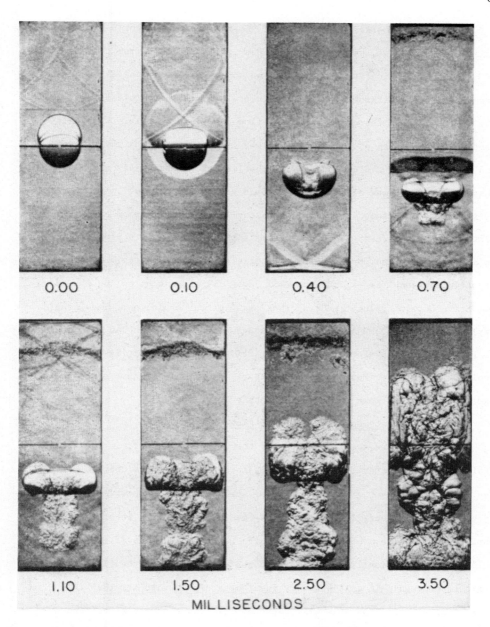

0.00   0.10   0.40   0.70

1.10   1.50   2.50   3.50

MILLISECONDS

Figure 1-20. Interaction of a Shock Wave and a Flame of Initially
Roughly Spherical Shape. Pressure Ratio of Incident Shock Wave 1.3;
Stoichiometric Butane-Air Mixture Ignited at Center of Combustion
Chamber, 8.70 msec Before Origin of Time Scale
[Markstein (1958)]

Heinrich (1974) has some very impressive movies which also show this effect. In his case, he propagated a spherical flame from the center of a square box which had a burst disc on one side. After the flame had propagated some distance in the box, the pressure rose sufficiently so that the disc burst and the gas impulsively flowed out of the hole. When this happened, the region of the flame that was furthest away from the orifice was accelerated in a direction toward the lighter product gases and became unstable. His movies clearly show a rapid growth of convolutions only on this side of the flame due to Taylor instability. This particular experiment is quite important because it shows that if the flame is propagating in a building and the pressure rises sufficiently to cause rapid release in one direction, the gas motions will cause flame convolution and a higher effective burning velocity during that phase of the explosion. Thus, the Taylor instability phenomenon is an aerodynamic effect which can contribute to the production of locally higher effective burning velocities during an accident situation. More recently, Solberg, et al. (1979) have observed the same effect during some large scale venting experiments.

5. Turbulence Generators in the Flow

There are two major types of turbulence generators in the flow that the flame generates and then passes through. The first of these is due to boundary layer growth on interior surfaces. Pressure waves generated by the flame cause flow ahead of the flame and this flow interacts with surfaces to produce a boundary layer which becomes turbulent if the flow velocity is sufficiently high. Under these circumstances, the flame will start to propagate in this boundary layer at a higher velocity than the bulk space velocity and this will cause oblique flame sheets to be generated which, in turn, causes an increase in the effective overall flame propagation rate. When this happens, the rate of pressure rise inside the enclosure becomes more rapid. The excellent experimental work of Utriew, et al. (1965) and Utriew and Oppenheim (1967) on the transition to detonation in hydrogen-oxygen mixtures shows that at late time just before a detonation is formed, the flame in the tube takes on a conical shape whose leading edge is traveling down the wall boundary layer at a very high velocity. In this case, the flame area is extremely large compared to the frontal area of the tube and the rate of pressure rise due to the flame propagation is extremely rapid compared to the initial rate of pressure rise in the system. The China Lake experiments [Lind and Whitson (1977)] showed essentially the same type of behavior at the surface of the pad. Here, as was discussed earlier, the flame traveled preferentially along the turbulent boundary layer generated by gas motion over the pad.

The effect of obstacles in the flow ahead of the flame is more difficult to evaluate. If flame propagation forces the flow to go through the obstacle region at high velocity, very high turbulence levels can be generated. Unfortunately, there has been very little experimental work on situations where the flame encounters a free standing obstacle such as a piece of process equipment or a pipe rack either in a building or in the open. Under these circumstances, one would expect that the turbulence generated by the interaction would be much less because the gas could take a path around the obstacle rather than through it.

There have been a few definitive studies of flame acceleration by obstacles of simple shape or placement. Dörge and Wagner (1975) centrally ignited an explosive mixture under essentially constant pressure conditions. However, the ignition point was surrounded by a grid which completely enclosed the ignition region. Flame propagation occurred at the normal burning velocity until the flame reached the grid. Then, if the Reynolds numbers of the grid were higher than approximately 60, the flame that propagated after passage through the grid exhibited a higher burning velocity. This is shown in Figure 1-21. The reason for the threshold value of 60 is that based on the velocity of the flow just prior to the flame reaching the grid, a wire Reynold number of less than 60 will not generate a turbulent wake, while a wire Reynolds number greater than 60 will. Note that the

68

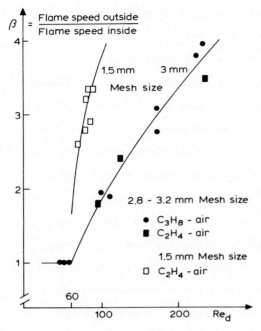

Figure 1-21. Increase in Flame Velocity Due to Turbulence Generated
by Flame Propagation Through a Grid. Re_d Based on Maximum Flow Velocity
at Moment that Flame Contacted the Grid and the Wire Diameter

Beta is the Ratio of Measured Flame Velocity Outside to Measured Flame
Velocity Inside the Grid
[Dörge and Wagner (1975)]

flame velocity outside of the grid region increases with increased Reynolds
number. In a later work, it was found that at higher values of Re_d, the
value of β asymptotically approached the value of 6 for most hydrocarbon-
air mixtures. With acetylene, they found this asymptote to be as large as
12.

　　　　Recently, two relatively simple confined obstacle experiments
have been performed. In the first of these by Lee, et al. (1979), a cylindrical flame in a methane-air mixture was propagated between two parallel plates and the obstacle consisted of a spiral rod, radiating from the
ignition line and attached to either or both plates. They observed a normal
space velocity of about 6 m/sec with no obstacles and could obtain effective
space velocities of up to 130 m/sec with the proper pitch to the spiral and

relative blockage area. In the second experiment reported by Eckhoff, et
al. (1980) and Moen, et al. (1981), a 2.5 m diameter by 10 m long tube,
open at one end, was filled with propane-air mixture and ignited with many
simultaneous sparks at the closed end. Washer shaped baffles were placed
in the tubes with holes that gave blockage areas of 16 percent, 50 percent,
and 85 percent. In different experiments, one, three, six, or nine equally
spaced baffles were placed over the length of the tube. Pressure records
were obtained at a number of stations. It was found that six baffles yielded the highest internal overpressures, and in one case (six baffles, 85 percent blockage), one gauge near the open end of the tube recorded an overpressure higher than the calculated constant volume adiabatic overpressure.
The implication of these experiments is that there is a preferred obstacle
spacing which will produce a more violent acceleration of the flame system.
Also, note that these experiments show that dynamic processes can cause the
system to be pressurized in some local region before combustion occurs there
and then very rapid combustion can generate pressures which are locally super
adiabatic.

6. Combustion Instability

Recently it has been discovered that combustion instability in a
large vented enclosure can cause the rate of pressure rise and the maximum
pressure at the second peak to be larger than would be predicted using proper modeling techniques and small scale experiments. Yao (1974) reports
the presence of a second peak during combustion venting. However, both
Zalosh (1980) and Solberg, et al. (1981) report that the second pressure
peak in a vented vessel, which occurs after the product gases are venting,
shows a high frequency oscillation of growing amplitude and this causes
the second peak to be larger than predicted. The frequency of the oscillation that they observe is related to the fundamental acoustic frequency of
the chamber.

The source of these oscillations is undoubtedly the result of

the same combustion instability that is observed in many continuous com-
bustions [Strehlow (1968a) and Putnam (1971)]. In those cases, it has
been shown that the criterion first stated by Rayleigh (1878) is operative. That is when, in the chamber, there is some inherent physical coupling of a fluctuating local heat release rate and a fluctuating local
acoustic pressure oscillation which causes the integral

$$\oint \delta P\, \delta Q dt = (+) \tag{1-36}$$

to be positive (as shown), an acoustic oscillation will grow in amplitude
at an exponential rate until nonlinear behavior stops its growth. This
linear acoustic instability can grow very rapidly depending on the nature
of the δP-δQ coupling in the system.

Either or both of two possible explanations for the unique behavior of larger vessels can be put forward. It may be that the lower
fundamental frequency of a large chamber couples more readily to the flame
processes because one half cycle is about equal to the time required to
burn through the preheat zone of the flame. Secondly, since the entire
combustion-venting process is slower in larger vessels [see the discussion
of Equation (1-35)], it may be that, in a small vessel, the combustion-
venting process is completed before the acoustic amplification becomes
serious. In any case, the presence of these oscillations in large vessel
explosions definitely contributes to the rate of pressure rise and under
venting conditions can cause a sizeable increase in the maximum pressure
that is observed.

Enclosure Explosion Dynamics

1. Long Pipe

Consider a tube with a very large L/D (say, of the order of one
hundred) containing a combustible mixture whose normal burning velocity is
high relative to the velocity of sound of the unburned gas. With these
circumstances, ignition at a closed end can cause the flame to accelerate

to such an extent that a detonation eventually appears in the tube [Utriew
and Oppenheim (1967), Utriew, et al. (1965), and Oppenheim (1972)]. See
Figure 1-22 for a schematic drawing of the initial behavior. At first, a
weak pressure wave is generated from the initial flame growth, then the
flame growth rate slows because the flame contacts the walls, the Taylor
instability mechanism becomes operative and the flame again starts to accelerate. By this time, the flow ahead of the flame is of sufficient velocity that a turbulent boundary layer is being produced along the walls of
the tube. Under these circumstances, the flame is observed to burn along
the boundary layer, producing, near the later part of flame propagation, an
elongated cone-shaped flame front, which has a very large surface area relative to the cross-sectional area of the tube. This means that the effective
burning velocity becomes extremely high during this portion of the process.
This, in turn, generates strong pressure waves propagating towards the leading
shock wave in the system (see Figure 1-23) and eventually this shock becomes sufficiently strong so that it triggers a homogeneous explosion in

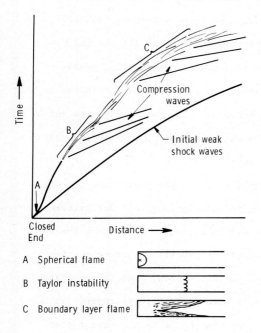

Figure 1-22. Flame Acceleration in a Long Straight Tube (Early Time)

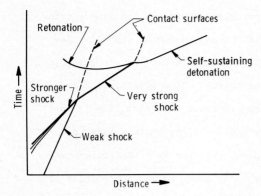

Figure 1-23. Detonation Initiation Caused by Flame Acceleration
in a Long Tube (Late Time)

some portion of the tube behind the lead shocks. This extremely high pres-

sure rise in an already hot combustible mixture leads to the propagation of

a detonation wave. Typically, the onset of detonation is observed to occur

some distance behind the lead shock, but also some distance ahead of the

leading edge of the flame. This detonation propagates toward the leading

shock and, after reaching it, produces a CJ detonation in the fresh undis-

turbed gas in the rest of the tube. It also produces a "retonation" wave,

which propagates back away from the initiation point until all the gas be-

tween the initiation point and the flame is consumed. Typically, pressures

in the neighborhood of the initiation region are extremely high compared to

what one would expect for a steady CJ detonation. This is because the ini-

tial detonation is occurring in a mixture which has been preheated and pre-

compressed by a relatively strong shock wave.

There is another possible complication to the initiation process

in a system of this type. If the pipe is relatively long, but not extremely

long, and, if the acceleration processes are sufficiently slow, then the en-

tire system can be pressurized by the combustion process before detonation

is initiated. Under these circumstances, the detonation, when it propagates,

is traveling in a gas which is already at a relatively high pressure. It has

been estimated that a localized pressure, at some point in a piping system

that contains a transition to detonation could reach values as high as 240
times the initial pressure due to initial prepressurization and eventual reflection of the detonation wave from the end of the tube.

Finally, one more important effect must be mentioned before we
leave the discussion of simple deflagration to detonation transition in
tubes. It has been observed that surface roughness and tube bends, etc.,
have a marked effect on the transition distance. In particular, extremely rough surfaces and bends, etc., shorten the distance to transition very
markedly. This means, of course, that it is very difficult to apply these
simple concepts of initiation to actual plant situations, because pipe
roughness, elbows and T-joints in ordinary plant piping systems can cause
a more rapid transition than would be observed in a smooth straight pipe.

If a long pipe is of relatively small diameter and the flame has
a relatively low burning velocity, one observes the behavior shown in Figure 1-24. Here, heat transfer to the wall is sufficiently rapid that after
the flame has propagated some distance (usually about 10 to 20 diameters),
the flame flow system travels down the tube at constant velocity and propagates in a quasi-steady manner. This is because heat transfer completely
cools the gas to near the wall temperature some distance behind the flame
and the hot gas column length stops growing and, therefore, stops pushing
unburned gas ahead of the flame. If this state is reached before the flame
system can generate a turbulent boundary layer on the tube wall, the flame
just does not accelerate.

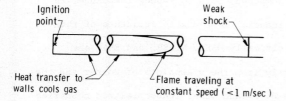

Figure 1-24. Flame Propagation - No Initiation

The boundary between the occurrence of flame accelerations and the occurrence of a quasi-steady flame have not been investigated at this time. Nevertheless, it should be an important safety consideration.

2. <u>Room Explosions (Complex Geometries)</u>

Astbury, et al. (1970), Astbury and Vaughan (1972) and Astbury, et al. (1972) report on the investigation of explosives in room-sized enclosures. They found that in a situation when ignition occurred in one room and the flame traveled to a second room, there was more severe explosion damage in the second room. Heinrich (1974) performed an interesting series of experiments in which two chambers were connected with a tube. Ignition in one chamber led to pressurization of the second chamber before the flame reached it and turbulence generated by the flow into the second chamber and separation at the entrance led to an enhanced rate of combustion in the second chamber and caused super adiabatic pressures to be generated in the second chamber.

The propagation of flames through complex structures is itself complex, and can lead to both locally super adiabatic pressures and to local transition to detonation. More evidences of this will be presented in Chapter 2.

COMBUSTION AND EXPLOSIVES SAFETY

Introduction

The safe production, transport and use of dangerous exothermic substances is everyone's business. Management and insurance companies are primarily interested in reducing the number and severity of incidents involving combustible and exothermic compounds both to reduce the danger to their personnel and to reduce the financial loss due to incidents of this type. Government agencies (now specifically OSHA, NIOSH, and EPA) are interested in safety aspects primarily from the viewpoints of worker and the neighboring community. Because the U. S. and other industrialized countries have this concern, a large body of information has been developed relative to the properties of dangerous compounds and to safe practices when handling these substances. In the United States, the primary source of good practice recommendations for the classification and handling of dangerous materials is supplied by the National Fire Protection Association (NFPA). This organization annually publishes a compilation of their officially adopted technical recommendations, which is called the National Fire Codes (NFC). The NFC (1980) contains 16 primary volumes plus two supplemental volumes. Additionally, they publish pocket editions of NFPA standards which are revised and updated at irregular intervals. One of the two addenda contains NFPA 70 [NFPA (1981)], the National Electrical Code, which has been officially adopted by OSHA as the basis for their regulations. Article 500 of that code entitled "Hazardous (classified) Locations" sets standards for electrical equipment design and operation in locations which can contain combustible gases, vapors or dusts. In addition, the American Standards Institute (ANSI) publishes many standards for handling hazardous materials.

We are not directly interested here in discussing the details of code regulations and good practices as recommended by the NFPA bulletins or ANSI standards or pocket guides. The quantity of this material is large and it ranges over a wide variety of situations, design recommendations, etc. We are, however, interested in the types of testing techniques that are used for producing the different classifications of materials and current developments relative to the improvement of these testing techniques and codes. It turns out that the NFPA good practice guides, even if adopted by the U. S. Government as regulations, usually do not reflect our best understanding based on current research results. The process of updating the NFPA codes is (as it must be) a relatively slow process because it must represent a concensus and, therefore, must go through committee evaluation. Thus, the current codes normally reflect the best practice as it was interpreted about five to ten years prior to the date of the code. This is why we will discuss not only testing techniques recognized by the code but also new techniques that have not as yet been codified. As an example of how the codes are out of date, note that in the dust classification of NEC 500, plastic dusts are not included in the listing but that in NFPA 70 [NFPA (1981)] they are.

Recently, there has been considerable interest in the development of new testing techniques and in the preparation of new ways to evaluate compounds that are dangerous, as well as the development of new preventive measures to mitigate against damage caused by explosive processes when they occur. Additionally, there are a number of compilations of properties of hazardous compounds and these are continuously being updated as new information becomes available. In the following subsections, a number of standardized testing techniques, primarily those published by the American Society for Testing Materials (ASTM), as well as some new research techniques for evaluating the relative hazard of different groups of compounds will be summarized. Furthermore, their relation to the codes and recommended good practice will be highlighted, research on preventive measures will be discussed, and compilations of properties of hazardous compounds and mixtures will be presented.

Vapors and Gases

1. ### Ignition Energy and Quenching Distance

For the experimental detonation of an ignition energy, high-performance (usually air-gap) condensers are used so that the majority of the stored energy will appear in the spark gap. The stored energy may be calculated using the equation $E = CV^2/2$, where C is the capacity of the condenser and V is the voltage just before the spark is passed through the gas. The spark must always be produced by a spontaneous breakdown of the gap because an electronic firing circuit or a trigger electrode would either obviate the measurement of spark energy or grossly change the geometry of the ignition source. It has been found experimentally that for this type of spontaneous spark up to 95 percent of the stored energy appears in the hot kernel of gas in less than 10^{-5} sec. The losses are thought to be due primarily to heat conduction to the electrodes. Since the total stored energy can be varied by changing either the capacity or the voltage, the electrode spacing (which is proportional to the voltage at breakdown) may be varied as an independent parameter. Two problems now arise: if the electrode spacing is too small, the electrodes will interfere with the propagation of the incipient flame and the apparent ignition energy will increase. If, however, the spacing is too large, the source geometry will become essentially cylindrical and the ignition energy will again increase because the area of the incipient flame is greatly increased in this geometry. For such cylindrical ignition, the important quantity is the energy per unit length of the spark, not the total energy. This condition is shown schematically in Figure 1-25a. The fact that the increase in minimum ignition energy is due to quenching at small electrode spacings may be confirmed by using electrodes flanged with electrically insulating material. Figure 1-25b shows the effect of electrode flanging on the ignition energy at small separations. Note that below a certain spacing, as the spacing is reduced, the spark energy required to ignite the bulk of the gas sample rises very much more rapidly with

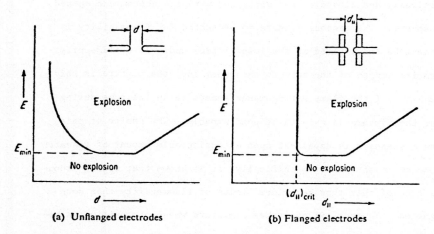

(a) Unflanged electrodes　　　　　**(b) Flanged electrodes**

Figure 1-25.　Quenching Distance and Minimum-Ignition-Energy
Measurement With a Capacitor Spark

flanged tips. This critical flange spacing is defined as the flat-plate quenching distance of the flame. Therefore, this simple experiment may be used to determine the flat-plate quenching distance and the minimum ignition energy for spherical geometry, and may also be used to evaluate whether the ignition source is essentially spherical or cylindrical. The latter piece of information is obtained by determining the effect of electrode separation on E_{min} at large separations.

Quenching distances may also be measured by quickly stopping the flow through a tube of the desired geometry when a flame is seated on the exit of the tube. If the flame flashes down the tube, the minimum dimensions are above the quenching distance for that mixture. Quenching distances that have been measured between two flat plates using this technique agree quite well with the quenching distances measured using the flanged electrodes in the spark-ignition experiment described above.

2. Flammability Limit Determination

The United States Bureau of Mines in Pittsburgh, Pennsylvania, has identified one particular technique for determining flammability limits as being their "standard" technique [Zabetakis (1965) and Coward and Jones (1952)]. In this technique, a 51 mm internal diameter tube 1.8 m long is mounted vertically and closed at the upper end with the bottom end opened to the atmosphere. The gaseous mixture to be tested for flammability is placed in the tube and ignited at the lower (open) end. If a flame propagates the entire length of the tube to the upper end, the mixture is said to be flammable. If the flame extinguishes somewhere in the tube during propagation, the mixture is said to be nonflammable. The choice of this technique as a standard is dependent upon a considerable amount of research that has gone on in the past. Specifically, it is known that upward propagation in a tube of this type has wider limits of flammability than downward propagation. In fact, if one takes a mixture whose composition is between that of the upward and downward propagation limit in this tube, and ignites it at the center of a large vessel, one finds that the flame propagates to the top of the vessel and extinguishes there. Only a portion of the material in the vessel is burned and a fair portion of the fuel-air mixture is left unburned. In other research, it has been found that, as the tube becomes smaller, the combustible range becomes narrower until one reaches the quenching diameter discussed in the previous section. At that point, there is no mixture of that fuel with air which will propagate a flame through the tube. Fifty-one millimeters was chosen as the diameter for the standard tube because this is the diameter at which a further increase in tube diameter causes only a slight widening of the limits that are measured.

The flammable limits of any particular fuel are of considerable importance to industrial safety. For example, a good rule of thumb relative to ventilation systems is that they should ingest sufficient air so that any flammable mixture that is ingested will be diluted to a fuel concentration of less than one-fourth of the flammable limit concentration for that particular fuel.

Flammability limits are not absolute, in the sense that they are unique physical properties such as burning velocity, minimum ignition or energy, or quenching distance, for example. The limit value that is obtained is definitely dependent upon the apparatus, geometry, and flow behavior of the system. The occurrence of a flammability limit is thought to be due to a combination of heat loss from the flame region due to various types of transport, plus the fact that under these limit situations, the flame temperature is so low that the kinetic mechanism changes.

As an example of other investigations that have yielded different values for the flammability limits, we refer to Sorenson, et al. (1975). In that experiment, the limit was determined by using a rich pilot flame to hold a lean limit flame. These authors found a lower limit of 4.3 percent methane in air. This should be compared to a limit of 5.25 percent methane for upper propagation in a standard flammability tube. In another piece of

work, Cubbage and Marshall (1972) discovered that the presence of a large
stoichiometric pocket in a vessel that contained small quantities of methane yielded combustion involvement of the external methane mixture at a
methane concentration of about 4.8 percent. Finally, Hardesty and Weinburg
(1974), by use of a very special burner, were able to burn methane in air
at a concentration of approximately 2.4 percent.

3. Flash Point Determination

Conceptionally, the flash point of a flammable liquid is the
lowest temperature at which the liquid is in equilibrium with air, such
that when a free flame is placed in the vapor phase above the liquid, a
flame will propagate through the vapor phase. The actual flash points
that are determined experimentally are found, however, to be very dependent upon the technique that is used. A good deal of information relative
to flash point determination is summarized in Mullins and Penner (1959).

Unfortunately, there are many "standard" techniques for measuring the flash point of a pure liquid or liquid mixture. In addition to
the closed cup equilibrium method which is most accurate, there are also
"open cup" methods in which there is no top on the cup and the density of the vapor is supposed to keep the partial pressure of the vapor
near the liquid close to its vapor pressure. The American Society of Testing Methods (ASTM) has certified at least six different types of apparatus
and procedures for flash point determination, and the International Standard group has specified its own procedures (see Table 1-6). The problem
is not simple because viscous liquids can be heated only slowly, some liquids tend to "skin" and mixtures of liquids always change their composition as they evaporate. The problem is further complicated by the fact
that different regulatory agencies at the local, state and federal level
have adapted different ASTM apparatus and procedures as standards.

Recently, there has been some new research directed towards the
understanding of the relation between the flash points and the flammability behavior of flammable liquid-air mixtures. Specifically, Affens and

Table 1-6. Some "Standard" Techniques for
Measuring Flash Points

Designation	Common Name	Comments
ASTM D-1310	Tag Open Cup	Open top -- poor because it is difficult to get an equilibrium mixture above liquid.
ASTM D-92	Cleveland Open Cup	Same as for tag open cup (for flash points above 80°C).
ASTM D-56	Tag Closed Cup	Equilibrium method for liquids with viscosities less than 9.5 centistokes at 25°C.
ASTM 93	Pensky-Martins Closed Tester	Used for viscous and liquids that "skin."
ASTM D-3243	Seta Flash Closed Tester	Used for jet fuels 2 ml sample electric heating.
ASTM D-3278	Seta Flash Closed Tester	For flash points between 0°C and 110°C viscosity less than 150 stokes at 25°C.
International Standards Group	Seta Flash Tester	ISO 3579 flash/no flash method, ISO 3580 definitive flash method.

co-workers [Affens, et al. (1977a and 1977b), Affens (1966), and Affens
and McClaren (1972)] have looked at the flammability properties of hydrocarbon solutions in air and have developed a "flammability index" for
such mixtures based on the application of Raoult's law, Dalton's law and
the principle of Le Chatelier's rule governing the flammability of vapor
mixtures. Additionally, Gerstein and Stine (1973) have discussed an interesting anomaly for the case when an inert inhibitor is present. Essentially, they determined the reason why such a mixture can exhibit no flash
point using the standard test procedure but still be flammable. This particular anomaly can be important when one is attempting to inert a hydrocarbon-air mixture with an inhibitor.

82

4. Autoignition Temperature (AIT)

Most techniques that are used in practice to measure the autoignition temperature of a flammable liquid involve the addition of a small
specified quantity of the liquid to a flask of some specified size that
contains air and is held at some high temperature by an external heating
source. In general, the operator watches for a flash of light after introduction of the sample. If a flash is observed within five minutes, autoignition is said to have occurred. As one can imagine, the autoignition
temperature is a very apparatus-dependent quantity. A comprehensive review of the status of autoignition temperature measurements was presented
by Setchkin (1954). More recently, Hilado and Clark (1972) have again reviewed the techniques for determining autoignition temperature and present
a compilation of data for more than three hundred organic compounds. They
report that the most widely accepted apparatus and procedure for determining autoignition temperature is that described in ASTM 2155 [ASTM (1970)].
In this apparatus, a small amount of the sample is injected into a heated
200 millimeter Erlenmeyer flask made of borosilicate glass. The operator
is expected to observe the flask for five minutes after injection of the
sample. If no flash is observed, the temperature is below the autoignition temperature. If a flash is observed, the temperature is above the
autoignition temperature for that particular compound. It must be noted
in passing that the specification of vessel size, shape and wall material
in ASTM 2155 are all required if one is to obtain a reasonably reproducible
value for the autoignition temperature of a flammable liquid. This is because each of these variables has an effect on the measured AIT.

It appears that the determination of autoignition temperature
will always be rather imprecise and apparatus-dependent. Nevertheless, the
relative values of the autoignition temperature of various compounds yields
an indication of allowable surface temperatures of equipment and apparatus
when exposed to air-vapor mixtures of these compounds. Thus, these rather

imprecise and apparatus-dependent autoignition temperatures are useful from a practical standpoint. Also, because of the difficulty of determining a precise autoignition temperature, there is little need for new research in this particular area of combustion safety.

5. Maximum Experimental Safe Gap (MESG)

The need to determine the quantity called the maximum experimental safe gap (MESG) is predicated on a philosophy relative to the design of electrical equipment which is to be operated in a hazardous atmosphere containing combustible vapors or gases. Specifically, the philosophy of all regulations relative to such equipment is that one should assume, unless pressurized purged enclosures are used, that the combustible gases or vapors will be able to enter the enclosure that contains the electrical apparatus. Since the sparks that are produced by the equipment will normally have sufficient energy to cause ignition of the combustible mixtures, the enclosures must be designed such that they are strong enough to contain an internal explosion without rupturing and, also must have clearance gaps to the outside, which are small enough such that venting of the hot products inside the chamber will not cause ignition of the combustible mixture outside the chamber. This phenomenon, i.e., ignition by venting hot product gases under pressure, is uniquely different than the flame quenching process that was described in an earlier section. Here, the hot product gases are being forced through the gap at high velocity and the critical gap widths at which propagation in the surrounding combustible mixture will just not occur are found to be approximately one-half the normal parallel plate quenching distance for most hydrocarbon-air mixtures.

The measurement of a maximum experimental safe gap, therefore, involves the construction of a specific apparatus for this type of test. As it turns out, at the present time, three uniquely different designs have been used rather extensively for MESG testing. These are the 20 milliliter bomb shown in Figure 1-26, the 8 liter bomb shown in Figure 1-27,

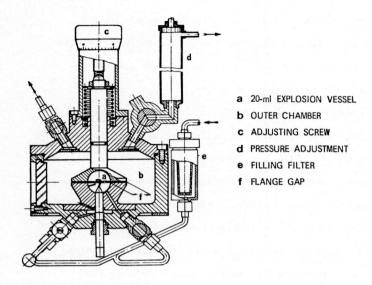

a 20-ml EXPLOSION VESSEL
b OUTER CHAMBER
c ADJUSTING SCREW
d PRESSURE ADJUSTMENT
e FILLING FILTER
f FLANGE GAP

Figure 1-26.   20 ml Explosion Vessel

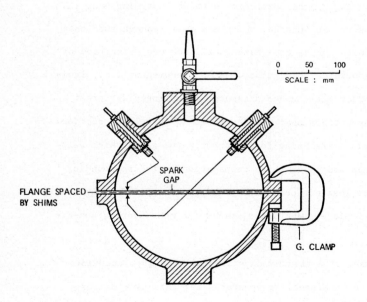

Figure 1-27.   8 Liter Explosion Vessel. (SMRE) 25.4 mm (1 inch)
Flanges, Assembly: 4 or 6 G Clamps

and the Underwriter's Laboratories Westerburg apparatus, which is described
in Dufour and Westerberg (1970) and Alroth, et al. (1976). In principle,
each of these apparatus contains a chamber which is the ignition chamber
and this contains an opening to the outside which consists of a gap of
fixed length and width with an adjustable thickness. In each case, the
apparatus is constructed sturdily enough so that the gap thickness will
not change due to internal pressure during an experiment. Strehlow, et
al. (1978b) showed that because of the large differences in the details of
the construction of these three pieces of apparatus, the dynamic response
during an experiment varies markedly from apparatus to apparatus. The 20
ml apparatus is so small that the total time for exhausting the system with
a nominal burning velocity of 40 cm/sec is of the order of 10 msec. Furthermore the pressure in the apparatus never rises above approximately two atmospheres because of the large gap area relative to the volume of the chamber. In the 8 liter vessel, the absolute size and the ratios of gap size
to the volume of the chamber are such that exhausting time is approximately
100 msec and the internal pressure rises up to 3 to 4 atmospheres. However,
in the Westerberg apparatus, the chamber has a volume of 28.315 l and the
gap has a very narrow width (only 10 cm). This means that in this chamber,
the internal pressure almost reaches that which one would expect from a
closed bomb and the exhausting time approaches one second.

These differences in response time and pressure in the three
vessels has caused large differences in the measured MESG from apparatus to
apparatus for a few oxygen containing hydrocarbons. In all cases, the
MESG for the UL apparatus is lower than that measured in other MESG equipment. This is primarily due to the long duration of the jet in the UL
apparatus and the small relative size of the acceptor vessel. In a number
of cases, retesting in the UL apparatus reduced the anomolous behavior.

The effect of chamber volume and gap length on the measured
MESG for a number of compounds was studied extensively in Britain and Germany over the past 15 years [PTB (1967), Nabert (1967a and 1967b), Lunn

and Phillips (1973), Phillips (1971) and Phillips (1972a, 1972b, and 1973)].

Based on this work, the international standard apparatus for MESG is now

the 20 ml bomb. In the United States, the Westerberg apparatus is used as

the standard. Additionally, in the United States, because electrical sys-

tems are housed in conduit, there is the requirement that one must also

test the compound by igniting it some distance away from the test volume

in a one-inch pipe. This produces a runup to detonation condition which,

in the industry, is called "pressure piling." This pressure piling can

lead to much higher dynamic pressures at the gap than one would observe in

a typical bomb situation. The U. S. testing technique also allows for

tests in which the combustible gas in the enclosure is turbulent, thus caus-

ing a more rapid rate of pressure rise. In designing U. S. electrical proof

equipment, these are normally taken into account.

Phillips and co-workers [Phillips (1972a, 1972b, and 1973)] have

constructed a theoretical model for the safe gap ignition process, which

quite faithfully reproduces the experiment observations. Experimentally,

it is observed that the hot gases escaping through the gap do not carry

a flame with them, but instead, mix with the cold combustible gases exter-

nal to the gap in a highly intense turbulent mixing zone some distance

from the exit of the gap. If conditions are right (i.e., if the gap is

large enough) after some delay time, this region "explodes" and a flame

propagates away from it through the rest of the combustible mixture in

the surrounding enclosure. If the gap is small enough, the temperature in

this region simply drops monotonically with time and no ignition is ob-

served. The Phillips theory predicts this behavior quite well. It is in-

teresting to note that Phillips showed that this jet mixing region behav-

ior is similar to the temporal behavior of a vessel explosion as illustrat-

ed in Figure 1-3. In this case, however, the critical variable is not the

concentration of the reactants but the gap width, and the heat loss term is

not due to losses to the wall but instead is due to the entrainment of

cold combustible gases. Nevertheless, the same delay to "explosion" (or in this case, flame propagation) and a predicted characteristic rate of temperature rise due to the rate of the chemical reaction is observed.

The National Electric Code defines four classifications for gases or vapors. These are A, B, C, and D. Class A contains only the compound acetylene. The standard compound for Class B is hydrogen, and that for Class C is diether ether. Gasoline is the standard compound which defines the safe gap for Class D compounds usage. For example, if a measured safe gap for a particular compound lies between that of diether ether and that of gasoline, it is considered to be a Class C compound and can be used only in locations where the equipment has been classified as Class C equipment. In order to obtain a classification of this type, the equipment itself must be tested in an explosive environment using diether ether as the fuel. If the compound has a safe gap which is larger than that of gasoline, it will be considered to be Class D, and any equipment must be tested with gasoline as a fuel before it can be used in an atmosphere that contains that particular fuel.

6. Inductive Spark Ignition

In ordinary industrial environments, most frequently a spark is obtained by breaking a circuit that contains a 110, 220, or 440 volt, 60 cycle, electric current. Testing to determine ignition currents required under such conditions has been done rather extensively in Germany [PTB (1967)] and has led to the concept of minimum igniting current. It has been found experimentally that the minimum ignition current produced during a break circuit spark is related to the maximum experimental safe gap for that particular compound in the same way that minimum ignition energy is related to quenching distance, that is $I_{min} = C(MESG)^2$. There are essentially no tests of inductive sparks performed in the United States.

Mists

In many situations of catastrophic fuel release, a mist of the fuel is formed in air, particularly if the fuel is a higher hydrocarbon. The most important difference between a fuel mist-air mixture and a fuel vapor-

air mixture is the fact that the upper flammability limit for mist is higher than that for a vapor. This is particularly true when the molecular weight
of the fuel is large and its volatility is low. This is because for small
droplets of low vapor pressure fuels, the mist will evaporate almost completely before a flame reaches the location of the droplets. In this circumstance, a fuel mist-air mixture will burn essentially like a premixed
fuel vapor mixture, and the flame will look somewhat like a premixed laminar flame. However, as the molecular weight of the fuel increases and
volatility decreases, the rate of evaporation of the fuel droplets as they
pass through the preheat zone of the flame decreases markedly and the
flame becomes a series of diffusion flames surrounding individual droplets. Under these circumstances, flame propagation can occur even though
all the fuel is not consumed in each droplet. Thus, the rich limit for a
fuel mist is much broader than it is for a pure vapor. Weatherford, et
al. [Wimer, et al. (1974), Weatherford (1975), Weatherford and Schakel
(1971) and Weatherford and Wright (1975)] have performed a series of experiments on the relative flammability and rate of fire spread during ignition and combustion of fuel mist-air mixtures. Very little else has
been done.

Combustible Dusts

1. Ignition Energy

The apparatus that is used for measuring dust cloud ignition
energy is similar to that used for measuring the ignition energy of gases
and vapors. After the dispersion of the dust into the bomb by a burst of
air, a spark from a condenser discharge is passed through the mixture and
the minimum ignition energy is determined by the usual technique. Dorsett,
et al. (1960) have described the technique. For most dusts, ignition energies lie in the range of 5 to 60 millijoules using this technique.

Recently, Eckhoff (1975 and 1976) and Eckhoff and Enstad (1976)
have pointed out that the ignition process for a dust cloud is more complex than it is for a vapor or gaseous fuel. Specifically, they found

that if the spark current is limited by an external resistance, the total energy needed for ignition that is stored in the capacitor is much less than if there is no external resistance. In this case, a good deal of the energy that is stored in the capacitor ends up as heat energy in the resistor and, therefore, the actual energy delivered to the gas is considerably lower than that delivered by a purely capacitance spark. Also, "long" electrical sparks appear to be more effective as dust explosion initiators than "short" sparks.

2. Explosion Severity

During the period 1958 to 1968, Jacobson, Nagy and co-workers [Jacobson, et al. (1961, 1962, and 1964), Nagy, et al. (1964a, 1964b, 1965, and 1968), Dorsett and Nagy (1968), and Dorsett, et al. (1960)] performed a standardized series of tests on a large variety of combustible dusts that are important to industry. For these experiments, they used the Hartmann apparatus which is a 75 cubic inch stainless steel cylindrical chamber, 13 inches long and mounted vertically. After placing a quantity of the test dust in the bottom of the chamber, a burst of air is injected to disperse the sample uniformly throughout the chamber. A relatively high energy spark is used to ignite the dust mixture. If ignition occurs, both the maximum pressure and the rate of pressure rise are measured from the pressure time history of the explosion. Experiments are performed at a number of dust levels ranging from 0.1 to 2.0 gm/m^3. With this technique, the minimum concentration for combustion and the optimum concentration for maximum rate of pressure rise can be determined.

Additionally, the minimum ignition temperature of a moving dust cloud is determined using the Godbert-Greenwald [Godbert and Greenwald (1936) and Godbert (1952)] furnace. The furnace is mounted vertically and the dust is passed through the furnace as a cloud from above using a burst of air escaping from a 100 psig reservoir. The temperature of the furnace is raised in successive experiments until the dust ignites and a flame is seen to issue from the furnace. Based on these standardized ex-

perimental techniques, two quantities have been defined relative to dust
explosion behavior. One of these is called explosion severity. It is
the product of the maximum explosion pressure produced in the Hartmann
bomb multiplied by the maximum rate of pressure rise in the Hartmann bomb.
To normalize explosion severity to a standard dust, the authors chose to
use float dust of Pittsburgh seam coal as a standard dust. Thus, the ratio of this product for the dust in question to the same product for
Pittsburgh seam dust yields the "explosion severity," a dimensionless ratio. The authors also define a quantity called the ignition sensitivity
of the dust. This is, again, normalized to Pittsburgh seam coal dust.
It is a product of the ignition temperature as determined using the standard furnace multiplied by the minimum ignition energy and the minimum dust
concentration which will just burn. In this case, the ratio is taken to
be that for Pittsburgh seam coal dust divided by the test dust product.
A large value of the explosion severity index and the ignition sensitivity
index indicates a dust which is much more dangerous than Pittsburgh seam
dust both in the severity of the explosion produced and its ease of ignition. The authors further define an index of explosibility, which is the
product of the ignition sensitivity and the explosion severity indices.
It should be pointed out that, for comparisons to be made, all the numbers
that are used in calculating these ratios must be generated in the same
apparatus in the same manner. Since so many tests with different dusts
have been performed in the Hartmann bomb, it has become, at least in the
United States, a standard test apparatus for dust explosion studies.
Hertzberg, et al. (1979) have modified this test procedure.

Recently, Eckhoff (1977) has presented some ignition energy
studies and Burgoyne (1978) has discussed testing techniques relative to
dust explosibility. Bartknecht (1980) has reported experiments in a one
cubic meter bomb as well as in a twenty liter and one liter bomb, which
are similar to the Hartmann apparatus. Theoretically, as they and Bartknecht have pointed out, the product of dP/dt multiplied by the volume of

the vessel to the one-third power should be a constant independent of the
size of the vessel. This constant is based on the same fundamental concept discussed earlier for gas explosions in low L/D vessels and is called
K_{St} by Bartknecht. Bartknecht has shown that the value of the constant
for a large vessel can be as much as 10 to 15 times larger than that for
a small vessel, particularly when the value of dP/dt_{max} in the small vessel (Hartmann bomb) is high. This is shown in Figure 1-28. He also points
out the fact that if one delays ignition after dispersing the dust, the
value of dP/dt_{max} drops precipitously while the maximum pressure obtained
in the bomb stays relatively constant. The reason for this is rather

obviously that the turbulence level in the bomb drops with time after dispersion, and that ignition of the mixture while it is highly turbulent
will produce a higher flame velocity, and therefore, a more rapid rate
of pressure rise in the bomb.

Bartknecht, in his studies, has come to the conclusion that one
should classify dusts into three categories, based on the value of K_{St} measured in the laboratory. In his classification scheme, Class I dusts are

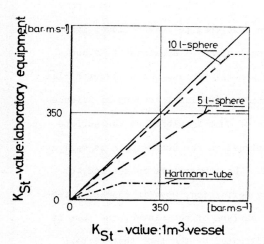

Figure 1-28. Comparison of K_{St}-Values Measured in Laboratory
Equipment With Those Obtained in a Large Vessel
(Averaged Straight Lines)

least dangerous and Class III dusts are most dangerous. Venting or suppression device response requirements are then based on the classification of the dust. This will be discussed further in Chapter 3.

Even though the situation is not completely resolved at the present time, the explosibility constant, K_{St}, is an important quantity because it can be used to estimate the relative danger of handling various types of dusts. NFPA 68 [NFPA (1978b)] has adopted Bartknecht's approach to venting that is summarized in Bartknecht (1980).

3. Layer Ignition Temperature

Dusts have the property that they settle out of the air and form surface layers on all horizontal upper surfaces. If housekeeping is poor, these dust layers can become quite thick. Under situations where the dust settles on a surface which is heated artificially (for example, the surface of a motor housing, or the reflector shield above an incandescent light bulb), the dust can be heated until spontaneous ignition of the dust layer occurs. This generally represents only a fire hazard, but if the smoldering dust layer is disturbed in the right manner, it could represent an explosion hazard, because it could act as an ignition source for a dust cloud.

Standardized tests have been developed to measure layer ignition temperature of dusts. The most usual piece of apparatus is a hot plate, which has been modified so that a metal surface which is quite flat reaches a temperature which is quite uniform [National Academy of Sciences (1981)]. At some time, t = 0, a sample of dust is placed on the surface in a container, and thermocouples are placed in the dust sample as well as on the surface of the hot metal platen. The temperature in the body of the dust is monitored with time, and by definition, when it reaches a temperature which is 50°C higher than the surface of the platen, the dust is said to have ignited. The test is complicated by the fact that certain dusts melt, and other dusts swell or froth when they are heated in this manner. Therefore, the ignition temperature is not always clear cut.

Nevertheless, layer ignition temperatures quite frequently are reported as a separate property of the dust and do have meaning relative to the behavior of the dust in contact with hot surfaces in an industrial environment. A new recommended procedure will appear as National Academy of Sciences (1981).

    4. Conductivity

        In general, dusts can be characterized as either being conductive or nonconductive. Most grain dusts, plastic dusts, etc., have very large resistivities and are considered to be nonconductive. Virtually all metal dusts have very high conductivities and are considered to be conductive. In the region between these two extremes, some graphite dusts have intermediate conductivities. However, one can consider most dusts to be either conductive or nonconductive. Safety in plant operations requires a different set of techniques whether or not the dust is conductive. This is because conductive dusts can form electrical paths when they settle on electrical apparatus. Thus, ignition can be effected by the fact that the dust is there, as contrasted to the case for a vapor mist or for a nonconductive dust where the combustible itself does not form a conducting path for the electric current. The electrical code, therefore, recognizes conductive dusts as a special hazard.

        There is, as yet, no standard test for conductivity. However, one is being suggested in National Academy of Sciences (1981). The usual procedure for determining the conductivity of a relatively nonconductive dust is to place the dust between two conducting metal bars mounted on a plate which has a very high electrical resistivity. The dust is spooned off so that the sample has a fixed thickness, length and height. Then the current between the two electrodes is measured for different applied voltages. DC voltages are preferred in this test. The resistivity of most nonconductive dusts appears to be slightly affected by the voltage in the range from ten to a thousand volts across a gap of approximately 10 mm. It has been noted, of course, that the moisture content of the dust is very impor-

94

tant to its measured conductivity. If the dust is very conductive, such as a metal dust, this steady state technique will not work because the conductivity is so high that the current will be very large. It has been suggested that any dust that has been prepared from a conductive solid (such as a metal dust) can be considered to be conductive without further testing.

Explosives and Propellants

Four techniques are usually used to determine the sensitivity of explosives and propellants so that safe storage and handling procedures can be specified. These are thermal decomposition studies at elevated temperatures, the impact test, the card gap test, and the test for minimum charge diameter.

The principles of an isothermal decomposition study were outlined in the first section of this chapter under the title vessel explosions. Referring to Figure 1-29, one finds that for low thermostat temperatures, the central temperature of the sample will rise to a temperature only slightly above the thermostat temperature. As the thermostat temperature is raised in subsequent experiments, the temperature difference between the center of the

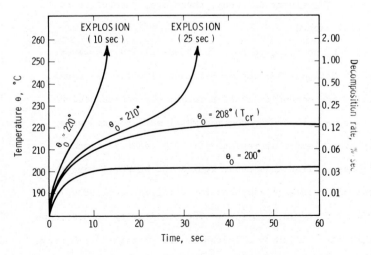

Figure 1-29. Heating Curves (Temperature Versus Time) in an Experiment Using an Isothermal Calorimeter
[U. S. Army Material Command (1972)]

sample and the thermostat will increase until the central temperature reaches some critical value. At bath temperatures higher that this, the sample will always explode after some induction delay time. When the bath temperature is raised to values higher than this critical value, the delay time to explosion will decrease. As outlined in the first section, the information on critical temperature and delay time can be used to deduce the kinetic parameters for the explosive.

In adiabatic calorimetry, the calorimeter is heated externally in such a way that it exactly matches the sample temperature at any time. Under these conditions, all samples of exothermic substances will eventually explode. However, since at low temperatures, the time to explode will be very long, one usually performs these experiments by increasing the calorimeter temperature in increments until at some specific increment, self-heating is seen to occur. At that point, the calorimeter is switched to adiabatic operation and the entire temperature-time history to explosion is observed. The equations for adiabatic explosion are applicable. A recent, interesting application of this technique is outlined by Townsend (1977).

In the impact test, a hammer is allowed to fall on a striker plate and a small sample of the substance under test is placed below the striker plate on a heavy metal anvil. The height of the drop is varied and the height at which 50 percent of the drops produce ignition of the sample is recorded as a measure of the ignition sensitivity of the explosive or propellant. This test is not very precise or reproducible, but it does give a measure of how easy it is to ignite a particular material by shock impact. Rarely, if ever, does the ignition produce a detonation in the sample. However, propellants as well as explosives usually exhibit ignition in this apparatus. Typical examples of this test and variations in design for both solid and liquid explosives are discussed by Macek (1962) and U. S. Army Material Command (1972).

The card gap test is used to determine initiation sensitivity of high explosives and propellants. A typical card gap test set-up is shown in Figure 1-30, as standardized by the Naval Ordnance Laboratory. In this test, the cards are 0.25 mm sheets of cellulose acetate or lucite. It is assumed that detonation was initiated in the sample if the mild steel plate has a hole punched in it after the experiment. The inert gap width is increased in successive tests until the witness plate is only distorted during an experiment. The principle here is that the shock wave transmitted through the cards is attenuated and that for each propellant or explosive, there is some minimum shock strength which will just cause transition to detonation in the sample. Thus, the larger the card gap length, the more shock sensitive the explosive. This technique has limited utility because if the minimum charge diameter of the explosive is greater than the experimental charge diameter, the explosive will appear to be extremely insensitive even though in larger sizes it may be sensitive to initiation.

The test for minimum charge diameter also has many variations. Usually, long cylindrical bare charges are used to determine the smallest

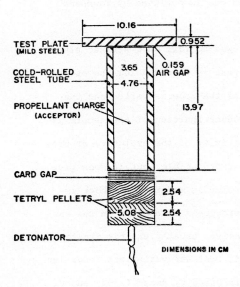

Figure 1-30. Charge Assembly and Dimensions for
NOL Gap Test

diameter that will just allow a constant velocity detonation to propagate
without decay. At, or just above, the minimum diameter, the measured
detonation velocity is always somewhat less than Chapman-Jouguet because
of the influence of the strong rarefaction fan that occurs because the
high pressure product gases are surrounded by atmospheric pressure air.

HAZARDOUS MATERIALS DATA AND HAZARD EVALUATION

Data

There are a number of different ways that compounds are rated according to their relative hazardousness. Simple ratings have been supplied
by the Manufacturing Chemist Association (MCA) in the form of chemical safety data sheets. Also, the National Fire Protection Association in NFPA 325M
[NFPA (1977b)] supplies simple hazardous ratings for a number of compounds
relative to fire hazard toxicity and reactivity. This last scheme has a
simple scale, from 0 to 4. Compounds rated 0 are considered to be perfectly safe while those rated 4 are highly dangerous. In general, substances
are placed in one category or another, based on accumulated experience gained
while handling the substance.

Recently, there have been a number of attempts to find a logical way,
based on the thermodynamic and reaction kinetic properties of different substances and mixtures, to somehow rate these substances. Stull (1977) discusses a technique for determining relative activity based on a nomograph
which combines a decomposition temperature of the substance and the activation energy of the decomposition process. He shows, for a few simple compounds, that relative ratings obtained from his nomograph agree with the
relative NFPA ratings of these compounds. Tsang and Domalski (1974) and
Domalski (1977) present a very complete compendum of methods for estimating self reaction hazards and they also discuss their current status and
interrelationships. They discuss testing techniques in some depth and various predictive schemes such as CHETAH, CRUISE, and TIGER. Dow Chemical
Company (1973) also presents a technique for evaluating relative hazard.

Comparisons are also made as to the relative efficiency of rating schemes based on (1) the enthalpy of polymerization, (2) the maximum enthalpy of decomposition, (3) the heat of combustion less the heat of decomposition, (4) the oxygen balance of the compound, and (5) a modified maximum enthalpy of decomposition of the compounds. It appears that the correlations are not really very good, and that further work along these lines must be done if an unambiguous thermodynamic-thermokinetic technique for evaluating unknown compounds is to be obtained.

There are many compilations of the properties of hazardous materials. These include Nabert and Schön (1963 and 1970), National Academy of Sciences (1973, 1975a, 1975b, 1978, and 1979), McCracken (1970), Sax (1965), Chemical Dictionary (1971), Weast (1979), Loss Prevention Handbook (1967), MERCK Index (1960), Coward and Jones (1952), Zabetakis (1960), NFPA 325 M [NFPA (1977b)], and NFPA 49 [NFPA (1975)].

Codes and Procedures

1. Codes and Good Practices

The National Fire Protection Association National Fire Codes (1980) and the American National Standards Institute (ANSI) standards have been mentioned before. They are, by far, the largest set of guidelines available in the United States relative to the manufacture, transport, storage, handling, and use of hazardous materials. In addition, there have been a number of discussions and descriptions of good plant design, such as those presented by ISA (1972), Russell (1976), Le Vine (1972a and 1972b), Institute of Chemical Engineering (1979), Rasbash (1970 and 1973), and the Oil Insurance Association (1974) as well as books that discuss safety aspects of plant design such as Wells (1980), Bodurtha (1980), Hammer (1980), and Koll and Ross (1980).

2. Procedures

Even though the codes and standards listed in the previous section cover most every conceivable situation, it is worthwhile to summarize the basic approach to good combustion safety using Figure 1-31 as an out-

line. This figure gives three parallel ways to reduce the risk of manufacturing, transporting, storing, or using hazardous combustible materials once
the material is considered "tolerable" (an example of an untolerable material would be nitrogen trichloride, which explodes if you shake the flask
that it is in). The three approaches should be considered as parallel approaches, all of which should be used, in proper balance, to maximize safe
plant operation. Some of these are discussed in NFPA 69 [NFPA (1978a)] and
they will be briefly reviewed here. They start on the left edge of Figure
1-31 with passive techniques. These will be discussed first. Then active

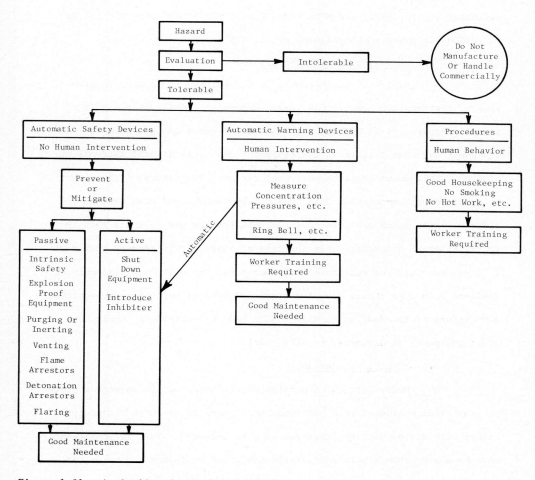

Figure 1-31. An Outline for Combustion Safety

techniques which are either automatic or require human intervention will be
discussed. Finally, procedural matters, including worker training, will be
discussed.

a. Passive Techniques

- ### Intrinsic Safety

In certain circumstances, it is possible to have a piece of
electrical equipment labeled "intrinsically safe." Under these circumstances, none of the sparks that can be produced by this particular piece
of equipment will be strong enough to cause ignition of any flammable vapor mixtures that are available. The principles of intrinsic safety are
discussed in ISA (1972) and NFPA 493 [NFPA (1978c)], and some partial applications in mines in Litchfield, et al. (1980).

- ### Explosion Proof Equipment

Explosion proof equipment is designed to be used in atmospheres that contain combustible gases, vapors or dusts. In the case of
gases or vapors, it is assumed that the gas or vapor can get into the
equipment and that a spark in the equipment can cause a local internal explosion. This means that the equipment designed for vapors or gases must
be strong enough to sustain this internal explosion, and also must have
clearance gaps which are less than the MESG for the gas or vapor that is
in the external atmosphere. In the case of dusts, it is assumed that dust
will not get into the enclosure and, therefore, the enclosure need only be
designed such that its surface temperatures never exceed the layer ignition
temperature of the dust in question. The proper location for explosion
proof equipment is discussed in NFPA (1981).

- ### Purging or Inerting

Under certain circumstances, the only way to operate a piece
of electrical equipment in a hazardous atmosphere is to purge the equipment
either with air or with an inert gas such as nitrogen. This is considerably
more expensive than simply using explosion proof equipment. However, if
the electrical equipment that must operate in the hazardous atmosphere is a

very sensitive piece of equipment, such as instruments for measuring concentrations or flow in a process line, the instrument itself could not survive
the internal explosion that is tolerated in an explosion proof enclosure.
Under these circumstances, each electrical enclosure must be purged with an
inert gas or air and be kept under a positive pressure at all times so that
all leakage from the enclosure is outward. Guidelines for this approach
are given in NFPA (1978a and 1974) and by Ecker, et al. (1974).

- Explosion Release (Venting)

The proper venting of buildings or process vessels in line
or ducts is discussed in Chapter 3, as well as NFPA 68 [NFPA (1978b)].

- Flame and Detonation Arresters

Another approach to stopping a flame from propagating a long
distance down a line into a process vessel or bag house is to place flame
arresters or flame traps in the line. A design of such flame traps is discussed by Cubbage (1959), Komamiya (1969), Howard and Russell (1972), and
in a Ministry of Labor report (1965). Detonation arresters are difficult
to construct and cause a much more severe pressure drop in the line. The
successful operation of such a detonation arrester is discussed by Sutherland and Weigert (1973). In this case, an acetylene detonation traveled
through a seven mile pipe towards the process equipment and was successfully stopped by a liquid bubble arrester placed in the line before it entered
the equipment.

- Flaring

When there is an upset in the plant operation which requires
the dumping of considerable quantities of combustibles, it is standard
practice to lead these gases to a tall stack and burn them at the top of
the stack as they escape. Schwartz and Keller (1977) discuss flare stack
operation and good practice.

b. Active Techniques

Active techniques used in safety devices are techniques
which require that an abnormality such as high pressure or temperature be

sensed by an automatic device and that this sensor then automatically shuts down the plant or introduces inhibitors to stop a combustion explosion, for example. Bartknecht (1981) describes suppression devices in some detail.

If automatic devices are used, they require the use of good preventive maintenance to ensure that they do not malfunction.

c. Automatic Warning Devices

Automatic warning devices also sense abnormal behavior but they require the intervention of a trained worker to take proper corrective action. Thus, in this case, not only is a preventive maintenance program for the sensor required, but proper training for the worker is also necessary.

d. Procedures

Here, proper worker training is absolutely essential. The person who is being trained must understand the need for good housekeeping, smoking, and hot work permit regulations imposed by management, and must somehow be motivated to act in a responsible manner. The best way to accomplish the needed worker awareness is to have a broad-based safety program where routine and emergency procedures are discussed with the workers at regular intervals. The worker will become concerned and know what to do only if management shows its concern.

3. Risk Analysis

Risk analysis is a relatively new technique that is being employed to evaluate the hazardous consequences of an accident occurring during any particular current operation or planned operation. The concept of acceptable risk has been discussed by Lawrence (1976). It normally involves the construction of a scenario or series of scenarios of expected sequences of events if an accident involving a hazardous material were to occur. The analysis is usually very specific. A good example of such a specific analysis is the Eichler and Napadensky (1977) evaluation of the effects of an accidental vapor cloud explosion on nuclear plants located near transportation routes. In this evaluation, they specifically tried to determine the safe standoff distance between a transportation route which handles combustibles and a nuclear plant, such that the overpressure from any blast wave at the plant site would be one psi or less. An interesting compilation of a number of different early risk analysis approaches has been presented in National Academy of Sciences (1977). This document discusses risk analysis for the water transportation of hazardous materials. Specifically, in its Appendix F, 12 separate approaches to risk analysis that had been attempted prior to that time are reviewed.

More recently, the concept of risk management has emerged [Griffiths (1981) and Slovic, et al. (1979)], and fault tree analysis has become a mature approach to risk management [Fleming, et al. (1979), Prugh (1980), and Allen and Rao (1980)]. There is even a new journal entitled Risk Analysis (1981). It appears as though any company that has maintained adequate records of previous accidents can use fault tree analysis to restructure their priorities towards risk.

CHAPTER 1

LIST OF SYMBOLS

A	pre-experimental factor
AIT	autoignition temperature
a	air or oxidizer concentration; sound speed
C	carbon; capacitance
C_p	specific heat at constant pressure
C_v	specific heat at constant value
C_1, C_2	specific heat at constant volume
c_1, c_2, $\cdots$ c_n	volume percents of fuels
D, D_1, D_2	concentration-dependent terms
D_{cr}	critical value of D
d_o	tube quenching distance

$d_{\|\|}$	parallel plate quenching distance
E	Arrhenius constant
E_f	flame energy
E_{min}	minimum ignition energy
F	fuel concentration
f	fuel concentration
H	hydrogen
h	heat transfer coefficient
h_1, h_2	enthalpies
I_{min}	minimum ignition energy
i, j	subscripts for chemical species and type of reaction
K_g	Bartknecht constant for a gas
k_u	thermal conductivity
L/D	ratio of length to diameter
LEL	lower explosion limit
L_1, L_2, ... L_n, L_m	volume percent lean limits
m	mass
N	nitrogen
n, m	exponents
O	oxygen; oxidizer concentration
P	a product
P_f	flame pressure
P_1, P_2	pressures
Q	heat of reaction per mole of product
q	nondimensional energy
R	universal gas constant
S	sulfur
S'	effective burning velocity
S_s	space velocity
S_u	burning velocity
s	vessel surface area

t	time
t_{ign}	ignition time
U_{CJ}, M_{CJ}, A_{CJ}, P_{CJ}	Chapman-Jouguet parameters
UEL	upper explosion limit
u, v, w, x, y	subscripts indicating mole fractions of elements in a fuel
u_1, u_2	flow velocities
V	vessel volume; voltage
V_1, V_2	specific volumes
β	time delay
ΔH_c	heat of combustion
ΔH_f	heat of formation
ΔH_r	heat of reaction
δP	differential pressure
δQ	differential heat
η_o	preheat zone thickness
γ	ratio of specific heats C_p/C_v
λ	a constant
ν_{ij}	stoichiometric coefficient
ϕ	fuel equivalence ratio
ψ	a ratio of burning velocities
ρ, ρ_1, ρ_2	densities
ρ_b	density of burnt gas
ρ_u	density of unburned gas
τ	time delay before ignition
θ	temperature
θ_f	flame temperature
θ_{ign}	ignition temperature
θ_o	initial temperature
θ_u	temperature of unburned gas
θ_2	critical temperature

CHAPTER 2

FREE-FIELD EXPLOSIONS AND THEIR CHARACTERISTICS

GENERAL DEFINITION OF AN EXPLOSION

Before discussing explosions, let us first define them. The dictionary definitions of an explosion are: 1) bursting noisily, 2) undergoing a rapid chemical or nuclear reaction with the production of noise, heat, and violent expansion of gases, and 3) bursting violently as a result of pressure from within. From our somewhat more scientific viewpoint, we quote from Strehlow and Baker (1976) another general definition of an explosion:

"In general, an explosion is said to have occurred in the atmosphere if energy is released over a sufficiently small time and in a sufficiently small volume so as to generate a pressure wave of finite amplitude traveling away from the source. This energy may have originally been stored in the system in a variety of forms; these include nuclear, chemical, electrical or pressure energy, for example. However, the release is not considered to be explosive unless it is rapid enough and concentrated enough to produce a pressure wave that one can hear. Even though many explosions damage their surroundings, it is not necessary that external damage be produced by the explosion. All that is necessary is that the explosion is capable of being heard."

The definition just given refers to explosions in air. Damaging explosions can, of course, occur in other media -- water or earth. However, our present concern with accidental explosions essentially precludes underwater or underground explosions, because the vast majority of explosions in

Table 2-1. Explosion Types
[Adapted from Strehlow and Baker (1976)]

Theoretical Models	Natural Explosions	Intentional Explosions	Accidental Explosions
Ideal point source	Lightning	Nuclear weapon explosions	Condensed phase explosions
Ideal gas	Volcanoes	Condensed phase high explosives	Light or no confinement
Real gas	Meteors	Blasting	Heavy confinement
Self-similar (infinite source energy)		Military	Combustion explosions in enclosures (no prepressure)
Bursting sphere		Pyrotechnic separators	Gases and vapors
Ramp addition (spark)		Vapor phase high explosives (FAE)	Dusts
Piston		Gun powders/propellants	Pressure vessels (gaseous contents)
Constant velocity		Muzzle blast	Simple failure (inert contents)
Accelerating		Recoilless rifle blast	Combustion generated failure
Finite stroke		Exploding spark	Failure followed by immediate combustion
Reaction wave		Exploding wires	Runaway chemical reaction before failure
Deflagration		Laser sparks	Runaway nuclear reaction before failure
Detonation		Contained explosions *	BLEVE's (Boiling Liquid Expanding Vapor Explosion) (pressure vessel containing a flash-evaporating liquid)
Accelerating waves			External heating
Implosion			Immediate combustion after release
			No combustion after release
			Runaway chemical reaction
			Immediate combustion after release
			No combustion after release
			Unconfined vapor cloud explosions
			Physical vapor explosions

* Contained vessel explosions such as those used in gas and dust explosion research, and explosions in internal combustion engine cylinders are examples.

these media are deliberate ones conducted for military purposes or peaceful
purposes such as blasting.

There are actually many types of processes which lead to explosions
in the atmosphere. Table 2-1 contains a listing of explosion sources, including natural explosions, intentional explosions and accidental explosions. The list is by type of energy release and is intended to be comprehensive. Included in Table 2-1 are theoretical models of sources used
for characterizing and studying explosions. Usually, these models are tractable idealizations of real processes.

We will draw rather freely in this chapter on three other references
besides the Strehlow and Baker (1976) paper already mentioned; they are
Baker (1973), Baker, et al. (1974b), and Strehlow (1980a).

"IDEAL" EXPLOSIONS

Source Properties

Some of the characteristics of explosions in air can be strongly affected by the character of the sources which drive the blast wave. In our
definition of an explosion, we stated that energy must be released over a
sufficiently small time and in a sufficiently small volume for the process
to be classed as an explosion. The general properties of an explosion
source which govern the strength, duration and other characteristics of the
resulting blast wave must then include at least its total energy, E; its
energy density, E/V; and the rate of energy release, i.e., its power. Four
of the sources listed in Table 2-1 have such a high energy density and power
that they produce blast waves which are essentially identical at and below
shock pressure levels which cause complete destruction. It has been found
that the blast wave produced by these "ideal" explosions can be correlated
entirely by a single parameter, the total source energy, irrespective of
the energy density or the power of the source. These four "ideal" explosion sources are point source, nuclear weapon, laser spark, and condensed
phase explosives. Nuclear weapon and laser spark explosions are included

here for completeness. They will not be discussed extensively in this text.

Point source explosions represent a mathematical approximation which has important theoretical implications. Explosions produced by condensed phase explosives are very important to the discussion of accidental explosions primarily because most experimental data on ideal blast waves have been accumulated by detonating condensed phase (high) explosives.[†] These explosives actually produce ideal blast waves because they are high density materials and, therefore, have a high energy content per unit volume. In military applications, however, the energy per unit mass or weight is a useful indicator of the explosives relative efficacy, and therefore, is usually reported along with the density of the explosive.

The following discussion will introduce a number of high explosives used for military and commercial purposes. Most are solid at room temperature, while some are liquids and some are semi-solids (gels). A typical military solid explosive is TNT. Typical blasting explosives are the various dynamites and commercial blasting agents, while nitroglycerin is a potent liquid explosive which is seldom used by itself but is used as an ingredient in dynamites and high-energy propellants. The range of energy densities for most condensed explosives is surprisingly small, as can be seen by comparing their energy density to that of TNT (TNT equivalency), the second column of Table 2-2. Probably the most energetic condensed explosive mixture on the basis of energy per unit mass (but a highly unstable one) is a stoichiometric mixture of liquid hydrogen and liquid oxygen, at 16,700 kJ/kg. This value can be used to refute the claims of anyone who states that he has discovered a chemical explosive more than 3.7 times as powerful as TNT (which has a specific detonation energy of 4520

[†]Refer to Baker (1973) for a detailed discussion of condensed phase explosions.

Table 2-2. Conversion Factors (TNT Equivalence)
for Some High Explosives

Explosive	Mass Specific Energy, E/M, kJ/kg	TNT Equivalent, $(E/M)_x/(E/M)_{TNT}$	Density, Mg/m^3	Detonation Velocity, km/s	Detonation Pressure, GPa
Amatol 80/20 (80% ammonium nitrate, 20% TNT)	2650	0.586	1.60	5.20	----
Baronal (50% barium nitrate, 35% TNT, 15% aluminum)	4750	1.051	2.32	----	----
Comp B (60% RDX, 40% TNT)	5190	1.148	1.69	7.99	29.5
RDX (Cyclonite)	5360	1.185	1.65	8.70	34.0
Explosive D (ammonium picrate)	3350	0.740	1.55	6.85	----
HMX	5680	1.256	1.90	9.11	38.7
Lead Azide	1540	0.340	3.80	5.50	----
Lead Styphnate	1910	0.423	2.90	5.20	----
Mercury Fulminate	1790	0.395	4.43	----	----
Nitroglycerin (liquid)	6700	1.481	1.59	----	----
Nitroguanidine	3020	0.668	1.62	7.93	----
Octol 70/30 (70% HMX, 30% TNT)	4500	0.994	1.80	8.48	34.2
PETN	5800	1.282	1.77	8.26	34.0
Pentolite 50/50 (50% PETN, 50% TNT)	5110	1.129	1.66	7.47	28.0
Picric Acid	4180	0.926	1.71	7.26	26.5
Silver Azide	1890	0.419	5.10	----	----
Tetryl	4520	1.000	1.73	7.85	26.0
TNT	4520	1.000	1.60	6.73	21.0
Torpex (42% RDX, 40% TNT, 18% Al)	7540	1.667	1.76	----	----
Tritonal (80% TNT, 20% Al)	7410	1.639	1.72	----	----
C-4 (91% RDX, 9% plasticizer)	4870	1.078	1.58	----	----
PBX 9404 (94% HMX, 3% nitrocellulose, 3% plastic binder)	5770	1.277	1.844	8.80	37.5
Blasting Gelatin (91% nitroglycerin, 7.9% nitrocellulose, 0.9% antacid, 0.2% water)	4520	1.000	1.30	----	----
60 Percent Straight Nitroglycerin Dynamite	2710	0.600	1.30	----	----

NOTE: The values for mass specific energy and TNT equivalence in this table are based on reported experimental
values for specific heats of detonation or explosion. Calculated values are usually somewhat greater
than those given in the first column of this table. Dobratz (1981) gives many calculated, and some experimental, values for high explosives.

kJ/kg). Most condensed explosives also have the property of exploding at

essentially a constant rate for a given initial density and explosive type,

called the detonation velocity D. Typically, detonation velocities range

from about 1.5 km/s for some blasting explosives to 8 km/s for military explosives. The detonation process is discussed in some detail in Chapter 1 of this text, by Cole (1965), and in the Engineering Design Handbook (1972). Table 2-2 lists the properties of a number of common high explosives.

Blast Wave Properties

As a blast wave passes through the air or interacts with and loads a structure, rapid variations in pressure, density, temperature and particle velocity occur. The properties of blast waves which are usually defined are related both to the properties which can be easily measured or observed and to properties which can be correlated with blast damage patterns. It is relatively easy to measure shock front arrival times and velocities and entire time histories of overpressures. Measurements of density variations and time histories of particle velocity are more difficult, and no reliable measurements of temperature variations exist.

Classically, the properties which are usually defined and measured are those of the undisturbed or side-on wave as it propagates through the air. Figure 2-1 shows graphically some of these properties in an ideal wave [Baker (1973)]. Prior to shock front arrival, the pressure is ambient pressure p_o. At arrival time t_a, the pressure rises quite abruptly (discontinuously, in an ideal wave) to a peak value $P_s^+ + p_o$. The pressure then decays to ambient in total time $t_a + T^+$, drops to a partial vac-

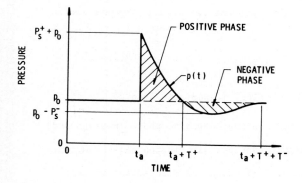

Figure 2-1. Ideal Blast Wave Structure

112

uum of amplitude P_s^-, and eventually returns to p_o in total time $t_a + T^+ +$ T^-. The quantity P_s^+ is usually termed the peak side-on overpressure, or merely the peak overpressure. The portion of the time history above initial ambient pressure is called the positive phase of duration T^+. That portion below p_o, of amplitude P_s^- and duration T^- is called the negative or suction phase. Positive and negative specific impulses[+], defined by

$$i_s^+ = \int_{t_a}^{t_a + T^+} [p(t) - p_o]\, dt \qquad (2\text{-}1)$$

and

$$i_s^- = \int_{t_a + T^+}^{t_a + T^+ + T^-} [p_o - p(t)]\, dt \qquad (2\text{-}2)$$

respectively, are also significant blast wave parameters.

In most blast studies, the negative phase of the blast wave is ignored and only blast parameters associated with the positive phase are considered or reported.[++] (The positive superscript is usually dropped.) The ideal side-on parameters almost never represent the actual pressure loading

---

[+] Specific impulse, i, is the total impulse, I, per unit area.

[++] There is, however, some recent evidence that negative phase impulses and secondary shocks may be quite significant for distributed blast sources. [See Esparza and Baker (1977a and 1977b) and later discussions in this chapter.]

applied to structures or targets following an explosion. So a number of
other properties are defined to either more closely approximate real blast
loads or to provide upper limits for such loads. (The processes of reflection and diffraction will be discussed in Chapter 3.) Properties of
free-field blast waves in addition to side-on pressure which can be important in structural loading are:

- Density ρ

- Particle velocity u

- Shock front velocity U

- Dynamic pressure $q = \rho\, u^2/2$

The dynamic pressure, q, is often reported as a blast wave property, because of its importance in drag or wind effects and target tumbling. In
some instances, drag specific impulse i_d, defined as

$$i_d = \int_{t_a}^{t_a + T} q\ dt = \frac{1}{2} \int_{t_a}^{t_a + T} \rho\, u^2\ dt \qquad (2\text{-}3)$$

is also reported.

Although it is possible to define the potential or kinetic energy
in blast waves, it is not customary in air blast technology to report or
compute these properties. For underwater explosions, the use of "energy
flux density" is more common [Cole (1965)]. This quantity is given approximately by

$$E_f = \frac{1}{\rho_o\, a_o} \int_{t_a}^{t_a + T} [p(t) - p_o]^2\ dt \qquad (2\text{-}4)$$

where ρ_o and a_o are density and sound velocity in water ahead of the shock.

At the shock front, a number of the wave properties are interrelated through the Rankine-Hugoniot equations. The two of these three equations most often used are based on conservation of momentum and conservation of energy across the shock [Baker (1973)]:

$$\rho_s \ (U - u_s) = \rho_o \ U \qquad (2\text{-}5)$$

$$\rho_s \ (U - u_s)^2 + P_s = \rho_o \ U^2 + P_o \qquad (2\text{-}6)$$

In these equations, subscript s refers to peak quantities immediately behind the shock front, and the total, or absolute, peak pressure is given by:

$$P_s = P_s + P_o \qquad (2\text{-}7)$$

The Point Source Blast Wave

A "point source" blast wave is a blast wave which is conceptually produced by the instantaneous deposition of a fixed quantity of energy at an infinitesimal point in a uniform atmosphere. There have been many studies of the properties of point source waves, both for energy deposition in a "real air" atmosphere and for deposition in an "ideal gas" ($\gamma = 1.4$) atmosphere. Deposition in water has also been studied [Cole (1965)]. Point source blast wave studies date to the second World War [Bethe, et al. (1944), Taylor (1950), Brinkley and Kirkwood (1947), and Makino (1951)]. They have been summarized by Korobeinikov, et al. (1961), Sakurai (1965), Lee, et al. (1969), and Oppenheim, et al. (1971), and will be briefly reviewed here.

Essentially, there are three regions of interest as a point source wave propagates away from its source. The first is the "near-field" region where pressures in the wave are so large that external pressure (or counter pressure) can be neglected. In this region, the wave structure admits to a self-similar solution and analytic formulations are adequate [Bethe, et al. (1947), Sakurai (1965), Bach and Lee (1970), and Oppenheim, et al. (1972b)]. This region is followed by an intermediate region, which is of extreme practical importance because the overpressure and impulse are sufficiently high in this region to do significant damage. But this region does not yield to an analytical solution, and therefore, must be solved numerically [von Neumann and Goldstine (1955) and Thornhill (1960)]. There have been approximate techniques developed to extend the analytical treatment from the near field. These have been summarized by Lee, et al. (1969). The intermediate region is followed in turn by a "far-field" region which yields to an analytic approximation such that if one has the overpressure time curve at one far field position, one can easily construct the positive overpressure portion of the curve for other large distances.

In the far field region there is theoretical evidence that an "N" wave[+] must always form and that the blast wave structure in the positive impulse phase is unaffected by the interior flow and is self-sustaining [Bethe, et al. (1947) and Whitham (1950)]. However, experimentally it is difficult to determine if such an "N" wave actually exists because atmospheric nonhomogenieties tend to round the lead shock wave [Warren (1958)]. This is why a distant explosion is heard as a "boom" rather than a sharp report.

Scaling Laws

Scaling of the properties of blast waves from explosive sources is a common practice, and anyone who has even a rudimentary knowledge of

---

[+]An "N" wave has a time history somewhat like the letter N. There are two equal shocks, with a linear decay from overpressure P_s^+ to underpressure P_s^-.

blast technology utilizes these laws to predict the properties of blast
waves from large scale explosions based on tests on a much smaller scale.
Similarly, results of tests conducted at sea level ambient atmospheric
conditions are routinely used to predict the properties of blast waves
from explosives detonated at high altitude conditions. Baker (1973) and
Baker, et al. (1973), summarize the derivation of scaling laws for scaling of blast wave properties; here we will state the implications of these
laws that are most commonly used.

The most common form of blast scaling is Hopkinson-Cranz or "cube
root" scaling. This law, first formulated by Hopkinson (1915) and
independently by Cranz (1926), states that self-similar blast waves are
produced at identical scaled distances when two explosive charges of similar geometry and of the same explosive, but of different sizes, are detonated in the same atmosphere. It is customary to use as a scaled distance
a dimensional parameter, $Z = R/E^{1/3}$, where R is the distance from the center of the explosive source and E is the total energy of the explosive.
Figure 2-2 shows schematically the implications of Hopkinson-Cranz blast
wave scaling. An observer located at a distance R from the center of an

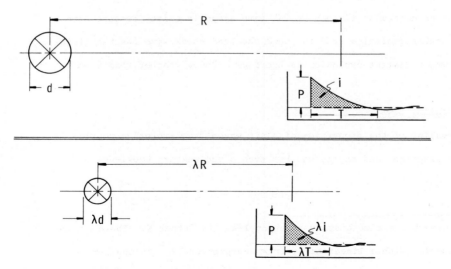

Figure 2-2. Hopkinson-Cranz Blast Wave Scaling

explosive source of characteristic dimension d will be subjected to a
blast wave with amplitude P, duration T, and a characteristic time history. The integral of the pressure-time history is the impulse i. The
Hopkinson-Cranz scaling law then states that an observer stationed at a
distance λR from the center of a similar explosive source of characteristic dimension λd detonated in the same atmosphere will feel a blast wave
of "similar" form with amplitude P, duration λT and impulse λi. All characteristic times are scaled by the same factor as the length scale factor
λ. In Hopkinson-Cranz scaling, pressures, temperatures, densities and
velocities are unchanged at homologous times. The Hopkinson-Cranz scaling
law has been thoroughly verified by many experiments conducted over a
large range of explosive charge energies. A much more complete discussion of this law and a demonstration of its applicability is given in
Chapter 3 of Baker (1973).

The blast scaling law which is almost universally used to predict
characteristics of blast waves from explosions at high altitude is that
of Sachs (1944). A careful proof of Sachs' law has been given by Sperrazza (1963). Sachs' law states that dimensionless overpressure and
dimensionless impulse can be expressed as unique functions of a dimensionless scaled distance, where the dimensionless parameters include
quantities which define the ambient atmospheric conditions prior to the
explosion. Sachs' scaled pressure is the ratio of blast pressure to ambient atmospheric pressure,

$$\bar{P} = (P/p_o) \qquad (2-8)$$

Sachs' scaled impulse is defined as

$$\bar{i} = \frac{ia_o}{E^{1/3} p_o^{2/3}} \qquad (2-9)$$

118

where a_o is the ambient sound velocity. These quantities are a function of dimensionless scaled distance[†], defined as

$$\bar{R} = \frac{Rp_o^{1/3}}{E^{1/3}} \tag{2-10}$$

The primary experimental proof of Sachs' law is given by Dewey and Sperrazza (1950).

Scaled Side-On Blast Wave Properties

Some standard conversion factors (TNT equivalence) for calculating equivalence of high explosive charges as given in Baker (1973) and Suppressive Shields (1977) are repeated here in Table 2-2. With these factors and the scaled distance $\bar{R} = Rp_o^{1/3}/E^{1/3}$, we have plotted dimensionless peak overpressure, $(p_S - p_o)/p_o$, versus $\bar{R}$ on a log plot in Figure 2-3 over a very short range, with data taken from a number of published sources. It is interesting to note that the overall disagreement between these sources is approximately a factor of $\pm$ 2. This was first observed by Baker (1973), and his curve, which is based on experimental data for Pentolite (50/50), is seen to represent a good average of the other curves. Figure 2-4 which covers a much larger overpressure-scaled distance region, shows the overall extent of scatter. In this curve, the shaded regions represent the total range covered by other curves. Figure 2-5 show the impulse curve, from Baker (1973), fitted to test data from various sources.

There has been controversy about the far-field behavior of the wave. Baker (1973) opts for a 1/R dependence, while Bethe, et al. (1947), Thornhill (1960), and Goodman (1960) state that the dependence should be proportional to $1/R (\ln R)^{1/2}$, and Porzel (1972) states that experimental data

[†]The quantity $(E/p_o)^{1/3}$ has dimensions of length and is sometimes called a characteristic length for an explosive source.

119

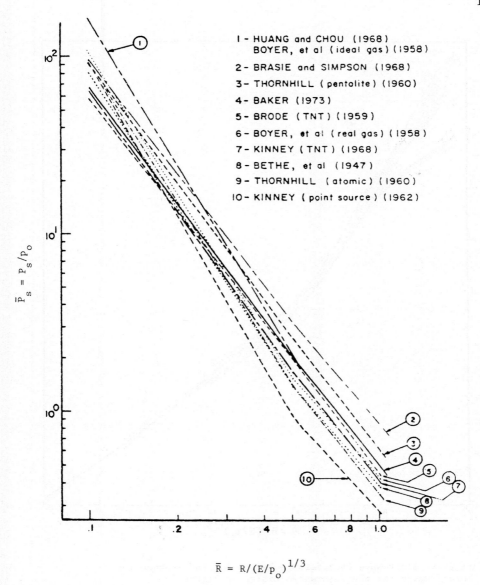

$$\bar{R} = R/(E/p_o)^{1/3}$$

Figure 2-3. Blast Wave Overpressure Versus Scaled Distance Taken
From a Number of References (Small Range)
[Strehlow and Baker (1976)]

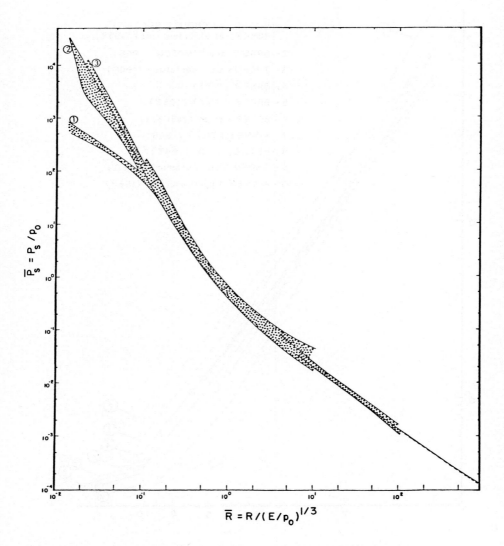

Figure 2-4. Blast Wave Overpressure Versus Scaled Distance for
Blast Waves for the References Listed on Figure 2-3.
(1) High Explosives; (2) Nuclear Explosions; (3) Point Source.
The Dotted Line in the Lower Right is for a $1/[\bar{R} \, (\ln \bar{R})^{1/2}]$
Dependence. The Solid Line is for a $1/\bar{R}$ Dependence.
[Strehlow and Baker (1976)]

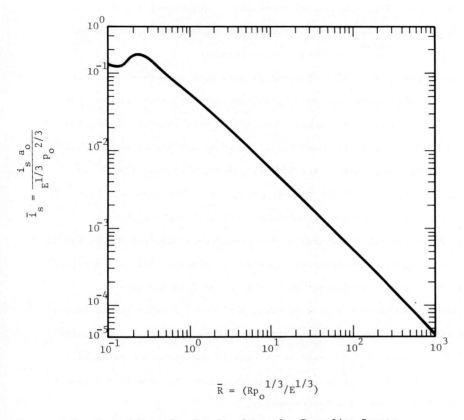

$$\bar{R} = (Rp_o^{1/3}/E^{1/3})$$

Figure 2-5. Scaled Specific Impulse Curve for Pentolite Bursts
[Baker (1973)]

show a $1/\bar{R}^{4/3}$ dependence. For comparison we have drawn both the $1/\bar{R}$ (ln $\bar{R}$)$^{1/2}$ and $1/\bar{R}$ dependences on Figure 2-4 for $\bar{R} > 10^2$ as a dotted line and solid line, respectively, to show how small the differences in far-field behavior really are. The question is actually moot for two reasons. Firstly, Warren (1958) has found a spread of measured overpressures in the far-field of about a factor of three, this undoubtedly due to refraction and focusing effects in the real atmosphere. He also found that the lead shock disappears in the far-field and is replaced by a slower pressure rise. This is also to be expected and is due to the nonuniformity of the atmosphere. Secondly, very little damage is done in the far-field, and

122

therefore, it has little practical importance. Techniques for evaluating far-field focusing effects due to atmospheric winds and temperature gradients will be discussed elsewhere in this chapter.

Many more side-on blast parameters than peak overpressure are reported in the literature for chemical and nuclear explosive sources, over varying ranges of scaled distance. Some of the more complete literature sources are Goodman (1960), Baker (1973), Swisdak (1975), or Kingery (1966) for chemical high explosives; and Kingery (1968), Glasstone (1962), or Glasstone and Dolan (1977) for nuclear explosives. The conversion factors in Table 2-2 cover some of the extant chemical high explosives. One can easily establish conversions for other known high explosives by merely comparing measured heats of explosion to those of TNT or Pentolite. One must take care in comparing compiled data for free-air sources to data for sources exploded on a good reflecting surface -- usually ground bursts. Surface burst data for chemical explosives such as TNT [see Kingery (1966)] will correlate well with free-air burst data if one assumes a reflection factor of about 1.8. That is, surface bursts appear to come from free-air bursts having 1.8 times the source energy.

If the ground acted as a perfect reflector with no energy being dissipated as ground shock or in cratering, the reflection factor for source energy would be exactly two, because all of the energy which would have driven an air shock in the lower hemisphere of a free-air burst now helps drive a shock into the hemisphere above the ground surface. The difference between the perfect reflection factor of two and the empirically observed factor of 1.8 for surface bursts on the ground is a rough measure of the amount of energy dissipated by cratering and ground shock.

Because TNT is so often used as the basis for comparing blast waves from high explosive and nuclear sources, standard curves for properties under sea level atmospheric conditions can prove quite useful. A set of such curves for TNT bursts in free air to small scale is included as Figures 2-45, 2-46, and 2-47 at the end of this chapter, while the same curves are repeated to a larger scale on foldout pages at the back of the book.

Wave Properties - Energy Distribution

One of the most important properties which determine the behavior
of an explosion process is the energy distribution in the system and how
it shifts with time as the pressure wave propagates away from the source.
Initially all the energy is stored in the source in the form of potential
energy. At the instant when the explosion starts, this potential energy
is redistributed to produce kinetic and potential energy in different regions of the system; the system now includes all materials contained within either the lead characteristic or lead shock wave that propagates away
from the source. The system is unsteady, both because new material is
continually being engulfed by the lead wave front, and because the relative distribution of energy in various forms and in various parts of the
system shifts with time.

To consider this problem in more detail, we will idealize the system to some extent. Assume 1) that the explosion is strictly spherical
in an initially homogeneous external atmosphere that extends to infinity,
2) that the source of the explosion consists of both energy-containing
material (source material) and inert confining material, and that during
the explosion process these materials do not mix appreciably with each
other or with the outside atmosphere, and 3) that shock wave formation
is the only dissipative process in the surrounding atmosphere. With
these assumptions, the source potential energy is distributed among a
number of distinct forms at various times and locations as the explosion
process proceeds. These are:

1. Wave Energy

The propagating wave system contains both **potential** energy

$$E_p = \int\limits_V \rho C_V \, (\theta - \theta_o) \, dV \tag{2-11}$$

and kinetic energy

$$E_k = \int\limits_V \frac{1}{2} \rho u^2 \, dV \qquad (2\text{-}12)$$

where V is the volume of the atmosphere enclosed by the lead characteristic or lead shock wave. This volume does not include the volume occupied by the products of explosion or the confinement material. Furthermore, at late time when the kinetic energy of the source and confining material are zero and the wave amplitude is such that shock dissipation is negligible, the total wave energy $(E_T = E_p + E_k)$ in the system must remain constant with time. This far-field wave energy should, therefore, be a unique property of each explosion process.

   2. Residual Energy in the Atmosphere (Waste Energy)

        In most explosions, a portion of the external atmosphere is traversed by a shock wave of finite amplitude. This process is not isentropic and there will be a residual temperature rise in the atmosphere after it is returned to its initial pressure. This residual energy will also reach some constant value at a later time. This was first called "waste" energy by Bethe, et al. (1944).

   3. Kinetic and Potential Energy of the Fragments (or Confining
        Material)

        Initially, the confining material will be accelerated and will also store some potential energy due to plastic flow, heat transfer, etc. Eventually, all this material will decelerate to zero velocity and will store some potential energy, primarily as increased thermal energy in the fragments.

   4. Kinetic Energy of the Source Material

        In any explosion involving a finite volume source, the source material or its products will be set into motion by the explosion process. This source material kinetic energy will eventually decay to zero as all motion stops in the near field.

5. Potential Energy of the Source Material

The source originally contained all the energy of the explosion as potential energy. As the explosion process continues, a portion of the source potential energy is redistributed elsewhere, and a portion normally remains in the source as high temperature products. While it is true that this stored energy eventually dissipates by mixing, this process is relatively slow compared to the blast wave propagation process, and for our purposes one can assume quite accurately that the residual energy stored in the products approaches a constant value at late time.

6. Radiation

Radiated energy is quickly lost to the rest of the explosion system and reaches a constant value quite early in the explosion process.

Figure 2-6 is a schematic showing how energy is redistributed in

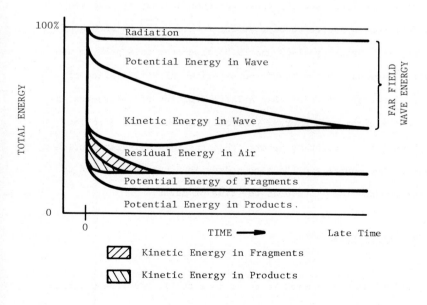

Figure 2-6. Energy Distribution in a Blast Wave as a Function
of Time After the Explosion (Schematic)
[Strehlow and Baker (1976)]

a blast wave as time increases. Note that at late time, when the wave is
a far-field wave, the system contains potential and kinetic wave energy,
residual potential energy (waste energy) in the atmosphere, potential energy in the fragments and potential energy in the products. Also, in
general, some energy has been lost to the system due to radiation. However,
radiation losses represent an important fraction of the total source energy
only for the case of nuclear explosions. A few general statements can be
made about Figure 2-6.

Only a fraction of the total energy which is initially available
actually appears as wave energy in the far-field. The magnitude of this
fraction relative to the total energy originally available must depend on
the nature of the explosion process itself. For example, the TNT equivalence of nuclear explosions is about 0.5 to 0.7 of that which one would
expect on the basis of the total energy available [Lehto and Larson (1969),
Thornhill (1960) and Bethe, et al. (1944)]. More to the point, in accidental explosions, the source has a sizable volume and normally releases
the stored energy relatively slowly; thus, for such explosions, one would
expect the effectiveness factor to be a strong function of the nature of
the release process. Brinkley (1969 and 1970) discusses this problem using a theoretical approach but presents no experimental verification of
the thesis that slow release means that a large fraction of the source energy is lost to the blast wave energy. Recently, Adamczyk and Strehlow
(1977) have shown for the bursting sphere, that the terminal distribution
of energy between the source and the surrounding is a strong function of
the sphere pressure and temperature at the time of burst.

ATMOSPHERIC AND GROUND EFFECTS

General

Ideal explosions are assumed to occur in a still, homogeneous atmosphere and to be unaffected by the presence of a ground surface. Real

conditions in the atmosphere and real surface effects can modify the wave in various ways.

Atmospheric Effects

Variations in initial ambient temperature and pressure can affect the blast wave so that noticeably different waves would be recorded from explosions on a high mountain or mesa than from explosions near sea level, or from explosions occurring on a hot summer day versus a cold winter day. These effects are, however, quite adequately accounted for if the Sachs scaling law, described earlier, is used to predict the wave properties. For very large explosions such as detonations of multi-megation nuclear weapons, the vertical inhomogeneity of the atmosphere will cause modification of an initially spherical shock front [Lutzsky and Lehto (1968)]. Changes in relative humidity and even heavy fog or rain have been found to have insignificant effects on blast waves [Ingard (1953)].

The more significant atmospheric effects which induce non-ideal blast wave behavior are unusual weather conditions which can cause blast focusing at some distance from the source. A low-level temperature inversion can cause an initially hemispherical blast front to refract and focus on the ground in an annular region about the source [Grant, et al. (1967)]. Severe wind shear can cause focusing in the downwind direction. This effect is discussed by Baker (1973) and Reed (1973). Structural damage from accidental explosions has been correlated with these atmospheric inhomogeneities [Siskind (1973), Siskind and Summers (1974), and Reed (1968)], and claims for damage from explosive testing were reduced when firings were limited to days when no focusing was predicted [Perkins, et al. (1960)]. A handbook on how to perform such calculations is available [Perkins and Jackson (1964)].

Ground Effects

Ground effects can also be important. If the ground acted as a perfectly smooth, rigid plane when explosions occurred on its surface, then it would reflect all energy at the ground plane and its only effect on the blast wave would be to double the apparent energy driving the wave. In actuality, surface bursts of energetic blast sources usually dissipate some energy in ground cratering and in ground shock, so that only partial reflection and shock strengthening occurs. A good "rule of thumb" given earlier is to multiply the effective source energy by a factor of 1.8 if significant cratering occurs. For sources of low energy density such as gaseous mixtures, very little energy enters the ground, and the reflective factor of two is a good approximation. This is equivalent to doubling the energy and then using free-air data correlations to predict blast wave properties.

A ground surface which is irregular can significantly affect the blast wave properties. Gentle upward slopes can cause enhancement, while steep upward slopes will cause formation of Mach waves and consequent strong enhancement. Downward slopes or back surfaces of crests cause expansion and weakening of shocks. These effects are usually quite localized, however, and "smooth out" quite rapidly behind the irregularities. Even deliberate obstructions such as mounded or revetted barricades produce only local effects [Wenzel and Bessey (1969)].

NONIDEAL EXPLOSIONS - GENERAL BEHAVIOR

Introduction

Most of the accidental explosion sources listed in Table 2-1 have much lower energy density than ideal sources, have a finite time of energy release, and also require some confinement to produce an explosion. Nuclear reactor runaway occurs at a much slower rate and in much more diluted fissionable material than does a nuclear weapon explosion. Chemical reactor vessel explosions usually involve an unwanted progression (i.e., runaway) of an internal chemical reaction caused at least partially by the effects of confinement, until the burst pressure of the vessel is exceeded. Gaseous and dust explosions in enclosures will usually cause only fires in the absence of some confinement. Simple pressure vessel failures for non-reacting gases can be strong explosion sources, but again can only produce strong blast waves if there is substantial vessel confinement prior to rupture. In the entire list of accidental explosions, the only nonideal sources not strongly dependent on confinement are unconfined vapor cloud explosions, and perhaps, physical explosions.

In the following parts of this section, source behaviors that produce nonideal blast waves will be discussed in a general way. Experimental and theoretical justification will be presented as appropriate. Then, in the following section, research on specific types of accidental explosions that are amenable to numerical calculations, analysis or experimentation will be discussed. In the final section of this chapter, a number of actual accidental explosions will be discussed to illustrate as many of the different types as possible.

Spherical Waves, Energy Density and Deposition Time

A number of systematic studies show that there are two primary properties of a spherically symmetric source region which lead to nonideality of the blast wave. If one assumes that the source region can be modeled as a region where energy is deposited within some finite spatial volume and during some finite time period, it is found that both of these non-idealities in the source lead to nonideality of the blast wave.[†]

Consider the two idealized limiting cases shown in Figure 2-7. Here, a quantity of energy, Q, is added to a spherical region of finite size in two limiting ways. If it is added very slowly compared to the

---

[†]All explosions (energy releases) occur in a finite time interval over a finite spatial volume; however, the rate of energy release may be rapid enough, and the energy density large enough to "appear" as a point source at distances farther than the distance where the overpressure is sufficient to be completely destructive.

time it takes a sound signal to travel across the sphere, the pressure will not rise and the work done on the surroundings is $E_S = P_o (V_f - V_o) =$ NR $(\theta_f - \theta_o)$ (see Figure 2-7b). In this case, there will be no blast wave even though work was done on the surroundings.

In the other limiting case illustrated in Figure 2-7c, the energy, Q, is added very rapidly and causes the pressure to rise to the pressure P_2 before any motion occurs. In this case $Q = NC_V (\theta_2 - \theta_o)$, and therefore, since the gas is an ideal polytropic gas,

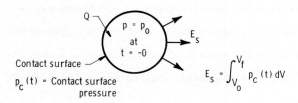

Contact surface

$p_c (t)$ = Contact surface pressure

$$E_S = \int_{V_o}^{V_f} p_c (t) \, dV$$

(a) Schematic

$Q = NC_p (\theta_f - \theta_o)$
$E_B = NC_v (\theta_f - \theta_o)$
$E_S = NR (\theta_f - \theta_o)$
$\dfrac{E_S}{Q} = \dfrac{\gamma - 1}{\gamma}$

(b) Constant pressure energy addition

$Q = NC_v (\theta_2 - \theta_o)$
at $t = -0$

$\dfrac{E_S}{Q} = \dfrac{1}{q} \left[(1+q) - (1+q)^{1/\gamma} \right]$

where

$q = \dfrac{Q}{NC\theta_o}$

Q = constant

$q \to 0 \quad V_o \to \infty \quad \dfrac{E_S}{Q} \to \dfrac{\gamma - 1}{\gamma} \quad ; \quad q \to \infty \quad V_o \to 0 \quad \dfrac{E_S}{Q} \to 1$

(c) Constant volume - Isentropic expansion

Figure 2-7. Energy Addition to a Constant Gamma Ideal Gas

$$Q = \frac{(P_2 - P_o) V}{\gamma - 1} \tag{2-13}$$

Equation (2-13) is the formula given by Brode (1955) for the energy stored in a bursting sphere. If one nondimensionalizes this expression with the initial energy in the spherical volume, one obtains an expression for the energy density in the sphere.

$$q = \frac{NC_V (\theta_2 - \theta_o)}{NC_V \theta_o} = \frac{\theta_2}{\theta_o} - 1 = \frac{P_2}{P_o} - 1 \tag{2-14}$$

If one calculates the fraction of this energy that is transmitted to the surroundings for the idealized process where the sphere expands slowly against a counter pressure equal to its instantaneous pressure, one finds that

$$\frac{E_s}{Q} = \frac{1}{q} [(1 + q) - (1 + q)^{1/\gamma}] \tag{2-15}$$

This is the formula given by Brinkley (1969) and Baker (1973) for the effective quantity of energy stored in the sphere expressed as a fraction of Brode's energy. The quantity E_s represents the maximum amount of work that can be done on the surroundings for any sphere burst. Thus, in the limit as $q \to \infty$ for fixed Q (point source), $E_s/Q \to 1$; while in the limit $q \to 0$ for fixed Q, $E_s/Q \to (\gamma - 1)/\gamma$. Note from above that for constant pressure (slow) addition, the energy transmitted to the surroundings is given by the equation

$$E_s/Q = (\gamma - 1)/\gamma \tag{2-16}$$

irrespective of the source volume. This is exactly the limit for the infinitely rapid addition of energy to a very large volume, even though the

two processes are quite different. Notice that in both of these limit
cases, i.e., for q → 0, there would be no blast.

A second quantity which is of importance to non-ideal behavior is
the time it takes to add the energy to the source region. This time can
be made dimensionless by dividing it by the time it takes a sound wave to
travel from the center of the source region to the edge before the heat
addition has occurred:

$$\tau = \frac{t_e \, a_o}{r_o} \tag{2-17}$$

In Equation (2-17), τ is the effective dimensionless time of energy addition, t_e is the effective time of energy addition, a_o is the velocity of
sound in the source region prior to the addition of energy, and r_o is the
radius of the source region.

To determine the importance of τ as well as other parameters, Strehlow and co-workers have studied bursting spheres [Strehlow and Ricker
(1976)], the homogeneous ramp addition of energy with a number of characteristic time periods [this essentially models a spark, Adamczyk (1976) and
Strehlow (1976)], constant velocity and accelerating flames propagating
from the center of the source region [Strehlow, et al. (1979)], and constant
velocity flames propagating from the edge of the source region toward the
center [implosions, Strehlow and Shimpi (1978)]. Schematic drawings of
these types of energy addition for comparison to the bursting sphere case
are shown in Figure 2-8. The calculations were performed using a lagrangian, finite-difference, artificial viscosity, computational scheme originally supplied by Oppenheim (1973) and modified to allow for heat addition.

These studies were performed in a systematic manner for a large number of cases for each type of energy addition shown in Figure 2-8. In
general, it was found that shock overpressure and energy scaled positive

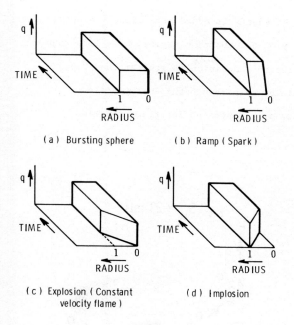

(a) Bursting sphere   (b) Ramp (Spark)

(c) Explosion (Constant velocity flame)   (d) Implosion

Figure 2-8.   Energy Addition to a Spherical Source Region
(Idealized)

impulse in the near-field are markedly affected by the nature of the source region. However, it was found that if the energy density in the source region is high enough and if the characteristic time for energy addition in the source region, τ, is short enough, far-field equivalence in overpressure is obtained for these nonideal sources. It has been found that bursting spheres with energy densities above about 8 [Equation (2-14)] produce a blast wave with far-field equivalency in overpressure [Ricker (1975)]. Also, for a constant velocity flame propagating from the center of a source region, an effective normal burning velocity of about 45 m/s is required if far-field equivalency in overpressure is to be obtained [Strehlow, et al. (1979)]. Additionally, it was found that energy scaled positive impulse, $\bar{\i}$, is conserved irrespective of the nature of the source region at energy scaled distances above 0.3.

There are other general behaviors of the blast wave produced by non-

134

ideal sources that are important. If energy is added sufficiently slowly,
as when the effective flame velocity of a combustion explosion is less than
45 m/s or when the effective time for ramp addition is greater than 0.5,
no shock is produced in the near-field, even though the energy-scaled
positive impulse is conserved. This is illustrated for flame addition
in Figure 2-9.

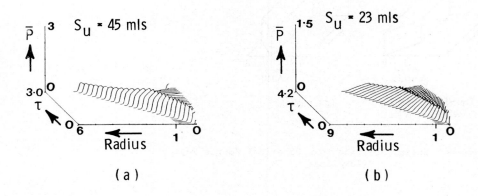

Figure 2-9. Change in Overpressure Distribution of a Blast Wave
Produced by a High Velocity and Low Velocity Combustion Wave

(a) High Velocity Wave (Notice Shock Structure)
(b) Low Velocity Wave (Notice Absence of a Shock in the Field)
[Strehlow, et al. (1979)]

Another behavior which is typical of low energy density sources
is that the negative phase of the blast wave is sizable when compared to
the positive phase. This was first noted by Rayleigh (1878). This is
markedly different than the blast wave produced from ideal sources. In
an ideal source blast wave there is a negative phase, but it generally
is very small when compared to the positive phase, and damage generated
by the negative phase impulse is usually quite small. Figure 2-10 shows
pressure time curves at various locations for a bursting sphere of relatively low energy density. In this graph R = 1.0 is the initial location
of the edge of the source region (and not the energy scaled radius). Note

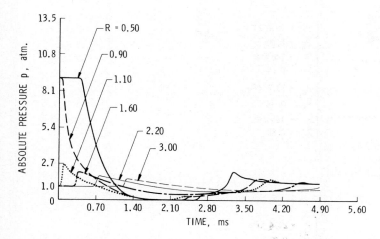

Figure 2-10.   Pressure Time Variations in the Flow for a Bursting Sphere With
q = 8.   Note the Very Large Negative Phase Region
[Shimpi (1978)]

the large negative phase for values of R which lie outside the source re-

gion.   This large negative phase for low energy density source explosions

can cause markedly different damage patterns than one would observe if the

same source energy were concentrated in a high explosive charge.   Also,

note the existence of a second shock following the negative phase of the

blast wave.   The same effects are seen when the source contains a gas phase

combustion wave or is modeled as the ramp addition of energy.

Experiments have verified this result for bursting spheres contain-

ing gases such as nitrogen, argon, etc.   Figure 2-11 is taken from Esparza

and Baker (1977a) and shows exactly the same blast wave behavior for the

bursting of frangible (glass) spheres.

Nonspherical Waves

1.   Two-Dimensional Analyses

Explosions are seldom spherically symmetric.   For example, a

fuel-air cloud in which the fuel is heavier than air will be rather flat,

instead of hemispherical.   The blast wave from a pressure vessel bursting

into two pieces will not be spherically symmetric, although it may be axi-

symmetric.   A way of dealing with asymmetric explosions is given by Chiu,

136

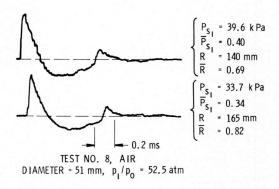

$$P_{S_I} = 39.6 \ kPa$$
$$\overline{P}_{S_I} = 0.40$$
$$R = 140 \ mm$$
$$\overline{R} = 0.69$$

$$P_{S_I} = 33.7 \ kPa$$
$$\overline{P}_{S_I} = 0.34$$
$$R = 165 \ mm$$
$$\overline{R} = 0.82$$

0.2 ms

TEST NO. 8, AIR
DIAMETER = 51 mm, p_I/p_0 = 52.5 atm

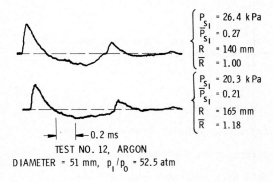

$$P_{S_I} = 26.4 \ kPa$$
$$\overline{P}_{S_I} = 0.27$$
$$R = 140 \ mm$$
$$\overline{R} = 1.00$$

$$P_{S_I} = 20.3 \ kPa$$
$$\overline{P}_{S_I} = 0.21$$
$$R = 165 \ mm$$
$$\overline{R} = 1.18$$

0.2 ms

TEST NO. 12, ARGON
DIAMETER = 51 mm, p_I/p_0 = 52.5 atm

Figure 2-11. Examples of Pressure Time Histories Produced by the Burst
of a Frangible Glass Sphere Containing High Pressure Air or Argon
[Esparza and Baker (1977b)]

et al. (1976). The theory is based upon Whitman's "ray-shock" theory and

the Brinkley-Kirkwood theory. By assuming a geometrical shape for the

shock, the geometric and dynamic relationships in the theory can be de-

coupled to give a set of ordinary differential equations. The theory was

applied to the burst of a pressurized ellipsoid. It was also applied to

the blast wave from an exploding wire, and the results agree with experi-

mental observations. Williams (1974) has also presented an approximate

analytical technique for determining the blast wave from a nonspherical

source.

In another recent treatment, Strehlow (1981) applied acoustic

monopole source theory to the calculation of the maximum overpressures

that can be generated by the deflagration of an unconfined cloud of arbitrary shape. The most interesting result of his study is that, for even
reasonably high deflagration speeds, the blast wave amplitude was found
to decrease rapidly with increased aspect ratio (length-to-depth ratio)
of the cloud. These results will be discussed in greater detail in the
following section under the topic, Unconfined Vapor Cloud Explosions.

2. Two-Dimensional Numerical Techniques

For asymmetric explosions which can be studied as two-dimensional phenomena (usually axisymmetric), there are several types of numerical schemes that can be used. The schemes require that constitutive
relationships or equations-of-state be known, but otherwise they are general. The schemes do, however, require a computer and usually a computer
with large memory and good speed. Because the computational fluid, thermal and chemical science field is expanding rapidly and because many computation codes are preliminary or proprietary, it is possible only to
describe a few schemes, and those only briefly.

The Particle-In-Cell (PIC) method is a finite-difference scheme
devised for compressible flow problems. In this scheme, the flow field is
divided into a grid or Eulerian mesh of cells. In selected grid cells are
a system of particles whose distribution and particle mass describe the
initial concentration of the source material or fluid. The equations of
motion without the transport terms are solved for the cells. The transport
terms are accounted for by allowing the particles to move as tracers among
the cells according to the velocities of the particles [Baker (1973)].
The details can be found in Harlow (1957) and Harlow, et al. (1959). The
PIC method has largely been supplanted by more efficient (and more accurate) schemes, but it has considerable historical significance.

The Fluid-In-Cell (FLIC) method is somewhat similar to PIC, except that massless particles are used as markers or tracers only. There
are two basic numerical steps for each timestep calculation. First, the
intermediate values for the velocities and specific internal energy are

calculated for cells, and this step includes the effects of acceleration caused by pressure gradients. In the second step, the transport effects are calculated [Gentry, et al. (1966)]. This scheme has been expanded and modified considerably since Gentry, et al. (1966) laid the foundation, and is, in various codes, used to solve a variety of blast wave problems.

Amsden (1973) first describes two numerical schemes which have been used for evaluating explosions. These schemes are referred to as:

- ICE - The Implicit Continuous-fluid Eulerian method. It can be applied to one-, two- or three-dimensional, time-dependent problems.

- ICED-ALE - A combination of ICE and another scheme, ALE. There are three phases to the method: The first phase is a typical explicit Lagrangian calculation. In the second phase, an iterative scheme provides advanced pressures for the momentum equation and advanced densities for the mass equation. This eliminates the need for a sound-signal propagation velocity criterion for stability, and thus allows larger timesteps. In the third phase, the computational grid is rezoned. "Marker" particles can be introduced to locate a free surface or to observe the overall flow.

Other finite-difference techniques are described in Roache (1972) and by Ramshaw and Dukowicz (1979), Fry, et al. (1976), Glaz (1979), Stein, et al. (1977), Nelson (1973), Benzley, et al. (1969), Wilkins (1969), Swegle (1978), W. E. Johnson (1971), Daly, et al. (1964), Schmitt (1979), Oran, et al. (1978), and Thompson (1975).

Still another technique applicable to two-dimensional explosions is the Finite Element Method. In the finite element method, the structural or fluid region of interest is subdivided into "elements" of fairly general shape, each having several nodal (corner) points. An interpolation function is prescribed over each element, and the parameters of the interpolation function are the unknowns. The two most commonly used forms of the

finite element method are the variational method and the residual method.
In the variational method, an extremal principle is employed to derive the
equations that are to be solved; the extremal principle may be the minimi-
zation of energy or viscous dissipation, for example. For the residual method, it is required that the difference between an approximate solution and
the true solution be small; this results in a general equation to be solved.
After formulating the general method, a particular form of interpolation
function is substituted into the equations to obtain numerical results. Descriptions of the method and its applications are given in Norrie and de
Vries (1973), Oden, et al. (1974), G. R. Johnson (1977 and 1978), and Hallquist (1979).

An application of the finite element method to time-dependent
compressible flow with shock waves (which is necessary for the calculation of the flowfield near an explosion) is described by Bowley and Prince
(1971). They studied transonic flow through cascades and channels.

In summary, one can say that a number of numerical methods for
calculating two-dimensional blast problems seem promising, but few have
been exercised enough to judge their relative merits or to generate prediction curves.

In concluding, the numerical computation field is becoming more
important because very complex problems become tractable. Improvements
can be expected as better mathematical models are devised, and as needed
basic data from careful laboratory or field experiments become available,
to better define supporting relationships, such as equations of state.

3. Experimental Measurements

Most real blast sources are nonspherical, and can be of regular geometry such as cylindrical or block-shaped, or can be quite irregular in shape. To date, few experiments have been run for other than the
cylindrical geometry of solid explosive sources. For cylinders, the wave
patterns have been shown [Wisotski and Snyer (1965), Reisler (1973)] to
be quite complex, as shown in Figure 2-12. The pressure-time histories

140

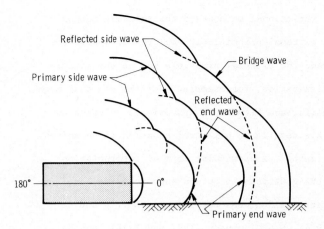

Figure 2-12.   Schematic Wave Development for Cylindrical Charges
[Reisler (1972)]

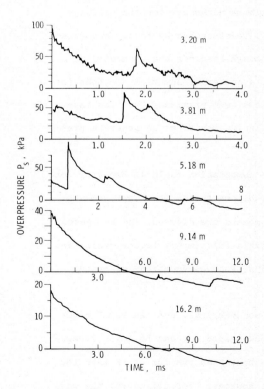

Figure 2-13.   Pressure Time Records From Cylindrical Charges
Along Charge Axis
[Reisler (1972)]

exhibit multiple shocks, as shown in Figure 2-13, and decay in a quite different manner in the near-field than do spherical waves.

Other types of nonspherical behavior also occur and have been studied experimentally. Gun muzzle blast or recoilless rifle backblast generates waves which consist of essentially single shocks, but these shocks have highly directional properties. This type of asphericity is particularly pronounced behind recoilless rifles, where the shock is driven by supersonic flow of propellant gases expanding through a nozzle [Baker, et al. (1971)].

The asphericity caused by reflection from a nearby reflecting surface, such as a flat wall on the ground, has been studied extensively. Schematically, the shock fronts for strong waves behave as shown in Figure 2-14, where C is the explosive charge, I indicates successive locations of the incident shock front, and R indicates reflected wave fronts. The dashed line ρ is the trace of the "triple point" u. This Mach reflection process will be described in more detail in Chapter 3.

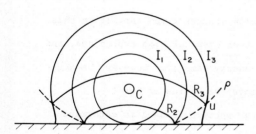

Figure 2-14. Reflection of Strong Shock Waves
                [Baker (1973)]

4. Summary

The above instances are only a few examples of nonspherical behavior. To reiterate, close to most real blast sources behavior is usually nonspherical. Fortunately, asymmetries smooth out as the blast wave progresses, and "far enough" from most sources, the wave will become a spherical wave. To determine how far is "far enough," one must rely on analysis or experiment for the particular situation.

Source Energy and Scaling

We have noted earlier that the total energy E is an important property of an explosion source. It enters all of the blast scaling laws, and is the most important parameter in describing and assessing the severity of an explosion. Unfortunately, this quantity has often proved quite difficult to estimate for accidental explosions and nonideal explosions, although it is well known for given masses or volumes of chemical high explosives.

It has been common practice in the evaluation of blast wave damage from accidental explosions (and planned explosions, too) to express the severity of the damage caused by the explosion in terms of its "TNT equivalency," i.e., in terms of how many pounds, kilograms, or kilotons of TNT would do an equivalent amount of damage. What one is in fact doing is equating the blast wave strength to that from an equivalent mass of TNT. That this is not an exact procedure should be clear from the comparisons of blast waves from TNT and other explosion sources given earlier in this chapter. It is reasonably correct for other chemical high explosives, and in the far-field for sources with lower energy density. However, because of the prevalence of use of TNT equivalence, we review it here.

After an accident, the blast damage pattern is used to determine the weight of TNT which would be required to do the observed amount of damage at that distance from the center of the explosion. If the explosion is chemical in nature, one then usually attempts to determine a percent TNT equivalence by determining a maximum equivalent TNT weight of the fuel or chemical by calculating either the heat of reaction of the mixture or the heat of combustion of the quantity of that substance which was released. Zabetakis (1960), Brasie and Simpson (1968), Burgess and Zabetakis (1973), Strehlow (1973a), and Eichler and Napadensky (1977) have all followed this approach based on the TNT equivalence concept for high explosives, where

relative damage is directly correlatable to the relative heats of explosion of different explosives measured in an inert atmosphere. The formulas are

$$(W_{TNT})_{calc} = \frac{\Delta H_c \cdot W_c}{4.520 \times 10^6} \qquad (2\text{-}18)$$

and

$$\% \text{ TNT} = [(W_{TNT})_{blast}/(W_{TNT})_{calc}] \times 100 \qquad (2\text{-}19)$$

where in Equation (2-18) $(W_{TNT})_{calc}$ = the equivalent maximum TNT weight,
kg; ΔH_c = heat of combustion of the hydrocarbon (or heat of reaction of
the exothermic mixture), J/kg; W_c = weight of hydrocarbon or reaction mixture available as an explosive source, kg; and 4.520×10^6 = heat of explosion of TNT, J/kg. In the same vein, Dow Chemical Company (1973) in their
safety and loss prevention guide, advocate evaluating the relative hazard
of any chemical plant operation by first calculating a ΔH of reaction or
explosion for the quantity of material which is being handled and then multiplying this basic number by factors based on other known properties such
as the substance's sensitivity to detonation.

In certain cases, nonideal explosions can be scaled more precisely
than either Sachs scaling, which is strictly energy scaling, or Hopkinson-
Cranz scaling, which also takes into account the orientation and explosive
behavior in the source region. It is true that in contrast to Sachs' law,
the Hopkinson-Cranz law can and has been used for scaling some close-in
effects of nonideal explosions, particularly for certain asymmetric sources.

Esparza and Baker (1977a) have recently proposed a more general scaling law for nonideal explosions in air which includes explosion source parameters describing details of the nonideal source. Extending the work,

we can derive a rather general law for such explosions:

$$\bar{P} = \left(\frac{P}{P_o}\right)$$

$$\bar{t}_a = \left(\frac{t_a \, a_o \, P_o^{1/3}}{E^{1/3}}\right) = f_j\left[\left(\frac{P_1}{P_o}\right), \frac{R \, P_o^{1/3}}{E^{1/3}}, \gamma_1, \left(\frac{a_1}{a_o}\right), \left(\frac{\dot{E}}{E^{2/3} \, a_o \, P_o^{1/3}}\right), \ell_i\right]$$

$$(2\text{-}20)$$

$$\bar{T} = \left(\frac{T \, a_o \, P_o^{1/3}}{E^{1/3}}\right) = f_j\,(\bar{P}_1, \bar{R}, \gamma_1, \bar{a}_1, \dot{\bar{E}}, \ell_i)$$

$$\bar{i} = \left(\frac{i \, a_o}{E^{1/3} \, P_o^{2/3}}\right)$$

where t_a is arrival time of the overpressure, T is duration of the overpressure, γ_1 is ratio of specific heats for gas in source, P_1 is internal absolute pressure of source, a_1 is sound velocity in source, $\dot{E}$ is energy release rate, ℓ_i are enough length ratios to define source geometry and direction from source, subscripts o and 1 refer to ambient and source conditions respectively, and a bar over a symbol signifies a nondimensional quantity.

Equation (2-20) is essentially an extension of Sachs' law with additional parameters to describe the source. This way of expressing a scaling law implies that the scaled quantities on the left-hand side of the equation are dependent variables, each of which are functions of the six scaled quantities on the right. The functional dependence f_j is different for each of the dependent variables, and it must be determined by experiment or analysis. Experimental data in Esparza and Baker (1977a and 1977b) at least partially confirm this law for bursting pressure spheres, and show typical functions f_j for scaled overpressure, arrival time, duration, positive phase and negative phase impulses, and second shock properties. Some other variations on these scaling laws are discussed in Chapter 3 of Baker (1973).

BLAST WAVE CALCULATIONS AND EMPIRICAL DATA FITS

Bursting Spheres

1. General Discussion

When a pressurized, gas-filled vessel bursts, a shock wave propagates away from the surface of the vessel. This shock wave may be strong
enough to cause damage or injury.

In most of the studies of gas vessel bursts, the vessel is assumed to be frangible, i.e., the vessel shatters into a large number of
fragments when it bursts. The effect of these fragments on the blast wave
is ignored, except when their kinetic energy is subtracted from the total
energy in the vessel. Boyer, et al. (1958) measured the blast waves from
shattering glass spheres pressurized with air, helium, and sulfur hexa-
fluoride (SF_6). Esparza and Baker (1977a and 1977b) measured the blast
waves from shattering glass spheres pressurized with air, argon, and Freon
vapor. They found that the negative specific impulse is almost as large
as the positive specific impulse for this type of wave. This was unexpect-
ed because, for high explosives, the negative impulse is usually negligible
compared to the positive impulse.

There have also been several numerical studies. Boyer, et al.
(1958) used a finite-difference computer program to calculate the flowfield near bursting spheres pressurized with air, helium, and SF_6. Their
numerical results agreed with their experiments. Chou, et al. (1967) and
Huang and Chou (1968) also developed a numerical program to generate data
for a bursting pressurized sphere. They used the Hartee method-of-charac-
teristics with Rankine-Hugoniot jump conditions across the shocks. This
allows shock locations to be more clearly defined than Boyer could obtain
with his method.

2. Results of Systematic Numerical Studies

Baker, et al. (1975) supplemented and extended the work reported
by Strehlow and Ricker (1976) to include many more cases of spherical pressure vessel bursts. The same finite-difference computer program (CLOUD)

[Oppenheim (1973)] was used. Pressures in the "numerical" spheres ranged from 5 to 37,000 atmospheres. Temperatures ranged from 0.5 to 50 times that of the surrounding atmosphere. The ratio of specific heats of the gas in the vessel was varied; 1.2, 1.4, and 1.667 were used to investigate this dependence.

In the analysis that was used for the overpressure and specific impulse calculations, the effects of the containing vessel and its fragments were ignored; that is, all of the energy of the gas in the vessel was put into the flowfield, rather than into the fragments as kinetic energy. Also, the surrounding atmosphere was assumed to be air. All fluids were assumed to obey equations of state for perfect gases. To determine the overpressure and impulse, one must know the initial absolute pressure p_1, absolute temperature θ_1, and the ratio of specific heats γ, of the gas in the vessel. The conditions of the atmosphere into which the shock wave propagates also must be known. These are the atmospheric pressure p_o, the speed of sound a_o, and the ratio of specific heats γ_o. These atmospheric values were assumed constant for all the sphere explosions that were calculated. Table 2-3 gives the initial sphere conditions for the cases that were run to generate the data used in this analysis.

a. Overpressure Calculation

The overpressure-versus-distance relationship for a bursting gas vessel is strongly dependent upon the pressure, temperature, and ratio of specific heats of the gas in the vessel. For high pressures and high temperatures relative to the air outside the vessel, the overpressure behavior is much like that of a blast wave from a high explosive.

To facilitate comparisons and use of these graphs, results are plotted in scaled (dimensionless) form, where $\bar{P}_s$ is scaled side-on overpressure, $\bar{R}$ is energy-scaled distance based on Brode's formula [Equation (2-13)], and $\bar{i}_s$ is energy-scaled impulse. On the $\bar{P}_s$ versus $\bar{R}$ graph

Table 2-3. Initial Conditions Assumed for
Pressure Sphere Bursts

Case	$\dfrac{P_1}{P_o}$	$\dfrac{\theta_1}{\theta_o}$	γ_1
1	5.00	0.500	1.400
2	5.00	2.540	1.400
3	5.00	10.000	1.400
4	5.00	50.000	1.400
5	10.00	0.500	1.400
6	10.00	50.000	1.400
7	100.00	0.500	1.400
8	100.00	50.000	1.400
9	150.00	50.000	1.400
10	500.00	50.000	1.400
A	94.49	1.000	1.400
B	94.49	1.167	1.200
C	94.49	0.840	1.667
11	37000.00	0.500	1.400
12	37000.00	5.000	1.400
13	37000.00	10.000	1.400
14	1000.00	1.000	1.400
15	1000.00	4.000	1.667
16	1000.00	0.500	1.400
17	5.00	5.000	1.400

(Figure 2-15), which is a smoothed version of the actual numerical results, the curves for higher pressures and temperature are located near the high explosive curve. The curves for lower pressures and temperatures lie farther from the high explosive curve.

When an idealized sphere "bursts," the air shock has its maximum overpressure right at the contact surface between the sphere gas and the air. Since, initially, the flow is strictly one-dimensional, the shock tube relationship between bursting pressure ratio and shock pressure can be used to calculate the pressure in the air shock at time $t = +0$. This relationship is given in Equation (2-21) [Liepman and Roshko (1967)].

$$\frac{P_1}{P_o} = \frac{P_{so}}{P_o} \left\{ 1 - \frac{(\gamma_1 - 1)\,(a_o/a_1)\left(\dfrac{P_{so}}{P_o} - 1\right)}{\sqrt{(2\gamma_o)\left[2\gamma_o + (\gamma_o + 1)\left(\dfrac{P_{so}}{P_o} - 1\right)\right]}} \right\}^{\left(\frac{-2\gamma_1}{\gamma_1 - 1}\right)} \tag{2-21}$$

In this equation, p_{so}/p_o is the dimensionless air shock pressure at the instant of burst, p_1/p_o is the dimensionless sphere pressure, and a_o/a_1 is the velocity of sound ratio. Note that the sphere dimensionless shock overpressure is $\bar{P}_{so} = (p_{so}/p_o - 1)$. Usually, it is the unknown in Equation (2-21), and one must solve the equation by iteration to obtain p_{so}/p_o.

The fact that the curves on Figure 2-15 are roughly parallel, particularly at lower pressures, allows one to construct a graphical proced-

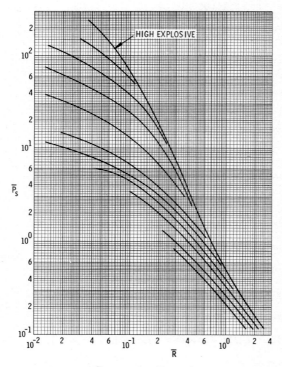

Figure 2-15. $\bar{P}_s$ Versus $\bar{R}$ for Overpressure Calculations, Bursting Sphere (Smoothed from Code Calculations)

ure for estimating the $\bar{P}_s$, $\bar{R}$ relation for any sphere burst. It was shown earlier through dimensional analysis that the blast field of any ideal gas sphere burst could be characterized by specifying the properties of the sphere and the surrounding atmosphere p_1/p_o, $Rp_o^{1/3}/E^{1/3}$, a_1/a_o, γ_1 and γ_o. To use these concepts, simply locate the initial shock pressure $\bar{P}_{so}$ and sphere radius $\bar{R}$, on Figure 2-15. Once this is done, the overpressure radius relationship can be obtained at any larger radius by following or paralleling the curves in Figure 2-15 [Strehlow and Ricker (1976)].

The energy-scaled radius of the sphere, $\bar{R}$, can be determined by noting that

$$E = \frac{P_1 - P_o}{\gamma_1 - 1} V_1 = \frac{4\pi}{3} \frac{P_1 - P_o}{\gamma_1 - 1} r_1^3 \qquad (2\text{-}22)$$

and since

$$\bar{R}_1 = \frac{r_1 P_o^{1/3}}{E^{1/3}} \qquad (2\text{-}23)$$

then

$$\bar{R}_1 = \left[\frac{3(\gamma_1 - 1)}{4\pi \left(\frac{P_1}{P_o} - 1 \right)} \right]^{1/3} \qquad (2\text{-}24)$$

The air shock pressure at $\bar{R} = \bar{R}_1$ can be obtained by solving Equation (2-21) using p_1/p_o, a_1/a_o, γ_1 and γ_o. Thus, all the dimensionless quantities which theory says are required to specify the blast field have been used. To simplify calculations, Equation (2-21) has been solved for sphere gases with a gamma of 1.4 (typically air, N_2, O_2, H_2, etc.) and 1.667 (typically He, Ar, Ne, etc.). These solutions are plotted in Figures 2-16 and 2-17.

150

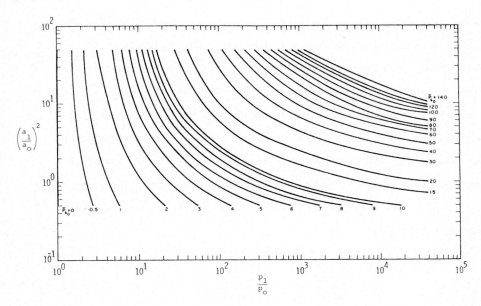

Figure 2-16. Vessel Temperature Versus Vessel Pressure for Constant Values of $\bar{P}_{s_o}$, $\gamma_1 = 1.4$

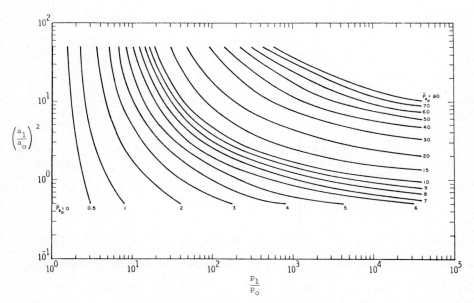

Figure 2-17. Vessel Temperature Versus Vessel Pressure for Constant Values of $\bar{P}_{s_o}$, $\gamma_1 = 1.667$

b. Specific Impulse Calculation

The $\bar{i} = i\, a_o/p_o^{2/3}\, E^{1/3}$ versus $\bar{R}$ data for some of the runs of Table 2-3 are plotted for several sets of initial conditions in Figure 2-18. For $\bar{R}$ less than about 0.5, the behavior is not clear, and a maximum i was chosen for the $\bar{i}$ versus $\bar{R}$ relationship in Figure 2-19. For $\bar{R}$ greater than about 0.5, all of the curves lie within about 25 percent of the high explosive (pentolite) curve. The pentolite curve was, therefore, chosen as representative of an $\bar{i}$ versus $\bar{R}$ curve in this region.

Therefore, for the burst of a pressure vessel, the $\bar{i}$ versus $\bar{R}$ relationship in Figure 2-19 or 2-20 should be used. For $\bar{R}$ in the range of 10^{-1} to 10^{0}, the $\bar{i}$ versus $\bar{R}$ curve in Figure 2-20 is more convenient. This is an enlargement of the upper left corner of Figure 2-19. These curves are accurate to about $\pm$ 25 percent. For a given distance, $\bar{R}$ is

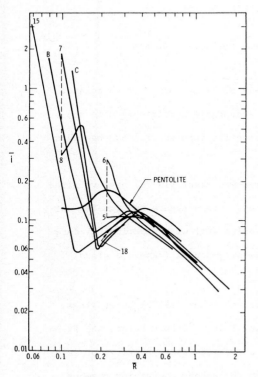

Figure 2-18. $\bar{i}$ Versus $\bar{R}$ for Bursting Spheres

152

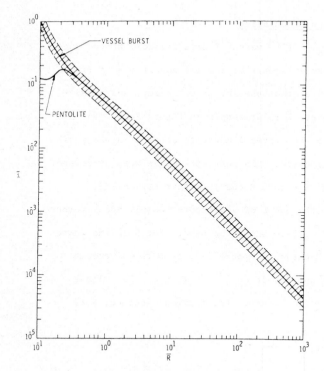

Figure 2-19. $\bar{i}$ Versus $\bar{R}$ for Pentolite and Gas Vessel Bursts

calculated, and $\bar{i}$ is read from Figure 2-19 or 2-20; then i is calculated.
Alternatively, one can choose a maximum acceptable specific impulse and
find the minimum distance at which the specific impulse is less than this
value.

Some experiments of Esparza and Baker (1977a) have verified
for at least a few cases that the blast wave overpressure and impulse calculated using the above technique agree quite well with experimental data.
They used argon and air as the sphere gases, and used frangible glass
spheres in the experiment.

In a second study, Esparza and Baker (1977b) burst glass
spheres that contained a flash-evaporating Freon., Before burst, the Freon
was held either as a liquid at slightly above its vapor pressure, or as a
gas at slightly below its pressure. When the sphere containing gaseous

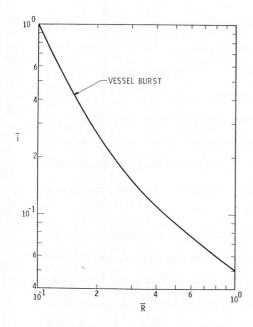

Figure 2-20. $\bar{i}$ Versus $\bar{R}$ for Gas Vessel Bursts, Small $\bar{R}$

Freon was burst, the blast wave was quite strong, but the shape was more
like a decaying sine wave than that of a shock followed by a rarefaction
region. Three distinct pressure peaks were observed. Evidently, the expanding Freon passed through the coexistence line on expanding, and partially condensed during the burst. When the sphere contained liquid phase
Freon, the blast wave was very weak, almost negligible compared to an air
burst at the same initial pressure. Evidently, the flash evaporation
process is so slow that the gas that is generated cannot contribute significantly to a blast wave. In all cases studied, however, the liquid
Freon was below its homogeneous nucleation temperature, i.e., $< 0.9\ \theta_c$
when θ_c is the critical temperature. The effect of bursting a sphere with
liquid above the homogeneous nucleation temperature is not known.

Vented Chambers

It is not unusual for an explosion to occur inside a structure that
is vented and attenuates the blast wave outside the structure. This type

of structure may have been constructed with blast wave attenuation in mind, or it may have been strongly built for other reasons. We discuss here the current state of knowledge for blast waves emitted in such explosions.

    1. Vented Explosions of High Explosives

        Esparza, et al. (1975) have correlated data on blast overpressure and specific impulse outside of suppressive structures with a high explosive detonated inside. A suppressive structure is a vented building which reduces the external blast pressure and impulse, fragment hazard, and thermal effects due to an explosion inside.

        The blast wave properties outside the structure depend upon the weight of explosive W (or energy), the volume of the structure, and the effective vent area ratio α_e. W is usually expressed in pounds or kilograms of TNT. It can be expressed for other explosives if one knows the total energy E (see Table 2-2). Then

$$W_{TNT} \ (kg) = \frac{E \ (J)}{4.520 \times 10^6} \quad (J/kg \ of \ TNT) \tag{2-25}$$

        The effective vent area ratio is calculated for the walls and roof of the structure. For single layer panels, the vent area ratio is the vent area divided by the total area of the walls and roof. For multi-layer panels,

$$\frac{1}{\alpha_e} = \sum_{i=1}^{N_p} \frac{1}{\alpha_i},$$

where N_p is the number of panels. If the structure consists of panels made of angles, zees, louvres, or interlocked I-beams, α_e is determined by using the information in Figure 2-21. Panels of nested angles which

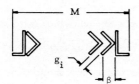

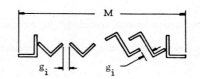

$$A_{vent} = \ell \sum_1^n g_i/N$$

ℓ = length of element

β = projected length of angle

$$N = \begin{cases} 2 \text{ if } (g_i/0.707) \simeq \beta \\ 4 \text{ if } (g_i/0.707) \simeq 2\beta \end{cases}$$

A_{wall} = LM

L = length of wall

α_e = A_{vent}/A_{wall}

$$A_{vent} = \ell \sum_1^n g_i$$

n = number of openings

A_{wall} = LM

α_e = A_{vent}/A_{wall}

(a) NESTED ANGLES

(b) SIDE-BY-SIDE ANGLES OR ZEES

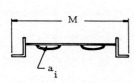

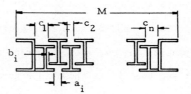

$$A_{vent} = \sum_1^n a_i/2$$

a_i = open area of louvre

A_{wall} = LM

α_e = A_{vent}/A_{wall}

$$A_{v_1} = 2\ell \sum_1^n a_i$$

$$A_{v_2} = A_{v_3} = 2\ell \sum_i^n b_i$$

$$A_{v_4} = 2\ell \sum_1^n c_i$$

$$\alpha_1 = A_{v_1}/A_w, \ldots$$

$$1/\alpha_e = \sum_{n=1}^n 1/\alpha_n$$

(c) LOUVRES

(d) INTERLOCKED I-BEAMS

Figure 2-21. Definition of Effective Area Ratio For
Various Structural Elements
[Esparza, et al. (1975)]

156

have approximately one opening per projected length are about twice as
efficient as a perforated plate in breaking up the side-on pressure as
it vents (N = 2). For closer nested angles such that there are about two
openings per projected length, the angles seem to be four times as efficient as a perforated plate (N = 4).

A model analysis was performed in which it was found that

$$P_s = f_1 \ (Z, \frac{X}{R}, \ \alpha_e)$$

$$(2\text{-}26)$$

$$\frac{i_s}{W^{1/3}} = f_2 \ (Z, \frac{X}{R}, \ \alpha_e)$$

where P_s is the side-on overpressure, i_s is the side-on specific impulse,
X is a characteristic length of the structure, R is the distance from the
center of the structure, and $Z = R/W^{1/3}$.

The P_s and i_s data for several suppressive structures with perforated panels, angles, zees, louvres, and interlocking I-beams have been
curve fit and the fit is shown in Figures 2-22 and 2-23 (S is the standard
deviation). The curves should only be used where Z, R/X, and α_e lie within the ranges shown on the graphs.

2. Vented Explosions of Combustible Gases and Dusts

Much more common than accidental explosions of high explosives
in vented chambers are accidental internal explosions in structures or manufacturing facilities involving mixtures of combustible gases with air, or
suspensions of combustible dusts in air. The enclosures may be designed
with vents which blow out or open at low differential pressures, or may be
naturally vented by open doors or by windows which also fail at low differential pressures. But, in spite of extensive experimental and analytic
study of pressures caused within the chambers, there are essentially no
data and no verified prediction methods for the characteristics of blast

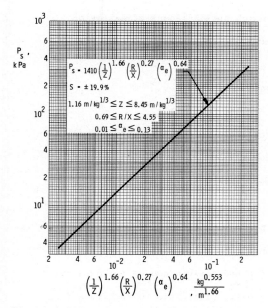

Figure 2-22.   Curve Fit to Side-on Pressures Outside
Suppressive Structures
[Esparza, et al. (1975)]

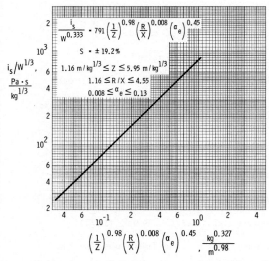

Figure 2-23.   Curve Fit to Scaled Side-on Impulse Outside of
Suppressive Structures
[Esparza, et al. (1975)]

waves emitted from vented chambers in which these classes of explosion occur. Principles of venting to preserve the integrity of a structure are discussed in Chapter 3.

Unconfined Vapor Cloud Explosions

1. General

A number of reviews - or overviews - of unconfined vapor cloud explosions have been prepared. Strehlow (1973b), Brown (1973), Munday (1976), Anthony (1977a), Eichler and Napadensky (1977), Davenport (1977a and 1977b), Wiekema (1980), and Gugan (1978) all discuss the general problem of unconfined vapor cloud explosions in some detail. Davenport's reports are quite thorough reports of incidents that have occurred and definitely show that massive release of a hydrocarbon can be followed either by no ignition, ignition with fire only, or ignition with fire and an explosion which produces damaging blast waves. Source behaviors which can lead to an explosion once ignition occurs are discussed in Chapter 1. In particular, the SWACER mechanism first studied by Knystautas, et al. (1979) is discussed in detail. This now appears to be the most viable mechanism for transition to detonation in a nominally unconfined cloud.

2. Free Cloud Detonation and Blast Wave Behavior

There is considerable information in the literature (both open and restricted) on blast overpressures from intentional detonation of fuel-air clouds because this type of Fuel-Air Explosion (or FAE) has military applications. Much of the literature regarding such planned explosions of FAE weapons is classified for government security reasons, although some work is reported in unclassified reports such as Kiwan (1970a, 1970b, and 1971), Kiwan and Arbuckle (1975), Kiwan, et al. (1975), Bowen (1972), Robinson (1973), and Axelson and Berglund (1978). The fuels used in the FAE weapon are, according to unclassified sources [Bowen (1972), Robinson (1973)], ethylene oxide, methylacetylene/propadiene/propylene (MAPP), normal-propyl-nitrate, or propylene oxide. In FAE weapons, the fuel is usually carried in canisters and explosively dispersed, and then detonated by one or more explosive detonators after some delay to allow time for mixing of the fuel with the air. The FAE weapon clouds are reported to be roughly disc-shaped or toroidal-shaped, with diameters considerably greater than their heights [Kiwan and Arbuckle (1975), Kiwan, et al. (1975), Robinson (1973)]. Reported values for maximum overpressures within the cloud in

Table 2-4. Heat of Combustion of Combustible Gases Involved In Vapor Cloud Accidents
[Adapted from Eichler and Napadensky (1977)]

Material	Formula	Low Heat Value (MJ/kg)	e_{HC}/e_{TNT} *
Paraffins			
Methane	CH_4	50.00	11.95
Ethane	C_2H_6	47.40	11.34
Propane	C_3H_8	46.40	11.07
n-Butane	C_4H_{10}	45.80	10.93
Isobutane	C_4H_{10}	45.60	10.90
Alkylbenzenes			
Benzene	C_6H_6	40.60	9.69
Alkylcyclohexanes			
Cyclohexane	C_6H_{12}	43.80	10.47
Mono Olefins			
Ethylene	C_2H_4	47.20	11.26
Propylene	C_3H_6	45.80	10.94
Isobutylene	C_4H_8	45.10	10.76
Miscellaneous			
Hydrogen	H_2	120.00	28.65
Ammonia	NH_3	18.61	4.45
Ethylene Oxide	C_2H_4O	26.70	6.38
Vinyl Chloride	C_2H_3Cl	19.17	4.58
Ethyl Chloride	C_2H_5Cl	19.19	4.58
Chlorobenzene	C_6H_5Cl	27.33	6.53
Acrolein	C_3H_4O	27.52	6.57
Butadiene	C_4H_6	46.99	11.22
HC Groups (est)	------	44.19	10.56

*Based on a heat of detonation of 4187 kJ/kg for TNT, which differs from the experimental value reported earlier in this chapter.

ethylene oxide-air detonations are about 2 MPa. Since the heat of combustion of most hydrocarbons is much larger than the heat of explosion of TNT (see Table 2-4), FAE weapons are very effective.

Fuel-oxygen mixtures have been used as blast simulants for small yield nuclear weapons in work sponsored by the U. S. Defense Nuclear Agency. This type of simulation is well summarized by Choromokos (1972). Methane-oxygen mixtures form the detonable gas, which was contained in large Mylar balloons prior to detonation. Small-scale and large-scale tests were conducted with tethered spherical balloons up to 33.5 m (110 ft) in diameter and hemispherical balloons up to 38.1 m (125 ft) in diameter. The latter balloon was successfully detonated with a TNT equivalent yield of 18,000 kg (20 tons).

There have also been many calculations of cloud detonation and the blast wave produced by cloud detonation. Specific calculations have been reported by Kiwan (1970a, 1970b and 1971), Shear and Arbuckle (1971), Oppenheim, et al. (1972a), Kiwan and Arbuckle (1975), Lee, et al. (1977), and Sichel and Foster (1979).

3. Spherical Cloud Deflagration

A systematic study of the effect of normal burning velocity on the blast wave produced by central ignition of a spherical cloud has been performed by Strehlow, et al. (1979). They also calculated the blast wave properties for a centrally ignited spherical detonation and bursting sphere for comparison. In addition, they studied the effect of flame acceleration on the blast wave structures.

The energy density, q, for all these calculations was 8, and the product gas in the source region was assumed to have an effective ratio of specific heats of $\gamma_1 = 1.2$. This is known to model quite faithfully most stoichiometric hydrocarbon-air explosions (see Chapter 1). For numerical stability reasons, the energy addition wave had to have a finite thickness. A number of computer runs were made to determine the optimum thickness which would lead to stable computation but also lead to a proper

self-similar solution for the wave structure prior to the time that the
wave reached the end of the source region. They found that choosing a wave
width of 0.1 r_o, where r_o is the total initial radius of the source region,
met both of these requirements. All of the runs that they reported have a
wave width of 0.1 r_o. For all cases of wave addition of energy, the normal
burning velocity divided by the local velocity of sound before the energy
addition started was considered to be a reference Mach number M_{su}. Reference Mach numbers that were used ranged from 0.034 up to 8.0 [for this energy density, the upper theoretical Chapman-Jouguet (CJ) wave Mach number is
5.2]. Typical output from low velocity subsonic wave addition of energy
to high velocity supersonic addition is shown in Figure 2-24. The overpressure scaled distance curve for all runs of practical importance, including detonation and open burst, is shown in Figure 2-25. In this figure, curves are given for a bursting sphere for an energy density of 8

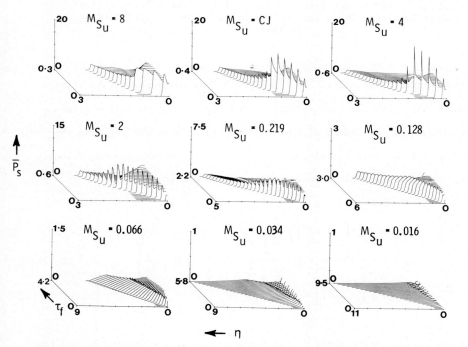

Figure 2-24. Blast Wave Structure for Constant Velocity Flames (Explosions).
$\eta = r/r_o$ When r_o is the Initial Source Radius. τ_f is Defined as $t\, a_o/r_o\ \sqrt{8}$

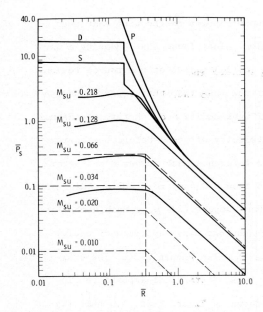

Figure 2-25. Maximum Wave Overpressure Versus Energy Scaled Distance
for Deflagrative Explosions With an Energy Density of q = δ and
Various Normal Burn Velocities. The Curve Labeled P is for Pentolite.
All Other Solid Curves Were Calculated Using the CLOUD Program. The
Curve Labeled D is for Detonation and S is for a Sphere Burst of the
Same Energy Density. The Origin of the Dashed Lines is Discussed in
the Text.

and the reference Pentolite. In Figure 2-25, the horizontal dashed lines

represent the predicted flame overpressures for very low velocity flames

calculated using the theory of Taylor (1946) for piston motion where the

piston motion has been transformed to an equivalent normal burning velocity

for a flame system. This transformation yields the equation

$$
\bar{P}_f = \frac{2\gamma_o \left[1 - \dfrac{\rho_2}{\rho_1}\right] \left(\dfrac{\rho_1}{\rho_2}\right)^2 M_{su}^{\;2}}{1 - \left[1 - \dfrac{\rho_2}{\rho_1}\right]^{2/3} \cdot \left(\dfrac{\rho_1}{\rho_2}\right)^2 M_{su}^{\;2}} \left[1 - M_{su} \dfrac{\rho_1}{\rho_2}\right] \tag{2-27}
$$

where ρ_1 is the density ahead of the flame and ρ_2 is the density behind the

flame and the overpressure is assumed to be so low that the ratio ρ_2/ρ_1 can be taken as the zero pressure rise density change across the flame. The vertical dashed line on Figure 2-25 represents the final radius of the cloud based on energy scaled radius for an energy density of 8 when the energy is added so slowly that there is no pressure rise. The dashed lines outside of this vertical line are predicted overpressures based on the equation $\bar{P}_s \propto 1/\bar{R}$. It can be seen that the agreement is quite good and that Taylor's acoustic theory adequately predicts the blast behavior for low burning velocity spherical explosions.

In Figure 2-26, the energy-scaled positive impulse is plotted against energy-scaled radius for all cases. It is not known why the impulses for flames are about 20 to 30 percent low. However, they fall within the error bars of Figure 2-19.

Numerical calculations were also made for the case of accelerating flames. In this case, they arbitrarily chose both an initial flame

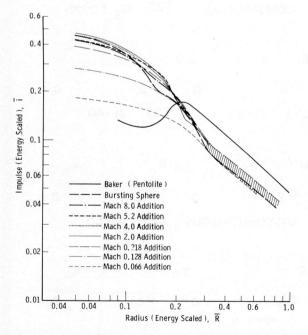

Figure 2-26. Energy Scaled Impulse Versus Energy Scaled Radius for Constant Velocity Flames, a Bursting Sphere and Reference Pentolite [Luckritz (1977)]

velocity for the central region and a terminal flame velocity for the
edge region and connected these with a region of constant acceleration
from the initial to the final velocity. The program was set up such that
both the width and the location of the acceleration region could be changed from run to run. Nine cases were run. These are summarized in Figure
2-27. The cases in the upper-right and left corner of this figure are the
reference cases for constant velocity flame propagation throughout the
source region. The graph labeled continuous was for continuous acceleration from this low velocity to the higher velocity over the entire source
region. The curve labeled discontinuous represents the discontinuous acceleration of the wave from the lower to the higher velocity at cell 25.
(There are 50 spherical shells in the source region.) Those labeled 1-10,
11-20, etc., represent cases which were run when the cell regions 1-10,
11-20, etc., contained the accelerating flame. Figure 2-28 shows the

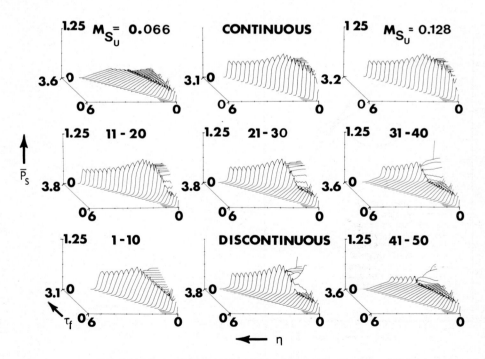

Figure 2-27. Blast Wave Structure for Accelerating Flames. η and τ_f are
Defined in Figure 2-25

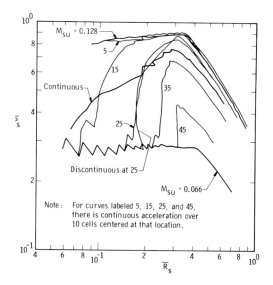

Figure 2-28. Effect of Flame Acceleration on
              Dimensionless Overpressure

scaled overpressure versus energy-scaled radius for these nine runs. Note
from this figure that the maximum overpressure for any acceleration process
is always less than that produced by the highest velocity flame in the system. Thus, it appears that, for a strictly spherical flame, acceleration
processes per se do not cause overpressure and one simply has to know the
highest effective flame velocity if one is to be able to predict the overpressure produced by the source region flame.

These calculations also indicate that there must be a high effective burning velocity if a spherical flame is to cause significant blast
damage. For example, from Figure 2-25 we see that an overpressure of 0.3
bar is produced by a flame Mach number M_{su} of 0.666, or an effective normal burning velocity of 23 m/s. Since hydrocarbon air mixtures have normal burning velocities in the range of 0.40 to 1 m/s, a considerable acceleration of the flame must occur by some mechanism if ignition of an unconfined fuel-air cloud is to produce a damaging blast wave by deflagrative
combustion.

4. The Blast Wave Produced by Deflagration in Clouds of Arbitrary Shape

It is well known that in an accident situation, the vapor cloud that is produced by a massive release of a combustible is never spherical (or even hemispherical) in shape. Clouds are usually roughly of pancake shape or cigar shape and are known to have aspect ratios that range up to 50 to 1. Also, clouds are usually edge ignited. The exception is ignition during an attempt to restart a stalled automobile or truck engine.

Strehlow (1981) has applied the monopole source theory of spherical acoustics [Stokes (1849)] to the deflagrative combustion of a number of nonspherical clouds. Simply stated, a monopole source in spherical geometry generates an overpressure which is determined uniquely by the first derivative with respect to time of the rate of mass addition in the source region

$$p - p_o = \ddot{m} \left[t - \frac{r}{a_o} \right] \bigg/ 4\pi r \qquad (2\text{-}28)$$

where r is the radius and the argument $t - r/a_o$ represents the locus of an acoustic wave traveling away from this simple source region. For deflagrative combusion of a cloud, the effective source of mass is the volume increase produced by combustion. This is converted to mass by multiplying by the air density ρ_o, i.e.,

$$\dot{m}(t) = \rho_o \dot{V}(t) = \rho_o \left(\frac{V_b - V_u}{V_u} \right) \left[S_u(t) \cdot A_f(t) \right] \qquad (2\text{-}29)$$

where $S_u(t)$ is the normal burning velocity and $A_f(t)$ is the effective flame area. We note from Equation (1-33) that for constant pressure combustion, $(V_b - V_u)/V_u = q/\gamma$, where q is the energy density of the combustion source. Using this and the definition of the velocity of sound for an ideal gas $a_o^2 = \gamma p_o/\rho_o$, we obtain

$$\bar{P} = \frac{q}{4\pi \, a_o^2 \, r} \frac{d}{dt} \left[S_u(t) \cdot A_f(t) \right] \qquad (2\text{-}30)$$

from Equations (2-28) and (2-29), where $\bar{P}$ is the acoustic overpressure.
Note that Equation (2-30) states that a constant velocity flame which has
a constant flame area will not generate acoustic overpressure. Equation
(2-30) shows that one must either have an accelerating flame or an ever
growing flame area before a pressure wave travels away from the source region.

When Equation (2-30) is applied to a constant velocity spherical
flame, one obtains the same analytic expression for the pressure at the
flame as was derived using the spherical piston approach of Taylor (1946)
[Equation (2-27)]. The reason why a constant velocity spherical flame generates overpressure is because its area is increasing at an increasing rate
at all times while combustion is occurring.

Strehlow (1981) made a number of simplifying assumptions and
calculated the effect of aspect ratio for two cloud shapes on the maximum
overpressure that would be generated outside the cloud. The assumptions
used in his analysis are that the cloud's energy content or total
volume is constant, that the burning velocities are the same and that
the distance from the observer to the cloud center is the same. The quantity $\bar{P}/\bar{P}_{spH}$ is the overpressure expected from a cloud with the spherical
shape and aspect ratio divided by the overpressure generated by a centrally
ignited spherical cloud.

Strehlow (1981) also noted that Equation (2-30) contains no adjustable constants. Thus, one can calculate either flame accelerations or
rates of flame area increase necessary to produce a specified overpressure
at a specified radius. Table 2-5 was constructed by assuming that a 10 kPa
overpressure is to be generated by deflagration combustion occuring 100 meters
away from the observer. Notice the extremely large values of flame acceleration
rates or rates of flame area increase that are required to produce a damag-

Table 2-5. Requirements to Generate 10 kPa at 100 m Radius

(1) $dS_u/dt = 0$		(2) $dA_f/dt = 0$	
S_u, m/s	$\dfrac{dA_f}{dt}$, m^2/s	A_f, m^2	$\dfrac{dS_u}{dt}$, m/s^2
1	1.7×10^7	100	1.7×10^5
10	1.7×10^6	10,000	1.7×10^3
100	1.7×10^5	1,000,000	17.0

ing blast wave. Thus, the application of simple source acoustic theory to deflagrative combustion of an unconfined cloud shows that:

- It is very difficult to produce a damaging blast wave by deflagrative combustion once the flame is not completely surrounded by a combustible mixture. This has also been verified experimentally by Thomas and Williams (1966).

- The maximum overpressure produced, other circumstances being equal, is proportional to the ratio of the minor dimension of the cloud to the distance to the observer.

- Blast pressure is rather uniformly distributed in all directions, i.e., the blast is roughly spherical.

- The cloud must be very large if a damaging blast wave is to be produced.

- Spherical flame propagation calculations such as those of Kuhl, et al. (1973) and Strehlow, et al. (1979) greatly overestimate blast pressures from deflagrative combustion following edge ignition of clouds with large aspect ratios.

From various accident accounts of the sequence of events that led to the production of a damaging blast wave after delayed ignition of a massive spill of combustible material it appears that:

- There is a threshold spill size below which blast damage does not occur. Gugan's (1980) documentation of incidents shows that blast damage has been observed for spills of less than 2000 kg but more than 100 kg only for the fuels H_2, H_2-CO mixture, CH_4 and C_2H_4. Blast damage has been recorded only for spills greater than 2000 kg for all other fuels.

- In the majority of cases where blast damage occurred, fire was present for a considerable period before the blast occurred.

- In many cases, damage is highly directional.

These observations when coupled with the results of simple source acoustic theory for deflagrative combustion lead to the following conclusions:

- There should be a size threshold below which blast will not occur as long as ignition is "soft," i.e., does not directly trigger detonation.

- The fact that fire is present early after ignition indicates that massive flame accelerations are necessary to lead to blast wave formation. Since the flame must have burned through the cloud edge by the time the blast is produced, simple acoustic source theory must be operative for even high deflagration velocities. Thus, the blast must arise from some sort of effectively supersonic combustion process or from very rapid increases in effective surface area of the flame.

- Simple acoustic source theory for deflagrative combustion shows that deflagrative processes per se cannot produce highly directional effects. However, it is well known that detonative combustion of a cloud does produce highly directional blast wave effects.

The net result of this work and the discovery of the SWACER effect [Lee and Moen (1980)] must lead to the conclusion that the damaging blast waves observed in vapor cloud explosions are caused by accelerations of the flame to either detonation or at least to high supersonic velocity.

5. Blast Wave Overpressure Estimation

The previous discussion in this section on the blast wave from vapor cloud explosions shows that one must assume detonation of a portion of the cloud if one is to make a conservative estimate of the damage potential. Unfortunately, it is not sufficient to use a spherical cloud because FAE calculations have shown, and one would expect that full CJ detonation pressures can be reached in all regions where a detonable cloud is present irrespective of the aspect ratio of the cloud. Because of this,

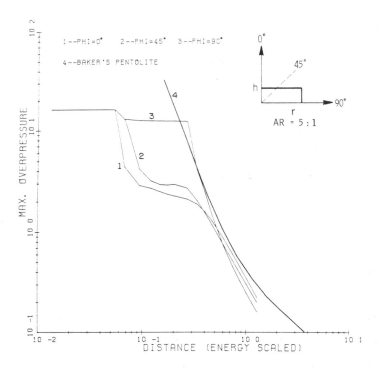

Figure 2-29. Maximum Overpressure Plotted Versus Energy Scaled Distance for a Pancake Shaped Cloud of Aspect Ratio 5 With Central Ignition of Detonation. Energy Density = 8.
[Raju (1981)]

one cannot use Figure 2-25 which was calculated for a spherical cloud to estimate the near-field blast potential of a pancake cloud with any realistic aspect ratio.

Recently, Raju (1981) has calculated the overpressure and impulse from the detonation of pancake clouds with aspect ratios of 5 and 10. His results are shown in Figures 2-29 through 2-32. These results are for free clouds with the stated aspect ratio.

Note that in the direction of the major axis of the pancake, both the overpressure and positive phase impulse are considerably higher than those predicted using either the pentolite curve or the spherical cloud detonation curve on Figures 2-25 and 2-26. Also, note in the far-field, both the overpressure and positive impulse fall below that predict-

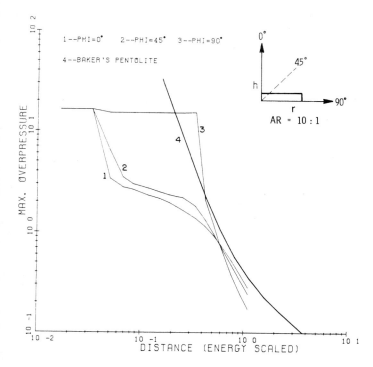

Figure 2-30. Maximum Overpressure Plotted Versus Energy Scaled
Distance for a Pancake Shaped Cloud of Aspect Ratio 10 With
Central Ignition of Detonation. Energy Density = 8.
[Raju (1981)]

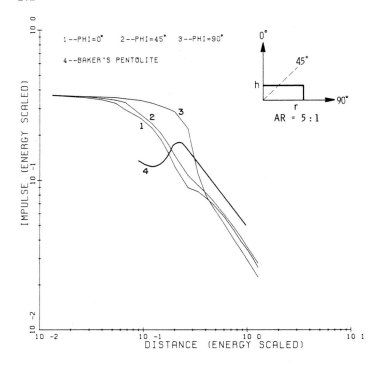

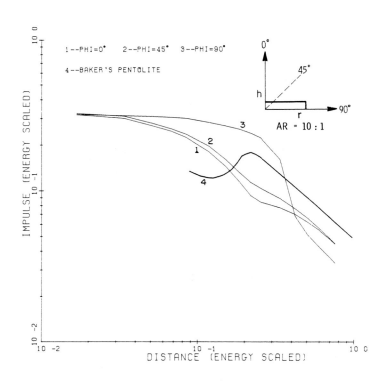

ed by the spherical cloud curves, or in the 0° or 45° direction. This observation is in agreement with the calculation of the blast wave structure
from a bursting ellipsoid [Chiu, et al. (1977)].

The overpressures and impulses presented in Figures 2-29 through
2-32 can be used to estimate the expected overpressures and impulses in the
near-field for pancake shaped clouds.

Physical Explosions

Physical explosions are also called vapor explosions. They can occur when two liquids at different temperatures are violently mixed, or when
a finely divided hot solid material is rapidly mixed with a much cooler
liquid. No chemical reaction is involved -- instead, a physical explosion
occurs when the cooler liquid is converted to vapor at so rapid a rate that
it cannot vent, and shock waves are formed.

There have also been attempts to estimate the source energy for such
physical or vapor explosions. Some investigators simply use the quantity
Q obtained by calculating the total heat which could be released by cooling
the hotter fluid or solid from its initial temperature to the initial temperature of the cool liquid which vaporizes to generate the explosion. A
phase change from liquid to solid is usually involved in the cooling of the
hotter material. This is equivalent to assuming that $E_s/Q = 1$, and gives
an artificially high prediction for source energy. Anderson and Armstrong
(1974) compute Q for several real or simulated physical explosions. They
assume constant volume heat addition followed by isentropic expansion,

---

Figure 2-31. Energy Scaled Impulse Plotted Versus Energy Scaled
Distance for a Pancake Shaped Cloud of Aspect Ratio 5 With
Central Ignition of Detonation. Energy Density = 8.
[Raju (1981)]

Figure 2-32. Energy Scaled Impulse Plotted Versus Energy Scaled
Distance for a Pancake Shaped Cloud of Aspect Ratio 10 With
Central Ignition of Detonation. Energy Density = 8.
[Raju (1981)]

identical to the second case of Adamczyk and Strehlow (1977), and a constant volume heat addition followed by expansion with equilibrium between the two liquids involved. The latter process releases more energy than the former.

Because phase changes are involved, Anderson and Armstrong (1974) use different energy equations than do Adamczyk and Strehlow (1977). They show that

$$Q = m \left[C_{pf} (\theta_{in} - \theta_f) + h_{gf} + C_{ps} (\theta_f - \theta_{fin}) \right] \qquad (2\text{-}31)$$

where m is the mass of hot (liquid) material, C_{pf} is the specific heat of hot liquid, C_{ps} is the specific heat of hot solid, θ_{in} is the initial temperature of hot material, θ_{fin} is the final temperature of cold fluid, θ_f is the freezing temperature of hot material, and h_{gf} is the heat of fusion of hot material.

Equations for the constant volume heat addition followed by equilibrium or isentropic expansion are, of course, more complex. Anderson and Armstrong (1974) show several example calculations for hot liquids mixed with cold liquids, such as molten uranium oxide (UO_2) in molten sodium (Na), molten steel in water, and molten aluminum in water. Their results are summarized in Figure 2-33. In this figure, the term "expansion with fluid equilibrium" gives the upper curves while "adiabatic expansion" gives the lower curves. The adiabatic expansion is probably a more realistic upper limit, and the curves labeled 1, 3 and 5 in the figure should not be used for estimating blast source energy. The expansion process in an explosion must be rapid, and equilibrium conditions cannot be maintained. For conditions simulating a nuclear reactor accident with molten aluminum fuel cladding falling into water, Anderson and Armstrong (1974) compute:

$$Q/m = 2025 \text{ J/g Al}$$

$$E_s(\text{equilibrium})/m = 550 \text{ J/g Al}$$

$$E_s(\text{adiabatic})/m = 310 \text{ J/g Al}$$

This gives ratios of E_s/Q of $0.153 \leq E_s/Q \leq 0.272$. Similarly, they compute for simulation of a steel foundry accident with molten steel reacting with water:

$$Q/M = 822 \text{ J/g steel}$$

$$E_s(\text{equilibrium})/m = 280 \text{ J/g steel}$$

$$E_s(\text{adiabatic})/m = 160 \text{ J/g steel}$$

This gives $0.195 \leq E_s/Q \leq 0.341$. As the ratios of E_s/Q indicate, physical explosions appear to be less efficient than gaseous explosions in converting source energy to blast waves.

Pressure Vessel Failure for Flash-Evaporating Fluids

Many fluids are stored in vessels under sufficient pressure that they remain essentially liquid at the vapor pressure corresponding to the storage or operating temperature for the particular liquid. Examples are the fuels propane or butane which are normally stored at "room" temperature, methane (LNG) which must be stored at cryogenic temperatures, refrigerants such as ammonia or the Freons which are also stored at room temperature, and steam boiler drums. If a vessel containing such fluids fails, the resulting sudden pressure release can cause expansion of vapor in the ullage space above the liquid and partial flash evaporation of the liquid. If expansion is rapid enough, the vapor may drive a blast wave into the surrounding air.

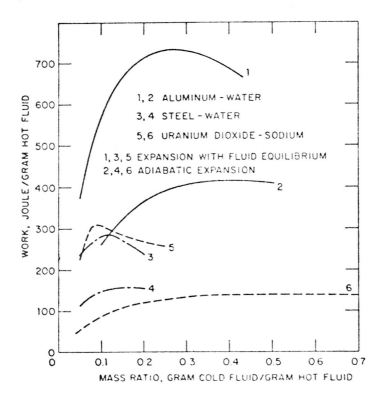

Figure 2-33. Work Versus Mass Ratio for Vapor Explosions
[Anderson and Armstrong (1974)]

Because the properties of flash-evaporating fluids differ from perfect gases, the methods for estimating blast yield for gas vessel bursts may not apply. If this is true, then the complete thermodynamic properties of the fluid in the vessel as functions of state variables such as pressure, specific volume, temperature, and entropy can be used to estimate blast yield.

For any expansion process from thermodynamic state 1 to thermodynamic state 2, the specific work done is defined as

$$e = u_1 - u_2 = \int_1^2 p \, dv \qquad (2\text{-}32)$$

where u is the internal energy, and v is specific volume. Assume that an isentropic expansion process occurs after vessel burst. This process is shown schematically in a pressure-specific volume (p-v) diagram in Figure 2-34, and in a temperature-entropy (θ-s) diagram in Figure 2-35. The particular initial state 1 shown in these two figures lies in the superheated vapor region, and so does the final state 2 after isentropic expansion to ambient pressure p_o. The cross-hatched area in Figure 2-34 is the integral of Equation (2-32), and, therefore, represents the specific energy e. Also, shown in the two figures are the saturated liquid and saturated vapor lines,

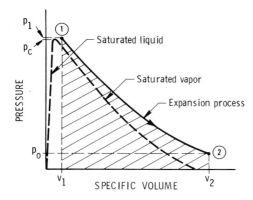

Figure 2-34.   p-v Diagram of Expansion

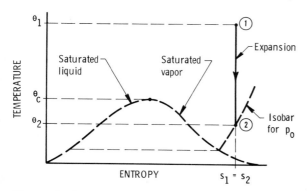

Figure 2-35.   θ-s Diagram of Expansion

which bound the wet vapor region. As is always true for flash-evaporating
fluids, the functional relationship between pressure and specific volume is
quite complex and the integral in Equation (2-32) cannot be obtained analytically. But, fortunately, there are tables of thermodynamic properties
available for many fluids, and the internal energy u or enthalpy h defined
as

$$h = u + pv \qquad (2-33)$$

are tabulated for the entire wet vapor region and the superheat region, as
functions of pressure and specific volume, or temperature and entropy. When
an initial or a final state falls within the wet vapor region, an important
parameter is the quality of the vapor, defined as

$$x = \frac{v - v_f}{v_g - v_f} = \frac{s - s_f}{s_g - s_f} = \frac{u - u_f}{u_g - u_f} = \frac{h - h_f}{h_g - h_f} \qquad (2-34)$$

where subscript f refers to fluid (saturated liquid state) and subscript g
refers to gas (saturated vapor state). Also, within the wet vapor region,
a given pressure uniquely defines a corresponding temperature, and vice
versa.

In bursts of vessels containing flash-evaporating fluids, three combinations of state variables are possible at states 1 and 2. These are:

Case 1: Superheated vapor at state 1 and at state 2 (as for the process shown in Figures 2-34 and 2-35).

Case 2: Superheated vapor at state 1 and wet vapor at state 2.

Case 3: Wet vapor (including both saturated liquid and saturated vapor) at state 1, and wet vapor at state 2.

The process of estimating e and total blast yield E is basically the
same for any of these combinations, but, depending on where state 1 lies,

the procedure for entering the thermodynamic tables differs somewhat. The
basic procedure is as follows:

Step 1: Estimate the initial state variables, including p_1, v_1, s_1,
u_1, or h_1.

Step 2: Assume isentropic expansion to atmospheric pressure p_a, i.e.,
$s_2 = s_1$. Determine v_2, u_2, or h_2.

Step 3: Calculate specific work e from Equation (2-32).

Step 4: Calculate total blast yield E by multiplying e by mass m of
fluid initially present in the vessel.

In Step 4, we use the basic definition of specific volume to obtain
the mass m of fluid from the known vessel volume V_1,

$$m = V_1/v_1 \qquad (2-35)$$

and compute E from

$$E = m \, (u_2 - u_1) \qquad (2-36)$$

Let us describe the differences in the three cases enumerated.
In Cases 1 and 2, the initial state conditions must be obtained from superheat tables for the fluid, usually entering with knowledge of the pressure
and temperature together. For Case 1, superheat tables are also used for
$p_2 = p_0$, $s_2 = s_1$, to obtain the final state conditions; while in Case 2, the
saturated vapor tables must be used with the definition of final quality x_2,
determined from final entropy s_2, being the most important factor. In Case
3, all values are found in the saturated vapor table, with initial quality
x_1 usually being determined from a real or fictitious initial specific volume. This case is probably the most common for flash-evaporating fluid
vessel bursts. The fictitious initial specific volume for a vessel which
is partially filled is obtained simply from Equation (2-35) by using m as
the mass of liquid in the vessel of volume V_1.

Some tables of thermodynamic properties for fluids which can be used to estimate blast yields by the process just described are the ASHRAE Handbook of Fundamentals (1972) for refrigerants, Keenan, et al. (1969) for steam, Din (1962) for a number of fluids including fuels such as propane and ethylene, and Goodwin, et al. (1976) for ethane. In many instances, these tables do not include internal energy u directly, but instead include h, p, and v. One then has to use Equation (2-33) to calculate u.

ACCIDENTAL EXPLOSIONS

Types of Accidental Explosions

The term "accidental explosions" includes a large spectrum of different possible occurrences, and each specific accidental explosion is at least slightly different from any other explosion. See NTSB (1972a) and Loss Prevention (1967) for discussions of a number of well investigated accidental explosions and Ordin (1974) for a discussion of hydrogen explosions. It is, nevertheless, possible to group accidental explosions into a number of major categories, each of which has rather distinctive characteristics, some of which are unique to that category (see Table 2-1). Unfortunately, the extent of understanding of the circumstances which lead to the explosion and the mechanism of the basic explosion process in each category is highly variable from category to category; some are very well understood, others are not as well understood. Also, some of the categories have the potential to yield much more disastrous explosions than others. Thus, even though all categories will be mentioned in what follows, the depth of discussion of each will be determined both by the current depth of understanding and by the overall potential societal impact of each category of explosion (i.e., its damage potential).

1. ### Detonation of Condensed Phase Systems

   a. ### Light or No Confinement

   As was indicated earlier, condensed phase detonation produces a blast wave that is essentially ideal. In most cases where there is an ac-

cidental detonation of such substances, there is confinement and this will
attenuate the blast wave somewhat. However, if one knows the quantity of
material involved and the degree of confinement, estimates of blast and fragment damage can be made in a rather straightforward manner [Engineering Handbook (1972), Baker (1973), and Swisdak (1975)]. Propellants are known to exhibit lower yields [Napadensky, et al. (1975)].

Such detonations can and have occurred during the manufacture,
transport, storage and use of high explosives and propellants. Additionally,
in the chemical industry this type of accidental explosion can occur in
chemical reactors, distillation columns, separators, etc., if some unwanted and highly sensitive substance is accidentally allowed to concentrate
in such an apparatus. As an example, Jarvis (1971a and 1971b), Freeman
and McCready (1971a and 1971b), and Keister, et al. (1971) report on the
detailed investigation of the explosion of a butadiene distillation column
due to the detonation of an accidental accumulation of a high concentration of vinyl acetylene in the column.

There have been a number of truly disastrous high explosive
detonation accidents. They have occurred where large quantities of these
substances are handled or stored either in bulk or in containers that are
in relatively close contact. For example, in 1921 a congealed mass of 2 x
10^6 kg of ammonium nitrate-sulphate double salt was being broken up and
crushed for use as fertilizer in Oppau, Germany [Lees (1980b)]. Unfortunately, dynamite was being used to fragment the pile. In one particular
shot, the pile detonated as a unit. This disaster killed an estimated 430
people, left a crater 130 m in diameter and 60 m deep and caused severe
damage to a radius of 6 km (Figure 2-36). Three official inquiries were
conducted, but details of the exact events that led up to the disaster
were never resolved because all the people associated with the incident
were killed. Up until the time of this disaster, it was thought that this
salt, even though exothermic, could not support a detonation. Even today,
laboratory scale testing would indicate that the substance is nondetonable

Figure 2-36. Aerial View of Damage From Oppau Explosion

because it has a minimum charge diameter for detonation which is much larger than the size usually used for sensitivity tests.

One other major explosion which is worthy of note is the
Texas City disaster of April 16, 1947 [Wheaton (1948) and Feehery (1977)].
The ship, *Grand Camp*, had in its hold 4500 tons of ammonium nitrate when
it caught fire. The fire accelerated out of control and eventually the remaining ammonium nitrate detonated. All houses within a one-mile radius
were destroyed, and it was estimated that 516 people were killed (many of
these were listed as missing). Property damage was estimated to be $67
million (in 1947 dollars). Figures 2-37 and 2-38 show some of the damage
from this disaster.

Large-scale accidental explosions of high explosives and
munitions loaded with high explosives escalated with the widespread use of
these materials in World War I. In fact, the Quantity-Distance (Q-D) tables

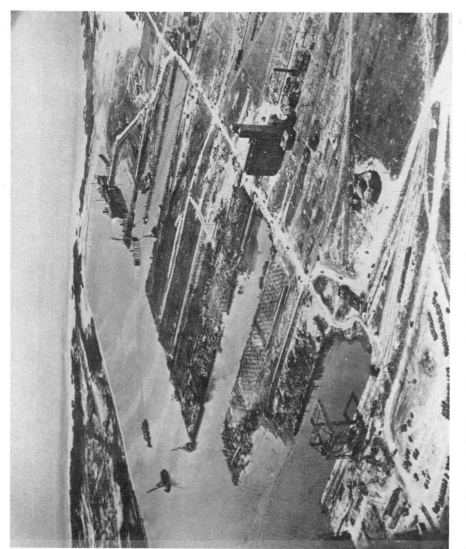

Figure 2-37. Aerial View of Damage in Texas City Disaster, April 16, 1947

184

Figure 2-38. Damage to Monsanto Chemical Plant in Texas City Disaster

adopted in the United States in an attempt to improve explosives safety
were originally based on a compilation and study of a number of such truly
disastrous explosions [Assheton (1930)].

   b.   Heavy Confinement

      The blast damage produced by heavy confinement is much less
than that produced by a high explosive when the confinement is weak. Design criteria for heavy confinement has already been discussed in this
chapter. Also, most design bunkers are designed with attenuation of the
blast wave in mind [Fugelso, et al. (1974)].

   2.   Combustion Explosions in Enclosures; (Not Pressurized), Gaseous
        or Liquid

   a.   Fuel Vapor

      These explosions usually occur when a fuel leaks into an en-
closure, mixes with air to form a combustible mixture and this mixture contacts a suitable ignition source that was present before the leak occurred.
The exceptions are ship or storage tank explosions where the space above
the stored fuel is in the explosive range. In these cases, the accidental
introduction of an appropriate ignition source will cause an explosion.

      There are two distinct limit behaviors for enclosure explo-
sions. If the enclosure has a length-to-diameter ratio of about $L/D = 1$,
and if the interior is not too cluttered with equipment, partitions, etc.,
the enclosure will usually suffer a simple overpressure explosion. The rate
of pressure rise in this case is relatively slow and the weakest windows or
walls will fail first. If this happens in a simple frame building, the ceiling will rise and all walls will fall out at about the same time. A good
example of the results of such an explosion is shown in Figure 2-39 taken
from Carroll (1979). In a steel container like a ship hold or boiler, the
enclosure will try to become spherical until a tear or rip vents the contents. While these explosions usually cause extensive damage to the enclosure [Senior (1974), and Tonkin and Berlemont (1974)], the blast wave that
they produce is ordinarily quite weak. This is because, in general, build-

186

Figure 2-39. Overpressure Explosion in a House. Intact Walls Have
Fallen Out and Roof is Lying in the Basement
[Carroll (1979)]

ings, ships or boilers are not very strong and they vent at overpressures which are typically very low (7 to 70 kPa). Thus, such an explosion acts as a very low energy density blast source.

The other limit behavior occurs in enclosures which have a large L/D or contain a large number of obstacles such as large pieces of equipment of internal partitions. In this case, after ignition, flame propagation causes gas motion ahead of the flame and this gas motion generates turbulence and large scale eddy folding at places where the flow separates from the obstacles. This, in turn, causes rapid increases in effective flame area, which causes a more rapid rise of pressure and further turbulent/eddy flame interactions. These processes may lead to gas-phase detonations in some locations in the enclosure. In these locations, the internal pressures can become very high (ca 1.5 MPa) very rapidly (< 1 ms), and highly localized massive damage can occur. It should be noted that this limit behavior is characterized by the location of maximum damage not being near the location of the source of ignition. In fact, maximum damage usually occurs as far from the ignition source as is possible within the enclosure. These explosions usually produce strong blast waves and high velocity fragments, and therefore, can cause more damage to the surroundings than simple overpressure explosions.

Almost any enclosure which is not designed to contain explosions can explode if filled with a combustible vapor or gas and ignited. A few examples will illustrate the wide variety of behaviors that are possible. Ostroot (1972) has documented over 200 combustion explosions in gas and oil fired furnaces or boilers caused by various types of malfunction or upset. These are primarily simple overpressure explosions. Halversen (1975) presents examples of tanker explosions. The majority of tanker explosions occur in spaces with low L/D and are simple pressure explosions as described above. However, some tanker explosions do produce damage that is highly localized and typical of high L/D enclosures. The results of a detonative tanker explosion are shown in Figure 2-40. The ship was being ballasted and

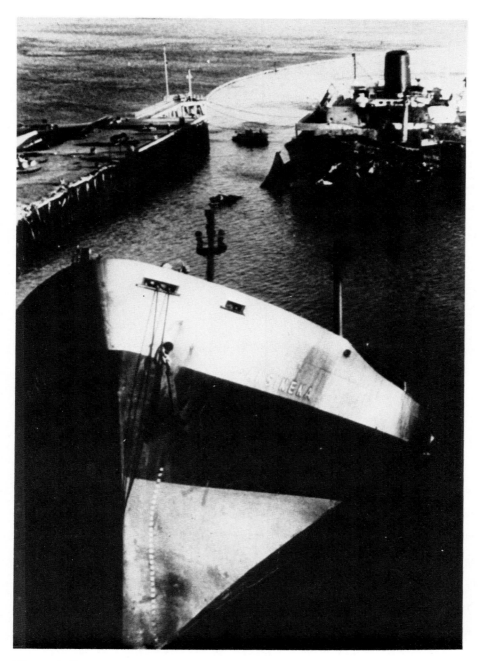

Figure 2-40. The Remains of the Liberian Tanker S. S. Sansinena After an Explosion in Her Hold on December 16, 1976, in Los Angeles Harbor

there was very little wind. A cloud of combustible vapor from the hold formed on deck and was ignited by some unknown source. The resulting flash fire entered the hold and undoubtedly progressed from a deflagration to a detonation. Witnesses reported that the entire tank deck and midship deck-house rose about 250 m into the air. Windows were broken up to 4 km away and severe damage was reported as far as 2 km away. The explosion killed six, left three missing and presumed dead, injured 58 and caused $21.6 million in damage [USCG (1977)]. While this explosion was due to external ignition of escaping fuel vapor during ballasting, many tanker and supertanker explosions are ignited by static electric sparks generated by the high pressure water spray that is used for cleaning [USCG (1974) and Owen (1979)] (see Figure 2-41).

Commercial buildings also quite often explode in a manner that indicates that significant wave propagation and flame accelerations occurred during the event. A good example of such a building explosion is discussed in NTSB (1976) (see Figure 2-42).

b. Dusts

Dust explosions in enclosures can be quite disastrous. Contrary to some commonly held beliefs, virtually all organic dusts, plus certain inorganic or metallic dusts, are combustible in air and can explode in an enclosure.[†] However, the sequence of events that leads to dust explosions is different than that which leads to most gas or vapor explosions. In order for a dust cloud to be explosible, the dust concentration must be so high that the optical extinction distance is of the order of 0.2 m. Such dust clouds are essentially opaque and usually contain concentrations well above those tolerable by man. Therefore, these conditions do not normally exist in the work place except possibly inside ducting or process equipment. The

---

[†]A dust is usually defined as material that will pass a standard 200 mesh screen (diameter < 76μm).

Figure 2-41. Tanker Kong Haakon VII After Fuel Vapor Explosion

normal sequence of events in a dust explosion is for a small explosion to
occur first in some piece of equipment. This causes the equipment to rupture and throw burning dust into the work place. Then, if the work place
is dirty, the gas motion and equipment vibration that result from the first
explosion causes the layered dust throughout the installation to become air-
borne. This dust is the fuel for a disastrous second explosion that travels
through the work place and causes major damage. In another typical sequence,
a pile of dust starts to smolder either by spontaneous combustion or because
it is covering a hot object (e.g., motor housing or lamp fixture). A worker

Figure 2-42. An Internal Natural Gas Explosion in the Elevator Shafts
(Located on the Left-Hand Face) of a 25-Story Commercial Building Blew
Out All the Bricks Surrounding the Shafts and Virtually All the Windows.
The Accident Occurred in April 1974 in New York City
[NTSB (1976)]

finds the fire and attempts to put it out with either a chemical extinguish-
er or a water hose. This stirs up a large dust cloud, a portion of which is
already burning, and an explosion results.

    Dust explosions in enclosures exhibit the same two limit behaviors as gas or vapor explosions. A volumetric explosion in an enclosure with a low L/D ratio will cause a simple overpressure burst of the enclosure. On the other hand, in structures with large L/D ratios, flame accelerations can occur, resulting in effective propagation velocities as high as detonation velocities. In this case, the damage can be severe and localized. Also, in this case, missiles can be thrown for considerable distances and the external blast wave may be quite strong.

The phenomena of dust explosions in enclosures probably have a longer history than vapor or gas explosions. This is because gases and vapors were not used extensively as fuels until quite recently. Dust explosions occur primarily in boilers, the chemical industry, the pharmaceutical industry [Nickerson (1976)], coal mines, grain elevators, and feed and flour mills. Bibliographies of the literature of grain dust explosions have been prepared by Aldis and Lai (1979) and Cardillo and Anthony (1979). General texts on dust explosions have been prepared by Palmer (1973) and Bartknecht (1981).

Coal mine explosions have been occurring ever since the start of the industrial revolution. Most industrial nations have extensive research programs to investigate the processes that occur during such explosions [Rae (1973), Cybulski (1975), Richmond and Leibman (1975), and Richmond, et al. (1979)], and the dynamics of the explosion process is reasonably well understood. Ignition is usually effected by the release and ignition of a pocket of methane during mining. Then an explosion propagates for long distances through the mine as a coal dust explosion because the work place is always quite dirty (it is expensive to remove all the coal dust in a mine). Cybulski (1975) has compiled world-wide statistics that show that 135 coal mine explosion disasters have occurred over the period 1900 to 1951; he defines a disaster as an explosion that has killed 50 or more persons. He shows a total of 20,448 deaths for these disasters, an average of 151 per disaster. His statistics also show that on the average, 117 persons per year were killed by coal mine explosions in the U. S. over the period 1931 to 1955.

This is much less than the average of 330 persons per year that were killed in the 1901 to 1930 period.[†] Unfortunately, coal mine explosions, large and small, have continued up until the present day and will undoubtedly occur in the future.

Dust explosions in grain elevators and various milling industries have also been a problem for a long time. Price and Brown (1922) review incidents that date back to 1876. The story is the same: layered dust is picked up by a primary explosion and triggers a disastrous secondary explosion. Approximately 30 to 40 grain elevator explosions occur in the U. S. each year. In December 1977, however, two very disastrous explosions occurred within five days of each other [Lathrop (1978)]. The explosions at the Continental grain facility in Westwego, Louisiana, on December 22, 1977, killed 36 people and caused a capital loss of about 30 million dollars. The explosion at the Farmers Export grain elevator in Galveston, Texas, on December 27, 1977, killed 18 and caused a capital loss of 24 million dollars. The damage from this latter explosion is apparent in Figure 2-43. These two explosions also significantly reduced the U. S. capacity to export grain.

Dust explosions in the chemical and pharmaceutical industries are generally confined to process equipment. The primary reason for this is that the product is expensive and dust control is sophisticated. Nevertheless, if the equipment is not protected, significant damage can occur [Bartknecht (1981)].

3. Pressure Vessels (Gaseous Contents)

    a. Simple Failure (Inert Contents)

Fortunately, there are very few simple pressure vessel failures. They can occur primarily in one of two ways. Either, the vessel is

---

[†]The reduction in average deaths per year is certainly a consequence of improved practices resulting from improved knowledge and understanding of these explosions and ways to mitigate their effects.

Figure 2-43. Aerial View of Farmers Export Elevator, Galveston,
Texas, After 1977 Explosion

faulty in construction or has corroded or been stressed due to mistreatment. Under these circumstances, the vessel can fail at any time. The other failure mode involves heating the vessel with either a faulty electrical connection [Anonymous (1970)] or with an external fire. In this case, the pressure of the gases increase and the strength of the vessel decreases until rupture occurs. The maximum blast can be calculated using the techniques for a frangible vessel. However, the actual blast will undoubtedly be less intense because the ductile vessel will tear relatively slowly. The primary fragments produced by this type of explosion can be quite dangerous (see Chapter 6).

Explosions of weak pressure vessels can also occur. As an example, a large scale utility boiler will explode when the internal pressure rises by as little as 10 to 15 kPa (1.5 to 2 psig). This pressure rise can result from a combustion explosion in the boiler as described in previous sections of this chapter, or it can result from a massive steam leak caused by the rupture of a large tube or header. In this latter case, steam enters the hot boiler at such a rate that the normal openings are insufficient to vent the pressure rise. In both cases, the boiler first tends to become round and may simply bulge without any of its surfaces being breeched. If the incident is more serious, sections of the boiler wall can be blown away. Generally, this type of incident does little damage to the surroundings, even though damage to the boiler can be quite severe.

b. Combustion Generated Failure

Compressed air lines are quite susceptible to a combustion generated explosion. Here the fuel is oil or char mainly on the wall (see Chapter 1). Such explosives have been discussed by Brown (1943), Smith (1959), Munck (1965), Fowle (1965), Chase (1966 and 1967), Burgoyne and Craven (1973), and Craven and Greig (1968). These explosions invariably show the highly localized damage patterns typical of explosions in vessels with high L/D ratios. One incident is known to have occurred where, because of a process upset, compressor oil vapor mixed in a pipe with oxygen-enriched air and detonated. In many places, several meter lengths of hydraulic grade pipe split into long filaments, especially following pipe elbows.

A unique pipeline explosion caused by the exothermic decomposition (and subsequent detonation) of high pressure acetylene is described by Sutherland and Wegert (1973). In this case, the detonation traveled through approximately seven miles of pipe. However, design was adequate and the pipe did not burst. Also, a detonation arrestor stopped the propagation before the detonation could enter the process units.

c. Failure Followed by Immediate Combustion of Contents

The blast damage for this type of explosion is similar to that for type (a) above. The major difference is the production of a fireball whose size depends on the quantity of gaseous fuel that is released. Fortunately, when a pressurized gaseous fuel is stored in a tank, much less can be stored per unit volume than can be stored if the fuel is liquified. Thus, these fireballs are much less severe than those produced in BLEVE's that will be discussed in category 4.

d. Runaway Chemical Reaction Before Failure

The explosion of chemical reactors occurs primarily because the controlled reaction that is occurring is exothermic and some type of upset occurs in the process control (e.g., too much catalyst present, loss of adequate cooling, inadequate stirring, etc.). The explosion process itself is discussed in Chapter 1. An explosion of this type should be contrasted to an explosion that occurs because of the detonation of the contents of the container. In the case of a runaway chemical reactor, the pressure buildup is considerably slower and the vessel usually ruptures in a ductile mode. If the contents are gaseous, the rupture is equivalent to a pressure vessel burst. If the contents are liquid and above the flash evaporation temperature (this is commonly true), the explosion acts like a BLEVE. Vincent (1971) and Dartnell and Ventrone (1971) describe, respec-

tively, the catastrophic rupture of a nitroaniline reactor and a runaway reaction in a vessel storing impure Para Nitro Meta Cresol. The explosion of
runaway chemical reactors are generally of interest only to the chemical
industry even though, in isolated incidents, public interests may be involved.

e. Runaway Nuclear Reactor Before Failure

A large number of scenarios have been constructed for the
consequences of a nuclear reactor runaway and/or core meltdown. Some of
these include the possible catastrophic breaching of the containment vessel by either an internal combustion explosion [Palmer, et al. (1976)] or
by a simple pressure burst. Nuclear reactors are constructed in such a
way that it is impossible for an accidental upset to produce anything re-
motely resembling an atomic bomb explosion. The most serious accident
scenario involves core meltdown, melting through the reactor vessel and/
or containment structure, and physical explosion by mixing with cooler
liquids in the environment. Keep in mind, however, that if a nuclear reactor runaway were to cause the containment vessel to be breached in a
catastrophic manner (explode), the hazard caused by the explosion itself
would be moot. This is because the attendant release of long lived radio-
active material would cause such a disruption to the locale that the explosion damage, no matter what its extent, would be considered minor. Fortunately, no example of such an accident can be supplied because none have
occurred. This is certainly an unequaled and enviable safety record for
the first 20 years of the commercialization of a new technology.

4. BLEVE's (Boiling Liquid Expanding Vapor Explosions)

This particular type of explosion results from a very specific
sequence of events [Walls (1978)]. It occurs upon rupture of a ductile
vessel containing a liquid whose vapor pressure is well above atmospheric
pressure. The sequence of events is as follows. For some reason, the ductile tank tears. Because the tank is ductile, the tearing process is relatively slow and a small number of large fragments are produced. Furthermore,

because the metal fragments are backed by a liquid which can evaporate very
rapidly (flash evaporate), the fragments can be thrown very large distances.
In this situation, relatively high initial velocities can be produced because
the backward thrust of the gases produced by the flash evaporating liquid im-
parts momentum to the tank pieces. The blast wave produced by such a rupture
is usually minimal. As has been shown by Esparza and Baker (1977b), in this
case, flash evaporation from the bulk fluid is such a slow process that very
little pressure rise is produced. As a matter of fact, one can make a first
good estimate of the maximum blast wave strength if one knows the volume of
the free vapor space above the flash evaporating liquid at the instant of
tank burst. This technique of blast estimation was discussed in a previous
section of this chapter.

    a. External Heating

       An added complication occurs if the liquid in the tank is
combustible and if the tank BLEVE is caused by heat from an external fire.
In these cases, the BLEVE produces a buoyant fireball whose duration and
size are determined by the total weight of fluid enclosed in the tank at
the instant that the BLEVE occurs. Such a fireball is shown in Figure 7-1.
If the tank is relatively large, the radiation from this fireball can cause
burns to exposed surfaces of skin and ignition of nearby combustible materials. This is discussed in more detail in Chapter 1.

       By far the most spectacular and dangerous BLEVE's have occur-
red in railroad transportation accidents involving tank cars containing
combustible high vapor pressure liquids such as liquid petroleum gas (LPG),
propane, propylene, butane, vinyl chloride, etc. A typical sequence of
events in this case starts with the derailment of a freight train carrying
a number of tank cars which are coupled together. The tank cars are piled
into a disorganized jumble, the piping on one car ruptures, or perhaps the
tank is perforated by a coupler, and the gas that is escaping ignites,
causing a torch which heats neighboring tank cars. This causes safety
valves on those tank cars to open, producing more torches. The heat trans-

Figure 2-44.   Section of Tank Car from Crescent City BLEVE
              After Rocketing

fer from these torches to the involved tank cars eventually causes one of
them to BLEVE. This rearranges the pile, producing at the same time rocketing pieces of the tank car (see Figure 2-44), a minor blast wave, and a
fireball. The fires continue and the rearranged tank cars BLEVE one at a
time, sometimes over a period of three to four hours [NTSB (1972b)] and
sometimes several days. There have been accidents with as many as six
BLEVE's recorded.

        Siewert (1972) documented 84 such accidents and showed that
the distances traveled by the fragments could be plotted as a straight line
on a log normal probability scale and that 95 percent of the pieces, irrespective of their size, are found within a radius of 700 m (2100 ft). The
fireball from the BLEVE of one U. S. railroad tank car (33,000 U. S. gallon
capacity) was seen to contact initially the surface of the earth at a radius of about 60 m (200 ft) and ignite combustible material up to a radius
of about 350 m (1000 ft). Thus, even though blast damage may not be severe,
an accident of this type can be catastrophic. Frequently, a fire in a

chemical plant will cause storage tanks or drums to BLEVE during the fire.

b. Chemical Reactor Runaway

Chemical reactors that contain a high pressure liquid phase undergoing an exothermic chemical reaction can BLEVE as can hot water tanks where the water is stored at pressures above atmospheric, i.e., at temperatures above 212°F.

5. Unconfined Vapor Cloud Explosions

An unconfined vapor cloud explosion is also the result of a unique sequence of events [Strehlow (1973b)]. The first event is the massive spill of a combustible hydrocarbon into the open atmosphere at or above ground level. This can occur in a chemical plant complex, in a transportation situation, or in a remote area as when a pipeline bursts. In all cases, once the spill has occurred, one of four things can happen. 1) The spill may be dissipated harmlessly without ignition. 2) The spill may be ignited immediately upon release. In this case, only a fire results; there is generally no explosion. 3) The spill can disperse over a wide area. After a time delay, this combustible cloud is ignited and a large fire ensues. 4) The sequence of events as listed in case 3) above occurs except that after the fire starts to burn, the flame propagating through the cloud somehow accelerates sufficiently to produce a dangerous blast wave. Davenport (1977a and 1977b) has documented, with statistics, the occurrence of cases 1), 3) and 4). Case 2) is usually considered as a separate case because immediate ignition generally poses no blast hazard. In a recent issue of Progress in Energy and Combustion Science, the vapor cloud explosion problem has been analyzed in some depth by Lewis (1980a), Brisco and Shaw (1980), Cox (1980), Lewis (1980b), and Cave (1980). Gugan (1978) also has discussed vapor cloud explosions in a general manner and documents many of them. He also includes (incorrectly) some BLEVE's and liquid phase detonation in his incident list.

Unconfined vapor cloud explosions can be both very spectacular and very dangerous. This is because the leak is into the open air, and with the right meteorological conditions, truly large clouds of combustible mixture can be produced before ignition occurs. By far, the most expensive (and most extensively investigated) unconfined vapor cloud explosion occurred at the Nypro Chemical Plant near Flixborough, England, in June 1974 [Parker, et al. (1975) and Tucker (1975 and 1976)]. The investigation revealed that a 0.5 m (20 in.) diameter temporary connecting pipe which had been installed with two out-of-line expansion bellows failed, causing the release of approximately 45 tons of cyclohexane at a pressure of 850 kPa (125 psi) and a temperature of 155°C through two 0.7 m (28 in.) diameter stub pipes. This fuel flash-evaporated and produced a sizeable cloud throughout the plant area. Ignition probably occurred at a furnace in the hydrogen plant some distance from the release point. Ignition was "soft" and there was a large fire in the plant area before flame acceleration processes occurred, producing a blast wave which extensively damaged the plant and houses up to one mile away. Twenty-eight people were killed and 89 were injured as a result of this incident. The damage to the plant and its environs was estimated to be about 100 million dollars.

A good example of a transportation accident involving an unconfined vapor cloud explosion is the East St. Louis incident of January, 1972 [NTSB (1973a) and Strehlow (1973a)]. In that case, rail cars were being sorted to form trains in the East St. Louis yards of the Alton and Southern Gateway railroad using a "humping" operation. In this operation, cars are released at the top of a hill and switches are thrown to direct the car to the proper train. There are brakes along the track to slow the car down once it reaches the proper location. A tank car containing propylene was diverted to one manifest and the car that had been diverted before it was an empty hopper car. The empty hopper car did not travel the full length of the track and had not yet coupled up with the main train. The brakes did not work properly for the propylene car and it hit the hopper car at high velocity. The hopper car jumped the tracks and its coupling punctured the tank of the propylene car. The two traveled down the tracks about 500 m spilling liquid propylene along the tracks. The cloud ignited in a caboose some distance from where the cars finally came to rest. After an initial fire, there was a severe explosion which injured 176 people and caused approximately 7.6 million dollars in damage to non-railroad property.

Another incident, one that occurred in Franklin County, Missouri, in 1970 [NTSB (1972a)] is unique because it appears that the vapor cloud detonated as a unit [Burgess and Zabetakis (1973)]. In this case, an underground pipeline burst and the propane that was contained in the pipe at 7 MPa (1000 psi) formed a fountain above the pipeline. The fuel-air mixture that was thus formed flowed downwind and started to fill a large valley. When the cloud height was about 6 m (20 ft), a pump house at the other end of the valley, constructed of concrete block, was demolished by an internal explosion and evidently caused a "hard" ignition which directly triggered detonation of the cloud. The detonation was followed by a fireball which occurred because most of the fuel-air mixture was too rich to detonate. Overall damage was very slight because of the low population density in the area.

The TNT yield of the Franklin County, Missouri, incident was calculated by Burgess and Zabetakis (1973) to be approximately 7.5 percent based on the combustion energy of the total quantity of fuel that escaped. Flixborough was determined to be approximately 5 percent by Sadee, et al. (1977) and East St. Louis was about 0.2 of one percent [Strehlow (1973a)], again based on the combustion energy of all the material that spilled. As a rough rule of thumb, we suggest estimating the expected damage potential from an unconfined vapor cloud explosion by assuming that the TNT equivalence for blast equals about 2 percent of the combustion energy of the fuel from the maximum spill that could occur, with an upper limit of about 10 percent (see Chapter 7 for methods for estimating fireball properties and effects).

There have, of course, been many instances where a spill ignited and did not produce a damaging blast wave [NTSB (1973b and 1973d) and Raj, et al. (1975)]. One of the largest spills of this type that is known occurred in Griffith, Indiana, in 1974 [Office of Pipeline Safety (1974) and Schneidman and Strobel (1974)]. An 0.46 m (18 in.) pipe connected to a 45,000 m^3 (283,000 barrel) underground reservoir which was filled with liquid butane at about 19.6 kPa (135 psig) opened at a T section with the opening pointing vertically. This nozzle blew for seven hours before the plume ignited. The odorized gas could be smelled 24 km (15 miles) downwind in Gary, Indiana. Upon ignition, the flame flashed back to the source and torched without producing a damaging blast wave.

6. Physical Vapor Explosions

Physical vapor explosions occur when two liquids at different temperatures are violently mixed or a finely divided hot solid material is rapidly mixed with a much cooler liquid. No chemical reactions are involved; instead, the explosion occurs when the cooler liquid is converted to vapor at such a rapid rate that localized high pressures are produced. These explosions have occurred in the steel and aluminum industry when molten metal is poured into a container that contains moisture. They have also been observed when liquid natural gas which contains approximately 10 percent of a hydrocarbon higher than methane is spilled on water. In this case, the cold liquid is liquid natural gas rather than the water. There is some indication that the catastrophic explosion of island volcanoes, such as Krakatoa in 1883 and Surtsey in 1963, has been due to a physical explosion when sea water is mixed intimately with hot magma [Gribbin (1978)].

Some examples of vapor explosions, from Fauske (1974), are given in Table 2-6. It is clear that physical explosions constitute an important or potentially important class of damaging explosion accidents.

A number of authors including Fauske (1974), Anderson and Armstrong (1974), Reid (1976), Witte, et al. (1970), Witte, et al. (1973), Board, et al. (1974), and Ochiai and Bankoff (1976) have discussed experi-

204

Table 2-6. Examples of Vapor Explosions
[Fauske (1974)]

Related Areas	Hot Fluid	Cold Fluid
Nuclear		
SPERT-1, SL-1, BORAX-1*	Aluminum	Water
Non-Nuclear		
Foundry Industry	Steel or Slag	Water
Aluminum Industry	Aluminum	Water
Kraft Paper Industry	Smelt (Na_2CO_3 and Na_2S)	Water
LNG Industry	Water	LNG
Volcanic Eruptions or Submarine Explosion (Krakatoa, etc.)	Lava	Water

* These are code names for nuclear reactor runaway experiments and accidents.

ments and theories which examine physical vapor explosions. Recently, a number of authors have postulated that when conditions are right, physical explosions can propagate in a detonative manner [Board, et al. (1975), Rabie, et al. (1979), and Hall and Board (1979)]. But, unfortunately, there is little agreement regarding the actual physical mechanisms which can initiate vapor explosions, reasonable methods for estimating explosive energy release, or the role of confinement or lack thereof in escalating such explosions or their effects.

CHAPTER 2

EXAMPLE PROBLEMS

Blast Pressures and Impulses From Bursting Gas Spheres

A spherical pressure vessel of radius 1.0 m (3.3 ft) containing air (γ_1 = 1.4) bursts in a standard sea level atmosphere. The inside gas pressure is 1.013 x 10^6 Pa (147 psi) and the temperature is 300°K (80°F). There are no reflecting surfaces nearby. Find the peak overpressure and specific impulse at a distance of 5.0 m (16.4 ft) from the center of the source.

1. Solution for Peak Overpressure

$\bar{R}_1$ and $\bar{R}$ for the distance of interest are calculated. $\bar{P}_{so}$, the starting peak overpressure is obtained from Figure 2-16. The correct curve is located in Figure 2-15 and $\bar{P}_s$ is read from the graph for the $\bar{R}$ of interest. From Equations (2-24) and (2-22):

$$\bar{R}_1 = \cfrac{1}{\left[\cfrac{\cfrac{4\pi}{3}\left(\cfrac{P_1}{P_o} - 1\right)}{\gamma_1 - 1}\right]^{1/3}} = \cfrac{1}{\left[\cfrac{\cfrac{4\pi}{3}\left(\cfrac{1.013 \times 10^6 \text{ Pa}}{1.013 \times 10^5 \text{ Pa}} - 1\right)}{1.4 - 1}\right]^{1/3}} = 0.2197$$

$$\bar{R} \text{ (at } r = 5.0 \text{ m)} = \cfrac{r}{\left[\cfrac{4\pi\, r_o^3}{3}\cfrac{\left(\cfrac{P_1}{P_o} - 1\right)}{\gamma_1 - 1}\right]^{1/3}} =$$

$$\cfrac{\cfrac{5.0 \text{ m}}{1.0 \text{ m}}}{\left[\cfrac{4\pi}{3}\cfrac{\left(\cfrac{1.013 \times 10^6 \text{ Pa}}{1.013 \times 10^5 \text{ Pa}} - 1\right)}{1.4 - 1}\right]^{1/3}} = 1.099$$

For P_1/P_o = 10 and T_1/T_o = 1, a_1/a_o = 1; therefore, $\bar{P}_{so} \approx 1.7$ (Figure 2-16). Looking at Figure 2-15, this point ($\bar{R}_1$, $\bar{P}_{so}$) falls near the third curve from the bottom. Following this curve, for $R = 1.099$, $\bar{P}_s =$ 0.26. Since $\bar{P}_s = (p_s - p_o)/p_o$, then $(p_s - p_o) = \bar{P}_s \, p_o = (0.26) \, (1.013 \times 10^5$ Pa$) = 2.6 \times 10^4$ Pa.

2. <u>Solution for Specific Impulse</u>

The $\bar{R}$ of interest has been calculated above. Read $\bar{i}$ for this $\bar{R}$ from Figure 2-19.

For $\bar{R} = 1.099$, $\bar{i} = 0.046$ (Figure 2-19). Since

$$\bar{i} = \frac{i \, a_o}{p_o^{2/3} \, E^{1/3}}$$

[see Equation (2-18)],

$$i = \bar{i} \, \frac{p_o^{2/3} \, E^{1/3}}{a_o}$$

From Equation (2-22),

$$E = \left(\frac{P_1 - P_o}{\gamma_1 - 1}\right) V_i = \left(\frac{1.013 \times 10^6 \text{ Pa} - 1.013 \times 10^5 \text{ Pa}}{1.4 - 1}\right) \times$$

$$\frac{4\pi}{3} (1.0 \text{ m})^3 = 9.55 \times 10^6 \text{ J}$$

$$i = \frac{(0.046) \, (1.013 \times 10^5 \text{ Pa})^{2/3} \, (9.55 \times 10^6 \text{ J})^{1/3}}{331 \text{ m/s}} = 64 \text{ Pa} \cdot \text{s}$$

<u>Blast Parameters for TNT Explosions</u>

This set of example calculations uses the large, fold-out metric figures, Figures 2-45, 2-46 and 2-47 at the back of the book.

A 27 kg spherical TNT explosive charge is detonated under sea level

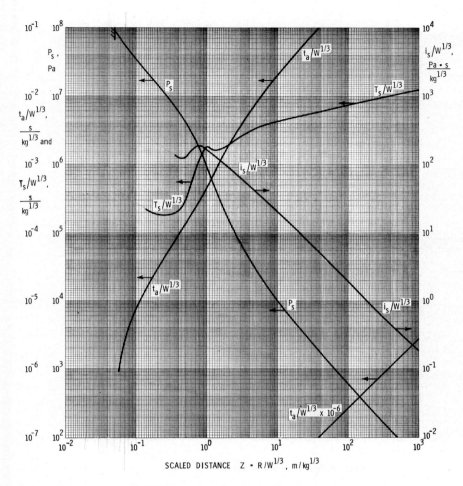

Figure 2-45. Side-On Blast Parameters for TNT

atmospheric conditions, but not near a reflecting surface such as the ground. Find both side-on and reflected blast wave properties at standoffs R of 30 m and 6 m from the charge center.

Calculate Z value for 30 m standoff:

$$Z = \frac{R}{W^{1/3}} \left(\frac{m}{kg_{TNT}^{1/3}} \right) = \frac{30}{27^{1/3}} = 10 \; \frac{m}{kg_{TNT}^{1/3}}$$

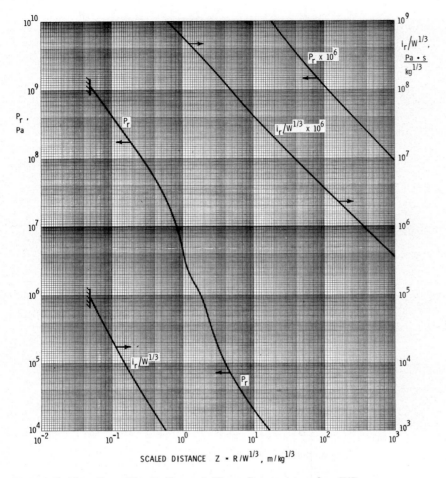

Figure 2-46. Normally Reflected Blast Parameters for TNT

Calculate Z value for 6 m standoff:

$$Z = \frac{R}{W^{1/3}} \left(\frac{m}{kg_{TNT}^{1/3}} \right) = \frac{6}{27^{1/3}} = 2 \; \frac{m}{kg_{TNT}^{1/3}}$$

Obtain Hopkinson-Cranz blast parameters directly from Figures 2-45 through 2-47.

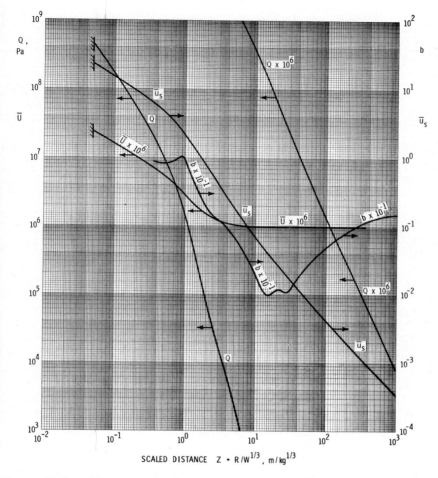

Figure 2-47. Additional Side-On Blast Parameters for TNT

For $Z = 10 \text{ m/kg}_{TNT}^{1/3}$,

$$P_s = 1.0 \times 10^4 \text{ Pa}$$

$$\frac{i_s}{W_{TNT}^{1/3}} = 2.0 \times 10^1 \frac{\text{Pa} \cdot \text{s}}{\text{kg}_{TNT}^{1/3}}$$

$$i_s = \left(2.0 \times 10^1 \frac{Pa \cdot s}{kg_{TNT}^{1/3}}\right) \times W_{TNT}^{1/3} =$$

$$\left(2.0 \times 10^1 \frac{Pa \cdot s}{kg_{TNT}^{1/3}}\right) \times 27^{1/3} \, kg_{TNT}^{1/3}$$

$$i_s = 6.0 \times 10^1 \, Pa \cdot s$$

$$\frac{t_a}{W_{TNT}^{1/3}} = 2.0 \times 10^{-1} \frac{s}{kg_{TNT}^{1/3}}$$

$$t_a = \left(2.0 \times 10^{-1} \frac{s}{kg_{TNT}^{1/3}}\right) \times W_{TNT}^{1/3} =$$

$$\left(2.0 \times 10^{-1} \frac{s}{kg_{TNT}^{1/3}}\right) \times 27^{1/3} \, kg_{TNT}^{1/3}$$

$$t_a = 6.0 \times 10^{-1} \, s$$

$$\frac{T_s}{W_{TNT}^{1/3}} = 4.2 \times 10^{-2} \frac{s}{kg_{TNT}^{1/3}}$$

$$T_s = \left(4.2 \times 10^{-2} \frac{s}{kg_{TNT}^{1/3}}\right) \times W_{TNT}^{1/3} =$$

$$\left(4.2 \times 10^{-2} \frac{s}{kg_{TNT}^{1/3}}\right) \times 27^{1/3} \, kg_{TNT}^{1/3}$$

$$T_s = 12.6 \times 10^{-2} \, s$$

For $Z = 2 \text{ m/kg}_{TNT}^{1/3}$,

$$P_s = 1.65 \times 10^5 \text{ Pa}$$

$$\frac{t_a}{W_{TNT}^{1/3}} = 1.85 \times 10^{-2} \frac{s}{kg_{TNT}^{1/3}}$$

$$t_a = \left(1.85 \times 10^{-2} \frac{s}{kg_{TNT}^{1/3}}\right) \times W_{TNT}^{1/3} =$$

$$\left(1.85 \times 10^{-2} \frac{s}{kg_{TNT}^{1/3}}\right) \times 27^{1/3} \, kg_{TNT}^{1/3}$$

$$t_a = 5.6 \times 10^{-2} \text{ s}$$

$$\frac{T_s}{W_{TNT}^{1/3}} = 1.85 \times 10^{-2} \frac{s}{kg_{TNT}^{1/3}}$$

$$T_s = \left(1.85 \times 10^{-2} \frac{s}{kg_{TNT}^{1/3}}\right) \times W^{1/3} \, kg_{TNT}^{1/3} =$$

$$\left(1.85 \times 10^{-2} \frac{s}{kg_{TNT}^{1/3}}\right) \times 27^{1/3} \, kg_{TNT}^{1/3}$$

$$T_s = 5.6 \times 10^{-2} \text{ s}$$

Figure 2-47 yields normally reflected overpressure P_r and scaled normally reflected specific impulse $i_r/W^{1/3}$.

For $Z = 10 \text{ m/kg}_{TNT}^{1/3}$,

$$P_r = 2.05 \times 10^4 \text{ Pa}$$

$$\frac{i_r}{W_{TNT}^{1/3}} = 3.3 \times 10^1 \frac{\text{Pa} \cdot \text{s}}{\text{kg}_{TNT}^{1/3}}$$

$$i_r = \left(3.3 \times 10^1 \frac{\text{Pa} \cdot \text{s}}{\text{kg}_{TNT}^{1/3}}\right) \times W_{TNT}^{1/3} =$$

$$\left(3.3 \times 10^1 \frac{\text{Pa} \cdot \text{s}}{\text{kg}_{TNT}^{1/3}}\right) \times 27^{1/3} \text{ kg}_{TNT}^{1/3}$$

$$i_r = 99 \text{ Pa} \cdot \text{s}$$

For $Z = 2 \text{ m/kg}_{TNT}^{1/3}$,

$$P_r = 7.3 \times 10^5 \text{ Pa}$$

$$\frac{i_r}{W_{TNT}^{1/3}} = 2.8 \times 10^2 \frac{\text{Pa} \cdot \text{s}}{\text{kg}_{TNT}^{1/3}}$$

$$i_r = \left(2.8 \times 10^2 \frac{\text{Pa} \cdot \text{s}}{\text{kg}_{TNT}^{1/3}}\right) \times W_{TNT}^{1/3} =$$

$$\left(2.8 \times 10^2 \frac{\text{Pa} \cdot \text{s}}{\text{kg}_{TNT}^{1/3}}\right) \times 27^{1/3} \text{ kg}_{TNT}^{1/3}$$

$$i_r = 8.4 \times 10^2 \text{ Pa} \cdot \text{s}$$

Figure 2-46 yields peak dynamic pressure Q, shock velocity $\bar{U}$, peak particle velocity $\bar{u}_s$, and decay constant b.

For $Z = 10 \text{ m/kg}_{TNT}^{1/3}$,

$$Q = 6.8 \times 10^5 \text{ Pa}$$

$$\bar{U} = 1.0$$

$$\bar{u}_s = 0.068$$

$$b \times 10^{-1} = 2.0 \times 10^{-2}; \quad b = 0.2$$

For $Z = 2 \text{ m/kg}_{TNT}^{1/3}$,

$$Q = 5.4 \times 10^7 \text{ Pa}$$

$$\bar{U} = 10.5$$

$$\bar{u}_s = 9.5$$

$$b \times 10^{-1} = 1.9 \times 10^{-1}; \quad b = 1.9$$

Blast Parameters for a Vapor Cloud Explosion

A total of 50 kg of propane is released and mixes with air in a hemispherical cloud on the ground. Predict blast wave parameters for side-on and reflected conditions for a deflagration with $M_{so} = 0.218$ at a distance R = 35.8 m.

To account for ground reflection, assume a reflectivity factor of 2. Thus, the 50 kg of propane as a hemisphere on the ground is equivalent to a free-air spherical source of 100 kg.

From Table 2-4, low heat value for propane is 46.4 MJ/kg, so

$$E = 100 \times 46.4 = 4.64 \times 10^3 \text{ MJ} = 4.64 \times 10^9 \text{ J}$$

Ambient conditions are $p_o = 1.013 \times 10^5$ Pa, and $a_o = 340$ m/s.

$$R_o = (E/p_o)^{1/3} = (4.64 \times 10^9 / 1.013 \times 10^5)^{1/3} = 35.8 \text{ m}$$

$$\bar{R} = R/R_o = 35.8/35.8 = 1.0$$

From Figure 2-25, $\bar{P}_s = 0.30$. From Figure 2-26, $\bar{i}_s = 0.34$.

$$P_s = p_o \times \bar{P}_s = 1.013 \times 10^5 \times 0.30 = 3.3 \times 10^4 \text{ Pa}$$

$$i_s = \bar{i}_s \frac{E^{1/3} \, p_o^{2/3}}{a_o} = \frac{0.34 \times (4.64 \times 10^9)^{1/3} + (1.013 \times 10^5)^{2/3}}{340} =$$

$$3.84 \times 10^3 \text{ Pa} \cdot \text{s}$$

From Equation (3-3),

$$\bar{P}_r = 2 \bar{P}_s + \frac{(\gamma + 1) \bar{P}_s^2}{(\gamma - 1) \bar{P}_s + 2\gamma} = 2 \times 0.3 + \frac{(1.4 + 1) \, 0.3^2}{(1.4 - 1) \, 0.3 + 2 \times 1.4} =$$

$$0.674$$

$$P_r = \bar{P}_r \times p_o = 0.674 \times 1.013 \times 10^5 = 6.83 \times 10^4 \text{ Pa}$$

From Equation (3-5),

$$i_r = \frac{P_r}{P_s} i_s = \frac{6.83 \times 10^4}{3.3 \times 10^4} \times 3.84 \times 10^3 = 7.94 \times 10^3 \text{ Pa} \cdot \text{s}$$

Blast from Electrical Circuit Breaker

Consider a power station switching room 10 x 15 x 9 meters in internal dimensions containing a door of 2 m^2 surface area and a circuit breaker in a 50 Hz power line that arcs and dissipates energy at a maximum rate of 50 MW, for a maximum duration of 1 sec with an initial rise time of 0.01 sec.

Use acoustic theory to calculate the blast wave overpressure and assume volumetric heating for the arc duration to calculate maximum static overpressure in the room. Also, make a first estimate of the flow velocity through the door, if it were open and acted as a vent.

Acoustic theory states that

$$p - p_o = \dot{m}(t)/4\pi r$$

where $\dot{m}(t)$ is the first derivative of mass flow rate and has units of kg/s^2. For this case, we can say that the arc is generating volume V at the density of the displaced air, and therefore, that

$$\dot{m}(t) = \rho_o \dot{V}(t)$$

At constant pressure (acoustic approximation)

$$P_o (V_2 - V_o) = nR (\theta_2 - \theta_o)$$

We do not know the volume of the arc, but if we assume n moles of gas are heated by the arc we get

$$\Delta E = nC_p \, (\theta_2 - \theta_o) \quad \text{(const. pressure)}$$

Substituting for $\theta_2 - \theta_o$ yields

$$P_o \, (V_2 - V_o) = \frac{R}{C_p} \Delta E$$

or

$$(V_2 - V_o) = \frac{\gamma - 1}{\gamma} \frac{\Delta E}{P_o}$$

Taking the second derivative with time yields:

$$\ddot{V}(t) = \frac{\gamma - 1}{\gamma} \frac{\Delta \ddot{E}}{P_o}$$

and substituting this expression to find the overpressure:

$$(p - P_o)_{shock} = \frac{\gamma - 1}{\gamma} \cdot \frac{\Delta \ddot{E}}{P_o} \cdot \frac{1}{4\pi r}$$

The effective heat capacity ratio in the arc will be much less than 1.4, so using 1.4 will produce a conservative answer. The rate of increase of power deposited in the gas, $\Delta E = 50$ MW/0.01 s or 5×10^9 J/s^2. Substitution yields

$$(p - P_o)_{shock} = \frac{0.4}{1.4} \cdot \frac{5 \times 10^9}{101325} \cdot \frac{1}{4\pi} \cdot \frac{1}{r}$$

or

$$(p - p_o)_{shock} = \frac{1122}{r} \text{ Pa}$$

Thus, one meter from the arc the overpressure will be

$$(p - p_o)_{shock} = 1.1 \text{ kPa}$$

This is not a damaging blast wave.

For the second part, assume that the room is sealed and pressurized by the deposition of 50 MW of power. The deposition of a quantity of energy ΔE will pressurize the room according to the following equation.

$$(p_2 - p_o) V = nR (\theta_2 - \theta_o) = \frac{\gamma - 1}{\gamma} \Delta E$$

Thus, the rate of pressure rise caused by the deposition of power is:

$$[(p_2 - p_o)_{static}]/s = \frac{0.4}{1.4} \frac{50 \times 10^6}{10 \times 15 \times 9} = 10,600 \text{ Pa/s}$$

Note that a more exact calculation could be made by assuming that the volume of the arc is a small percentage of the room volume and that the arc's energy causes isentropic compression of the remainder of the gas in the room. However, because of the relatively small pressure rise (ca 10 percent), the above calculation should be reasonably correct.

A first estimate of the maximum flow rate through the 2 m^2 door can be obtained by noting that the pressure rises at the rate of about 10^4 Pa/s. This means that 141 m^3 of air must leave the room per second through the door. Thus:

$$\frac{141}{2} = 70.5 \text{ m/s}$$

is the maximum velocity due to impulsive motion of the gas through the door.

Note that a more sophisticated approach here would be to assume quasi-static flow through the door and the standard orifice formula to determine a quasi-equilibrium pressure in the room, and thus, the flow velocity through the door. The rate of pressure rise due to arc heating is already known. One would calculate a rate of pressure drop due to mass flow through the door and determine the flow velocity at which this rate of pressure drop equals the rate of rise due to the arc. This would be the equilibrium room pressure of the room. If this pressure were above the pressure that exists at the time that the arc is quenched, the terminal pressure would be used to calculate a maximum flow rate. These would also be only approximate solutions. The full solution would involve solving the differential equation for rate of pressure rise including both the source term (the arc) and the loss term (flow through the door) to determine $P(t)$ for the room.

Also, note that this approach could be used to design adequate venting for the room, or to decide if the room were sufficiently strong to survive the accident.

CHAPTER 2

LIST OF SYMBOLS

$A_f(t)$	effective flame area
a_o	ambient sound velocity
a_1	explosion source sound velocity
C_{pf}	specific heat of hot liquid
C_{ps}	specific heat of cold liquid
C_v	specific heat at constant volume
E	energy in explosive source
$\dot{E}$	energy addition rate
E_B	terminal energy in source products

E_f	energy flux density
E_k	kinetic energy
E_p	potential energy
E_s	source energy, energy transmitted by source to surroundings
E_T	total wave energy
e	specific work in vapor expansion
f_i	one of a set of functions
h	enthalpy
h_f	fluid enthalpy
h_g	gas enthalpy
h_{gf}	heat of fusion of hot material
i_d	drag specific impulse
i_s, i_s^+	specific positive side-on impulse in blast wave
i_s^-	specific negative side-on impulse in blast wave
ℓ_i	dimensionless shape factors
M	mass of condensed high explosive
M_{su}	burning velocity Mach number
M_w	wave addition Mach number
m	mass of fluid
N, n	number of moles of mixture
P_s, P_s^+	peak side-on overpressure
P_s^-	peak side-on underpressure
P_{sph}	peak overpressure for a spherical acoustic source
$p_c(f)$	constant surface pressure
p_f	flame pressure
p_o	ambient atmosphere pressure
p_s	absolute shock front pressure
p_{so}	absolute shock pressure at the surface of a bursting sphere
$p(t)$	pressure-time history

P_1	initial absolute pressure in bursting vessel
Q	total heat addition
q	energy density; dynamic pressure
R	radial distance from center of spherical blast source; universal gas constant
$\bar{R}, \bar{i},$ etc.	barred quantities are dimensionless parameters related to the unbarred symbol
r	radial distance from origin for two-dimensional cloud
r_o	radius of explosion source
r_1	radius of bursting sphere
S_u	burning velocity
s	entropy
s_f	entropy of fluid
s_g	entropy of gas
T, T^+	blast wave positive phase duration
T^-	blast wave negative phase duration
t	time
t_a	shock wave arrival time
t_e	effective time for energy addition
U	shock velocity
u	particle velocity
u_f	fluid internal energy
u_g	gas internal energy
u_s	peak particle velocity behind shock front
u_1, u_2	internal energies in vapor
V	volume of explosive source
V_f	final volume
V_o	initial volume
V_1	initial volume of compressed vapor
v	specific volume
v_f	specific volume of fluid
v_g	specific volume of gas

W	weight of explosive charge
W_c	weight of fuel
W_{TNT}	weight of TNT explosive
X	characteristic dimension of vented chamber
x	quality of wet vapor
Z	distance scaling parameter
α_e	effective vent area ratio
β	projected length in a vent panel configuration
ΔH_c	heat of combustion
γ	ratio of specific heats
γ_o	ratio of specific heats in atmosphere
γ_1	ratio of specific heats in explosion source
λ	scale factor
λ_e	equivalent scaled distance parameter
ρ	density
ρC_v	heat capacity
ρ_E	density of condensed explosive
ρ_o	initial density
ρ_s	peak density behind shock front
ρ_1, ρ_2	densities
τ	dimensionless time for energy addition
θ	absolute temperature
θ_c	critical temperature of vapor
θ_f	freezing temperature of hot material; final temperature
θ_{fin}	final temperatures of cold fluid
θ_{in}	initial temperature of hot material
θ_o	initial ambient absolute temperature
θ_1	initial absolute temperature in blast source
θ_2	initial state temperature for isentropic expansion

CHAPTER 3

LOADING FROM BLAST WAVES

INTRODUCTION

The properties and characteristics of free-field air blast waves, i.e., waves propagating undisturbed from some explosion source through the atmosphere, are discussed at some length in Chapter 2. But, only in very special cases do the free-field properties represent true transient loads applied to structures or objects which the blast wave intercepts. The loads are strong functions of the orientation, geometry and size of the objects which the waves encounter. This chapter reviews the rather voluminous information found in the literature on the interaction of air blast waves with various objects, and the resulting transient pressure loads applied to the objects.

In the methods and results given here, it is generally assumed that the initial shock loads on solid objects can be decoupled from the response of the objects to the loads, and that the objects can be treated as rigid bodies which cause processes such as shock reflection and diffraction, and alteration of flow behind the shock front. There is a large density difference between the wave transmitting media, air, and most solids that the wave encounters. The large density difference between air and most solids, and the great mismatches in acoustic impedance render these assumptions tenable in most air blast loading problems. However, if one were concerned with shock loading and structural response for underwater or underground explosions, decoupling of loading and target motion or deformation would be an inappropriate assumption.

Loads applied by blast waves from sources located outside a structure or object, i.e., external blasts, are treated first. Infinite, flat surfaces are considered as a limiting case, and then finite objects are

considered. Next, loads from explosions occurring within a structure,
i.e., internal blasts, are discussed for vented and unvented explosions.

EXTERNAL LOADING

Reflected Waves With Normal Incidence

An upper limit to blast loads is obtained if one interposes an infinite rigid wall in front of the wave, and reflects the wave normally.
All flow behind the wave is stopped, and pressures are considerably greater than side-on. The pressure in normally reflected waves is usually
designated p_r, and the peak overpressure, P_r. The integral of overpressure over the positive phase, defined in Equation (3-1), is the reflected
specific impulse i_r. Durations of the positive phase of normally reflected waves are designated T_r. The parameter i_r has been measured closer to
high explosive and nuclear blast sources than have most blast parameters.

$$i_r = \int_{t_a}^{t_a + T_r} [p_r(t) - p_o] \, dt \qquad (3-1)$$

The Hopkinson-Cranz scaling law described in Chapter 2 applies to
scaling of reflected blast wave parameters just as well as it does to side-
on waves. That is, all reflected blast data taken under the same atmospheric conditions for the same type of explosive source can be reduced to a
common base for comparison and prediction. The scaling law for non-ideal
explosions also applied to reflected parameters, and, far enough from the
source that the total energy E is the predominant parameter; Sach's law
applies too.

The literature contains considerable data on normally reflected
blast waves from high explosive sources, usually bare spheres of Pentolite
or TNT [Goodman (1960), Jack and Armendt (1965), Dewey, et al. (1962),
Johnson, et al. (1957), Jack (1963), Wenzel and Esparza (1972)]. From

224

these sources it is possible to construct scaled curves for P_r and i_r for specific condensed explosives over fairly large ranges of scaled distance. Figure 2-46 gives these parameters over a large range in scaled distances. Measurements for reflected impulse extend in to smaller scaled distances, i.e., closer to the blast source, than do measurements of reflected pressure because a much simpler measurement technique suffices for impulse measurement [Johnson, et al. (1957) and Dewey, et al. (1962)]. Furthermore, reflected specific impulses can be predicted right to the surface of a condensed spherical explosive source, using a simple formula applicable in the strong shock regime given by Baker (1967).

$$i_r = \frac{(2M_T E)^{1/2}}{4\pi R^2} \tag{3-2}$$

where $M_T = M_E + M_A$ is the total mass of explosive M_E plus the mass of engulfed air M_A, and R is the distance from the charge center. Very close to the blast source, $M_E \gg M_A$, and Equation (3-2) gives a simple $1/R^2$ relation for variation of i_r with distance. This relation is also noted by Dewey, et al. (1962).

Unfortunately, for explosive sources other than bare spheres of solid high explosives, very little data exist for normally reflected overpressures and specific impulses. For shock waves weak enough that air behaves as a perfect gas, there is a fixed and well-known relation between peak reflected overpressure, P_r, and peak side-on overpressure, P_s [Doering and Burkhardt (1949) and Baker (1973)],

$$\bar{P}_r = 2\bar{P}_s + \frac{(\gamma+1)\bar{P}_s^2}{(\gamma-1)\bar{P}_s + 2\gamma} \tag{3-3}$$

where

$$\bar{P}_r = P_r / p_o$$

$$(3\text{-}4)$$

$$\bar{P}_s = P_s / p_o$$

At low incident overpressures ($\bar{P}_s \rightarrow 0$), the reflected overpressure approaches the acoustic limit of twice the incident overpressure. If one were to assume a constant $\gamma = 1.4$ for air, then for strong shocks the upper limit would appear to be $\bar{P}_r = 8\, \bar{P}_s$. But, air ionizes and dissociates as shock strengths increase, and γ is not constant. In fact, the real upper limit ratio is not exactly known, but is predicted by Doering and Burkhardt (1949) to be as high as 20. Brode (1977) has also calculated this ratio for normal reflection of shocks in sea level air, assuming air dissociation and ionization. His equation, given without noting its limits of applicability, is, for P_s in psi,

$$\frac{P_r}{P_s} = \frac{2.655 \times 10^{-3}\, P_s}{1 + 1.728 \times 10^{-4}\, P_s + 1.921 \times 10^{-9}\, P_s^2} + 2 +$$

$$(3\text{-}5)$$

$$\frac{4.218 \times 10^{-3} + 4.834 \times 10^{-2}\, P_s + 6.856 \times 10^{-6}\, P_s^2}{1 + 7.997 \times 10^{-3}\, P_s + 3.844 \times 10^{-6}\, P_s^2}$$

We have calculated this ratio, and have used Figure 2-40 to determine corresponding scaled distances, for a wide range of side-on overpressures. These calculated reflected overpressures reach the proper low pressure asymptote of twice side-on peak overpressure, and agree remarkably well with the empirical fit to data for P_r in Figure 2-46. They have, therefore, been used to establish the dashed portions of the curves for P_r in Figure 2-46. Above $P_s = 690$ kPa (100 psi) and standard atmosphere conditions, Equation (3-3) is increasingly in error and should not be used. Brode's equation gives a maximum reflection factor at the surface of a spherical HE charge at sea level of $P_r/P_s = 13.92$.

Equation (3-3) gives only peak pressures and, hence, little indication of time histories of reflected pressure, and, therefore, of reflected specific impulse. Lacking more accurate prediction methods, one can roughly estimate the reflected specific impulse if the side-on specific impulse is known or predictable, by assuming similarity between the time histories of side-on overpressure and normally reflected overpressure. This assumption gives

$$\frac{i_r}{i_s} \approx \frac{P_r}{P_s} \tag{3-6}$$

which, together with Equation (3-3) or (3-5), allows estimation of i_r.

Reflected Waves With Oblique Incidence

Although normally incident blast wave properties usually provide upper limits to blast loads on structures, the more usual case of loading of large, flat surfaces is represented by waves which strike at oblique incidence. Also, as a blast wave from a source some distance from the ground reflects from the ground, the angle of incidence must change from normal to oblique.

There have been many theoretical studies of oblique shock wave reflection from plane surfaces, and some experiments. The general physical processes are well described in Kennedy (1946), Baker (1973) and Harlow and Amsden (1970). We will summarize their work here and present curves which can be used to estimate some of the properties (mainly shock front properties and geometry) of obliquely reflected waves.

Oblique reflection is classed as either regular or Mach reflection, dependent on incident angle and shock strength. Geometries of these two cases are shown in Figures 3-1 and 3-2. In regular reflection, the incident shock travels through still air (Region One) at velocity U, with its

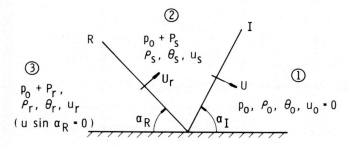

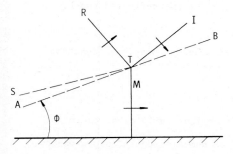

Figure 3-1.   Regular Oblique Reflection of a Plane Shock
From a Rigid Wall
[Kennedy (1946)]

Figure 3-2.   Mach Reflections From a Rigid Wall
[Kennedy (1946)]

front making the angle of incidence α_I with respect to the wall. Properties behind this front (Region Two) are those for a free air shock.   On
contact with the wall, the flow behind the incident shock is turned, because the component normal to the wall must be zero, and the shock is reflected from the wall at a reflection angle α_R that is different from α_I.
Conditions in Region Three indicate reflected shock properties.   A pressure transducer flush-mounted in the wall would record only the ambient
and reflected wave pressures (direct jump from Region One to Region Three)
as the wave pattern traveled along the wall, whereas, one mounted at a
short distance from the wall would record the ambient pressure, then the
incident wave pressure, and finally, the reflected wave pressure.   Some
interesting properties of this regularly reflected shock, given by Kennedy

(1946), are as follows:

1. For a given strength of incident shock, there is some critical angle of incidence $\alpha_{I\ crit}$, such that the type of reflection described above cannot occur for $\alpha_I > \alpha_{I\ crit}$.

2. For each gaseous medium, there is some angle α' such that for $\alpha_I > \alpha'$, the strength of the reflected shock is greater than it is for head-on reflection. For air (approximated as an ideal gas with $\gamma = 1.40$), $\alpha' = 39°23'$.

3. For a given strength of incident shock, there is some value for $\alpha_I = \alpha_{min}$ such that the strength of the reflected shock, P_r/p_o is a minimum.

4. The angle of reflection α_R is an increasing monotonic function of the angle of incidence α_I.

As noted in the discussion of regular oblique reflection, there is some critical angle of incidence, dependent on shock strength, above which regular reflection cannot occur. Ernst Mach [Mach and Sommer (1877)] showed that the incident and reflected shocks would coalesce to form a third shock. Because of the geometry of the shock fronts, they were termed the Mach V or Mach Y, with the single shock formed by the coalesced incident and reflected shocks normally called the Mach stem. The geometry of Mach reflection is shown in Figure 3-2. In addition to the incident and reflected shocks I and R, we now have the Mach shock M; the junction T of the three shocks is called the triple point. In addition, there is also a slipstream S, a boundary between regions of different particle velocity and different density, but of the same pressure. When α_I in Figure 3-1 exceeds $\alpha_{I\ crit}$, the Mach wave M is formed at the wall and grows as the shock systems move along the wall, with the locus of the triple point being a straight line AB.

Harlow and Amsden (1970) present a resume of theory and experiment on regular reflection and the limit of regular reflection (which is also

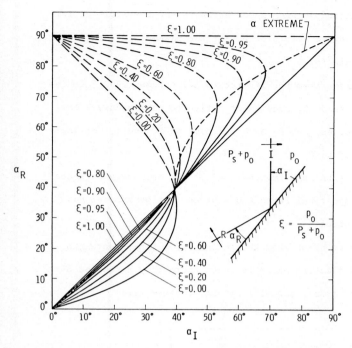

Figure 3-3. Angle of Incidence Versus Angle of Reflection for Shocks
of Different Strengths Undergoing Regular Reflection
[Harlow and Amsden (1970)]

the start of Mach reflection). Two useful curves from their paper are

given here. Figure 3-3 gives angle of reflection α_R as a function of an-

gle of incidence α_I in the regular reflection regime. The parameter ξ is

defined as

$$\xi = \frac{P_o}{P_s + P_o} \tag{3-7}$$

[Harlow and Amsden (1970) call ξ the shock strength, but it is, in fact,

the _inverse_ of the shock strength.] Inverting Equation (3-7), we also

have the relation

$$\bar{P}_s = \frac{P_s}{P_o} = \frac{1}{\xi} - 1 \tag{3-8}$$

230

Figure 3-4 shows the relation between shock strength and the limit of regular reflection. It was originally plotted for the parameter ξ, but we have added a separate scale for $\bar{P}_s$.

A final set of curves from the literature [Glasstone (1962)] is included as Figure 3-5, to allow prediction of reflected peak pressure for oblique shocks. These curves give $\bar{P}_r$ as a function of $\bar{P}_s$ and α_I for incident shock strengths $\bar{P}_s$ up to 4.76.

Some data exist [Wenzel and Esparza (1972)] for obliquely reflected strong shock waves from spherical Pentolite sources, but we know of no data for more diffuse sources.

The total process of blast wave reflection for a finite strength source located above a reflecting plane is quite thoroughly discussed by Kennedy (1946) and Baker (1973), and will not be repeated here.

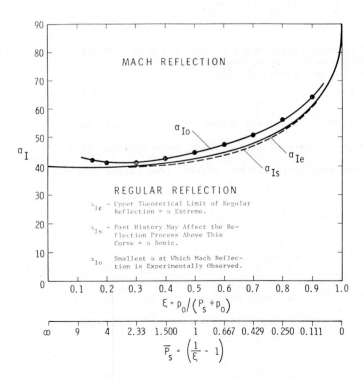

Figure 3-4.   Regions of Regular and Mach Reflection in the
α-ξ Plane, Where $\gamma = 1.4$
[Harlow and Amsden (1970)]

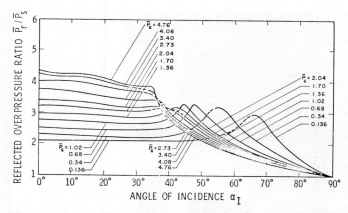

Figure 3-5. Reflected Overpressure Ratio as a Function of Angle of Incidence
for Various Side-On Overpressures
[Glasstone (1962)]

Diffracted Loading

Complex loading of a real target results during the diffraction of
the shock front around the target. Figure 3-6 shows photographs of the
diffraction process, while Figure 3-7 shows schematically, in three stages,
the interaction of a blast wave with an irregular object. As the wave
strikes the object, a portion is reflected from the front face, and the
remainder diffracts around the object. In the diffraction process, the
incident wavefront closes in behind the object, greatly weakened locally,
and a pair of trailing vortices is formed. Rarefaction waves sweep across
the front face, attenuating the initial reflected blast pressure. After
passage of the front, the body is immersed in a time-varying flow field.
Maximum pressure on the front face during this "drag" phase of loading is
the stagnation pressure.

We are interested in the net transverse pressure on the object as
a function of time. This loading, somewhat idealized, is shown in Figure
3-8 [details of the calculation are given by Glasstone (1962)]. At time
of arrival t_a, the net transverse pressure rises linearly from zero to
maximum P_r in time $(T_1 - t_a)$ (for a flat-faced object, this time is zero).

232

Figure 3-6.   Schlieren Pictures Showing Interaction of a
              Shock Wave With a Cylindrical Tank

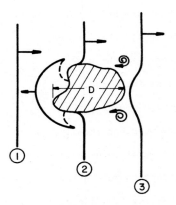

Figure 3-7.   Interaction of Blast Wave With Irregular Object

Pressure then falls linearly to drag pressure in time $(T_2 - T_1)$. This
time history of drag pressure, $q(t)$, is a modified exponential, with a
maximum given by

$$C_D Q = C_D \cdot \frac{1}{2} \rho_s u_s^{\,2} \tag{3-9}$$

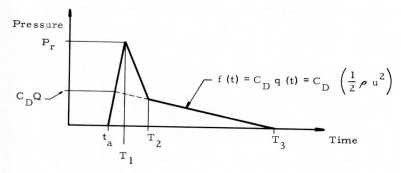

Figure 3-8. Time History of Net Transverse Pressure on Object During Passage of a Blast Wave

where C_D is the steady-state drag coefficient for the object; Q is the peak dynamic pressure; and ρ_s and u_s are peak density and particle velocity, respectively, for the blast wave. The characteristics of the diffraction phase of the loading can be determined if the peak side-on overpressure P_s or the shock velocity U are known, together with the shape and some characteristic dimension D of the object. The peak amplitude of the drag phase, $C_D Q$, can also be determined explicitly from P_s or u_s.

An approximate method of prediction of net diffracted blast loads such as in Figure 3-8 has been developed by Baker, et al. (1974b) and Baker, et al. (1975), for blasts with known values of P_s and i_s. The method utilizes an assumed time history of drag pressure known to be reasonably accurate for TNT and nuclear blasts, estimates of diffraction times based on shock tube experiments, drag coefficients from wind tunnel data, and reflected and stagnation blast front properties based on equations which are well known in blast physics.

Side-on overpressure is often expressed as a function of time by the modified Friedlander equation [see Chapter 1 of Baker (1973)]

$$p(t) = P_s (1 - t/T) e^{-bt/T} \qquad (3-10)$$

234

where T is the duration of the positive phase of the blast wave. Integrating this equation gives the impulse

$$i_s = \int_o^T p(t)\, dt = \frac{P_s T}{b} \left[1 - \frac{(1 - e^{-b})}{b} \right] \qquad (3\text{-}11)$$

The dimensionless parameter b is called the time constant which is a function of shock strength, and is discussed in detail in Chapter 6 of Baker (1973). The time constant is in Table 3-1, Figure 3-9, and Figure 2-39 for a range of shock strengths $\bar{P}_s$, where

$$\bar{P}_s = P_s/p_o \qquad (3\text{-}12)$$

and p_o is ambient air pressure. The peak reflected overpressure P_r and peak dynamic pressure Q are unique functions of P_s for a given ambient pressure p_o. For shocks of intermediate to weak strengths, $\bar{P}_s \leq 3.5$,

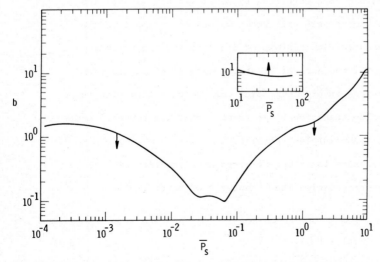

Figure 3-9. Blast Wave Time Constant b Versus Dimensionless Side-On Overpressure $\bar{P}_s$

Table 3-1. Blast Wave Time Constant Versus Dimensionless Side-On Overpressure

$\bar{P}_s$	67.90	37.200	20.4000	11.9000	7.2800	3.4600	2.0500	1.38000
b	8.98	8.750	9.3100	10.5800	7.4700	3.4900	2.0600	1.58000

$\bar{P}_s$	0.772	0.5060	0.1610	0.0616	0.0374	0.0261	0.01980
b	1.320	1.0500	0.3820	0.0984	0.1170	0.1110	0.14900

$\bar{P}_s$	8.70-3	3.91-3	2.48-3	1.41-3	2.42-4	1.153-4
b	0.3600	0.6440	0.8770	1.1400	1.6100	1.45000

these functions are [Baker (1973)]:

$$\bar{P}_r = 2\,\bar{P}_s + \frac{3\,\bar{P}_s^{\,2}}{4} \qquad\qquad (3\text{-}13)^{\dagger}$$

and

$$\bar{Q} = \frac{5}{2}\,\frac{\bar{P}_s^{\,2}}{7 + \bar{P}_s} \qquad\qquad (3\text{-}14)$$

where

$$\bar{P}_r = P_r/p_o, \qquad \bar{Q} = Q/p_o \qquad\qquad (3\text{-}15)$$

---

†This is obtained from Equation (3-3) for $\gamma = 1.4$, which is also used in determining Equation (3-14).

For the time history of drag pressure, a good fit to experimental data for TNT is a slightly modified form of that employed by Glasstone (1962).

$$q(t) = Q (1 - t/T)^2 e^{-bt/T} \qquad (3\text{-}16)$$

The procedure for determining the transverse loading blast parameters in Figure 3-9 which are independent of object size and shape is then as follows:

(1) Obtain P_s and i_s from data or computer analysis,

(2) Calculate $\bar{P}_s$,

(3) Read b from Table 3-1, Figure 3-9, or Figure 2-47,

(4) Solve Equation (3-11) for T, knowing P_s, i_s, and b,

(5) Substitute $\bar{P}_s$ in Equations (3-13) and (3-14) to obtain $\bar{P}_r$ and $\bar{Q}$,

(6) Obtain P_r and Q from Equation (3-15), and

(7) Substitute in Equation (3-16) for q(t), realizing that $T = T_3 - t_a$ in Figure 3-8.

The remaining quantities needed to define the time history of transverse pressure are dependent on the size and shape of the object. They are only well defined for objects of regular shape, such as right circular cylinders, flat rectangular strips, etc. Methods for estimating $(T_1 - t_a)$ and $(T_2 - T_1)$ are given by Glasstone (1962) for several such objects, and will not be repeated here. One does need to know, however, the shock front velocity U. This is a unique function of the shock strength $\bar{P}$, and is given by [see Chapter 6 of Baker (1973)]

$$\bar{U}^2 = 1 + \frac{6 \, \bar{P}_s}{7} \qquad (3\text{-}17)$$

Drag coefficients C_D are available from Hoerner (1958) for a variety of bodies over a wide range of flow velocities. Estimates for the subsonic flow range which applies over the shock strengths of interest to us are

Table 3-2. Drag Coefficients, C_D, of Various Shapes
[Source: Hoerner (1958)]

SHAPE	SKETCH	C_D
Right Circular Cylinder (long rod), side-on	Flow	1.20
Sphere		0.47
Rod, end-on	Flow	0.82
Disc, face-on	Flow or	1.17
Cube, face-on	Flow	1.05
Cube, edge-on	Flow	0.80
Long Rectangular Member, face-on	Flow	2.05
Long Rectangular Member, edge-on	Flow	1.55
Narrow Strip, face-on	Flow	1.98

given in Table 3-2. Melding these quantities, dependent on the size and shape of the body, to the previous ones which are derivable from side-on blast wave properties permits an estimate of the entire time history of transverse pressure, at least for bodies of regular geometry.[†]

INTERNAL BLAST LOADING

Shock Wave Loading

The loading from an explosive charge detonated within a vented or unvented structure consists of two almost distinct phases. The first phase is that of reflected blast loading. It consists of the initial high pressure, short duration reflected wave, plus perhaps several reflected pulses arriving at the chamber walls. These later pulses are usually attenuated in amplitude because of irreversible thermodynamic process, and they may be very complex in waveform because of the complexity of the reflection process within the structure, whether vented or unvented.

Maxima for the initial internal blast loads on a structure can be estimated from scaled blast data or theoretical analyses of normal blast wave reflection from a rigid wall, discussed earlier in this chapter.

Following initial shock wave reflection from the internal walls, the internal blast pressure loading can become quite complex in nature. Figure 3-10 from Gregory (1976) shows a stage in the loading for a cylindrical, vented structure. At the instant shown, portions of the cap, base and cylindrical surface are loaded by the reflected shock and the incident shock is reflecting obliquely from all three internal surfaces. The oblique reflection process can generate Mach waves, if the angle of incidence is great enough, and pressures can be greatly enhanced on entering corners or

[†]These methods, and Figures 3-8, 3-9, and Table 3-2, will be used again in Chapter 5 to help determine velocities of "secondary fragments," i.e., objects picked up and hurled by the blast waves.

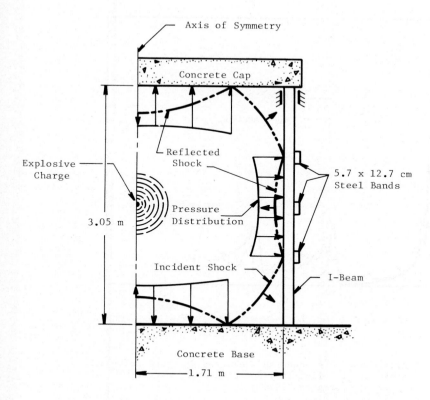

Figure 3-10.   Schematic Representation of Shock Reflections From
        Interior Walls of Cylindrical Containment Structure
                    [Gregory (1976)]

reflecting near the axis of a cylindrical structure. In box-shaped structures, the reflection process can be even more complex.

Following the initial internal blast loading, the shock waves reflected inward will usually strengthen as they implode toward the center
of the structure, and re-reflect to load the structure again. As noted
earlier, the second shocks will usually be somewhat attenuated, and after
several such reflections, the shock wave phase of the loading will be over.

The shock wave loading can be measured with suitable blast measuring systems, or it can be computed for systems possessing some degree of
symmetry. In a spherical containment structure, the loading can be relatively easily predicted for either centrally located or eccentric blast

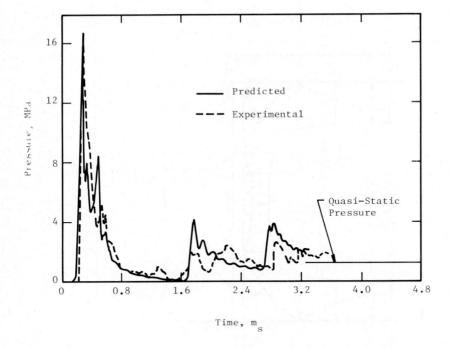

Figure 3-11. Comparison of Predicted and Measured Pressure Pulse at
Point on Sidewall of Cylindrical Containment Structure
[Gregory (1976)]

sources [Baker (1960), Baker, et al. (1966)]. In a cylindrical structure,
existing (but complex) two-dimensional computer programs can be used to
predict loads for blast sources on the cylinder axis [see Figure 3-11 from
Gregory (1976)]. For any more complex geometry such as a blast source located off-center in a cylinder, a box-shaped structure, presence of internal equipment, or structures of more complicated shapes, accurate computation of the details of the internal blast loading is not possible, and one
must rely on measurements. Kingery, et al. (1975) and Schumacher, et al.
(1976) contain most of the internal blast measurements for uniformly vented structures, for cubical and cylindrical geometries, respectively.

As just noted, the initial and reflected air shock loading on the
interior surfaces of suppressive structures is quite complex for all real
structural geometries. But, simplified loading predictions can often be

made rather easily from scaled blast data for reflected waves and several approximate equations. The first approximation we will use is to assume that the incident and reflected blast pulses are triangular with abrupt rises, i.e.,

$$
\left.\begin{aligned}
p_s(t) &= P_s \, (1 - t/T_s), \quad 0 \le t \le T_s \\[2em]
p_s(t) &= 0 \qquad\qquad\qquad, \qquad t \ge T_s
\end{aligned}\right\} \tag{3-18}
$$

and

$$
\left.\begin{aligned}
p_r(t) &= P_r \, (1 - t/T_r), \quad 0 \le t \le T_r \\[2em]
p_r(t) &= 0 \qquad\qquad\qquad, \qquad t \ge T_r
\end{aligned}\right\} \tag{3-19}
$$

The durations of these pulses are not the same as the actual blast wave durations T, but instead are adjusted to preserve the proper impulses, i.e.,

$$
T_s = \frac{2 \, i_s}{P_s} \tag{3-20}
$$

$$
T_r = \frac{2 \, i_r}{P_r} \tag{3-21}
$$

These two equations constitute our second simplifying approximation.

A third simplifying approximation is that the initial internal blast loading parameters are, in most cases, the normally reflected parameters, even for oblique reflections from the structure's walls (provided the slant range is used as the distance R from the charge center to the location on the wall). For strong shock waves, this assumption is almost exactly true up to the angle for limit of regular reflection of slightly greater than 39°, and for weak waves, the limit is as great as 70° (see Figure 3-5). For structure designs which are box-like with length-to-width and height-to-width ratios near one, shock reflections from the walls will be regular almost everywhere.

In enclosed structures such as suppressive structures, the initial shock wave reflects and re-reflects several times, as discussed earlier. In certain configurations and over limited areas of the inner surface, the reflected waves can "implode" or reinforce, but generally they are attenuated considerably before again striking the walls, floor or ceiling. For approximate estimates of the magnitudes of these reflected waves, it can be assumed that the second shock has half the amplitude and impulse of the initial reflected shock, the third shock has half the amplitude of the second shock, and that all later reflections are insignificant. These assumptions are summarized below in equational form:

$$
\left.
\begin{array}{l}
P_{r2} = P_{r1}/2 \\[2em]
i_{r2} = i_{r1}/2 \\[2em]
P_{r3} = P_{r2}/2 = P_{r1}/4 \\[2em]
i_{r3} = i_{r2}/2 = i_{r1}/4 \\[2em]
P_{r4}, \text{ etc.} = 0 \\[2em]
i_{r4}, \text{ etc.} = 0
\end{array}
\right\}
\qquad (3\text{-}22)
$$

Adjusted durations of the reflected pulses are the same as the initial pulses, i.e.,

$$T_{r3} = T_{r2} = T_{r1} \qquad (3\text{-}23)$$

The final assumption we will make is that the time of reverberation, i.e., the time delay between arrival of successive reflected shocks, is simply

$$t_r = 2 \, t_a \qquad (3\text{-}24)$$

Again, this is not strictly true, because the second and third shocks are weaker than the first, and, therefore, travel more slowly. But, the accuracy of this assumption is of the same order as the other assumptions.

Schematically, the complete set of simplifying assumptions is illustrated in Figure 3-12, which shows a typical simplified pressure loading at a point on the inner surface of a structure. (The later two reflected pulses are often ignored in estimating the internal blast loading because the pressures and impulses are much lower than in the initial pulse.) In the simplifications employed here, the combined loads from all three pulses are 1.75 times those from the initial pulse. One greater degree of simplification can then be employed for structures with response times much longer than the longest time in Figure 3-12. That is simply to combine all three pulses, and multiply the amplitude (and equally the impulse) by 1.75.

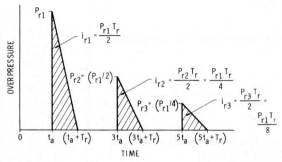

Figure 3-12. Simplified Internal Blast Pressure

Venting and Quasi-Static Pressures

High Explosive Source

When an explosion from a high explosive source occurs within a
structure, the blast wave reflects from the inner surfaces of the structure, implodes toward the center, and re-reflects one or more times.
The amplitude of the re-reflected waves usually decays with each reflection, and eventually the pressure settles to a slowly decaying level,
which is a function of the volume and vent area of the structure and the
nature and energy release of the explosion. A typical time history of
pressure at the wall of a vented structure is shown in Figure 3-13. The
process of reflection and pressure buildup in either unvented or poorly
vented structures has been recognized for some time, dating from World
War II research on effects of bombs and explosives detonated within enclosures. More recently, study of these pressures has revived because
of interest in design of vented explosion chambers. Esparza, et al.
(1975) summarizes some recent work in this area.

Weibull (1968) reports maximum pressures for vented chambers of
various shapes having single vents with a range of vent areas to volume

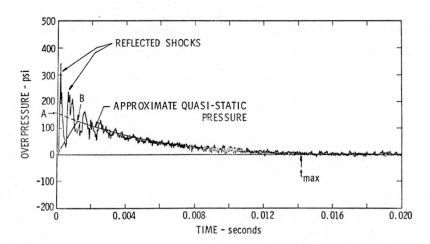

Figure 3-13.   Typical Time History of Internal Pressure at Inner
Surface of a Suppressive Structure
[Kingery, et al. (1975)]

of $(A_V/V^{2/3}) \leq 0.0215$. These maximum quasi-static pressures are shown by Weibull to be independent of the vent area ratio, and to be a function of charge-to-volume ratio (W/V) up to 5.00 kg/m^3. He fitted a single straight line to his data, but Proctor and Filler (1972) later showed that fitting a curve to the data, with asymptotes to lines related to heat of combustion for small (W/V) and to heat of explosion with no after-burning for large (W/V), was more appropriate. Additional data on maximum quasi-static pressures and on venting times have been obtained by Keenan and Tancreto (1974) and by Zilliacus, et al. (1974). Concurrent with experimental work which preceded applications to suppressive structures, Proctor and Filler (1972) developed a theory for predicting time histories of quasi-static pressures in vented structures. Kinney and Sewell (1974) did likewise, and also obtained an approximate formula for this time history. Converted to scaled parameters, this equation is:

$$\ell n \; \bar{P} = \ell n \; \bar{P}_1 - 2.130 \; \bar{\tau} \qquad (3\text{-}25)$$

Here, $\bar{P}$ and $\bar{P}_1$ are scaled absolute pressures given by

$$\bar{P} = P(t)/p_o \qquad (3\text{-}26)$$

$$\bar{P}_1 = (P_{qs} + p_o)/p_o \qquad (3\text{-}27)$$

with P_{qs} being the quasi-static pressure. The quantity $\bar{\tau}$ is a dimensionless time for venting given by

$$\bar{\tau} = \bar{A} \; \bar{t} = \left(\frac{\alpha_e A_s}{V^{2/3}} \right) \left(\frac{t a_o}{V^{1/3}} \right) \qquad (3\text{-}28)$$

In this equation, α_e is an effective vent area ratio to be discussed later,

A_s is internal surface area of the structure, V is internal volume of the structure, t is time, and a_o is sound velocity of air in the structure. The rationale for use of these scaled parameters is developed by Baker and Oldham (1975). Equation (3-25) gives a value for scaled venting time $\bar{\tau}_{max}$ of

$$\bar{\tau}_{max} = 0.4695 \; \ell n \; \bar{P}_1 \qquad\qquad (3-29)$$

The problem of blowdown from a vented chamber is also solved theoretically by Owczarek (1964), given initial conditions in the chamber but assuming isentropic expansion through the vent area.

In the suppressive structures program, sufficient data have been recorded to add significantly to the measurements for other types of vented or unvented chambers [Kingery, et al. (1975) and Schumacher, et al. (1976)]. In comparing such data with either previous data or theory, there are several questions raised by the general physics of the process and by the differences in venting through single openings in walls. Referring to Figure 3-13, one can see that the maximum quasi-static pressure is quite difficult to define because it is obscured by the initial shock and first few reflected shocks.

Obviously, several reflections must occur before irreversible processes attenuate the shocks and convert their energy to quasi-static pressure. It, therefore, seems inappropriate to call point A in Figure 3-13 the peak quasi-static pressure, although this is the point used by Kingery, et al. (1975) to compare with code predictions from Proctor and Filler (1972) and the Kinney and Sewell equation. A better approach is to allow some time before the maximum quasi-static pressure is established, such as point B in Figure 3-13.

Figure 3-13 also illustrates another problem inherent in reduction of vented pressure data, i.e., accurate determination of duration of this

pressure. When the pressure traces approach ambient, the shock reflections have largely decayed. But the pressure approaches the baseline nearly asymptotically, so that the duration is quite difficult to determine
accurately. A possible duration t_{max} is shown in the figure.

A dimensionless equation based on a scaling law for the explosion
venting process was formulated by Baker and Oldham (1975). In more general terms, the law states that

$$\left(\frac{p}{p_o}\right) = f_1 \left[\left(\frac{\alpha_e A_s}{v^{2/3}}\right), \left(\frac{P_1}{P_o}\right), \left(\frac{ta_o}{v^{1/3}}\right), \gamma\right] \tag{3-30}$$

Based on a theoretical analysis of chamber venting by Owczarek (1964), two
of the terms in Equation (3-30) can be combined, and $\bar{P} = (P/p_o)$ is a function of ratio of specific heats γ and a new scaled time

$$\bar{\tau} = \bar{A} \bar{t} = \left(\frac{\alpha_e A_s}{v^{2/3}}\right) \left(\frac{ta_o}{v^{1/3}}\right) \tag{3-31}$$

An alternate form of Equation (3-30) is then

$$\bar{P} = f_2 (\bar{P}_1, \bar{\tau}, \gamma) \tag{3-32}$$

The initial pressure $\bar{P}_1$ for structures with no venting or small venting
can be shown to be related to another scaling term,

$$\bar{P}_1 = f_3 (E/p_o V) \tag{3-33}$$

where E is a measure of total energy released by the explosion. For tests
with explosives of the same type and no change in ambient conditions, a
dimensional equivalent of Equation (3-33) is

$$P_1 = f_4 \, (W/V) \tag{3-34}$$

where W is charge mass (weight) and V is chamber volume. The scaled pressure-time histories during the gas venting process can be integrated to give scaled gas impulse $\bar{i}_g$. This parameter is defined as

$$\bar{i}_g = i_g \left(\frac{\alpha_e A_s a_o}{p_o V} \right) \tag{3-35}$$

Equations (3-25) and (3-29) can be shown to give

$$\bar{i}_g = \frac{1}{2.130} \left(e^{2.130 \, \bar{\tau}_{max}} - 1 \right) - \bar{\tau}_{max} \tag{3-36}$$

For a single layer structure, the vent area ratio α_e is the vent area divided by the total area of the wall. For a multi-layer wall, however, Baker, et al. (1975) assumed that

$$\frac{1}{\alpha_e} = \sum_{i=1}^{N} \frac{1}{\alpha_i} \tag{3-37}$$

This relationship has, at the moment, no theoretical proof. However, it does reach the appropriate limits for large and small number of plates, and provides a relative measure of venting for a variety of panel configurations. Definition of the individual values of α_i for each layer in a multi-layered vented panel requires careful study of the panel configuration and experimental verification. Specific formulas and methods for predicting α_e for various suppressive structure panels are presented by Esparza, et al. (1975), of which some are summarized in Figure 2-21.

In spite of complexities in the venting process, gas venting pressures and their durations can be predicted with reasonable accuracy, particularly if one differentiates between these relatively long term and low amplitude pressures and the internal blast pressures resulting from blast wave impingement and reflection. Figure 3-14 shows the simplified form for the gas venting pressures which can be assumed.

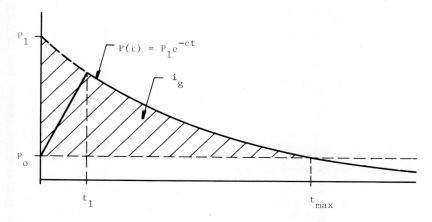

Figure 3-14. Simplified Gas Venting Pressure

In this simplified form, the gas venting pressure is assumed to follow the solid curve and rise linearly from zero time until it reaches, in time t_1, a curve which is decaying exponentially from an initial maximum value of P_1. The decay then follows the time history

$$P(t) = P_1 e^{-ct} \tag{3-38}$$

until it reaches ambient pressure p_o at time $t = t_{max}$. The exponential decay is shown to agree well with experiment [Kingery, et al. (1975), and Schumacher, et al. (1976)]. The cross-hatched area under the entire overpressure curve is defined as the gas impulse, i_g, and is given mathematically as[+]

[+]Referring to Figure 3-14, the value for i_g will always be greater than the true gas impulse because t_1 is never zero. Unless the structure is very well vented, the error from using Equation (3-39) is small.

$$i_g = \int_o^{t_{max}} [P(t) - p_o]\, dt$$

$$= \int_o^{t_{max}} (P_1 e^{-ct} - p_o)\, dt \qquad (3\text{-}39)$$

$$= \frac{P_1}{c} \left(1 - e^{-ct_{max}}\right) - p_o\, t_{max}$$

The time t_1 we will assume to be the end of the internal blast loading phase, shown earlier (e.g., Figure 3-12) to be

$$t_1 = 5\, t_a + T_r \qquad (3\text{-}40)$$

The maximum value for the overpressure, P_{QS}, in the gas venting phase of the loading is the static pressure rise which would occur in an unvented enclosure before heat transfer effects attenuate it. From data

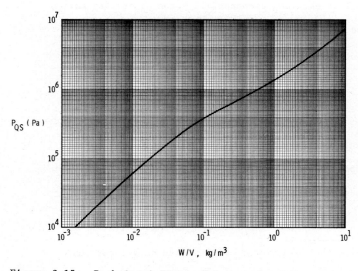

Figure 3-15. Peak Quasi-Static Pressure for TNT Explosion in Chamber

and analyses in several references, the curve of Figure 3-15 has been shown to yield good predictions of P_{QS}, if the quantity of explosive W and the internal volume of the structure V are known.

Gas venting parameters other than P_{QS}, or $P_1 = P_{QS} + P_o$, can be most easily predicted using plots or equations for some of the scaled parameters described earlier. The quantity c in Equation (3-38) is given with reasonable accuracy by

$$c = 2.130 \left(\frac{\alpha_e A_s a_o}{V} \right) \tag{3-41}$$

For air at standard sea level conditions, $a_o = 340$ m/sec, thus

$$c = 725 \frac{\alpha_e A_s}{V} \ (\text{sec}^{-1}) \tag{3-42}$$

for A_s in m^2 and V in m^3. As indicated, units of c in this last equation are sec^{-1}.

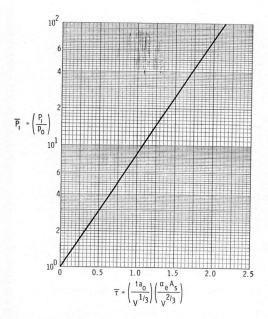

Figure 3-16.   Scaled Blowdown Duration Versus Scaled Maximum Pressure

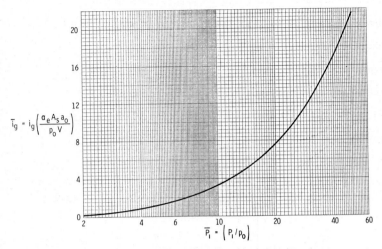

$$\bar{i}_g = i_g \left(\frac{\alpha_e A_s a_0}{p_0 V} \right)$$

$$\bar{P}_1 = \left(P_1 / p_0 \right)$$

Figure 3-17. Scaled Gas Pressure Impulse Versus Scaled Initial Pressure

Figures 3-16 and 3-17 give scaled durations of gas overpressure $\bar{t}_{max}$ and scaled gas impulse $\bar{i}_g$ as functions of scaled initial pressure $\bar{P}_1 = P_1 / p_0$. Equations for the scaled parameters can be inverted to give the corresponding dimensional quantities, as follow:

$$i_g = \bar{i}_g \left(\frac{p_0 V}{\alpha_e A_s a_0} \right) \qquad (3\text{-}43)$$

$$t_{max} = \bar{\tau}_{max} \left(\frac{V}{\alpha_e A_s a_0} \right) \qquad (3\text{-}44)$$

Self-consistent units must be used when "unscaling" using these equations.

The values for the vent area ratio α_e are obtained for typical suppressive structure designs in the manner shown in Figure 2-21.

Combustible Gas or Dust Mixtures With Air

The recent work in the suppressive shield program has generated acceptable prediction methods and curves to estimate pressures inside vented structures with high explosives detonated inside. But, except in design of munitions and explosive manufacturing plants, accidental explosions within structures are much more likely to occur with combustible gases or combustible dusts suspended in air. Industry has, in fact, been plagued for many years with explosions of this nature, and there is a voluminous literature dealing with internal gaseous and dust explosions, and the effects of venting on pressures generated in enclosures by such explosions.

Much of the work on dust explosion venting is summarized by Palmer (1973), Gibson and Harris (1976), and by Bartknecht (1978a and 1978b). Gibson and Harris (1976), Bartknecht (1978a), Anthony (1977b) and Bradley and Mitcheson (1978a and 1978b), also summarize effects of venting on gaseous explosions. The NFPA (1978b) has published an explosion venting guide based on work by the U. S. Bureau of Mines and work in Great Britain and Germany.

Indicative of studies of venting in the United States chemical industry are papers by Runes (1972) and Howard (1972). Analytic studies, sometimes including test data, have been made by Yao (1974), Munday (1976), Sapko, et al. (1976), Pasman, et al. (1974), and Bradley and Mitcheson (1978a). Baker, Esparza and Kulesz (1977) have summarized some of this work in a review paper.

Parameters assumed to be important in these studies are geometric ones such as shape and volume of the enclosure, the total vent area, heat of combustion Q_c of the dust or gaseous fuel, and the stoichiometric ratio, all of which determine the maximum pressure P and maximum rate of pressure rise $\dot{P} = dP/dt$. Some correlations of P and $\dot{P}$ have been attempted by different investigators with the ratio (A/V), a dimensionless "K factor" which is the ratio of vessel cross-section area A_c to vent area A, and to a cube root law which is similar to the Hopkinson-Cranz law discussed in Chapter 2. The most convincing correlations are those of Bartknecht (1978a and 1978b) and of Bradley and Mitcheson (1978a and 1978b). Bartknecht correlates data with a "cubic law" given by

$$\left(\frac{dp}{dt}\right)_{max} \cdot V^{1/3} = K \tag{3-45}$$

where K is a specific constant for a particular combustible gas or dust. Bradley and Mitcheson correlate maximum pressure rises in spherical vented chambers with a quotient of dimensionless ratios $\bar{A}/\bar{S}_o$, where $\bar{A}$ is the product of vent area and coefficient of discharge (K_d) divided by total sphere internal surface area (A_s), and $\bar{S}_o$ is the ratio of gas velocity U ahead of a flame front to the acoustic velocity in the unburned gas just after ignition:

$$\bar{A} = \frac{A\,K_d}{A_s}; \quad \bar{S}_o = \frac{U}{a_o} \tag{3-46}$$

Instances of severe damage to buildings by dust explosions are reviewed by Palmer (1973), and similar accidents for internal gaseous explosions are reviewed by Mainstone (1976) and by Strehlow and Baker (1976).

No careful scaling of explosion venting appears to have been done for industrial explosions, although analytic papers such as those of Yao (1974), Munday (1976), Sapko, et al. (1976), and Bradley and Mitcheson (1978a and 1978b) present their analyses or some prediction equations in dimensionless form. But, Baker (1977) conducted a model analysis of solid propellant burning in a vented chamber, and Baker, et al. (1980) followed with a similitude analysis for vented dust explosions. They concluded that:

    (1) "Replica" (or cube-root) scaling with a geometric scale factor λ should apply;

    (2) Geometry of model and full-scale chambers must be exactly maintained;

(3) Scaled pressure (P/p_o) should be a function of scaled vent
     area $(A/V^{2/3})$ and scaled total energy $(E/p_o V)$;

(4) Pressure rate scales according to $(\dot{P} V^{1/3}/a_o p_o)$, and is pro-
     portional to $1/\lambda$ for equal atmospheric pressure p_o and sound
     velocity a_o. Bartknecht's "cubic law," Equation (3-45) is
     consistent with this last conclusion.

Absolute maxima for internal pressures can be predicted for industrial explosions if the mass of weight of fuel and air in the chamber are known, and the chamber is assumed to be unvented. The heat of
combustion then raises the pressure to a known value in a constant volume process. But, the maxima so calculated are overconservative. Empirical fits to test data give the best current estimates of maximum
pressures, and many are reported in editions of the NFPA Guide for Explosion Venting, prior to 1975. But, unfortunately, it is common practice to plot these data as functions of the dimensional parameter (A/V)
so scaling of these test data to chambers of different size is question-
able. Anthony (1977b) and Baker, Esparza and Kulesz (1977) point out
that this practice is physically incorrect, and if one attempts to scale
up data from small scale experiments, vent areas larger than the total
surface area of a vessel can result.

Bartknecht (1978a and 1978b) presents a number of nomographs for
estimating vent areas which will reduce maximum pressures to specified
levels, for various combustible gases and dusts. Figure 3-18 is a typical set of such nomographs. Here, curves are given for various values
of reduced pressure and K_{ST}[+]. These curves are consistent with the third
scaling conclusion given by Baker (1977), i.e., scaled maximum pressure
is a unique function of scaled vent area ratio, for each class of dust.

[+]K_{ST} is Bartknecht's K parameter, Equation (3-45), for dusts [ST is an ab-
breviation for the German word for dust (staub)].

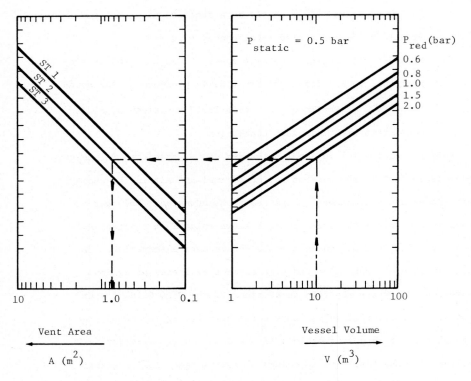

$P_{static} = 0.5$ bar

P_{red}(bar)

0.6
0.8
1.0
1.5
2.0

ST 1
ST 2
ST 3

10   1.0   0.1   1   10   100

Vent Area

$A \ (m^2)$

Vessel Volume

$V \ (m^3)$

Figure 3-18.   Explosion Venting Nomographs for Combustible Dusts
[Bartknecht (1978b)]

This can be seen in Figure 3-19, which is obtained by reading values from
Figure 3-18 and scaling them, using widely disparate values of vessel volume V. The series of lines on the right-hand side of Figure 3-18 collapse
to a single curve, and verify the use of the scaled vent area ratio, over
the range of validity of the curves. This range is noted in an inset in
Figure 3-19.

The general form of the pressure-time histories of unvented and
vented gas and dust internal explosions are shown in Figures 3-20 and
3-21. In some instances with gaseous internal explosions (but apparently

Figure 3-20.   Typical Pressure-Time History for Unvented Gas
or Dust Explosion
[Bartknecht (1978a)]

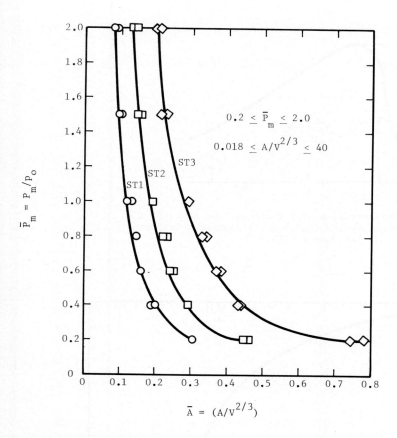

Figure 3-19.   Scaled Maximum Pressures for Dust Explosions
Versus Scaled Vent Area Ratio

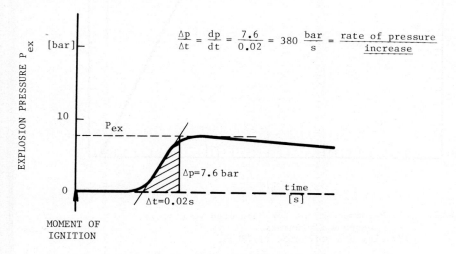

Figure 3-20.   Typical Pressure-Time History for Unvented Gas
or Dust Explosion [Bartknecht (1978a)]

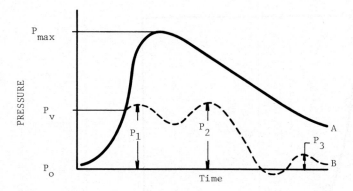

Figure 3-21. A Representation of a Pressure-Time History of an
Unvented (Curve A) and Vented (Curve B) Deflagrative Explosion
[Anthony (1977b)]

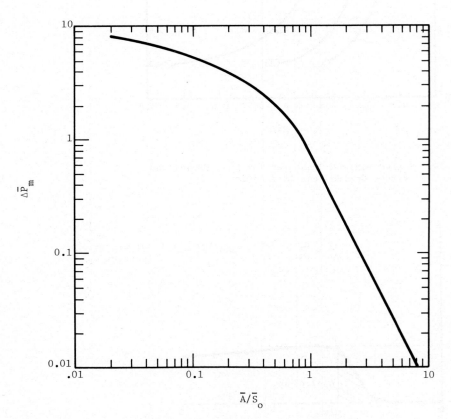

Figure 3-22.   Safe Recommendations for Uncovered Vent Areas,
Gaseous Explosions
[Bradley and Mitcheson (1978b)]

not with dust explosions), these relatively slowly varying pressures can be altered by transition to detonation, by turbulent effects during venting, and by rate of vent cover opening. The excellent papers by Bradley and Mitcheson (1978a and 1978b) include, for vented gaseous explosions, comparisons of their theoretical predictions of these pressure-time histories with experiments by many investigators. They also give upper limit curves of scaled maximum overpressure $\Delta \bar{P}_m$ versus the scaled venting parameter $\bar{A}/\bar{S}_o$ for both open vents and covered vents. These curves are reproduced here as Figures 3-22 and 3-23. Values of $\bar{S}_o$ for a number of fuel gases which can be used to enter these graphs are given in Table 3-3, also from Bradley and Mitcheson (1978b).

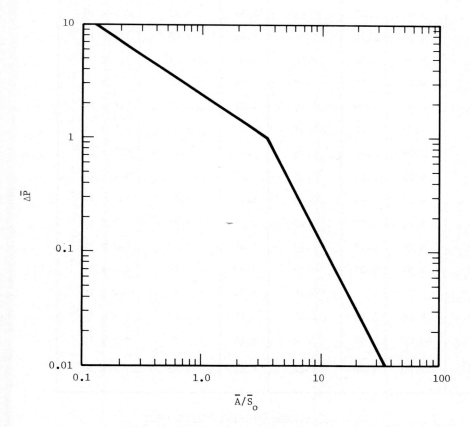

Figure 3-23. Safe Recommendations for Covered Vent Areas,
Gaseous Explosions
[Bradley and Mitcheson (1978b)]

The reader is cautioned that older work by the U. S. Bureau of Mines, and the editions of the NFPA Guide of Explosion Venting based on vent area predictions in the form of A/V can lead to gross <u>overestimates</u> of the vent areas needed for limiting maximum explosion pressures in

Table 3-3. Properties of Selected Gas-Air Mixtures at Initial Conditions of 1 Atmosphere and 298°K [Bradley and Mitcheson (1978b)]

Gas	Mole % In Air	ϕ	S_o m sec^{-1}	$\frac{\rho_{uo}}{\rho_{bo}}$	P_e Atmospheres	$\bar{S}_o$
CH_4	9.48	1.00	0.43	7.52	8.83	8.5×10^{-3}
C_2H_2	7.75	1.00	1.44	8.41	9.78	3.2×10^{-2}
C_2H_2	9.17	1.20	1.54	8.80	10.28	3.7×10^{-2}
C_2H_4	6.53	1.00	0.68	8.06	9.39	1.4×10^{-2}
C_3H_8	4.02	1.00	0.45	7.98	9.31	9.6×10^{-3}
C_3H_8	4.30	1.07	0.46	8.09	9.48	9.9×10^{-3}
C_3H_8	5.00	1.26	0.38	7.97	9.55	9.2×10^{-3}
C_3H_8	6.00	1.52	0.15	7.65	9.30	3.0×10^{-3}
C_5H_{12}	2.55	1.00	0.43	8.07	9.42	9.0×10^{-3}
C_5H_{12}	2.70	1.06	0.43	8.18	9.76	9.3×10^{-3}
C_5H_{12}	3.00	1.18	0.40	8.16	9.77	8.4×10^{-3}
C_5H_{12}	3.50	1.39	0.29	7.92	9.80	6.3×10^{-3}
$C_{16}H_{34}$			0.39	7.82	9.40	7.1×10^{-3}
H_2	29.50	1.00	2.70	6.89	8.04	4.4×10^{-2}
H_2	40.00	1.60	3.45	6.50	7.78	5.3×10^{-2}
TOWN GAS	25.00	1.40	1.22	6.64	8.03	1.9×10^{-2}

ϕ = Equivalence ratio = $\dfrac{\text{Actual Fuel/Air Volume Ratio}}{\text{Stoichiometric Fuel/Air Volume Ratio}}$

P_e = Theoretical closed vessel maximum explosion pressure

vented vessels. On the other hand, use of the voluminous data on maximum pressures and pressure rise rates obtained in the instrumented Hartmann apparatus can lead to serious underestimates of these parameters. Bartknecht's prediction curves and equations, and those of Bradley and Mitcheson are the most realistic available at present. Studies of effects of turbulence, vent opening time, and details of vessel shape on the vented explosion process are far from complete, however, and there is an almost complete lack of data or theory regarding pressure waves transmitted to the surroundings from vented gaseous and dust explosions.

It has recently been discovered by Zalosh (1979) that the venting of room-size enclosures leads to higher overpressures than one would predict using either the Bradley and Mitcheson (1978a and 1978b) or the Bartknecht (1977, 1978a) approach.[†] The reason for this is that the second pressure peak, which is produced while the combustion products are flowing through the vent, is considerably higher than that predicted by previous measurements or calculations for small vessels. From the pressure records (see Figures 3-24 and 3-25 for examples) it can be seen that when the second pressure peak occurs, there are significant oscillations of very high amplitude in the chamber. These oscillations correspond to the natural acoustic frequencies of the chamber and are obviously being driven by the combustion process. It is apparent that these oscillations are caused by a combustion-driven instability of the flame front during the later processes of combustion in the venting chamber. It is interesting that these large chambers have characteristic frequencies of the order of 100 to 1000 Hz and this corresponds approximately to the time it takes to burn through the preheat zone of a premixed gas flame.

[†] Additional large scale tests that confirm these observations have been performed and reported by Astbury, et al. (1970), Astbury and Vaughan (1972), Astbury, et al. (1972), Solberg, et al. (1979) and Eckhoff, et al. (1980).

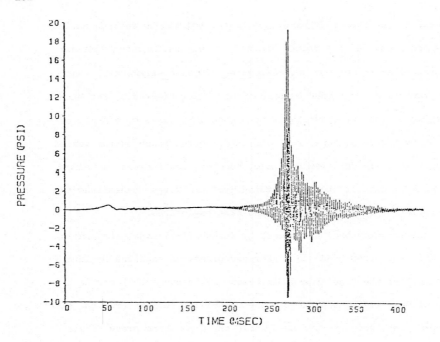

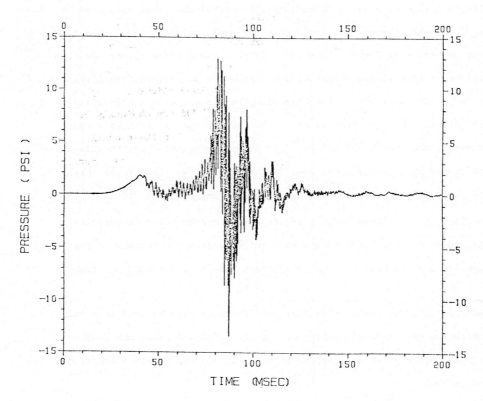

This is not true for small chambers where the characteristic frequency
of the chamber is in the 10,000 Hz range. Thus, it is apparent that for
large sized chambers, flame propagation processes can couple with chamber oscillations to produce destructive amplitudes during the latter part
of the venting. This means that the estimates given in previous sections
are not conservative for large size enclosures and that one may easily
produce internal overpressures that are larger than one would predict.
These effects are still under investigation.

In another recent development, there is some indication that combustion explosion venting of the vessel, when the initial pressure in the
vessel is higher than atmospheric pressure, leads to different venting
requirements than when the vessel is at atmospheric pressure. There is
a proposal for research by Factory Mutual Research and Fenwall Industries
[Yao and Friedmann (1977)] to study jointly this problem with a series of
relatively large tests. The tests are to be supported by a consortium of
chemical companies interested in this problem area. However, the fact
that serious consideration is being given to these tests implies that
there is a problem in the venting of a combustion explosion from a vessel
whose initial pressure is significantly above atmospheric.

One additional problem with combustion venting is the fact that
one should use, in general, either an orifice or a very short length of
pipe to direct the flow outside of the building. If a long pipe is used,
the flame acceleration processes in the pipe can cause significantly larger local overpressures than one would calculate or expect in the vessel.

Figure 3-24. Pressure Signature for Propane Test AP-5 With
2-1/2 Vents Deployed
[Zalosh (1979)]

Figure 3-25. Pressure Signature for Ethylene Test BE-1 With
Six Vents Deployed
[Zalosh (1979)]

This is because combustible material is ordinarily vented first, followed by a flame; and this flame can accelerate in the turbulent duct flow. Thus, venting a combustion explosion from an enclosure by means of a long duct is not recommended because explosion of the duct itself can cause significant damage.

RUNAWAY CHEMICAL REACTOR VESSEL

As discussed in Chapter 1, there are many circumstances in the process industry where exothermic reactions are carried out in reactor vessels at relative high pressures and elevated temperatures. The rule of thumb for safe operation of such reactors is that the external cooling system must be adequate to carry away the heat of reaction such that a runaway temperature excursion does not occur. However, even in the best planned systems, runaways do occur occasionally and good process plant design requires that the vessel be vented, usually with a passive, burst disk which is sized to relieve at some pressure above the operating pressure of the vessel. The burst disk is usually connected to a pipe system which carries the contents of the vessel to some location where the catastrophic dump can be handled in a safe manner. Quite frequently the burst disk is not placed on a separate line but is close to the vessel on a process line of the plant. Thus, both the burst disk orifice and the piping must be of sufficient size to both lower the pressure of the vessel while exothermic reaction is occurring and also handle the flow from other portions of the system which may be connected to the vessel. In general, venting calculations can be performed by making the assumption that the flow through the plumbing system is quasi-steady and that during blowdown the pressure in the reactor vessel is uniform throughout the vessel at any instance of time.

Consider the case of a vessel that contains a gas (which may be non-ideal), and where we need not consider two-phase flow equations. Following a technique suggested by Duxbury (1976), assume that the flow through the pipe vent system is a fanno flow. The theory of fanno flows tells us that when the reactor pressure is sufficiently high, the flow entering the pipe is subsonic and that flow continuously accelerates and reaches sonic velocity at the end of the pipe. At that point in the pipe, the flow is fanno "choked." Under these conditions, the pipe is carrying the maximum possible mass flow. The flow is accompanied by a pressure drop which is caused by frictional drag on the walls. Thus, the friction of the pipe wall must be induced to calculate properly the flow through the piping system. To perform the calculation, we must also know the equation of state of the gas of gaseous mixture which is flowing from the reactor. If the conditions are such that the gas cannot be accurately described by the perfect gas law, it is important to use a reasonably accurate nonideal equation of state. In addition, assume that head changes due to changes in pipe elevation are negligible. The energy equation may then be written in the form

$$h + \frac{G^2}{2\rho^2} = \text{constant} \tag{3-47}$$

where G is ρu, the mass flow through the pipe; h is enthalpy; and u can be interpreted as the average velocity across the cross-section of the pipe.

Additionally, the momentum equation must be solved to calculate the pressure drop due to viscous forces in the pipe. This equation has the form

$$\frac{dP}{dL} - \frac{G^2}{\rho^2} \frac{d\rho}{dL} + \frac{2fG^2}{\rho D} = 0 \tag{3-48}$$

where f is the Fanning friction factor which is related to Reynolds number and pipe roughness, and L is the coordinate in the pipe direction. For a short length of pipe, one may assume that the gas density has some average value for the last term on the left-hand side of this equation. With this assumption, Equation (3-48) can be written in finite difference form as follows:

$$P_1 - P_2 = \frac{2fG^2}{\rho_m} \cdot \frac{\Delta L}{D} + G^2 \left(\frac{1}{\rho_2} - \frac{1}{\rho_1} \right) \qquad (3\text{-}49)$$

Here, ρ_m is the mean value of ρ_1 and ρ_2 and ΔL is a short length of pipe. To solve Equations (3-47) and (3-49) simultaneously for the mass flow through the pipe, one must assume a mass flow and pressure level at the entrance of the pipe. This is done by assuming that the pipe entrance region is a convergent nozzle (a good vent pipe design should use such a nozzle to avoid the vena contracta effect). The flow in a convergent nozzle is essentially isentropic. Once the pipe area and a mass flow is assumed, then the isentropic relationships yield the pressure, temperature and density at the entrance of the pipe. The pipe is then broken into a number of short lengths (the ΔL's) and the state properties and flow Mach number of the gas along the pipe are calculated until either the flow Mach number becomes greater than unity or the end of the pipe is reached. If the flow Mach number becomes unity before the end of the pipe is reached, the assumed mass flow is too large. If the flow Mach number is not unity by the time the end of the pipe is reached, the assumed mass flow is too small. Iteration on the mass flow until the choked flow condition is achieved (Mach number equals 1.0) will then yield the initial rate of removal of the mass from the reactor vessel. This pipe flow calculation can be performed assuming that the gas is nonreactive. However, to determine the effect of mass removal from the vessel on the runaway reaction, the calculation must include the chemistry that is occurring in the vessel during the blowdown process. The reaction can be treated as having quasi-steady kinetics which is occurring uniformly throughout the

entire vessel. The best kinetic rates to use are those determined in
adiabatic calorimeter tests.

The actual behavior of the reacting system in the vessel is different before the burst disk opens and after it opens. The most conservative
assumption that one can make before the disk opens is that heat loss to
the walls is negligible; that is, that one has an adiabatic runaway. With
this assumption from the known kinetics of the reaction in the vessel and
the heat of the reaction, one can calculate the pressure-time curve for the
vessel, starting at time t = 0 when there has been no upset and continuing
until the pressure reaches the disk's burst pressure. After the burst
disk opens, one must continue to assume that the vessel contents are reactive and that the exothermic chemistry is still occurring. However,
now, instead of occurring at constant volume, it is occurring under a
situation where the specific volume in the vessel is dropping with time.
While the actual thermodynamic path process for this phenomenon is not
known, a good first approximation is the Hugoniot relationship applied
to a steady situation (where there is no flow). This essentially assumes
a straight line process path in the (p,v) plane and is exactly correct
when the specific volume of the gas remains constant (i.e., before blowdown). To calculate this stage of the process, the known mass flow out
of the vessel is used to determine a time at which some small percentage
of the vessel contents have left the vessel. This fixes a specific volume change. It also fixes an average value of the pressure and temperature of the gas in the vessel and the extent of the reaction during that
time. This information can be used to calculate the state of the gas at
the end of that time by simultaneously solving the Hugoniot relationship,
the equation of state and the reaction kinetic relationship for the substance in the vessel. Note that, depending upon the size of the vent that
was assumed, pressure in the vessel may continue to rise, stay relatively
constant or start to drop at the instant of burst. Obviously, the minimum safe design is one where the pipe is sized such that from the time of

burst, the pressure in the vessel drops continuously. Another criterion that can be used to evaluate overall safety is whether or not the pressure drop is rapid enough to cause the rate of the chemical reaction to slow down. This latter criterion is more conservative than pressure drop.

Relief venting of an exothermic runaway reaction when the reactor contains both the liquid and gas phase is more difficult to calculate. The major reason for this is that the flow through the venting ducts will be two-phase flow and the dynamics of two-phase flow are not as well understood as those for pure liquid or pure gaseous flows. Singh (1978 and 1979) has made some recommendations for calculational techniques when two-phase venting occurs.

CHAPTER 3

EXAMPLE PROBLEMS

Vented Pressure Parameters

An explosion of 50 kg of TNT occurs in a strong, vented chamber. The chamber has a volume $V = 30$ m^3, and a vent area $A_v = \alpha_e A_s = 10$ m^2. Atmospheric conditions are standard sea-level conditions of $p_o = 1.013 \times 10^5$ Pa and $a_o = 3.40 \times 10^2$ m/s. What are the vented pressure parameters?

From Figure 3-15,

$$\frac{W}{V} = 1.667 \ \frac{kg}{m^3}$$

$$P_{QS} = 1.9 \times 10^6 \ Pa$$

$$P_1 = P_{QS} + p_o = 1.9 \times 10^6 + 1.013 \times 10^5 = 2 \times 10^6 \ Pa$$

$$\bar{P}_1 = \frac{2 \times 10^6}{1.01 \times 10^5} = 19.8$$

From Figure 3-16,

$$\bar{\tau} = 1.39$$

From Figure 3-17,

$$\bar{i}_g = 7.50$$

$$t = \frac{\bar{\tau} V}{a_o A_v} = \frac{(1.39)(30)}{(3.40 \times 10^2)(10)} = 0.0123 \text{ s} = \underline{12.3 \text{ m/s}}$$

$$i_g = \frac{\bar{i}_g P_o V}{A_v a_o} = \frac{(7.50)(1.013 \times 10^5)(30)}{(10)(3.40 \times 10^2)} = \underline{670 \text{ Pa·s}}$$

Maximum Gas Explosion Pressure for Uncovered Vents

A propane-air mixture explodes in a cubical chamber with an uncovered vent opening. The vessel volume is $V = 50 \text{ m}^3$ and vent area is $A_v = 10$ m^2, while the propane (C_3H_8) is mixed at five percent volume ratio with the air. The vent is sharp-edged, with a discharge coefficient $K_D = 0.6$. Standard atmosphere conditions exist with $p_o = 1.013 \times 10^5$ Pa. What is the maximum gas explosion pressure?

Since the chamber is cubical, the internal surface area is $6(\sqrt[3]{50})^2$ $= 81.431 \text{ m}^2 = A_s$. From Equation (3-44),

$$\bar{A}_v = \frac{A_v K_d}{A_s} = \frac{(10)(0.6)}{81.431} = 0.07368$$

$$\frac{\bar{A}_v}{\bar{S}_o} = \frac{0.07368}{9.2 \times 10^{-3}} = 8.01$$

270

From Figure 3-22, an $\bar{A}_v/\bar{S}_o$ of 8.01 gives a $\Delta\bar{P}_m$ of 0.01 which, in turn, gives

$$\Delta P_m = (\Delta\bar{P}_m)\ (p_o)$$

$$= (0.01)\ (1.013 \times 10^5)\ Pa$$

$$= (1.013\ kPa)$$

Maximum Gas Explosion Pressure for Covered Vents

What is the maximum gas explosion pressure for the same vessel and same conditions as the previous problem if the vent is covered?

From Figure 3-23, an $\bar{A}_v/\bar{S}_o$ of 8.01 gives a $\Delta\bar{P}$ of 0.18 which, in turn, gives

$$\Delta P = (\Delta\bar{P})\ (p_o)$$

$$= (0.18)\ (1.013 \times 10^5)\ Pa$$

$$= (18.2\ kPa)$$

CHAPTER 3

LIST OF SYMBOLS

A	vent area
$\bar{A}$	dimensionless vent area ratio
A_s	internal surface area of vented chamber
a_o	ambient sound velocity
b	dimensionless time constant (initial decay rate)

C_D	drag coefficient
c	decay rate for gas venting
D	internal diameter of a pipe
E	total energy released by an explosion
f	Fanning friction factor
G	$= \rho u$, the mass flow through a pipe
h	enthalpy
i_g	gas specific impulse
i_r	reflected specific impulse
i_s	side-on specific impulse
K	constant related to type of combustible gas or dust
K_d	discharge coefficient for an orifice
K_{ST}	Bartknecht parameter for a combustible dust
M_A	mass of air
M_E	mass of explosive
M_T	mass of explosive plus engulfed air
P_m	maximum pressure within a vessel for an internal gas or dust explosion
P_r	peak reflected overpressure
$\bar{P}_r, \bar{P}_s, \bar{I}_r$, etc.	barred quantities are nondimensional groups corresponding to indicated dimensional quantities
P_1, P_2	pressures in fluid flow in a pipe
P	pressure
P_o	ambient pressure
P_r	pressure in reflected blast wave (absolute)
Q	peak drag pressure
R	distance from charge center
$\bar{S}_o$	dimensionless burning velocity
T, T_s	duration of positive phase of side-on blast wave
T_r	duration of positive phase of reflected overpressure
T_1, T_2, T_3	times describing diffracted shock loading

t	time
t_a	shock wave arrival time
t_r	reverberation time of reflected shock waves in a chamber
U	velocity of free-running shock
U_r	velocity of reflected shock
u	flow velocity
u_r	peak particle velocity behind reflected shock
u_s	peak particle velocity behind free-running shock
V	chamber or enclosure volume
W	explosive charge weight
α_e	effective vent area ratio
α_I	angle of shock incidence on a reflecting surface
α_R	angle of shock reflection from a reflecting surface
ΔL	short length of pipe
γ	ratio of specific heats
θ_r	peak absolute temperature behind reflected shock
θ_s	peak absolute temperature behind free-running shock
ξ	inverse of shock strength
ρ	fluid density
ρ_m	fluid density in pipe flow
ρ_o	ambient density
ρ_r	peak density behind reflected shock
ρ_s	peak density behind free-running shock
ρu	mass flow through a pipe
ρ_1, ρ_2	fluid densities in pipe flow
τ, τ_{max}	venting times

CHAPTER 4

STRUCTURAL RESPONSE: SIMPLIFIED ANALYSIS TECHNIQUES

In this chapter we will introduce the concepts of energy solutions and pressure-impulse diagrams. We will then use these simplified analytic tools to determine stresses, strains, and deflections in dynamically loaded structural elements.

One needs to know stresses, strains, and deflections in structural components to determine if buildings and other structures are damaged by some loading such as a blast load. All of the approaches used in this chapter give final states and not time histories, which is of most interest to designers who wish to know maximum stresses and deflections. The approaches are simplified ones, and have the advantage that results can be presented as nondimensional graphs. The existence of different domains for the response of structures will become apparent. Some parameters do not affect structural response in certain limiting domains. In addition, the effect of increase or decrease in all parameters on the stresses or deflections in a structure can be determined from simplified graphical computations whenever these simplified solutions are used.

Naturally, the solutions presented in this chapter involve engineering approximations and simplified representations of structural components of buildings. We will derive each approach and procedure so assumptions are understood and so the techniques can be extended to other applications. Whenever transient solutions or more complex structural computations using numerical techniques are required, Chapter 5 should be used.

AMPLIFICATION FACTORS

Sinusoidal Loading

The most basic dynamic system, usually studied in the first sessions of an engineering dynamics course, consists of a mass on a spring. In en-

274

gineering jargon, this is called a linear elastic, single-degree-of-freedom

system. Such a system is shown excited with a sinusoidal forcing function

in the insert to Figure 4-1. By substituting into Newton's second law (F =

ma), and solving the resulting differential equation, a solution for the

maximum deformation can easily be developed. It is presented graphically

in Figure 4-1. Basically, this figure is a plot of the maximum deflection

divided by the static deformation presented as a function of frequency ratio,

i.e., the frequency of excitation, $2\pi/T$, divided by the natural frequency, ω,

of the responding structure, $\sqrt{\frac{k}{m}}$. The ordinate is also called the dynamic

amplification factor.

The results presented in Figure 4-1 can be replotted by inverting the

abscissa and multiplying by 2π. Now, the results are presented as a function

of a duration ratio, the duration of loading, T, divided by the period of the

responding structure, $\sqrt{\frac{m}{k}}$. Figure 4-2 is this plot for sinusoidal excita-

tions. Resonance can still be seen. We cast these known results in this

format because of certain similarities and differences which can then be seen

in linear oscillators loaded with harmonic motion and by impact (or other

transient pulses).

Blast Loading

The problem of greater interest is a single-degree-of-freedom, elas-

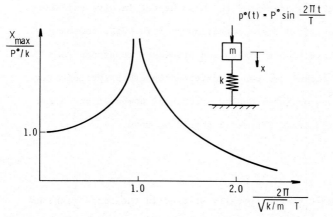

Figure 4-1. Dynamic Amplification Factor for Sinusoidal Excitation

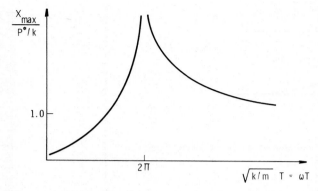

Figure 4-2. Amplification Factor Expressed as Duration Ratio

tic oscillator loaded with a blast wave or impact excitation, as in Figure
4-3. We will assume that the exponentially decaying forcing function seen
in Figure 4-3 is a mathematical approximation to an air blast wave. If one
takes the time integral for p*(t), the area under the curve, the total applied impulse, I, is obtained which equals P*T. This quantity called impulse is important to us in subsequent discussions. Ordinarily, the quantity T is located where half the impulse falls between time equal to zero
and T, and the other half of the impulse falls between time T and infinity.
This quirk arises mathematically because the exponential decay never does
return to a zero pressure; hence, no definable duration of loading exists.

For the system shown in Figure 4-3, the equation of motion (Newton's
Second Law) is

$$m \frac{d^2 x}{dt^2} + k\, x = P*\, e^{-t/T} \qquad (4\text{-}1)$$

For the initial conditions of no displacement and no velocity at time t =
0, the transient dynamic solution is given by:

$$\frac{x(t)}{(P*/k)} = \frac{(\omega T)^2}{[1 + (\omega T)^2]} \left[\frac{\sin \omega t}{\omega T} - \cos \omega t + e^{-\frac{(\omega t)}{(\omega T)}} \right] \qquad (4\text{-}2)$$

276

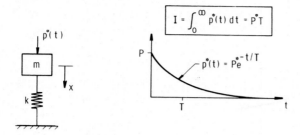

Figure 4-3. Linear Oscillator Loaded by a Blast Wave

where $\omega = \sqrt{\dfrac{k}{m}}$. Equation (4-2) involves three nondimensional numbers. In functional format it can be expressed as:

$$\frac{x(t)}{(P*/k)} = \psi\ [\omega T,\ \omega t] \tag{4-3}$$

If maximum motion is our only interest rather than transient solutions, Equation (4-2) must be differentiated with respect to time t, and the resulting velocity set equal to zero to obtain the scaled time (ωt_{max}) when x(t) is a maximum. This scaled time is given by:

$$0 = \frac{\cos \omega t_{max}}{\omega T} + \sin \omega t_{max} - \frac{e^{-\frac{\omega t_{max}}{\omega T}}}{\omega T} \tag{4-4}$$

Equation (4-4) in functional format is expressed by:

$$\omega t_{max} = \psi\ [\omega T] \tag{4-5}$$

Unfortunately, Equation (4-4) is a transcendental equation and cannot be solved explicitly for ωt_{max}; however, it can be solved by trial and error. Once ωt_{max} is obtained for specific values of ωT, then ωt_{max} can be substituted into Equation (4-2) to obtain $\dfrac{X_{max}}{P*/k}$ as a function of ωT. In functional format, the transient solution given by Equation (4-2) is reduced to:

$$\frac{X_{max}}{P*/k} = \psi \ (\omega T) \qquad\qquad (4\text{-}6)$$

The trial and error solution to Equation (4-6), or Equations (4-2) and (4-4) if you prefer, is shown by the heavy solid continuous line in Figure 4-4.

Figure 4-4 for a blast-loaded elastic oscillator is the counterpart to Figure 4-2 for the more familiar sinusoidally driven excitation. The ordinates and abscissi in both curves are identical; however, the shapes of the curves differ considerably. To understand how blast-loaded structures behave, we must study the meaning of results presented in Figure 4-4.

Notice that the solution presented in Figure 4-4 can be approximated by two straight line asymptotes, which are the light lines in the figure. For very large values of $\sqrt{\frac{k}{m}}\, T$ (greater than 40) and for very small values of $\sqrt{\frac{k}{m}}\, T$ (less than 0.4), these asymptotes are very accurate approximations to the general solution. The asymptotes can be computed by appropriate

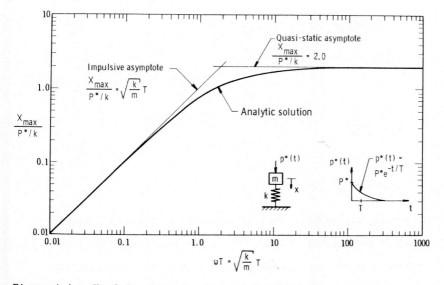

Figure 4-4. Shock Response for Blast-Loaded Elastic Oscillator

energy balance equations, which will be discussed later in this section.
For purposes of discussion, three different loading realms can be identified. The realm where $\sqrt{\frac{k}{m}}$ T is greater than 40, will be called the quasi-
static loading realm. In this domain, the maximum dynamic deflection is
twice the static deflection; hence, the word quasi-static. The origins of
a dynamic load factor of 2.0 used in many civil engineering problems and
codes comes from the response of structures to loads in this quasi-static
loading realm. The criterion of $\sqrt{\frac{k}{m}}$ T greater than 40 means that the duration of loading T relative to the period $\sqrt{\frac{m}{k}}$ of the loading structure is
very large. In other words, the load dissipates very little before the
maximum deformation is achieved. In this quasi-static loading realm, the
deformation depends only on the peak load P and the structural stiffness k.
The response is independent of the duration of loading T and the mass of
the structure m. As the time ratio $\sqrt{\frac{k}{m}}$ T becomes smaller, however, a dynamic load factor becomes less and less applicable, until finally other approaches are required.

For values of $\sqrt{\frac{k}{m}}$ T less than 0.4, the ordinate and abscissa in Figure 4-4 are equal. So, for small values of ωT:

$$\frac{X_{max}}{P*/k} = \sqrt{\frac{k}{m}}\ T \quad (\omega T \leq 0.4) \tag{4-7}$$

or

$$\frac{\sqrt{km}\ X_{max}}{I} = 1.0 \quad (\omega T \leq 0.4) \tag{4-8}$$

where I = P*T.

In this domain, the deformation is directly proportional to the product (P*T), which is the impulse I; hence, the expression "impulsive loading realm." Notice that the deformation depends only on the product (P*T)
or area under the load history in the impulsive loading realm. Any combina-

tion of peak loads and durations with the same impulse will result in the same maximum deformation in this domain. Now, both structural stiffness and structural mass influence the results. A dynamic load factor of 2 would be very conservative in this impulsive loading realm. Biggs' or Newmark's method of analysis, which is presented in Chapter 5, uses the concept of dynamic load factor in this domain, but with this factor equal to (ωT). The significance of small values of ωT, which is the criterion for this realm, is that the load is imparted to the structure and removed before the structure has had adequate time for undergoing significant deformation. In other words, the duration of loading is short relative to the response time of the structure in the impulsive loading realm.

Finally, a third loading realm exists for values of ωT between 0.4 and 40 which is a transition realm connecting the impulsive loading realm to the quasi-static realm. This realm will be termed the "dynamic loading realm" because the deformation here depends upon the entire load history. Unfortunately, no approximate idealizations can be applied in this domain. Here motion depends upon both pressure and impulse as well as structural stiffness and mass. Actually, computation of the two asymptotes, quasi-static and impulsive, yield an approximation to the entire shock response, Figure 4-4. One can use a French curve and the knowledge that the solution at the intersection of the asymptotes is approximately twice as great as the actual value of $X_{max}/(P*/k)$ to draw an approximate transition solution. In the dynamic loading realm, the duration of the applied load is of the same order of magnitude as the response time of the structure. Structural response solutions to transient pulses are analytically more complex in this, the dynamic loading realm.

Several analytical observations should be made that greatly ease computations when maximum structural deformations or stresses are to be computed in either the impulsive loading realm or quasi-static loading realm. The asymptotes for maximum deformation are easily computed using

energy procedures. The strain energy S.E. in this linear elastic system
is given by:

$$S.E. = (1/2) \ k \ X_{max}^{2} \qquad\qquad (4-9)$$

The maximum possible work Wk which could be imparted to the structure by a
constant force whose amplitude decreases insignificantly is:

$$Wk = P*X_{max} \qquad\qquad (4-10)$$

Equating Wk to S.E. yields the asymptote for the quasi-static loading realm.

$$P*X_{max} = (1/2) \ k \ X_{max}^{2} \qquad\qquad (4-11)$$

or

$$\frac{X_{max}}{P*/k} = 2.0 \qquad \text{(quasi-static asymptote)} \qquad\qquad (4-12)$$

Remember that the significance of the quasi-static loading realm is that
load durations are long with only small amounts of load dissipation before
the structure reaches its maximum deformation. This reasoning explains why
an upper bound to the maximum possible work imparted to the structure yields
excellent results when equated to the maximum structural strain energy that
must absorb this work. This principle of equating an estimated work to
strain energy is a key procedure in subsequent structural discussions, especially those that use energy balance procedures.

To compute the impulsive loading realm asymptote, we must estimate
the initial kinetic energy KE imparted to the structure. Remember that the
criterion for the impulsive loading realm is a loading duration so short

that little structural deformation occurs before the load is over. This
means that at time zero, an initial velocity is imparted to the structure
equal to $\frac{I}{m}$. The kinetic energy, K.E., associated with this velocity at
time zero, when no strain energy is stored in the structure, equals:

$$\text{K.E.} = (m/2) \; \frac{I}{m}^2 = \frac{I^2}{2m} \qquad (4\text{-}13)$$

This kinetic energy at time zero will eventually be absorbed as strain energy, which is given by Equation (4-9). Equating K.E. to S.E. gives the
asymptote for the impulsive loading realm.

$$\frac{I^2}{2m} = (1/2) \; k \; X_{max}^2 \qquad (4\text{-}14)$$

or

$$\frac{\sqrt{km} \; X_{max}}{I} = 1.0 \quad (\text{impulsive asymptote}) \qquad (4\text{-}15)$$

This principle of equating kinetic energy to strain energy for an estimate
of the impulsive loading realm asymptote is an equally important procedure
which will be used in subsequent structural discussions using energy balance procedures.

P*-I DIAGRAMS FOR IDEAL BLAST SOURCES

Elastic System

The graphical solution presented in Figure 4-4 is often replotted in
what is called a Pressure-Impulse or P*-I diagram. Figure 4-5 is this P*-I
diagram for Figure 4-4. Figure 4-5 is obtained by first inverting the ordinate in Figure 4-4 and replotting it as the new ordinate in Figure 4-5. Next,

282

the old abscissa in Figure 4-4 is multiplied by the new ordinate in Figure

4-5 to obtain the product (P*T) or I. This new scaled impulse is plotted as

the abscissa in Figure 4-5.[+]

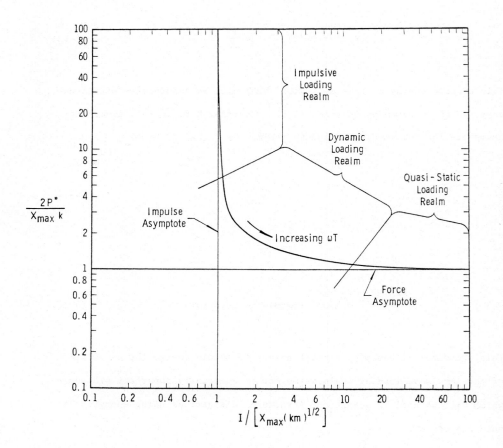

Figure 4-5. P*-I Diagram for Blast-Loaded Elastic Oscillator

[+]This figure is, in fact, a <u>force</u>-impulse, rather than a pressure-impulse

diagram. Later, in specific applications for blast-loaded structures, we

will use the symbols P and p to denote <u>pressures</u>, rather than the <u>forces</u>

used in this discussion, which are denoted as P* and p*.

The information contained in a P*-I diagram, Figure 4-5, is exactly the same as that presented in the shock amplification response curve, Figure 4-4. The major difference is that the shock amplification response plot emphasizes deformation or stress as functions of scaled time; whereas, the P*-I diagram emphasizes the combination of applied load and impulse for defining the threshold of damage.

Once some value of X_{max} is specified which constitutes the threshold of damage in a specific structure (specified values of k and m), then the rectangular hyperbola shaped curve in Figure 4-5 is an isodamage curve. This isodamage curve defines what combined values of applied load P* and impulse I yield this specified deformation. If larger loads and impulses are imparted in the region above and to the right of the curve in Figure 4-5, the structure will be damaged because deformations will be larger than threshold ones. Should smaller loads and impulses be imparted in the region below and to the left of the curve in Figure 4-5, the structure should be undamaged.

The three previously defined loading realms are still apparent in Figure 4-5. The vertical asymptote to Figure 4-5 is the impulsive loading realm asymptote; changes in the applied impulse I are required to depart from an isodamage contour, and changes in applied load P remain on the same isodamage contour. Similarly, the horizontal asymptote is the quasi-static loading realm asymptote, as changes in applied load only are required to move away from an isodamage contour in this domain.

Rigid-Plastic System

So far, our discussion has centered on shock response and P*-I diagrams for a linear elastic, single-degree-of-freedom system. Very similar curves also exist for structures undergoing plastic deformation. Figure 4-6 is an analytically derived P*-I diagram for the rigid, perfectly plastic, single-degree-of-freedom, Coulomb element in the insert to this figure [Baker, Westine and Dodge (1973)]. The same exponentially decaying forcing function is imparted to this system as to the previous elastic ex-

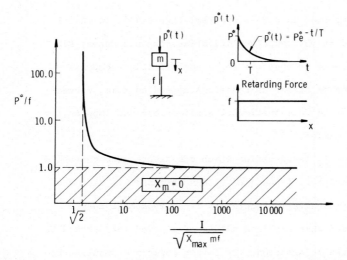

Figure 4-6. P*-I Diagram for Blast-Loaded Rigid-Plastic System

ample. The only difference is that a Coulomb friction element with retarding force f replaces the elastic spring in the previous example. Provided
the peak applied load P* exceeds the retarding force f, the system breaks
free and displaces. If P* does not exceed f, the system never moves. The
heavy curved line in Figure 4-6 is the analytical solution to this rigid-
plastic system. Notice that the nondimensional P*-I diagram still has the

general shape as the P*-I diagram for the elastic system. The existence of
all three loading realms is apparent. The impulsive loading realm asymptote
is still a vertical line, and the quasi-static loading realm asymptote is
still a horizontal line as in the previous example.

The same principles of equating the maximum possible work to the
strain energy for determining the quasi-static asymptote, and of equating
the kinetic energy to the strain energy for the impulsive loading realm
asymptote still apply. For this plastic system, the maximum possible work
Wk is given by:

$$Wk = P*X_{max} \qquad (4\text{-}16)$$

and the strain energy is given by:

$$S.E. = f\ X_{max} \qquad (4\text{-}17)$$

Equating the maximum work to the strain energy gives:

$$P*X_{max} = f\ X_{max} \qquad (4\text{-}18)$$

or

$$P*/f = 1.0 \quad (\text{quasi-static asymptote}) \qquad (4\text{-}19)$$

This asymptote is the same one given in Figure 4-6. Similarly, the impulsive loading realm asymptote is computed by calculating the kinetic energy K.E.

$$K.E. = 1/2m\left(\frac{I}{m}\right)^2 = \frac{I^2}{2m} \qquad (4\text{-}20)$$

while when equated to the strain energy, yields:

$$\frac{I^2}{2m} = f\ X_{max} \qquad (4\text{-}21)$$

or

$$\frac{I}{\sqrt{X_{max}}\ m\ f} = \sqrt{2} \quad (\text{impulsive asymptote}) \qquad (4\text{-}22)$$

This asymptote is also the same one given for the impulsive loading realm in Figure 4-6.

We have already stated that these principles of equating the maximum possible work to the strain energy to obtain the quasi-static asymptote, and of equating the kinetic energy to the strain energy to obtain the impulsive loading realm asymptote are very important. These principles apply whether elastic or plastic behavior is being studied. They will be used repeatedly as a first principle later in this chapter when energy procedures are used to obtain solutions for beams, plates, and columns.

Scatter in Experimental Results

When plastic deformation occurs, large experimental scatter can be observed in measured deformations on supposedly identical structures from supposedly identical tests. This phenomenon can be partially explained by returning to the rigid-plastic $P*$-I diagram shown in Figure 4-6. Note that the quasi-static loading realm in this plastic system has an asymptote at $P*/f = 1.0$. But $P*/f = 1.0$ is also the asymptote for no deflection at all, because the mass will not move unless $P*$ exceeds f. So, for $P*/f = 1.0^-$, no displacement occurs, but for $P*/f = 1.0^+$, large displacements can occur. (The large displacements at $P*/f = 1.0^+$ are unlimited ones because this quasi-static asymptote is independent of X_{max}. A bifurcation as in a buckling problem is an example.) Such a model indicates that large scatter will result in experimentally measured displacements in the quasi-static loading realm for a plastic system. The impulsive loading realm, on the other hand, is much better conditioned and should exhibit less scatter because its asymptote is a function of X_{max}.

Figure 4-7 shows dimensionless residual tip deformations for air blast loaded cantilever Inconel X beams. Everything was held constant in these tests except the standoff distances from an explosive charge. As can be seen in Figure 4-7, the deformations range from very small to very large (a δ/L of 1.0 corresponds to a beam being bent over flat with the ground). Usually, six beams were tested at any one standoff distance and the scatter (the ratio between the minimum and maximum deflections) was as large as a factor of 7.0. These tests are plotted on log-log paper because of the scatter. Although the results are not truly in the quasi-static loading

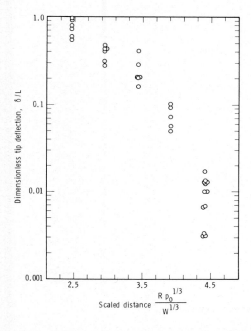

Figure 4-7. Damage to Inconel-X Cantilever Beams,
Dynamic Loading Realm

realm, they are in the dynamic loading realm or elbow of the curve in Figure 4-6. Complex computational procedures seeking great accuracy are not
worth the effort when results are found to scatter as in Figure 4-7.

In contrast to Figure 4-7, Figure 4-8 is the residual mid-span deformation for explosively loaded simply supported steel beams in the impulsive
loading realm. Everything in these tests was held constant except for the
applied specific impulse i^+. As can be seen in Figure 4-8, scatter was always less than a factor of 30 percent and these impulsively loaded beams can
have their deformations plotted on geometric paper rather than log paper.
The differences in boundary conditions or material types have nothing to do
with differences in scatter when Figures 4-7 and 4-8 are compared. In the

---

[+]The specific impulse i is the integral of a pressure-time function, rather
than a force-time function.

288

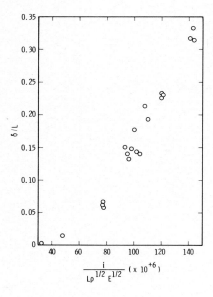

Figure 4-8. Damage to Simply Supported Steel Beams,
Impulsive Loading Realm

impulsive loading realm, plastic beam deformations are much better conditioned than beam deformations near the quasi-static loading realm asymptote.
The calculations presented graphically in Figure 4-6 qualitatively predict
this behavior. In reality, perfect plasticity does not exist; hence, finite
rather than infinite deformations have to exist even in the quasi-static
loading realm.

The _elastic_ model presented in Figure 4-5 exhibits no such ill-condi-
tioned structural response behavior because both asymptotes are a function of
X_{max}. One should also note that although displacement is poorly conditioned
in the quasi-static loading realm for a plastic system, the yield load (or
force) is sharply defined. Hence, although deformations can scatter, the
threshold loading level for initiating damage does not scatter appreciably.

The purpose of this discussion is to show that scatter can occur, and
this scatter may not be the cause of poor experimental technique but instead
results from certain physical phenomena, such as the "go - no go" condition

of the rigid perfectly plastic problem. Analytically, one can anticipate
this "ill-conditioning" of the displacement because, as mentioned previously, the one asymptote is independent of displacement. In addition, although
one experimentally measured quantity can scatter greatly (such as deformation), other experimentally measured quantities (such as yield force) do not
necessarily scatter as greatly. Quite often, for a plastically deforming,
blast-loaded structure, complex computational procedures to determine the
level of damage are not worth the effort because increases in computational
accuracy may be meaningless. As we have seen, scatter can be quite large in
experiments conducted in one loading realm, but much less pronounced for experiments conducted in another loading realm.

Experimentally Obtained P*-I Diagrams[†]

This concept of a P*-I diagram (or, more often, a P-i diagram) is
more than a mathematical one. Experimental test results obtained in a variety of structures can be plotted on a pressure-impulse diagram, and will result in approximate rectangular hyperbolas for some threshold or level of
damage. One of the most extensive data bases for plotting a P-i diagram
involves blast damage to typical homes and factory buildings, and comes from
bomb damage in Britain during World War II. Jarrett (1968) curve fitted an
equation with the format:

$$R = \frac{K \, W^{1/3}}{\left[1 + \left(\frac{7000}{W} \right)^2 \right]^{1/6}} \qquad (4\text{-}23)$$

for brick houses with constant levels of damage to relative explosive charge
weight W and standoff distance R. The constant K in Equation (4-23) changes
with various levels of damage. Equation (4-23) can be converted to a P-i
diagram because side-on blast wave overpressure and side-on specific impulse

[†]These diagrams appear often with impulse as the ordinate and pressure as
the abscissa. The first curve discussed here is an example.

290

at sea level ambient atmospheric conditions can be calculated (see Chapter 2) from R and W. Figure 4-9 is a side-on impulse versus overpressure diagram using Jarrett's curve fit to bomb damage houses to establish isodamage contours.

The existence of both the quasi-static and impulsive loading realms is apparent for the three different isodamage contours shown in this figure. Levels of damage increase as pressures and impulses increase in Figure 4-9. The British use this empirically obtained P-i diagram for brick houses to determine damage criteria for other homes, small office buildings, and light framed industrial buildings under the supposition that these other structures are similar.

Jarrett's Equation (4-23) or Figure 4-9, which is equivalent, is the basis for explosive safe standoff criteria used in the United Kingdom. The curves in Figure 4-9 are more useful than the equation for establishing damage thresholds in industrial explosions, because the curves do not depend on the concept of "TNT equivalence" under sea level ambient atmospheric conditions. When presented in the format of P versus i, the receiver is characterized, and the character of the source plus transmission of the

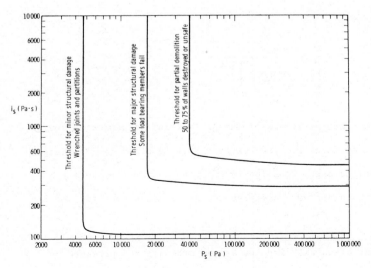

Figure 4-9. Impulse Versus Pressure Diagram for Constant Levels of Building Damage

shock is omitted from this representation. For example, as presented in Figure 4-9, the pressures and impulses can be the result of an explosion of a line source, an area source, or from numerous different nonideal energy releases at sea level or at various altitudes.

Throughout this book, the reader will find many different effects of blast on buildings and people presented in P-i diagrams. People respond to blast as a complex mechanical system, so being able to show primary blast lethality or loss of hearing as P-i diagrams should cause no surprise. This concept of a P-i diagram is an extremely important one and will be used extensively throughout the book so the reader should be certain that he understands this concept.

P*-I DIAGRAMS FOR NONIDEAL EXPLOSIONS

The loadings on a structure or structural element caused by an internal gas or dust explosion are, in many cases, different from the loadings produced by condensed explosives. Overpressures produced by condensed explosives are characterized by very short rise times and nearly exponential decay. For vapors or dust explosions, the loading is characterized by a finite rise time and a different time variation. In fact, many researchers performing gas or dust explosion experiments give as their results the peak

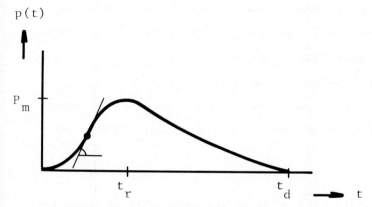

Figure 4-10. Schematic Overpressure-Time History in
Confined Gas or Dust Explosions

292

value of the overpressure, P_m, a maximum pressure rate, $\dfrac{dp_{max}}{dt_{max}}$, and, in case of venting, some duration, t_d, (see Figure 4-10).

This nonideal explosive loading can be described approximately by the following formula:

$$p(t) = P_m \left[t/t_r - \frac{1}{2\pi} \sin \frac{2\pi t}{t_r} \right] \qquad\qquad t \leq t_r \qquad\qquad (4\text{-}24)$$

$$= P_m \left(1 - \frac{t - t_r}{t_d - t_r} \right) e^{-\left(\frac{t - t_r}{t_d - t_r} \right)} \qquad\qquad t_r \leq t \leq t_d \qquad\qquad (4\text{-}25)$$

The maximum rate of pressure rise defined as the slope at $t = t_r/2$ will then be:

$$\frac{dp}{dt} \left(t = \frac{t_r}{2} \right) = \frac{2P_m}{t_r} \qquad\qquad (4\text{-}26)$$

For this loading, the influences of the finite rise time and the shape of the loading on structural response can be evaluated using a one-degree-of-freedom, elastic-plastic, spring-mass system.[†] The calculated elastic P*-I diagram representing a dust explosion, curve B, is shown in Figure 4-11 for a time ratio, t_r/t_d, equal to 0.4. For comparison, the P-I curve for an ideal explosion is shown as curve A.

Two important differences can be seen in Figure 4-11:

(1) The quasi-static loading realm pressure asymptote for a finite rise time loading is 1.0 (equivalent to a static loading); whereas, the

---

[†]Note that, again, we use force P*, rather than pressure P, when analyzing a simple mechanical system.

293

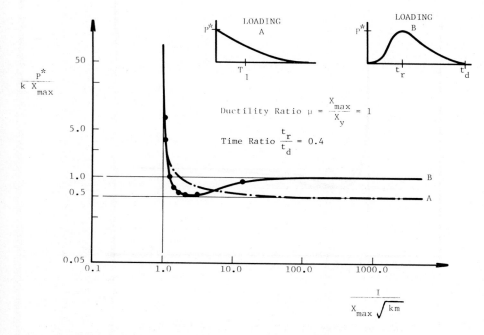

Figure 4-11. Comparison of Zero and Finite Rise Times
P*-I Diagrams

pressure asymptote for a loading with zero rise time is 0.50 (equivalent
to a dynamic load factor of 2.0).

(2) In the region $1.15 < \dfrac{I}{X_{max}\sqrt{km}} < 5.5$, the loading with finite
rise time is more severe than the loading with zero rise time. This behavior is produced by resonance between the loading rate and the structural frequency.

In the impulsive loading realm, both loadings have the same asymptote.

For nonideal explosions, results differ only slightly if the ratio
of t_r/t_d is shifted within reasonable limits. This is shown for an elastic system in Figure 4-12, where the curve in Figure 4-11 ($t_r/t_d = 0.40$)
is compared to a loading width of $t_r/t_d = 0.20$.

The influence of adding plasticity to the oscillator is shown in Figure 4-13. In this comparison, the loading is identical in both cases, a
gradual rising pressure pulse with a t_r/t_d ratio of 0.4. The oscillator re-

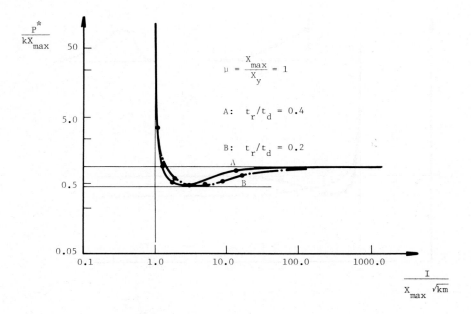

Figure 4-12.   Effect of Change in t_r/t_d for Finite Rise Time
Loading on P*-I Diagrams

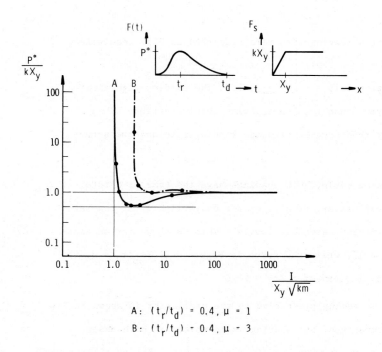

A: $(t_r/t_d) = 0.4$, $\mu = 1$
B: $(t_r/t_d) = 0.4$, $\mu = 3$

Figure 4-13.   Effect of Plasticity on P*-I Diagrams

sponds in an elastic, perfectly plastic manner as shown in the insert to

Figure 4-13. One case, Case A, is the threshold for the beginning of plas-

ticity; whereas, the other case, Case B, is allowed to deform plastically

until the ductility ratio μ (the ratio of maximum deformation divided by

the yield deformation) equals 3.0.[+] The effects of plasticity on the P*-I

diagram are: 1) to damp the dynamic overshoot which occurs elastically in

the elbow of the curve, and 2) to shift the threshold curve to the right

in the impulsive loading realm. As had already been noted when Figure 4-6

was discussed, no changes occur for plastic thresholds in the quasi-static

loading realm. The results in Figure 4-13 have not been expressed relative

to the yield deformation X_Y, a constant in both systems; whereas, X_{max} dif-

fers. The shift seen in the impulsive loading realm occurs because more

energy can be absorbed whenever the structure deforms plastically. In the

quasi-static loading realm, the work applied to the structure and the strain

energy being absorbed both increase linearly; hence, the net result is no

change in the quasi-static threshold.

A pressure vessel burst is another nonideal blast loading which dif-

fers from the loadings associated with a dust or gaseous vapor explosion or

a high explosive. The two principle differences between a blast from a

pressure vessel burst and a high explosive are: 1) a large amplitude, long

duration negative pressure (called the negative phase) occurs in vessel

bursts, and 2) there is a significant second positive phase in vessel bursts

(see Chapter 2). Figure 4-14 gives a recorded pressure history from a burst-

ing pressurized glass sphere.

In order to investigate the influence of a loading with a pressure-

time history shown in Figure 4-14, the elastic oscillator was excited (math-

ematically) with the forcing function depicted schematically in Figure 4-15.

To denote this function, we must now speak of three different times, shown

as t_{d1}, t_{d2}, and t_{d3}, all measured in Figure 4-15 from time zero. By de-

---

[+]Ductility ratio will be discussed further in Chapters 5 and 8.

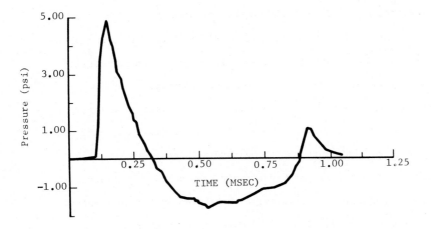

Figure 4-14. Recorded Pressure-Time History from a Bursting Sphere
[Esparza and Baker (1977b)]

$$t \le t_{d1}: \quad p(t) = P_m\left(1 - \frac{t}{t_{d1}}\right)$$

$$t_{d1} < t < t_{d2}: \quad p(t) = \gamma \, P_m \, \sin\left[\left(\frac{t - t_{d1}}{t_{d2} - t_{d1}}\right)\right] \pi$$

$$t_{d2} < t \le t_{d3}: \quad p(t) = \delta \, P_m\left(1 - \frac{t - t_{d2}}{t_{d3} - t_{d2}}\right)$$

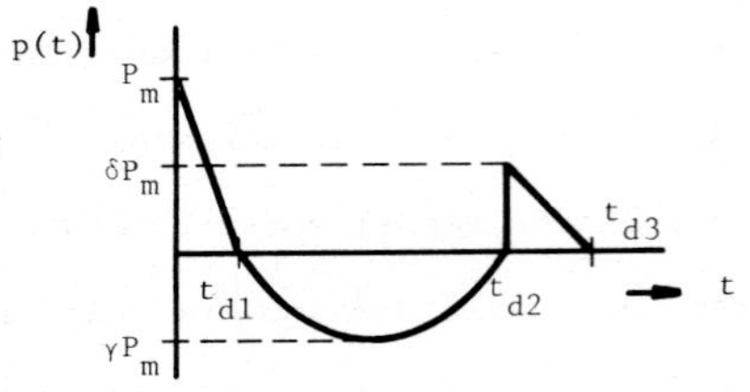

where $t_{d1} = \alpha \, t_{d3}$ and $t_{d2} = \beta \, t_{d3}$

Figure 4-15. Schematic of Blast Loading from Vessel Burst

fining time ratios α and β, plus relative amplitude ratios γ and δ as in
Figure 4-15, the pressure curve of Figure 4-14 was approximated reasonably
well with $\alpha = 0.24$, $\beta = 0.76$, $\gamma = -0.85$, and $\delta = 0.8$.

The response of an elastic oscillator system to a pressure vessel
burst is compared to that of a condensed explosive in Figure 4-16. The
impulse I is the total impulse with the negative phase subtracted from
that of the two positive phases. No major differences are found in either

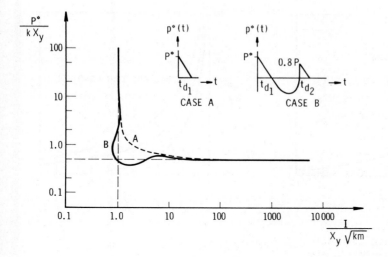

Figure 4-16. Effects of Pressure Vessel Burst on P*-I Diagram

the quasi-static or impulsive loading realms; however, in the dynamic loading realm, major differences occur. Although these influences as seen in
Figure 4-16 are for specific values of α, β, γ, and δ, in general the irregular shape in the elbow of the P*-I diagram is caused by phasing between
the response time and various loading durations. Only for extremes in the impulsive loading realm or quasi-static loading realm can the results of a
vessel burst be accurately predicted by a single triangular pressure pulse.

The last subject to be discussed about P*-I diagrams is the influence
on structural damage thresholds of the pulse shape after the pressure rise.
For either simple elastic or rigid-plastic systems, Abrahamson and Lindberg
(1976) computed the results for rectangular, triangular, and exponential
shaped pulses with zero rise times. Figure 4-17 shows a diagram for the
elastic system, and Figure 4-18 for the plastic one. Both curves show that
the quasi-static and impulsive loading realms are relatively unaffected for
the same critical levels of displacement. The major differences occur in
the elbows of these figures where the more rectangular loadings have a more
abrupt transition from one asymptote to the other.

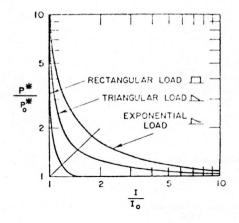
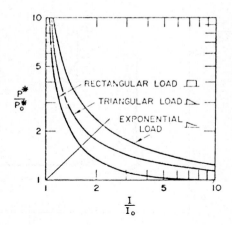

Figure 4-17. Critical Load Curves for an Elastic
Spring-Mass System
[Abrahamson and Lindberg (1976)]

Figure 4-18. Critical Load Curves for the Simple
Rigid-Plastic System
[Abrahamson and Lindberg (1976)]

ADDITIONAL P*-I DIAGRAM DETAILS

A popular method of using P-i diagrams is in conjunction with distance versus explosive charge weight (R-W) curves as overlays. The P-i
diagram defines a target's or structure's susceptibility to air blast.
Under sea-level ambient atmospheric conditions, the charge weight W and
standoff distance R uniquely determine, along with structural orientation
and geometry, the pressure and impulse imparted to a target. By creating
an overlay as illustrated conceptually in Figure 4-19, one graphically determines all combinations of energy releases and standoff distances that
are on the threshold for critically damaging some structure. In this illustrative example, by reading the results from Figure 4-19, it can be
seen that 1/8 kg of TNT at 0.43 m, 1/4 kg at 0.85 m, 1/2 kg at 1.05 m, 1
kg at 2.0 m, and 2 kg of TNT at 3.5 m all yield the same damage threshold
for the structure represented by the isodamage curve.

299

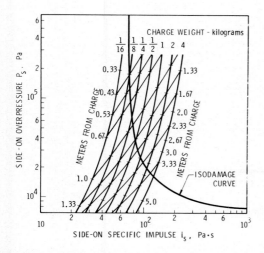

Figure 4-19.   Illustration of Overlays to a P*-i Diagram.
Incident (Side-On) Overpressure and Impulse From
Pentolite Spheres

One word of warning is in order about the use of P-i diagrams. Any
one diagram applies only to some single given mode of damage. For example,
the tail boom of a helicopter might fail in a gross overall bending mode
when large charges are located far away; however, this same tail boom might
experience localized panel failure when small explosive bursts occur close
to the structure. This example suggests that two, or conceivably more,
modes of failure can occur if large ranges in effects are being studied or
different damage mechanisms come into play. Each of the separate modes or
mechanisms of failure will have its own individual P-i diagram. The survivability of a complex target might have to be established by plotting
several P-i diagrams as in Figure 4-20. When such a composite plot is made,
the threshold of damage for all modes becomes the dashed line in Figure 4-20,
as the analyst usually wishes to predict or protect against all modes of
damage.

In the next chapter on numerical techniques, we will present Bigg's
approach, which is nothing more than a single-degree-of-freedom, elastic-

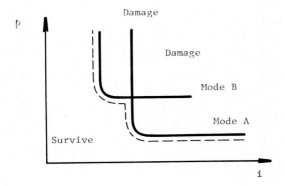

Figure 4-20.   P-i Diagram for Complex Targets

plastic oscillator loaded with various pulses. A figure, such as Figure 5-19, which is Bigg's solution, is an amplification diagram which corresponds exactly with Figure 4-4, our elastic solution. The quantity $\frac{Y_m}{Y_{el}}$ in Bigg's solution is our quantity $\frac{X_{max}}{P*/k}$ and $\frac{t_d}{T_n}$ is our $\sqrt{\frac{k}{m}}\,T$. A series of contours must be used in Figure 5-19 because plasticity as well as elasticity is included in Bigg's solution. The quantity $\frac{R_m}{F_1}$ in Figure 5-19 introduces the effects of plasticity into this solution. When $\frac{R_m}{F_1}$ equals 2.0 or more, Bigg's solution behaves elastically and his $\frac{R_m}{F_1}$ equal to 2.0 contour is our amplification diagram, Figure 4-4. Use of Bigg's approximate analysis procedure will be discussed in much greater detail in Chapter 5.

ENERGY SOLUTIONS

Designers generally wish to predict peak (maximum) bending stresses, shears, and deformations in blast-loaded structural components. They are seldom interested in an entire time history of a structure responding to a blast wave. Energy solutions, such as those about to be presented, are excellent for predicting these maxima; however, these energy solutions give no predictions of time histories.

We will begin by deriving some solutions for dynamically loaded

beams and plates in the quasi-static and impulsive loading realms. These
solutions will be compared to test results which demonstrate their validity. All of these solutions are based upon assumed deformed shapes, so we
will also discuss what happens when other shapes are used. Ultimately,
we will show how nondimensional P-i diagrams can be derived for beams,
plates, columns, and other structural elements when the principles used in
this section are properly combined.

To derive all of these energy solutions, one must follow certain
steps. These steps are:

1) Assume an appropriate deformed shape;

2) Differentiate the deformed shape to obtain strains;

3) Substitute the strains into the appropriate relationship for
   strain energy per unit volume;

4) Integrate the strain energy per unit volume over the volume of
   the structural element to obtain the total strain energy;

5) Compute the kinetic energy by substituting into $\frac{I^2}{2m}$;

6) Compute the maximum possible work by integrating over the loaded
   area for the pressure times the deflections;

7) Obtain the deformation in the impulsive loading realm by equating kinetic energy to strain energy;

8) Obtain the deformation in the quasi-static loading realm by
   equating work to strain energy; and

9) Substitute deformation into strain equation to obtain strains.

These steps will be used over and over again in the following illustrative
examples. If the reader wishes to apply these methods to other structural
elements, he should learn them as a procedure. In the following examples,
we are going to assume that the reader has had a college course in strength
of materials, thus, many basic structural relationships will be used without being derived or references cited.

Elastic Cantilever Beam

As a first problem, assume we wish to compute maximum deformation
and bending strain in an impulsively loaded cantilever beam. The beam has
a coordinate system with its origin at the root of the beam as shown in
Figure 4-21. We will assume that the deformed shape (Step 1) is given by:

$$w = w_o \left[1 - \cos \frac{\pi x}{2L} \right] \tag{4-27}$$

Selection of a meaningful deformed shape is a major step in this
solution. For an elastic cantilever, an acceptable deformed shape must
have zero deformation at the root of the beam, the maximum deformation at
the tip of the beam, no slope at the root, and no second derivative or
moment at the tip. The shape in Equation (4-27) meets all of these boundary conditions. Other deformed shapes are possible, but we will discuss
what they will do to an analysis after two solutions have been presented.

The next step (Step 2) is to differentiate this assumed deformed
shape twice to obtain curvature:

$$\frac{d^2 w}{dx^2} = \frac{\pi^2 w_o}{4 L^2} \cos \frac{\pi x}{2L} \tag{4-28}$$

If we assume that deflections are small (elastic Bernoulli-Euler bending),

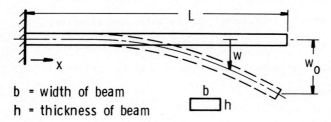

b = width of beam
h = thickness of beam

Figure 4-21. Cantilever Beam for Elastic Energy Solution

the strain energy is given by:

$$S.E. = \int_o^L \frac{M^2 \, dx}{2EI} = \frac{EI}{2} \int_o^L \left(\frac{d^2 w}{dx^2}\right)^2 dx \qquad (4\text{-}29)$$

where E is Young's modulus and I is the second moment of area.[+]

    Substituting Equation (4-28) for curvature in Equation (4-29) (Step 3), and integrating the resulting cosine squared function over the length of the beam (Step 4) gives:

$$S.E. = \frac{\pi^4 \, EI \, w_o^2}{64 \, L^3} \qquad (4\text{-}30)$$

    Next, we need to estimate the kinetic energy. This fifth step is obtained by:

$$K.E. = \Sigma_{beam} \, (1/2) \, m \, V_o^2 = \int_o^L 1/2 \, (\rho \, b \, h \, dx) \left[\frac{i \, b \, dx}{\rho \, b \, h \, dx}\right]^2 \qquad (4\text{-}31)$$

or

$$K.E. = \frac{i^2 \, b \, L}{2 \, \rho \, h} \qquad (4\text{-}32)$$

    Omitting steps 6 and 8 for now because we are not now calculating beam response in the quasi-static realm, and equating strain energy S.E. in Equation (4-30) to kinetic energy K.E. in Equation (4-32) (Step 7), gives the solution for deformation in the impulsive loading realm.

$$\frac{i^2 \, b \, L}{2 \, \rho \, h} = \frac{\pi^4 \, EI \, w_o^2}{64 \, L^3} \qquad (4\text{-}33)$$

[+]Note that the symbol I has two meanings; total impulse and second moment of area. Do not confuse these quantities.

Or, after assuming a rectangular section and substituting $(1/12)\ b\ h^3$ for I:

$$\frac{w_o}{L} = \frac{\sqrt{384}}{\pi^2} \left(\frac{L}{h}\right) \left[\frac{i}{h\ \sqrt{E\rho}}\right] \qquad (4-34)$$

But, for small deformations, the strain ε is related to the deformed shape through:

$$\varepsilon = \frac{M\ c}{EI} = -\frac{h}{2} \left(\frac{d^2 w}{dx^2}\right) \qquad (4-35)$$

where c is the distance from the neutral axis to the outer fibers of the beam. After substituting Equation (4-28), this gives:

$$\varepsilon = \frac{\pi^2\ h\ w_o}{8\ L^2} \cos\frac{\pi x}{2L} \qquad (4-36)$$

The maximum strain will occur at the root of the cantilever where the cosine function equals 1.0. After substituting Equation (4-34) for w_o in Equation (4-36) (Step 9), the maximum strain ε_m is obtained:

$$\varepsilon_m = 2.45\ \frac{i}{h\ \sqrt{E\rho}} \qquad (4-37)$$

This solution applies for strains in the impulsive loading realm. The solution states that strain is independent of beam length, a correct and interesting conclusion for the impulsive loading realm. Strain is independent of span in this realm because doubling the span doubles the kinetic energy going into the system; however, doubling the span also doubles the amount of beam material which is available for absorbing the strain energy. The net result is that beam span cancels out of this solution.

Deformations still depend upon span, but the maximum strain does not in this loading realm.

To demonstrate the validity of this solution, we can compare Equation (4-37) to experimental data generated by Baker, et al. (1958). In these tests, high explosive charges were detonated in the vicinity of 6061-T6 aluminum cantilever beams. Plotted in Figure 4-22 is the maximum bending strain as a function of $\dfrac{i}{L\sqrt{\rho E}}$ for beams with a length of 305 mm and a thickness of 1.3 mm. Some uncertainty exists in computing the impulse imparted to the beams because of air blast wave diffraction around the beams; hence, the test data are plotted as horizontal bars which cover the range of possible impulses. As can be seen in Figure 4-22, Equation (4-37) accurately predicts observed results for elastic cantilever beams in the impulsive loading realm.

Although we have no test data to demonstrate a solution for elastic cantilever beams in the quasi-static loading realm, this solution is easi-

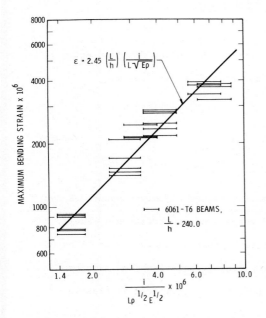

Figure 4-22. Elastic Response of Cantilevers, Impulsive Loading Realm
[Baker, et al. (1958)]

ly derived. The maximum possible work on the beam loaded by a pressure P_r (Step 6) is given by:

$$Wk = \int_0^L P_r \; b \; w \; dx = P_r \; b \; w_o \int_0^L [1 - \cos \frac{\pi x}{2L}] \; dx \qquad (4\text{-}38)$$

or

$$Wk = [1 - \frac{2}{\pi}] \; P_r \; b \; L \; w_o \qquad (4\text{-}39)$$

Equating Wk given by Equation (4-39) to strain energy S.E. (Step 8) given by Equation (4-30) and reducing the results gives the deformation solution for the quasi-static loading realm.

$$\left(\frac{w_o}{L}\right) = 2.865 \left(\frac{L}{h}\right)^3 \left(\frac{P_r}{E}\right) \qquad (4\text{-}40)$$

Substituting into Equation (4-36) and setting the cosine function equal to 1.0 gives the maximum strain in the quasi-static loading realm.

$$\varepsilon_{max} = 3.535 \; \frac{P_r \; L^2}{E \; h^2} \qquad (4\text{-}41)$$

This illustration shows that easily derived algebraic solutions can be developed by using assumed deformed shapes and conservation of energy. This approach is attractive for designers because it shows how doubling or halving a parameter will modify the resulting deformations and strains. But, as an analysis procedure, this approach is not limited to elastic solutions, as we now show.

Plastic Deformations In Beams

A second comparison will be made for plastic deformations in simply supported and in clamped-clamped beams which have been loaded impulsively. We will derive the rigid-plastic solution for simply supported beams, and tell the reader what the solution would be in clamped-clamped beams. Test results for beams with a variety of different materials and boundary conditions will be compared to our solution.

For a simply supported beam of total length L whose origin is at the mid-point, the assumed deformed shape will be a parabola given by:

$$w = w_o \left(1 - \frac{4\,x^2}{L^2} \right) \qquad (4\text{-}42)$$

This deformed shape has a maximum deflection and no slope at mid-span, plus no deflection and a maximum slope at the supports. Two derivatives indicates that this assumed deformed shape yields constant curvature.

The strain energy stored in a rigid-plastic beam is the yield moment times the angle through which it rotates $\left(\frac{d^2 w}{dx^2}\,dx \right)$ summed by integrating over the length of the beam. Because of symmetry, this strain energy can be obtained by doubling an integration over half of the beam.

$$S.E. = -2 \int_o^{L/2} M_y \frac{d^2 w}{dx^2}\,dx = \frac{16\,M_y\,w_o}{L^2} \int_o^{L/2} dx \qquad (4\text{-}43)$$

or

$$S.E. = \frac{8\,M_y\,w_o}{L} \qquad (4\text{-}44)$$

In the impulsive loading realm, the kinetic energy is given by:

$$K.E. = \Sigma_{beam} \frac{I^2}{2m} = 2 \int_o^{L/2} \frac{(i \; b \; dx)^2}{2 \; (\rho \; A \; dx)} \tag{4-45}$$

or

$$K.E. = \frac{i^2 \; b^2 \; L}{2 \; \rho \; A} \tag{4-46}$$

Equating K.E. in Equation (4-46) to strain energy in Equation (4-44) and simplifying gives:

$$\frac{i^2 \; b^2 \; L}{\rho \; M_y \; A} = 16 \left(\frac{w_o}{L} \right) \tag{4-47}$$

If we assume a rectangular cross section beam of width b and thickness h, then, after substituting $(1/4) \; \sigma_y \; b \; h^2$ for M_y (σ_y is the yield point) and b h for the area A, the impulse is related to the maximum deflection by the equation:

$$\frac{i^2 \; L}{\rho \; \sigma_y \; h^3} = 4 \left(\frac{w_o}{L} \right) \qquad \text{[S.S. beam]} \tag{4-48}$$

Although we leave it to the reader to derive, a similar solution for a clamped-clamped impulsively loaded beam would be given by:

$$\frac{i^2 \; L}{\rho \; \sigma_y \; h^3} = 8 \left(\frac{w_o}{L} \right) \qquad \text{[Clamped beam]} \tag{4-49}$$

Notice that the only difference between the solution for a simply supported beam and a clamped beam is in the numerical coefficient, i.e., 8 for a clamped beam and 4 for a simply-supported beam. All of the parameters L, ρ, σ_y, h, w_o, and i interrelate in a similar fashion. If one parameter is doubled or halved, the resulting response is changed by the same relative percentage in both beams. Both solutions can, therefore, be presented in the same equation by inserting a parameter N. This parameter N equals 1.0 for simply supported beams and 2.0 for clamped-clamped beams.

$$\frac{i^2 L}{N \rho \sigma_y h^3} = 4 \left(\frac{w_o}{L}\right) \tag{4-50}$$

Had other boundary conditions such as pinned-clamped or cantilever been studied, impulsively loaded, rigid-plastic beams also would have solutions as given by Equation (4-50), provided appropriate N values were used. We will take advantage of this observation later when nondimensional P-i diagrams are presented for structural components with a variety of different boundary conditions.

Figure 4-23 is a plot of experimental data taken by Florence and Firth (1965) on both clamped-clamped and simply supported beams with a half span to thickness ratio ℓ/h of 36. Because these investigators use half span ℓ rather than full span L, Equation (4-50) must be written as Equation (4-51) for purposes of comparison.

$$\frac{i^2}{N \rho \sigma_y h^2} = \left(\frac{h}{\ell}\right)\left(\frac{w_o}{\ell}\right) \tag{4-51}$$

All beams, whether pinned or clamped and made from 2024-T4 aluminum, 6061-T6 aluminum, 1018 cold-rolled steel, or 1018 annealed steel, were impulsively loaded using sheet explosive. The clamped beams did not develop significant extensional stresses, because the ends were allowed to move inwards at

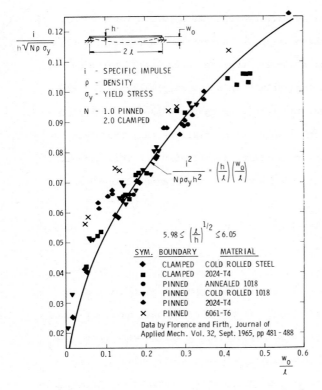

Figure 4-23. Beam Bending in the Impulsive Realm

the same time that the boundaries were constrained against rotation. Figure 4-23 demonstrates the validity of Equation (4-51) and this approximate rigid-plastic analysis procedure by comparing test results on six different beam systems to our solution.

Although we will not develop this solution, a rigid-plastic solution for beams in the quasi-static loading realm could also be derived by computing the maximum possible work and equating it to the strain energy which has already been estimated, Equation (4-44). Inherent in all these energy solutions is an assumed deformed shape. The next logical question is: "What influence do different deformed shapes have on solutions?"

Influence of Deformed Shape on Energy Solutions

To illustrate the influence of assumed deformed shape on computed structural strains and deformations, we performed elastic bending and plastic bending analyses on simply supported beams loaded with a uniform impulse. Although the details of each solution could be presented, these particulars contribute little to this assessment. The analyses use the same energy procedures which have already been presented several times. The results are, therefore, presented in tables, and the details are omitted.

For an elastic analysis, we evaluated computed results for three different deformed shapes. The first shape is a parabola, the second shape is the first mode sine wave, and the third shape is the static deformation for a uniformly applied load. All three solutions give similar results for the strain energy S.E., the maximum deformation w_o, and the maximum strain ε_{max}, except for numerical coefficients as can be observed in Table 4-1. The results are nondimensional, so numerical coefficients can be compared directly. For strain energy, deformation, or strain, only the numerical coefficients differ for the different deformed shapes. This difference is small, usually appearing in the second integer. So, reasonable engineering answers are obtained to within a few percent, regardless of which deformed shape is selected, provided the shape meets the boundary conditions. Our inability to know precisely the loads and the scatter which naturally occurs (illustrated in Figure 4-8), infer that exact choice of the "best" deformed shape is relatively unimportant.

Incidentally, when the elastic simply supported beam solution is compared to the so-called "exact" Bernoulli-Euler beam solution using a series expansion to predict both deflection and strain, the parabola gives the "exact" solution in the impulsive loading realm, and the static deformed shape gives the "exact" solution in the quasi-static loading realm. In an elastic cantilever beam, the static deformed shape also gives the "exact" answer in the quasi-static loading realm and the parabola gives the "exact" solution

Table 4-1. Impulsive Bending Solution For Elastic
Simply Supported Beam

Parameter / Deformed Shape $\dfrac{w}{w_o} =$	Parabola $4\left(\dfrac{x}{L}\right)^2$	First Mode $\sin\left(\dfrac{\pi x}{L}\right)$	Static Deformed Shape $\dfrac{16}{5}\left[\left(\dfrac{x}{L}\right) - 2\left(\dfrac{x}{L}\right)^3 + \left(\dfrac{x}{L}\right)^4\right]$
Strain Energy $\dfrac{(S.E.)\ L^3}{Eb\ h^3\ w_o^2} =$	2.666	2.029	2.048
Deformation $\dfrac{w_o\ \sqrt{\rho E}\ h^2}{i\ L^2} =$	0.4330	0.4964	0.4941
Strain $\dfrac{\varepsilon_{max}\ \sqrt{\rho E}\ h}{i} =$	1.732	2.449	2.372

in the impulsive loading realm. So, in the impulsive loading realm, beams deform with almost a constant curvature.

A second comparison between the effects of assumed deformed shapes was made for rigid-plastic bending in a simply supported beam loaded with a uniform impulse. One more deformed shape was added to the deformed shapes already used for the elastic comparisons. The fourth shape is a static hinge yielding plastically in the center of the beam, while the rest of the beam remains rigid. This deformed shape is a common one used in civil engineering plasticity studies; whereas, the other three deformed shapes distribute the deformation and the strain energy throughout a member. The stationary hinge concentrates the deformation and strain energy. Table 4-2 compares nondimensional numerical coefficients for strain energy, maximum plastic deformation, and maximum plastic strain.

Table 4-2. Impulsive Bending Solution For Plastic,
          Simply Supported Beam

Parameter $\dfrac{w}{w_o} =$	Parabola $4\left(\dfrac{x}{L}\right)^2$	First Mode $\sin\left(\dfrac{\pi x}{L}\right)$	Static Deformed Shape $\dfrac{16}{5}\left[\left(\dfrac{x}{L}\right) - 2\left(\dfrac{x}{L}\right)^3 + \left(\dfrac{x}{L}\right)^4\right]$	Stationary Hinge $2\left(\dfrac{x}{L}\right)$ for $0 \leq \dfrac{x}{L} \leq \dfrac{1}{2}$
Strain Energy $\dfrac{(S.E.)\ L}{\sigma_y\ w_o\ b\ h^2} =$	2.00	1.571	1.60	1.00
Deformation $\dfrac{w_o\ \rho\ \sigma_y\ h^3}{i^2\ L^2} =$	0.250	0.3183	0.3125	0.500
Strain $\dfrac{\varepsilon_{max}\ \rho\ \sigma_y\ h^2}{i^2} =$	1.00	1.571	1.500	No Meaning

Once again, similar answers are obtained regardless of which deformed shape is used. The stationary hinge yields no solution for strain because there is no "gauge length" associated with a concentrated hinge. The deformation is significantly larger for a stationary hinge than for the associated deformations from distributed deformed shapes. Dimensionally, the solution for the concentrated hinge is correct, but quantitatively, the deformations are larger because the strains are not distributed. For steel beam-like members, distributed deformed shapes are closer to reality and consequently give superior quantitative answers. In underreinforced concrete members, failure mechanisms often look like concentrated hinges; hence, the use of stationary hinges works better for concrete design.

The moral behind these illustrations is that either a first mode approximation or a static deformed shape work excellently. We recommend a first mode approximation for symmetric deformations, as in simply supported and clamped-clamped beams, because the resulting algebra is slightly easier. For nonsymmetric responses, as in a simply supported clamped beam, the static deformed shape may be easier to use. For other boundary conditions, we make no recommendations.

In all of the illustrations, the assumed deformed shape is less important than the effects of coupling. Supporting a flexible structural component on a flexible foundation has a much greater influence on structural response than the assumed deformed shape.

Biaxial States of Stress

Energy solutions can also be used to predict structural response in plates or other objects under biaxial rather than uniaxial states of stress. Closed-form classical plate solutions are very difficult. Generally, either the contributions from shearing stresses or the contributions from extensional behavior are left out so that a mathematical equation is obtained which can be solved. When an energy solution is used, extensional _and_ bending behavior can be analyzed for both normal and shearing stresses. The procedures to be followed are precisely those which have been used before. For a clamped rectangular plate, a possible deformed shape is:

$$w = \frac{w_o}{4} [1 + \cos \frac{\pi x}{X}] [1 + \cos \frac{\pi y}{Y}] \qquad (4-52)$$

The parameters X and Y are the half spans, and the origin for this coordinate system is at the center of the plate.

The kinetic energy imparted to the plate is obtained from:

$$K.E. = \frac{I^2}{2m} = 4 \int_o^X \int_o^Y \frac{i^2 (dx)^2 (dy)^2}{2 \rho h (dx) (dy)} \qquad (4-53)$$

or

$$K.E. = \frac{2 i^2 X Y}{\rho h} \qquad (4-54)$$

The strain energy is more complex and must be calculated from first principles. For either bending or extension, the strain energy per unit volume is computed from the stresses and strains, σ and ε:

$$\frac{S.E.}{Vol.} = \int_{Strains} \left[\sigma_{xx} \, d\varepsilon_{xx} + 2\sigma_{xy} \, d\varepsilon_{xy} + \sigma_{yy} \, d\varepsilon_{yy} \right] \qquad (4\text{-}55)$$

The subscripts xx and yy represent normal stresses and strains, and the subscript xy means shearing stresses and strains. In an elastic plate, the normal stresses equal $E\varepsilon$ and the shearing stresses equal $\dfrac{E}{2\,(1+\nu)}\,\varepsilon_{xy}$, so integrating the general solution, Equation (4-55), gives:

$$\frac{S.E.}{Vol.} = \frac{E}{2}\,\varepsilon_{xx}^{2} + \frac{E}{2\,(1+\nu)}\,\varepsilon_{xy}^{2} + \frac{E}{2}\,\varepsilon_{yy}^{2} \qquad [\text{Elastic Solution}] \qquad (4\text{-}56)$$

where E is Young's modulus and ν is Poisson's ratio.

In a rigid-plastic plate, the normal stress equals a constant σ_y and the shearing stress equals another constant $\dfrac{\sigma_y}{\sqrt{3}}$, if the von Mises yield theory is used. Substituting into Equation (4-55) and integrating for a rigid-plastic solution yields:

$$\frac{S.E.}{Vol.} = \sigma_y\,\varepsilon_{xx} + \frac{2\,\sigma_y}{\sqrt{3}}\,\varepsilon_{xy} + \sigma_y\,\varepsilon_{yy} \qquad [\text{Plastic Solution}] \qquad (4\text{-}57)$$

We will proceed with a plastic analysis because a plastically deformed plate will be used later to make comparisons with test results. The bending solution is obtained by substituting

$$\varepsilon_{xx} = -Z\,\frac{\partial^{2} w}{\partial x^{2}} \qquad (4\text{-}58a)$$

$$\epsilon_{yy} = -Z \frac{\partial^2 w}{\partial y^2} \qquad (4\text{-}58b)$$

$$\epsilon_{xy} = 2Z \frac{\partial^2 w}{\partial x \partial y} \qquad (4\text{-}58c)$$

for the strains and performing the indicated differentiations on Equation (4-52). A triple integration, over beam thickness and in the x and y directions, is required before the strain energy is obtained for bending.

Next, the strain energy associated with extensional behavior is obtained by substituting

$$\epsilon_{xx} = (1/2) \left(\frac{\partial w}{\partial x} \right)^2 \qquad (4\text{-}59a)$$

$$\epsilon_{yy} = (1/2) \left(\frac{\partial w}{\partial y} \right)^2 \qquad (4\text{-}59b)$$

$$\epsilon_{xy} = \left(\frac{\partial w}{\partial x} \right) \left(\frac{\partial w}{\partial y} \right) \qquad (4\text{-}59c)$$

into Equation (4-57) and performing, once again, the indicated differentiations on Equation (4-52). After substituting and integrating, the extensional strain energy is obtained.

We leave out the tedious algebra, since it does not illustrate any new principles. The final solution relating impulse to deformation is obtained by equating strain energy S.E. to kinetic energy K.E. After making algebraic reductions, the answer is given by:

$$\left[\frac{i\,X}{\sqrt{\rho\,\sigma_y}\,h^2}\right]^2 = \frac{\pi}{4}\left[1+\left(\frac{X}{Y}\right)^2\right]\left(\frac{w_o}{h}\right) + \frac{2}{\sqrt{3}}\left[\frac{X}{Y}\right]\left(\frac{w_o}{h}\right) +$$

(4-60)

$$\frac{3\pi^2}{64}\left[1+\left(\frac{X}{Y}\right)^2\right]\left(\frac{w_o}{h}\right)^2 + \frac{4}{\sqrt{3}}\left[\frac{X}{Y}\right]\left(\frac{w_o}{h}\right)^2$$

There are four terms on the right-hand side of Equation (4-60). The first of these is associated with normal bending stresses, the second with bending shear stresses, the third with normal extensional stresses, and the last with extensional shear stresses. So, the relative contributions from each type of strain energy can be related to the total solution.

Equation (4-60) can be compared to test data reported by Jones, et al. (1970) for rectangular plates with an aspect ratio $\frac{Y}{X}$ equal to 1.695. Both hot-rolled mild steel plates and 6061-T6 aluminum plates were loaded with sheet explosive and deformed plastically. Substituting for $\frac{Y}{X}$ in Equation (4-60) and rearranging yields $\frac{i_R^2\,Y^2}{\rho\,\sigma_y\,h^4}$ as a function of $\frac{w_o}{h}$ which can be compared directly with experimental results, Figure 4-24:

$$\frac{i_R^2\,Y^2}{\rho\,\sigma_y\,h^4} = 5.00\left(\frac{w_o}{h}\right) + 5.71\left(\frac{w_o}{h}\right)^2$$

(4-61)

Agreement is excellent when data are compared to Equation (4-61) as in Figure 4-24. The error for small values of w_o/h is probably caused by only a portion of the full span deforming when deformations are small.

So far, all comparisons have been for structural response in the impulsive loading realm. A quasi-static loading realm solution for clamped rectangular plate permanent deformations can be developed using Equation (4-52) as the assumed deformed shape. Equating S.E. to Wk in a plastic anal-

318

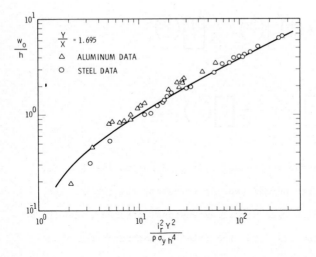

Figure 4-24. Predicted and Experimental Deformations in
Uniformly Impulsed Rectangular Plates

ysis eventually gives the relationship:

$$\frac{P_R \, X^2}{\sigma_y \, h^2} = \frac{\pi}{3} \left[1 + \left(\frac{X}{Y}\right)^2 \right] + \frac{8}{3\sqrt{3}} \left[\frac{X}{Y}\right] +$$

(4-62)

$$\frac{3\pi^2}{32} \left[1 + \left(\frac{X}{Y}\right)^2 \right] \left(\frac{w_o}{h}\right) + \frac{8}{\sqrt{3}} \left[\frac{X}{Y}\right] \left(\frac{w_o}{h}\right)$$

Test data by Hooke and Rawlings (1969) on clamped rectangular, mild-steel plates of various scaled thicknesses and aspect (X/Y) ratios can be used in comparisons with Equation (4-62). Figure 4-25 presents some of these data in a plot of scaled applied load $[PX^2/(\sigma_y h^2)]$ as a function of scaled permanent mid-span deflection (w_o/h) for rectangular plates with an (X/Y) of 1/2. In these tests, a rectangular plate was mounted in one face of a scaled box. A long duration load was applied pneumatically after a solenoid valve was activated to pressurize the box. Because this loading has a gradual

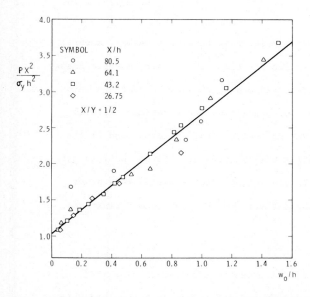

Figure 4-25.   Permanent Deformations in Quasi-Statically
Loaded Rectangular Plate

rise associated with it, the load factor is 1.0 and not 2.0. Our dynamic
solution given in Equation (4-62) was divided by 2.0 and plotted in Figure
4-25 after substituting for X/Y to obtain:

$$\frac{P_R\,X^2}{\sigma_y\,h^2} = 1.039 + 1.733\left(\frac{w_o}{h}\right) \qquad (4\text{-}63)$$

Agreement is excellent. The same type of agreement can also be shown for
square plates even though we do not present these comparisons.

Other beam and plate comparisons can be shown, but we stop with these
four because they include all our needs. Comparisons have been made on
structural components in the impulsive as well as quasi-static loading domains, for elastic as well as plastic response, for strains as well as deformations, and for biaxial (plate) as well as uniaxial (beam) states of stress.
These results demonstrate that energy techniques are design tools which are
very useful and easy to apply.

NONDIMENSIONAL P-i DIAGRAMS

Designers generally wish to predict peak bending stresses, shears, and deformations in all loading realms for blast-loaded structures. Energy solutions which have just been presented and which will be combined into entire P-i diagrams are excellent for predicting these maxima. We will present P-i diagrams for beams, extensional strips,[†] columns, and plates. For each solution, the final results will be presented first and then the relationships will be derived. By presenting information in this order, many readers not interested in the details can skip over the derivations.

Beam Solutions

Figure 4-26 is a nondimensional pressure-specific impulse (P-i) diagram for determining the maximum strain and deflection in beams loaded by a blast wave. The ordinate to Figure 4-26 is scaled impulse, the abscissa is scaled pressure, and the contours are for constant values of scaled strain. The blast wave is characterized by its peak applied pressure P and specific impulse i. These pressures and specific impulses are either side-on or reflected ones dependent upon the orientation of the building relative to the enveloping wave. In this graphical solution, we assume that the loading is uniform over the entire span of length L. The beam has a loaded width b, a mass density ρ, a cross-sectional area A, a total depth h, an elastic modulus E, a yield stress σ_y, a second moment of area I, and a plastic (not elastic) section modulus Z. Throughout this solution and subsequent ones, we will assume that the stress-strain curve is elastic perfectly plastic, without significant strain hardening or strain rate effects.

Different boundary conditions can be evaluated by inserting the appropriate nondimensional numbers, i.e., the appropriate Ψ coefficients, from the Table in Figure 4-26. Simply supported, clamped-clamped, clamped-pinned, and cantilever beams are all included in the graphical solution.

[†]An extensional strip has negligible bending stiffness.

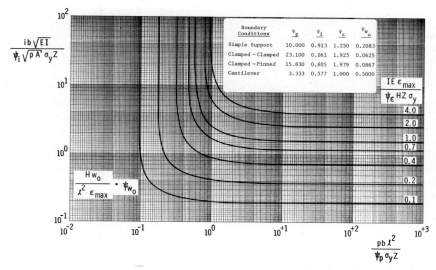

Figure 4-26. Elastic-Plastic Solution for Bending of Blast Loaded Beams

Each curve represents a specific maximum scaled strain, or a maximum scaled ductility ratio μ. No strain energy is absorbed in extensional or shear behavior; this solution is entirely a bending one. Any self-consistent set of units can be used because this solution is nondimensional. Once strain has been obtained graphically from Figure 4-26, the maximum deflection is obtained by solving the equation in the lower left-hand corner.

Figure 4-27 is a corresponding bending beam solution; however, it is valid for elastic response only. The major added benefit derived from Figure 4-27 is that it can be used to estimate the shear forces at the supports. For a Bernoulli-Euler beam, a plastically responding beam has no shear force at the instant of maximum deformation, because $\frac{dM}{dx}$ equals zero. Maximum shear is reached earlier in the response which cannot be estimated from an energy solution because it gives end states and not a transient solution. For an elastic solution, a maximum shear force V is reached when the beam is in its maximum deformed position. Provided the response is elastic, Figure 4-27 gives the same solution as an elastically responding beam in the more generalized elastic-plastic solution of Figure 4-26.

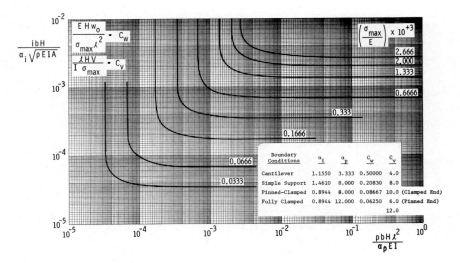

Figure 4-27. Stresses, Shears, and Deflections in Bending of Blast Loaded
Elastic Beams

Figures 4-26 and 4-27 were derived using conservation of energy principles. To demonstrate how these relationships were developed, we will
illustrate with an elastic, simply supported beam. First, a deformed shape
must be assumed. Assuming a deformed shape which corresponds to the static
deformed shape for a beam undergoing uniform loads gives:

$$w = \frac{16}{5} w_o \left[\frac{x}{L} - 2 \left(\frac{x}{L} \right)^3 + \left(\frac{x}{L} \right)^4 \right] \tag{4-64}$$

This deformed shape is then differentiated twice with respect to x so that
the elastic bending moment M can be obtained from $M = -EI \dfrac{d^2 w}{dx^2}$. This procedure gives for the bending moment

$$M = \frac{192}{5} \frac{EI \, w_o}{L^2} \left[\left(\frac{x}{L} \right) - \left(\frac{x}{L} \right)^2 \right] \tag{4-65}$$

The strain energy S.E. stored in a deformed beam can then be determined by substitution into S.E. $= \int_o^L \frac{M^2 dx}{2EI}$. Substitution gives:

$$S.E. = \frac{(192)^2\, EI\, w_o^2}{(50)\, L^4} \int_o^L \left[\left(\frac{x}{L}\right)^2 - 2\left(\frac{x}{L}\right)^3 + \left(\frac{x}{L}\right)^4 \right] dx \qquad (4\text{-}66)$$

or after completing the integration,

$$S.E. = 24.576\, \frac{EI\, w_o^2}{L^3} \qquad (4\text{-}67)$$

The asymptote which is impulse dependent is determined by equating the kinetic energy K.E. to the strain energy. The kinetic energy is given by:

$$K.E. = (1/2) m\, V_o^2 = \frac{I^2}{2m} \qquad (4\text{-}68)$$

Substituting ρAL for m and ibL for I gives for the kinetic energy:

$$K.E. = \frac{i^2 b^2 L}{2\rho A} \qquad (4\text{-}69)$$

Equating S.E. to K.E. gives the impulsive loading realm asymptote:

$$\frac{i^2 b^2 L}{2\rho A} = 24.576\, \frac{EI\, w_o^2}{L^3} \qquad (4\text{-}70)$$

Equation (4-70) relates applied impulse to deformation. To relate impulse to bending stress, we must use the moment-curvature relationships. The maximum moment as given by Equation (4-65) occurs at $\frac{x}{L} = 1/2$. The maximum moment is then computed to be:

$$M_{max} = \frac{192}{20} \frac{EI\ w_o}{L^2} \tag{4-71}$$

Substituting $\sigma_{max} = \dfrac{M_{max}\ h/2}{I}$ and solving for $\dfrac{w_o}{L}$ gives:

$$\frac{w_o}{L} = \frac{5}{24} \frac{\sigma_{max}}{E\ h} L \tag{4-72}$$

Finally, taking the square root of Equation (4-70) and substituting Equation (4-72) into Equation (4-70) to eliminate w_o gives the asymptote for the impulsive loading realm in terms of the maximum bending stress.

$$\frac{i\ b\ h}{\sqrt{\rho\ EI\ A}} = 1.461\ \frac{\sigma_{max}}{E} \tag{4-73}$$

Equation (4-73) is the impulsive loading realm asymptote plotted in Figure 4-27. The numerical coefficient 1.461 in Equation (4-73) is the α_i coefficient for a simply supported beam. In Equation (4-72), the number 5/24 is the C_w coefficient in Figure 4-27 to relate stress to deformations in a simply supported beam.

The quasi-static asymptote in Figure 4-27 is computed by calculating the maximum possible work Wk and equating this quantity to the strain energy. This maximum work is:

$$Wk = \int_o^L Pb\ (dx)\ w \tag{4-74}$$

After substituting Equation (4-64) for w:

$$Wk = \frac{16}{5} Pb \ w_o \int_o^L \left[\frac{x}{L} - 2 \left(\frac{x}{L} \right)^3 + \left(\frac{x}{L} \right)^4 \right] dx \qquad (4\text{-}75)$$

Or after integrating:

$$Wk = \frac{16}{25} Pb \ L \ w_o \qquad (4\text{-}76)$$

The strain energy has already been calculated as Equation (4-67). Equating S.E. to Wk gives the quasi-static loading realm asymptote.

$$\frac{16}{25} Pb \ L \ w_o = 24.576 \ \frac{EI \ w_o^2}{L^3} \qquad (4\text{-}77)$$

Equation (4-77) relates the applied pressure to deformation. To relate pressure to bending stress, we substitute Equation (4-72) for w_o and algebraically rearrange terms to obtain:

$$\frac{Pb \ h \ L^2}{EI} = 8.0 \ (\sigma/E) \qquad (4\text{-}78)$$

Equation (4-78) is the quasi-static loading realm asymptote plotted in Figure 4-27. The numerical coefficient 8.0 in Equation (4-78) is the α_p coefficient for a simply supported beam. The coefficient C_v relating maximum bending stress to the maximum shear force is obtained by differentiating the moment equation, Equation (4-65), with respect to x to obtain the shear force V with respect to deformation w_o.

$$V = \frac{dM}{dx} = \frac{192}{5} \ \frac{EI \ w_o}{L^3} \left(1 - \frac{2x}{L} \right) \qquad (4\text{-}79)$$

The maximum shear occurs at $x = 0$ or $x = L$. Setting $x = 0$ and substituting Equation (4-72) for w_0 gives:

$$V_{max} = 8.0 \; \frac{\sigma_{max} \; I}{L \; h} \qquad (4\text{-}80)$$

Equation (4-80) is the shear equation presented in Figure 4-27. The numerical value of 8.0 in Equation (4-80) is the C_v coefficient for a simply supported beam.

A transition for the dynamic loading realm was faired by using a hyperbolic tangent squared relationship which, from practical experience, seems to fit quite well. Notice that for small values

$$S.E. = Wk \cdot \tanh^2 \left(\frac{K.E.}{Wk} \right)^{1/2} \qquad (4\text{-}81)$$

of the argument, the hyperbolic tangent equals its argument and we obtain the impulsive loading realm asymptote of $S.E. = K.E.$ For large arguments the tanh equals 1.0, and we obtain the quasi-static loading realm asymptote of $S.E. = Wk$. Less than one percent error is introduced in the linearly elastic oscillator in Figure 4-4 when the square of the hyperbolic tangent is used as an approximation to the exact solution for that spring mass system. All of the nondimensional P-i diagrams which will be presented use Equation (4-81) as an approximate way of creating a transition between asymptotes.

These energy solutions, within the bounds of a Bernoulli-Euler, small deformation, beam solution, give exact answers for both strain and deformation in the quasi-static loading realm. These "exact" answers occur because the deformed shape is correct in this domain. In the impulsive loading realm, only approximate answers are given because the deformed shape is not quite right; however, the results are sufficiently accurate, especially when one realizes the uncertainties associated with the load. More accurate answers are obtained if a more accurate deformed shape is assumed. Actually, the interrelationship of one variable with another remains the same irrespective of the assumed deformed shape, as we have noted before. The only effect of using other deformed shapes is to slightly modify the numerical coefficients α_i, α_p, C_v, and C_w.

The same procedures can be used to compute the P-i diagram for cantilever, clamped-clamped, clamped-pinned, or beams with any other boundary condition. If the assumed deformed shape corresponds even approximately to a beam deflection curve and meets the correct boundary conditions, then fairly accurate answers will result. The only difference in the solutions of beams with different boundary conditions is that different numerical values arise in the α_i, α_p, C_v, and C_w coefficients.

Deriving P-i diagrams for an elastic-plastic beam is a much more complicated procedure. To illustrate this procedure we will solve for the impulsive loading realm asymptote in a uniformly loaded, elastic-plastic beam. First, one needs a constitutive relationship. We choose:

$$\sigma = \sigma_y \tanh \frac{E\varepsilon}{\sigma_y} \tag{4-82}$$

Stress in Equation (4-82) for small values of $\frac{E\varepsilon}{\sigma_y}$ is given by $E\varepsilon$, and for large values is given by σ_y. These limits are the correct ones for an elastic-plastic system. To determine the strain energy, we must return to Equation (4-55), which for uniaxial stress will be given by:

$$\frac{S.E.}{Vol.} = \int_{Strains} \sigma \, d\varepsilon = \int_o^\varepsilon \sigma_y \tanh \left(\frac{E\varepsilon}{\sigma_y}\right) d\varepsilon \tag{4-83}$$

328

or

$$\frac{S.E.}{Vol.} = \frac{\sigma_y^2}{E} \log \cosh \left(\frac{E\varepsilon}{\sigma_y}\right) \tag{4-84}$$

Next, we need a deformed shape which for a simply supported beam will be assumed to be:

$$w = w_o \sin \frac{\pi x}{L} \tag{4-85}$$

Strain is related to deformation through the strain curvature relationship which is:

$$\varepsilon = -Z \frac{d^2 w}{dx^2} = \frac{\pi^2 w_o Z}{L^2} \sin \frac{\pi x}{L} \tag{4-86}$$

Substituting Equation (4-86) into Equation (4-84) and integrating over the volume for a rectangular cross-section beam yields:

$$S.E. = \frac{4 \sigma_y^2 b}{E} \int_0^{h/2} \int_0^{L/2} \log \cosh \left[\frac{\pi^2 w_o E Z}{\sigma_y L^2} \sin \frac{\pi x}{L}\right] dZ\, dx \tag{4-87}$$

The best method of integrating Equation (4-87) is to make a transformation of variables. Defining $\bar{X}$ as $\frac{\pi x}{L}$ and $\bar{Z}$ as $\frac{\pi Z}{h}$ and substituting into Equation (4-87) gives:

$$S.E. = \frac{4 \sigma_y^2 b h L}{\pi^2 E} \int_0^{\pi/2} \int_0^{\pi/2} \log \cosh \left[\left(\frac{\pi w_o h E}{\sigma_y L^2}\right) \bar{Z} \sin \bar{X}\right] d\bar{Z}\, d\bar{X} \tag{4-88}$$

If the beam is impulsively loaded, the kinetic energy is given by:

$$\text{K.E.} = \frac{I^2}{2m} = \frac{i^2 \, b \, L}{2 \, \rho \, h}$$

(4-89)

Equating K.E. to S.E. and rearranging gives:

$$\left[\frac{\pi^2 \, i^2 \, E}{8 \, \rho \, \sigma_y^2 \, h^2} \right] = \int_o^{\pi/2} \int_o^{\pi/2} \log \cosh \left[\left(\frac{\pi \, w_o \, h \, E}{\sigma_y \, L^2} \right) \bar{Z} \sin \bar{X} \right] d\bar{Z} \, d\bar{X}$$

(4-90)

This solution for the impulsive loading realm asymptote must be solved numerically on a computer. Because the quantities $\bar{Z}$ and $\bar{X}$ are nondimensional variables of integration, the final solution can be presented as a plot of $\left[\dfrac{\pi^2 \, E \, i^2}{8 \, \rho \, \sigma_y^2 \, h^2} \right]$ versus $\left[\dfrac{\pi \, w_o \, E \, h}{\sigma_y \, L^2} \right]$. The quasi-static loading realm asymptote can also be obtained by equating work Wk to S.E. These quantities, when combined by using the approximation given by Equation (4-81), allow entire elastic-plastic P-i diagrams to be drawn as in Figure 4-26.

Figure 4-28 presents an interesting comparison by plotting the elastic-plastic solution which was just derived and elastic only and rigid-plastic only solutions based on the same deformed shapes. These comparisons show that for small loads, the elastic solution is adequate, and for large loads, the plastic solution applies. In the region of the transition from elastic to plastic, some difference can be seen which is caused by some regions of the beam being plastic while others are elastic. The insert to Figure 4-28 shows the hyperbolic tangent function for the stress-strain curve.

Several other observations should be noted based upon these energy solutions for beams. In the impulsive loading realm, maximum bending stress is independent of span L. This conclusion is mathematically correct and is caused by the span entering the strain energy and kinetic energy expressions to the same power. Deformations, however, do depend up-

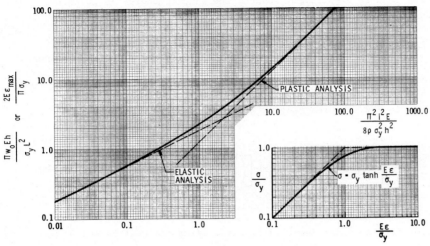

Figure 4-28.   Elastic-Plastic Solution for a Simply Supported Beam Subjected to a
Uniform Impulsive Loading

on span in both the impulsive and quasi-static loading realms. In the
impulsive loading realm, the response depends only on the impulse or area
under the applied pressure-time history as we should expect from earlier
discussions of simple mechanical systems. In the quasi-static loading
realm, response is independent of beam density and duration of the loading as should be expected from the same discussions of simple systems.

Whenever these relationships are used for design, especially in
concrete design, remember that maximum responses only and not time histories are being estimated. Waves usually propagate in dynamically loaded
beams. This means that maximum bending stresses and shears can occur at
various times anywhere throughout the beam and not just at the midspan or
supports. We strongly recommend placing stirrups and reinforcing throughout the entire span of concrete beams; otherwise, concrete beams may fail
rather than transmit these waves.

Finally, no factors of safety have been used in these calculations.
Appropriate factors of safety must be selected based on building codes or
what designers feel to be appropriate.

Extensional Behavior in Strips

When a member undergoes large deformations relative to its thickness
and is axially constrained or is simply quite thin, the principal mode of
energy dissipation can be extensional rather than bending. Figure 4-29
presents an elastic-plastic, one-dimensional, extensional solution. In an
extensional solution, one assumes that the ends are constrained from moving
together so that in-plane forces can be developed. The results presented
in Figure 4-29 are very similar to the previously presented bending solution in that contours of constant scaled maximum strain are presented on a
plot of scaled applied impulse and pressure. All loads are assumed to be
uniformly distributed over the member being loaded. After the strain has
been determined, the maximum deformation, the slope at the boundaries, and
the magnitude of the maximum anchoring force can all be determined using
Figure 4-29.

The symbols in Figure 4-29 are very similar to those used previously. The one new symbol is A, the cross-sectional area of the member. Other
symbols include the applied reflected or side-on pressure P, the applied
reflected or side-on impulse i, the loaded width b, the total span L, the

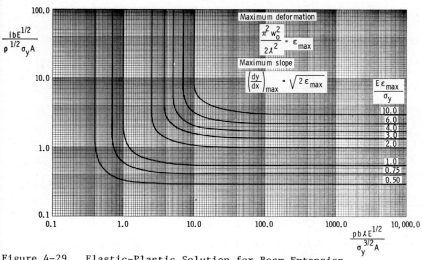

Figure 4-29. Elastic-Plastic Solution for Beam Extension

mass density ρ, the elastic modulus E, the yield point σ_y, the maximum strain ϵ_{max}, the maximum deformation w_o, and the maximum slope $\left(\dfrac{dw}{dx}\right)_{max}$. Any self-consistent set of units can be used, as all scaled quantities are nondimensional.

To derive the graphical solution presented in Figure 4-29, a deformed shape was assumed to be given by:

$$w = w_o \sin \frac{\pi x}{L} \qquad (4-91)$$

The extensional strain for small deformations is approximated by $\left(\dfrac{1}{2}\right)\left(\dfrac{dw}{dx}\right)^2$. Differentiating Equation (4-91) and substituting gives:

$$\epsilon = \frac{\pi^2 w_o^2}{2L^2} \cos^2 \left(\frac{\pi x}{L}\right) \qquad (4-92)$$

The maximum strain occurs when the cosine equals 1.0 or:

$$\epsilon_{max} = \frac{\pi^2 w_o^2}{2L^2} \qquad (4-93)$$

This equation is the relationship relating strains to deformation in Figure 4-29. If this solution is to be an elastic-plastic one, we need an elastic-plastic constitutive relationship. Equation (4-94) is assumed to be this relationship because it lets stress equal $E\epsilon$ for values of $\dfrac{E\epsilon}{\sigma_y}$ less than 0.5, and lets stress equal σ_y for values of $\dfrac{E\epsilon}{\sigma_y}$ greater than 2.0.

$$\sigma = \sigma_y \tanh \left(\frac{E\epsilon}{\sigma_y}\right) \qquad (4-94)$$

The strain energy per unit volume in an elastic-plastic system is the area

under the stress-strain curve. Integrating Equation (4-94) gives for the
strain energy per unit volume

$$\frac{\text{S.E.}}{\text{Vol.}} = \int_o^\varepsilon \sigma_y \tanh\left(\frac{E\varepsilon}{\sigma_y}\right) d\varepsilon \tag{4-95}$$

or

$$\frac{\text{S.E.}}{\text{Vol.}} = \frac{\sigma_y^2}{E} \log \cosh\left(\frac{E\varepsilon}{\sigma_y}\right) \tag{4-96}$$

Substituting Equation (4-92) for ε in Equation (4-96) and multiplying by
the differential volume A dx yields an integral for the strain energy:

$$\text{S.E.} = \frac{\sigma_y^2 A}{E} \int_o^L \log \cosh\left[\frac{\pi^2 E w_o^2}{2\sigma_y L^2} \cos^2\left(\frac{\pi x}{L}\right)\right] dx \tag{4-97}$$

Substituting a dimensionless variable $\bar{X}$ equal to $\frac{\pi x}{L}$, and substituting ε_{max}
for $\frac{\pi^2 w_o^2}{2 L^2}$ (Equation 4-93) finally gives as an integral for the strain energy:

$$\text{S.E.} = \frac{\sigma_y^2 A L}{\pi E} \int_o^\pi \log \cosh\left[\frac{E\varepsilon_{max}}{\sigma_y} \cos^2 \bar{X}\right] d\bar{X} \tag{4-98}$$

The asymptote can now be calculated as before. The impulsive loading realm
asymptote is obtained by equating kinetic energy K.E. to strain energy. The
kinetic energy is given by:

$$\text{K.E.} = \frac{I^2}{2m} = \frac{i^2 b^2 L}{2\rho A} \tag{4-99}$$

Equating Equations (4-98) and (4-99) plus rearranging terms gives:

$$\left[\frac{i\ b\ E^{1/2}}{\rho^{1/2}\ \sigma_y\ A}\right]^2 = \frac{2}{\pi} \int_0^\pi \log \cosh\left[\left(\frac{E\varepsilon_{max}}{\sigma_y}\right)\cos^2 \bar{X}\right] d\bar{X} \qquad (4\text{-}100)$$

A computer is needed to numerically integrate Equation (4-100) for various constant values of scaled strain $\dfrac{E\varepsilon_{max}}{\sigma_y}$. Equation (4-100) does show that the impulsive loading realm asymptote in functional format can be given by:

$$\frac{i\ b\ E^{1/2}}{\rho^{1/2}\ \sigma_y\ A} = \psi\left(\frac{E\varepsilon_{max}}{\sigma_y}\right) \qquad \text{(Impulsive Realm)} \qquad (4\text{-}101)$$

Equation (4-100) is plotted in Figure 4-29 as a series of constant scaled maximum strain asymptotes in the impulsive loading realm.

To obtain the quasi-static loading realm asymptote, we calculate the work Wk.

$$Wk = Pb\ w_o \int_0^L \sin \frac{\pi x}{L}\ dx \qquad (4\text{-}102)$$

or

$$Wk = \frac{2\ Pb\ L\ w_o}{\pi} \qquad (4\text{-}103)$$

Substituting Equation (4-93) for w_o in Equation (4-103), equating Equation (4-103) to Equation (4-98), and rearranging terms gives an equation for the quasi-static asymptotes.

$$\frac{Pb\ L\ E^{1/2}}{\sigma_y^{3/2}\ A} = \frac{(\pi/2)^{3/2}}{\left(\frac{E\varepsilon_{max}}{\sigma_y}\right)^{1/2}} \int_0^\pi \log\ \cosh\left[\left(\frac{E\varepsilon_{max}}{\sigma_y}\right)\cos^2\bar{X}\right] d\bar{X} \qquad (4\text{-}104)$$

A computer is also needed to numerically integrate Equation (4-104) for con-
stant values of $\frac{E\varepsilon_{max}}{\sigma_y}$. Equation (4-104) shows that the quasi-static loading
realm asymptote is functionally given by:

$$\frac{Pb\ L\ E^{1/2}}{\rho^{1/2}\ \sigma_y\ A} = \psi\left(\frac{E\varepsilon_{max}}{\sigma_y}\right) = \psi\ (\mu) \qquad \text{(Quasi-Static Realm)} \qquad (4\text{-}105)$$

Equation (4-105) with the proper functional format is plotted as the asymptotes to the quasi-static loading realm in Figure 4-29. An approximation
still had to be made to establish a transition between the impulsive and
quasi-static loading realms. The same hyperbolic tangent squared relationship, Equation (4-81), was used for this extensional solution as had been
used in the beam solutions.

The derivation of this elastic-plastic solution for beam extension
shows the complexity that is involved whether bending or extensional
solutions are being derived. In the earlier elastic or rigid-plastic solutions, the mathematics were easier and closed-form solutions were possible.
The previously derived elastic bending beam solution was an excellent example.

Buckling of Columns

Figure 4-30 shows a scaled pressure-impulse diagram for buckling of
an axially loaded elastic column. Different boundary conditions and the
possibility of having sidesway are accounted for in the α_p and α_i coefficients associated with pressure and impulse. The solid line in Figure 4-30
is the threshold separating unstable column response from stable. If the
nondimensional loads imparted to a column establish a point which is to the

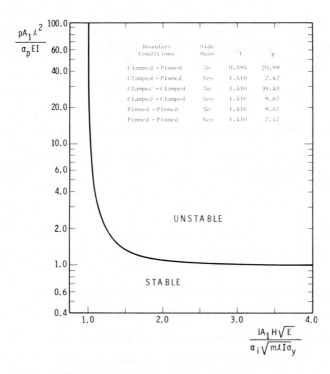

Figure 4-30. Buckling for Dynamic Axial Loads

left and/or below the threshold line, then the column should remain stable.
On the other hand, should these nondimensional loads establish a point above
and to the right of the threshold, large permanent, unstable deformation
should be expected. In developing this solution, energy procedures were
once again applied. The major new parameter is the mass (not weight) of
the overlying floor M. We assume that the mass of the column is insignificant relative to the mass of the heavy floor above. The parameters L, E,
I, σ_y, and h all pertain to the total span, modulus of elasticity, second
moment of area, yield point, and total depth of the column. The parameter
A_1 is the loaded area of the roof or floor over the column. Dead weight
effects are ignored in this solution as they are considered to be insignificant relative to the dynamic loads from the applied blast wave.

To derive a column solution, we again assume a deformed shape. If the column is simply supported without sidesway, a sine wave, as in Equation (4-106), is a good assumption

$$w = w_o \sin \frac{\pi x}{L} \qquad (4\text{-}106)$$

Differentiating Equation (4-106) twice and substituting into $M = -EI \dfrac{d^2 w}{dx^2}$ gives the moment

$$M = \frac{\pi^2 EI \, w_o}{L^2} \sin \frac{\pi x}{L} \qquad (4\text{-}107)$$

The strain energy is the integral $\displaystyle\int_o^L \frac{M^2 dx}{2EI}$ or

$$S.E. = 2 \int_o^{L/2} \frac{\pi^4 EI \, w_o^2}{2L^4} \sin^2 \left(\frac{\pi x}{L}\right) dx \qquad (4\text{-}108)$$

which, upon integrating, gives:

$$S.E. = \frac{\pi^4 EI \, w_o^2}{4L^3} \qquad (4\text{-}109)$$

The load on the column will act through a deflection δ equal to S-L, where L is the original length of the column. The differential length dS is given by:

$$dS = dx \sqrt{1 + \left(\frac{dy}{dx}\right)^2} \qquad (4\text{-}110)$$

Expanding Equation (4-110) using the binomial theorem and then integrating
to find the total length S gives:

$$S = \int_o^L dx \left[1 + (1/2) \left(\frac{dy}{dx} \right)^2 + \ldots \right] \tag{4-111}$$

Completing this integration and subtracting L from S to obtain δ gives, as
a first approximation:

$$\delta = (1/2) \int_o^L \left(\frac{dy}{dx} \right)^2 dx \tag{4-112}$$

We can now proceed to solve for the work which equals $P\,A_1\,\delta$ or:

$$Wk = P\,A_1\,\delta = \frac{P\,A_1}{2} \int_o^L \left(\frac{dy}{dx} \right)^2 dx \tag{4-113}$$

Substituting in the first derivative of Equation (4-113) to integrate gives:

$$WK = \frac{\pi^2\,P\,A_1\,w_o^2}{2L^2} \int_o^L \cos^2 \left(\frac{\pi x}{L} \right) dx \tag{4-114}$$

or, upon completion:

$$Wk = \frac{\pi^2\,P\,A_1\,w_o^2}{4L} \tag{4-115}$$

The quasi-static asymptote is obtained when the strain energy is
equated to the work:

$$\frac{\pi^4 \; EI \; w_o^2}{4L^3} = \frac{\pi^2 \; P \; A_1 \; w_o^2}{4L} \tag{4-116}$$

or:

$$\frac{P \; A_1 \; L^2}{EI} = \pi^2 \qquad \begin{array}{l}\text{(quasi-static asymptote} \\ \text{S.S. beam; no sidesway)}\end{array} \tag{4-117}$$

Equation (4-117) is the Euler beam solution with a dynamic load factor of 1.0 instead of 2.0. Because the vertical load $P \; A_1$ is independent of w_o, we have the classical, small deformation, Euler column instability. The factor α_p in Figure 4-30 is equal to π^2 for this pinned-pinned column without sidesway. The concept of effective column length with L being the distance between points of inflection can be applied in this analysis. A review of α_p for a pinned-pinned column with sidesway shows a column with only one quarter the strength because the effective length of the column is twice as long. Similarly, α_p for a clamped-clamped column without sidesway is four times stronger than the simply supported column because the effective length is halved.

To compute buckling in the impulsive loading realm, we need the kinetic energy imparted to the overlying mass m. This kinetic energy equals:

$$K.E. = (1/2) \; m \; V_o^2 = (1/2) \; m \left(\frac{i \; A_1}{m}\right)^2 \tag{4-118}$$

or:

$$K.E. = \frac{i^2 \; A_1^2}{2m} \tag{4-119}$$

Equating K.E. to S.E. gives the impulsive loading realm asymptote.

$$\frac{i^2 A_1^2}{2m} = \frac{\pi^4 EI w_o^2}{4L^3} \qquad (4\text{-}120)$$

Notice that, unlike the quasi-static loading realm result, the deformation w_o does not cancel out of Equation (4-120). This result means that "stable buckling" occurs in the impulsive loading realm. A certain quantity of kinetic energy is being put into the column, which strain energy can dissipate until the deformations are large enough to cause yielding. This observation means that we must use Equation (4-107) to obtain the maximum moment by setting $\sin \frac{\pi x}{L}$ equal to 1.0, and substitute into $\sigma = \frac{Mh}{2I}$ to relate the maximum bending stress (to be limited by σ_y) to the deformation w_o. This substitution gives:

$$\sigma_y = \frac{\pi^2 E h w_o}{2L^2} \qquad (4\text{-}121)$$

Substituting Equation (4-121) into Equation (4-120), rearranging terms algebraically, and taking the square root of the result finally gives:

$$\frac{\left(i A_1\right) \sqrt{E} h}{\sigma_y \sqrt{M L I}} = \sqrt{2.0} \qquad \begin{array}{l}\text{(Impulse asymptote S.S.}\\ \text{beam; no sidesway)}\end{array} \qquad (4\text{-}122)$$

The numerical coefficient $\sqrt{2.0}$ is the α_i coefficient in Figure 4-30. Other α_i coefficients must be computed independently. The static concept of effective length no longer applies in the impulsive loading realm; hence, it should not be used. In the impulsive loading realm, buckling does not occur in the classical sense; it is really a bending process. Permanent deformation does not occur until the column yields in the impulsive loading

341

realm. The same Equation (4-81) was used to estimate a transition between
the quasi-static and impulsive loading realms as has been used to approximate this transition in all earlier analyses.

PLATES

Figure 4-31 is an elastic scaled overpressure versus scaled specific
impulse diagram for initiating failure in a plate. By failure in a plate,
we mean that the bending stress has reached yield at one location only in a
brittle plate, and a complete failure mechanism has reached yield in a ductile plate. A ductile plate will undergo no significant permanent deformation
just because one location has yielded. In a ductile material, much higher
loads can be carried until such time as a series of yield lines form a

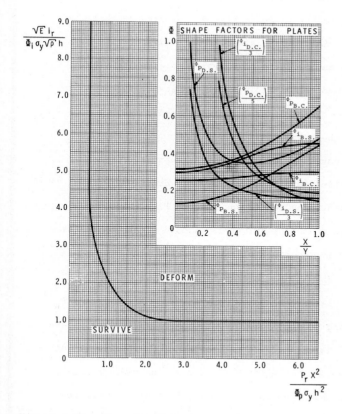

Figure 4-31. Normalized Load Impulse Diagram for Plates

collapse mechanism. On the other hand, a brittle plate will fracture as soon as yield is reached anywhere in the plate. Hence, a ductile plate will carry much more load than a brittle one. The graphical solution presented in Figure 4-31 accounts for these differences in material characteristics.

The insert to Figure 4-31 is entitled "shape factors for plates." The ϕ_i function for impulse and the ϕ_p function for pressure that are obtained from this insert account for: (1) the plate aspect ratio X/Y, (2) plate boundary conditions, and (3) plate ductility or brittleness. In this insert, the first subscript which is either i or p indicates whether this is a ϕ factor for an impulse i or a pressure p. The second subscript, which is either a B or a D, indicates whether the brittle B or ductile D solution is being applied. The third and last subscript, which is either an S or a C, indicates whether the plate is simply supported S or clamped C on all boundaries. As an example, the symbol $\phi_{i_{B.C.}}$ implies that this is the ϕ factor for impulse in a brittle plate which is clamped. The ϕ function is determined by reading the ordinate of the insert for an appropriate plate aspect ratio on the abscissa. In Figure 4-31, X and Y represent plate half spans, not the total spans.

After the ϕ factors have been determined, the main figure can be used to determine if a brittle plate fractures or if a ductile plate undergoes permanent deformation. The continuous line in Figure 4-31 is a threshold determined through an energy solution. Should the point describing the load fall above and to the right of the threshold line, the plate should deform or fracture. Should the loading be located below and to the left of the threshold, the plate should survive. The parameters i_R and P_R are the reflected impulse and pressure imparted to the plate. Plate parameters include ρ the mass density, h the total thickness, X the shorter of the two half spans, E the modulus of elasticity, and σ_y the yield point. All plates are assumed to be homogeneous, isotropic, flat slabs which respond in bending to uniformly applied pressures and impulses.

To derive the P-i diagram for a simply supported plate, we begin once again by assuming a deformed shape w as given by:

$$w = w_o \cos \frac{\pi x}{2X} \cos \frac{\pi y}{2Y} \qquad (4\text{-}123)$$

The strains are given in a bending plate by:

$$\varepsilon_{xx} = -z \frac{\partial^2 w}{\partial x^2}, \quad \varepsilon_{yy} = -z \frac{\partial^2 w}{\partial y^2}, \quad \text{and } \varepsilon_{xy} = 2z \frac{\partial^2 w}{\partial x \, \partial y} \qquad (4\text{-}124)$$

The strain energy S.E. per unit volume under a biaxial state of stress is:

$$\frac{S.E.}{Vol.} = \left(\frac{E}{2} \varepsilon_{xx}^2 + G \varepsilon_{xy}^2 + \frac{E}{2} \varepsilon_{yy}^2 \right) \qquad (4\text{-}125)$$

where G is the shear modulus equal to $\frac{E}{2(1+\nu)}$. Differentiating Equation (4-123), substituting in the strain relationships, and squaring yields:

$$\varepsilon_{xx}^2 = \frac{\pi^4 w_o^2 z^2}{16 X^4} \cos^2 \left(\frac{\pi x}{2X} \right) \cos^2 \left(\frac{\pi y}{2Y} \right) \qquad (4\text{-}126a)$$

$$\varepsilon_{yy}^2 = \frac{\pi^4 w_o^2 z^2}{16 Y^4} \cos^2 \left(\frac{\pi x}{2X} \right) \cos^2 \left(\frac{\pi y}{2Y} \right) \qquad (4\text{-}126b)$$

$$\varepsilon_{xy}^2 = \frac{\pi^4 w_o^2 z^2}{4X^2 Y^2} \sin^2 \left(\frac{\pi x}{2X} \right) \sin^2 \left(\frac{\pi y}{2Y} \right) \qquad (4\text{-}126c)$$

Substituting for G in Equation (4-125), 0.3 for ν, and Equations (4-126), results in a triple integral for the strain energy. Performing the required integration and collecting terms eventually yields:

$$\text{S.E.} = \frac{\pi^4}{6(4)^3} \frac{E \, w_o^2 \, h^3}{XY} \left[\left(\frac{Y}{X}\right)^2 + 3.08 + \left(\frac{X}{Y}\right)^2 \right] \qquad (4\text{-}127)$$

The kinetic energy imparted to the plate is easily determined. The kinetic energy is given by:

$$\text{K.E.} = \sum_{\text{plate}} \frac{m}{2} V_o^2 = 4 \int_0^X \int_0^Y \left[\frac{i^2 \, (dx)^2 \, (dy)^2}{2 \, \rho \, h \, (dx) \, (dy)} \right] = 2 \, i^2 \frac{XY}{\rho h} \qquad (4\text{-}128)$$

Equating K.E. to S.E. gives the impulsive loading realm asymptote of:

$$\frac{w_o}{Y} = \frac{16 \, \sqrt{3}}{\pi^2 \left[\left(\frac{Y}{X}\right)^2 + 3.08 + \left(\frac{X}{Y}\right)^2 \right]^{1/2}} \left(\frac{iX}{\sqrt{\rho E} \, h^2} \right) \qquad (4\text{-}129)$$

The strains are maximum in the center of a simply supported plate and at the outer fibers (i.e., at $\cos \frac{\pi x}{2X} = 1.0$, $\cos \frac{\pi y}{2Y} = 1.0$, and $z = h/2$) where

$$\varepsilon_{xx_{max}} = \frac{\pi^2}{8} \left(\frac{h}{X}\right) \left(\frac{Y}{X}\right) \left(\frac{w_o}{Y}\right) \qquad (4\text{-}130a)$$

$$\varepsilon_{yy_{max}} = \frac{\pi^2}{8} \left(\frac{h}{Y}\right) \left(\frac{w_o}{Y}\right) \qquad (4\text{-}130b)$$

$$\varepsilon_{xy} = 0 \qquad (4\text{-}130c)$$

Substituting for w_o/Y then gives:

$$\varepsilon_{xx_{max}} = \frac{2\sqrt{3}}{\left[1.0 + 3.08\left(\frac{x}{y}\right)^2 + \left(\frac{x}{y}\right)^4\right]^{1/2}} \left(\frac{i}{\sqrt{\rho E}\ h}\right) \tag{4-131a}$$

$$\varepsilon_{yy_{max}} = \frac{2\sqrt{3}}{\left[1.0 + 3.08\left(\frac{y}{x}\right)^2 + \left(\frac{y}{x}\right)^4\right]^{1/2}} \left(\frac{i}{\sqrt{\rho E}\ h}\right) \tag{4-131b}$$

But, in a plate, stress is related to elastic strain through Hooke's Law:

$$\sigma_{xx} = \frac{E}{(1-\nu^2)} \left(\varepsilon_{xx} + \nu\,\varepsilon_{yy}\right) \tag{4-132a}$$

$$\sigma_{yy} = \frac{E}{(1-\nu^2)} \left(\varepsilon_{yy} + \nu\,\varepsilon_{xx}\right) \tag{4-132b}$$

Hence, substituting for strains yields:

$$\frac{\sigma_{xx_{max}}}{E} = \left(\frac{3.81\ i}{\sqrt{\rho E}\ h}\right) \frac{\left[0.3 + \left(\frac{y}{x}\right)^2\right]}{\left[1.0 + 3.08\left(\frac{y}{x}\right)^2 + \left(\frac{y}{x}\right)^4\right]^{1/2}} \tag{4-133a}$$

$$\frac{\sigma_{yy_{max}}}{E} = \left(\frac{3.81\ i}{\sqrt{\rho E}\ h}\right) \frac{\left[1.0 + 0.3\left(\frac{y}{x}\right)^2\right]}{\left[1.0 + 3.08\left(\frac{y}{x}\right)^2 + \left(\frac{y}{x}\right)^4\right]^{1/2}} \tag{4-133b}$$

Next, we need a yield criterion for a biaxial state of stress. If we use the von Mises yield criterion, then:

$$\left(\sigma_{yy} - \sigma_{zz}\right)^2 + \left(\sigma_{zz} - \sigma_{xx}\right)^2 + \left(\sigma_{xx} - \sigma_{yy}\right)^2 = 2\,\sigma_y^2 \tag{4-134}$$

where σ_y is the yield stress under uniaxial loading.

In Equation (4-134), σ_{zz} equals zero; hence, Equation (4-134) becomes:

$$\left(\frac{\sigma_{yy}}{E}\right)^2 - \left(\frac{\sigma_{yy}}{E}\right)\left(\frac{\sigma_{xx}}{E}\right) + \left(\frac{\sigma_{xx}}{E}\right)^2 = \left(\frac{\sigma_y}{E}\right)^2 \tag{4-135}$$

Substituting Equations (4-133) in Equation (4-135) and gathering terms finally yields the impulsive loading realm asymptote for an elastic simply supported plate:

$$\left(\frac{3.81\ \sqrt{E}\ i}{\sigma_y\ \sqrt{\rho}\ h}\right) = \left[\frac{1.0 + 3.08\left(\frac{X}{Y}\right)^2 + \left(\frac{X}{Y}\right)^4}{0.79 + 0.11\left(\frac{X}{Y}\right)^2 + 0.79\left(\frac{X}{Y}\right)^4}\right]^{1/2} \left\{\begin{array}{l}\text{brittle S.S.}\\\text{plate}\\[6pt]\text{impulsive}\\\text{asymptote}\end{array}\right\} \tag{4-136}$$

This is the asymptote plotted in Figure 4-31 for a brittle simply supported plate in the impulsive realm. The ϕ_i function in Figure 4-31 accounts for the effects of the (X/Y) aspect ratio and any numerical constants. A simply supported plate is assumed to fail by shattering upon reaching yield anywhere in the plate.

Even though a ductile plate has reached yield at its center, this plate will neither rupture nor deform plastically. Before permanent plastic deformation can be observed, a ductile plate must form a collapse mechanism by the creation of yield lines. This behavior gives a ductile material more energy absorbing capabilities than its brittle counterpart. It is very difficult to extend this solution to an elastic-plastic one; however, we will do it by ignoring any localized plastic behavior so that a ductile material elastic solution estimate can be made. This approach is not rigorously correct; it is used to obtain acceptable engineering answers. In a simply supported plate, the yield lines which originate at the center eventually propagate out toward the plate corners. After reaching the corners, the plate collapses. In the corners at $x = X/2$ and $y = Y/2$, no normal stresses exist. Collapse is caused by shear stresses reaching the appropriate yield condition. In the corner, the shear stress is given by:

$$\varepsilon_{xy} = \frac{\pi^2 z w_o}{2 X Y} \tag{4-137}$$

Substituting $h/2$ for z and Equation (4-137) for $\frac{w_o}{Y}$ in Equation (4-129) gives:

$$\varepsilon_{xy} = \frac{4 \sqrt{3} i}{h \sqrt{\rho E} \left[\left(\frac{Y}{X}\right)^2 + 3.08 + \left(\frac{X}{Y}\right)^2 \right]^{1/2}} \tag{4-138}$$

The yield criterion for shear requires that $\tau_{xy} = \sigma_y / \sqrt{3}$. Setting this yield stress equal to $G\varepsilon_{xy}$ and rearranging terms finally gives:

$$\frac{\sqrt{E} \; i}{\sigma_y \sqrt{\rho} \; h} = \frac{1 + \nu}{6} \left[\left(\frac{Y}{X}\right)^2 + 3.08 + \left(\frac{X}{Y}\right)^2 \right]^{1/2} \left. \begin{cases} \text{ductile S.S.} \\ \text{plate} \\ \\ \text{impulsive} \\ \text{asymptote} \end{cases} \right\} \tag{4-139}$$

This is the asymptote plotted in Figure 4-31 for a ductile, simply supported plate in the impulsive loading realm.

Before the quasi-static asymptotes can be calculated for a simply supported plate, the maximum possible work must be estimated. This work Wk is given by the integral:

$$Wk = 4 \int_o^X \int_o^Y P \; w \; dx \; dy = 4 \, P \, w_o \int_o^X \int_o^Y \cos \frac{\pi x}{2X} \cos \frac{\pi y}{2Y} \, dx \, dy \tag{4-140}$$

348

or

$$Wk = \frac{16}{\pi^2} P \, w_o \, X \, Y \tag{4-141}$$

Equating Wk to S.E. gives the quasi-static loading realm asymptote of:

$$\frac{w_o}{Y} = \frac{6(4)^5 \, P \, X^2 \, Y}{\pi^6 \, E \, h^3 \left[\left(\frac{Y}{X}\right)^2 + 3.08 + \left(\frac{X}{Y}\right)^2 \right]} \tag{4-142}$$

Substituting Equation (4-142) into the strain Equations (4-130) and using Hooke's Law gives:

$$\frac{\sigma_{xx_{max}}}{E} = \frac{8.68 \left(\frac{P}{E}\right)\left(\frac{Y^2}{h^2} + 0.3 \frac{X^2}{h^2}\right)}{\left[\left(\frac{Y}{X}\right)^2 + 3.08 + \left(\frac{X}{Y}\right)^2 \right]} \tag{4-143}$$

$$\frac{\sigma_{yy_{max}}}{E} = \frac{8.68 \left(\frac{P}{E}\right)\left(\frac{X^2}{h^2} + 0.3 \frac{Y^2}{h^2}\right)}{\left[\left(\frac{Y}{X}\right)^2 + 3.08 + \left(\frac{X}{Y}\right)^2 \right]} \tag{4-144}$$

Finally, use of Equation (4-135) as a yield criterion, substitution of Equation (4-144) into Equation (4-135), and rearrangement of terms gives the quasi-static loading realm asymptote for brittle fracture in a simply supported plate.

$$\frac{8.68 \, P \, X^2}{\sigma_y \, h^2} = \frac{1.0 + 3.08 \left(\frac{X}{Y}\right)^2 + \left(\frac{X}{Y}\right)^4}{\left[0.79 + 0.11 \left(\frac{X}{Y}\right)^2 + 0.79 \left(\frac{X}{Y}\right)^4 \right]^{1/2}} \left. \begin{array}{l} \text{brittle S.S.} \\ \text{plate} \\ \\ \text{quasi-static} \\ \text{asymptote} \end{array} \right\} \tag{4-145}$$

As in the impulsive loading realm, the ductile plate in the quasi-static loading realm is assumed to form a collapse mechanism when the shears in the corners reach yield. The shear stress in the corners is given by Equation (4-137) in both loading realms. Substituting h/2 for z and Equation (4-142) for w_o/Y gives:

$$\varepsilon_{xy} = \frac{3(4)^5 \ P \ X \ Y}{2 \ \pi^4 \ E \ h^2 \left[\left(\frac{Y}{X}\right)^2 + 3.08 + \left(\frac{X}{Y}\right)\right]^2} \qquad (4\text{-}146)$$

Finally, setting the shear yield stress $\sigma_y/\sqrt{3}$ equal to $\left[\dfrac{E}{2 \ (1 + \nu)}\right] \varepsilon_{xy}$, substituting Equation (4-146) for ε_{xy}, and rearranging terms yields the asymptote for ductile, simply supported plates in the quasi-static loading realm.

$$\frac{P \ X^2}{\sigma_y \ h^2} = 0.09524 \left[\left(\frac{Y}{X}\right) + 3.08 \left(\frac{X}{Y}\right) + \left(\frac{X}{Y}\right)^3\right] \quad \begin{Bmatrix} \text{ductile S.S.} \\ \text{plate} \\ \\ \text{quasi-static} \\ \text{asymptote} \end{Bmatrix} \qquad (4\text{-}147)$$

This asymptote is also plotted in Figure 4-31 with the right-hand side of the equation the $\phi_{P_{D.S.}}$ function. The reader can now derive the ϕ_p and ϕ_i functions for clamped plates. The procedures are precisely the same; however, the algebra is even more tedious because separate segments of the plate must be integrated independently and summed to account for changes in sign at inflection points.

As can be seen, the same procedures are used to solve plate equations as beam equations. In the limit with Y/X = ∞, these plate equations do not yield the beam solutions. These variations are caused by the difference between plain stress versus plain strain solutions and by differences in the equations for assumed deformed shape.

350

Structural Response Times

Because we analyze structures by subdividing them into components
with rigid rather than elastic supports, structural periods can also be of
interest. Provided the periods of supporting backup structures are long
relative to the fundamental period of the structural component, uncoupled
structural response calculations are appropriate. The only computational
procedure which explicitly calculates periods of natural frequencies is
the coupled multi-history frame solutions that appear in the next chapter.
The graphical energy solutions which have been used in this chapter can all
be used to estimate the first fundamental period associated with that of

Table 4-3. Fundamental Periods for Various
Structural Components

Type of Structure	Equation
Elastic Beam	$\dfrac{\tau}{L^2} \dfrac{\sqrt{EI}}{\sqrt{\rho A}} = 3.63 \left(\dfrac{\alpha_i}{\alpha_p} \right)$
Elastic-Plastic Beam	$\dfrac{\tau}{L^2} \dfrac{\sqrt{EI}}{\sqrt{\rho A}} = 11.81 \left(\dfrac{\psi_i}{\psi_p} \right) \left[\dfrac{Pb \; L^2}{\psi_p \; \sigma_y \; Z} \right]^{0.302}$
Elastic-Plastic String	$\dfrac{\tau}{L} \dfrac{\sqrt{E}}{\sqrt{\rho}} = 1.57 \left[\dfrac{\sigma_y^{1/2} \; E^{1/2} \; A}{Pb \; L} \right]^{0.285}$
Buckling Column	$\dfrac{\tau \; h \; E^{3/2} \; I^{1/2}}{\sigma_y \; M^{1/2} \; L^{5/2}} = 2.72 \left(\dfrac{\alpha_i}{\alpha_p} \right)$
Plate Bending	$\dfrac{\tau}{\sqrt{\rho}} \dfrac{\sqrt{E} \; h}{x^2} = 5.436 \dfrac{\phi_i}{\phi_p}$

that of response by realizing this period occurs close to the elbows in the
nondimensionalized pressure-impulse diagrams. From this insight, Table 4-3
was created for estimating various structural periods.

All parameters in Table 4-3 are as used in each analysis for each
component. The α_i, α_p, ϕ_i, and ϕ_p components are also associated with each
solution and account for the influence of different boundary conditions.
The elastic solutions all result in relatively simple algebraic expressions;
however, the elastic-plastic beams or string solutions result in a little
more complicated format because of the added complexities caused by plasticity.

SUMMARY

In this chapter, we presented the concept of a pressure-impulse P-i
diagram. The P-i curve of approximate rectangular hyperbola shape applies
to elastic and plastic beams, plates, and columns. We have also shown
how P-i diagrams should be modified for the various shaped blast pressure
pulses associated with pressure vessel ruptures and dust explosions. By
using British bomb damage data from World War II, we illustrated that P-i
diagrams are not merely a mathematical concept, but apply to real structures.

Next, we demonstrated that energy solutions can be used as an easily
applied procedure for estimating the asymptote to the P-i diagram. The impulsive loading realm asymptote is obtained by equating kinetic energy to
strain energy, and the quasi-static loading realm is obtained by equating
the maximum possible work to strain energy. The applicability of these
energy solutions for estimating maximum deformations and strains in both
elastic and plastic beams and plates has been illustrated in a series of
comparisons with test results.

To aid in the design and analysis of beams, plates, extensional strips,
and columns, nondimensional pressure-specific impulse (P-i) diagrams are pre-

sented which can be used to solve for deflections and strains in blast-loaded structural components. Finally, to illustrate how a structure is analyzed by breaking it up into components and applying these nondimensional pressure-impulse diagrams, an example problem is given for blast loaded, framed, sheet metal buildings.

CHAPTER 4
EXAMPLE PROBLEM

This example problem will be worked using the English system of units with metric equivalency in brackets. The reason for departing from the metric system used throughout the rest of this book is that most civil engineering handbooks have structural section properties in English units. To use metric equivalency of section properties would cause confusion if this book did not conform.

Definition of the Problem

Figure 4-32 is a plan view of a framed, one-story, 13-foot (3.96 m) tall steel building with four square bays. All distances between columns

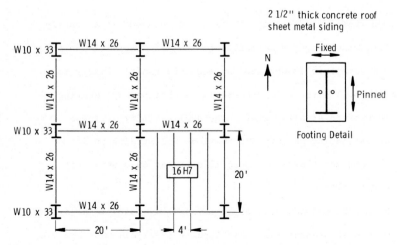

Figure 4-32. Plan View for Example Problem

are 20 feet (6.10 m). All roof beams are W 14 x 26, and 16H7 roof joists

run in the N-S direction on 4-foot (1.22 m) centers. All 13-foot (3.96 m)

tall columns are W 10 x 39, and are pinned-pinned in the E-W direction and

clamped-clamped in the N-S direction. All roof beams and joists are as-

sumed to be pinned-pinned. Properties from <u>Manual of Steel Construction</u>

(1975) for these steel chambers are given in Table 4-4.

The roof of the building is concrete 2.5 inches (6.35 cm) thick which

has a weight per unit area of 33.3 lb/ft^2 (1596 N/m^2). The wall of the

building is a corrugated sheet metal siding whose maximum span is the full

13 feet (3.96 m) from roof to floor. The properties of this siding also ap-

pear in Table 4-4. We will assume that all beams and the siding have a

yield stress σ_y of 33,000 psi (2.28 x 10^{+8} Pa).

The loading imparted to this building will be an external blast wave

moving west to east with a side-on peak overpressure P_s of 1.42 psi (9.79 x

10^{+3} Pa) and a side-on specific impulse i_s of 0.145 psi-sec (100.0 Pa·s).

The normally reflected peak overpressure P_r associated with this loading is

3.00 psi (2.07 x 10^{+4} Pa) and the normally reflected specific impulse i_r is

0.300 psi-sec (206.8 Pa·s).

In this problem, use the results developed and presented in this

chapter to determine if: 1) the blast load is "severe enough" to cause

damage, and, if it could cause damage, then perform a more detailed analy-

sis to see if: 2) the roof joists are adequate, 3) the roof beams are

adequate, 4) the columns will buckle, 5) the columns will be bent, and 6)

the siding will be deformed.

Approximate Analysis

Figure 4-9 for bomb damage to British brick homes can be used to

determine approximately if this blast load is a threat. Because Figure

4-9 is a side-on pressure-impulse diagram in the metric system, we plot

a P_s of 9,790 Pa and an i_s of 100.0 Pa·sec. This point falls just below

the impulse threshold for "Threshold for Minor Structural Damage - Wrenched

Table 4-4. Properties of Structural Members

A. W 14 x 26 Beam

Symbol	Description	Property	
		English	Metric
A	Cross-Section Area	7.67 in^2	4.95 x 10^{-3} m^2
h	Depth	13.89 in	0.353 m
I$_{xx}$	Second Moment of Area About Major Axis	244.00 in^4	1.02 x 10^{-4} m^4
I$_{yy}$	Second Moment of Area About Minor Axis	8.86 in^4	3.70 x 10^{-6} m^4
S$_{xx}$	Elastic Section Modulus About Major Axis	35.10 in^3	5.75 x 10^{-4} m^3
Z$_{xx}$	Plastic Section Modulus About Major Axis	40.00 in^3	6.55 x 10^{-4} m^3
-	Weight Per Unit Length	26.00 lb/ft	38.7 kg/m

B. W 10 x 33 Beam

Symbol	Description	English	Metric
A	Cross-Section Area	11.50 in^2	7.42 x 10^{-3} m^2
b	Flange Width	8.00 in	0.203 m
h	Depth	9.75 in	0.248 m
I$_{xx}$	Second Moment of Area About Major Axis	171.00 in^4	0.712 x 10^{-4} m^2
I$_{yy}$	Second Moment of Area About Minor Axis	36.50 in^4	0.152 x 10^{-4} m^2
S$_{xx}$	Elastic Section Modulus About Major Axis	35.00 in^3	5.73 x 10^{-4} m^3

Joints and Partitions." Because this technique is approximate, some minor damage should be expected.

Damage to Roof Joist

A roof joist responds in bending as a simply supported beam. Figure

Table 4-4. Properties of Structural Members (Cont'd)

B. W 10 x 33 Beam (Cont'd)

Symbol	Description	Property	
		English	Metric
Z_{xx}	Plastic Section Modulus About Major Axis	38.80 in^3	6.36×10^{-4} m^3
Z_{yy}	Plastic Section Modulus About Minor Axis	14.00 in^3	2.29×10^{-4} m^3
-	Weight Per Unit Length	33.00 lb/ft	49.1 kg/m

C. 16H7 Open-web Steel Joist

M_y	Yield Moment	413,000.00 in-lb	46,700.00 N·m
h	Depth	16.00 in	0.406 m
I_{xx}	Second Moment of Area About Major Axis	110.00 in^4	0.458×10^{-4} m^4
-	Weight Per Unit Length	10.30 lb/ft	15.4 kg/m

D. Corrugated Steel Siding

A/b	Cross-Section Area Per Unit Width	0.75 in^2/ft	1.59×10^{-3} m
I_{xx}/b	Second Moment of Area Per Unit Width	0.138 in^4/ft	1.88×10^{-7} m^3
S_x/b	Elastic Section Modulus Per Unit Width	0.149 in^3/ft	8.01×10^{-6} m^2
-	Weight Per Unit Area	3.40 lb/ft^2	163 N/m^2

4-26 is an elastic-plastic beam bending solution which can be used to evaluate the safety of the roof joists. We will assume that the concrete adds no structural strength (a conservative assumption), but adds mass when a 4-foot wide strip of concrete responds along with the roof joist. Because the roof will be enveloped with a side-on blast wave, P_s and i_s are the appropriate applied loads. To use Figure 4-26, we must compute the scaled overpressure

$$\frac{P\,b\,L^2}{\psi_p\,\sigma_y\,Z} = \frac{P\,b\,L^2}{\psi_p\left(M_y\right)} = \frac{(1.42)\,(4 \times 12)\,(20 \times 12)^2}{(10)\,(413,000)} = 0.951$$

and the scaled specific impulse

$$\frac{i\,b\,\sqrt{EI}}{\psi_i\,\sqrt{\rho A}\,\sigma_y\,Z} = \frac{i\,b\,\sqrt{E}\,\sqrt{I}\,\sqrt{g}}{\psi_i\,\sqrt{\rho g A}\left(M_y\right)} = \frac{(0.0145)\,(4 \times 12)\,\sqrt{30 \times 10^{+6}}\,\sqrt{110}\,\sqrt{386}}{(0.913)\,\sqrt{\frac{10.3}{12} + \frac{33.3 \times 4}{12}}\,(413,000)} = 0.602$$

Reading the scaled strain from Figure 4-26 gives a value of 0.3. Substituting for the parameters in the scaled strain allows us to estimate the strain from the explosion.

$$\frac{E\,I\,\varepsilon_{max}}{\psi_\varepsilon\,h\,Z\,\sigma_y} = \frac{I\,\sigma_{max}}{\psi_\varepsilon\,h\,M_y} = \frac{(110)\,\sigma_{max}}{(1.25)\,(16.0)\,(413,000)} = 0.3$$

Rearranging, we get:

$$\sigma_{max}\,(\text{blast}) = 22,500\ \text{psi}\ (1.55 \times 10^{+8}\ \text{Pa})$$

This maximum stress is caused by the blast load. In addition, the dead weight stress of the steel joist and concrete roof must be considered. This dead weight stress is given by:

$$\sigma = \frac{w\,L^2}{8\,S} = \frac{\left[\frac{10.3}{12} + \left(\frac{33.3 \times 4}{12}\right)\right]\,[20 \times 12]^2}{8\,[110/(16/2)]}$$

or

$$\sigma \text{ (dead load)} = 6,260 \text{ psi } (0.432 \times 10^{+8} \text{ Pa})$$

The sum of both stresses is 28,760 psi ($1.98 \times 10^{+8}$ Pa) which is less than the 33,000 psi ($2.28 \times 10^{+8}$ Pa) yield stress; thus, the roof joist will not yield.

Damage to Roof Beam

We also use the bending beam solution for the roof beam. One of the W 14 x 26 beams running E-W through the center of the building should be the heaviest loaded roof beam because the roof joists from both sides apply loads to this beam. Because beam bending again dominates, Figure 4-26 can be used. Using section properties from Table 4-4;

$$\frac{P \ b \ L^2}{\psi_p \ \sigma_y \ Z_{xx}} = \frac{(1.42) \ (20 \times 12) \ (20 \times 12)^2}{(10) \ (33,000) \ (38.8)} = 1.53$$

The scaled impulse is given by:

$$\frac{i \ b \ \sqrt{EI}}{\psi_i \ \sqrt{\rho A} \ \sigma_y \ Z} =$$

$$\frac{(0.0145) \ (20 \times 12) \ \sqrt{30 \times 10^{+6}} \ \sqrt{244} \ \sqrt{386}}{(0.913) \ \sqrt{\left(\dfrac{33.3 \times 20}{12}\right) + \left(\dfrac{10.3 \times 4 \times 20}{20 \times 12}\right) + \left(\dfrac{26}{12}\right)} \ (33,000) \ (38.8)} = 0.640$$

Interpolating for the scaled strain from Figure 4-26 gives a value of 0.34. Substituting for the parameters in the scaled strain allows us to estimate the stress from blasting

$$\frac{E \ I \ \varepsilon_{max}}{\psi_\varepsilon \ h \ Z \ \sigma_y} = \frac{(244) \ \sigma_{max}}{(1.25) \ (13.89) \ (38.8) \ (33,000)} = 0.34$$

or

$$\sigma_{max} \text{ (blast)} = 31{,}000 \text{ psi } (2.16 \times 10^{+8} \text{ Pa})$$

The dead load stress is given by:

$$\sigma = \frac{w\ L^2}{8\ S} = \frac{\left[\left(\dfrac{33.3 \times 20}{12}\right) + \left(\dfrac{10.3 \times 4 \times 20}{20 \times 12}\right) + \left(\dfrac{26}{12}\right)\right][20 \times 12]^2}{8\ [35.1]}$$

or

$$\sigma \text{ (dead load)} = 12{,}500 \text{ psi } (0.863 \times 10^{+8} \text{ Pa})$$

The sum of the dead load and blast load is 43,500 psi ($3.04 \times 10^{+8}$ Pa) which exceeds the yield stress of 33,000 psi ($2.28 \times 10^{+8}$ Pa). So, the roof beam will yield. If we sum the stresses, which is not rigorously correct because the system is no longer elastic, and divide by E to estimate the strain, the deformation w_o can be computed from the insert to Figure 4-26.

$$\frac{h\ w_o}{L^2\ \varepsilon_{max}} = \frac{(13.89)\ w_o}{(20 \times 12)^2 \left(\dfrac{43{,}500}{30 \times 10^{+6}}\right)} = \psi_{w_o} = 0.1747$$

or, rearranging:

$$w_o = 1.00 \text{ inches } (25.4 \text{ mm})$$

This deformation is small relative to the entire span, so the roof probably would still be usable. The roof beam will have yielded, however.

Damage to Columns

The most severely loaded column is the column in the center. We
will assume that sidesway can occur and that the column is fixed-fixed
about its weak axis and pinned-pinned about its strong axis. Using Figure
4-30 for buckling of a column about its weak axis, and properties for the
W 10 x 33 from Table 4-4, the scaled pressure is:

$$\frac{P \ A_1 \ L^2}{\alpha_p \ E \ I_{yy}} = \frac{(1.42) \ (20 \times 12)^2 \ (13 \times 12)^2}{(9.87) \ (30 \times 10^{+6}) \ (36.5)} = 0.184$$

The scaled impulse axis about the weak axis is:

$$\frac{i \ A_1 \ h \ \sqrt{E}}{\alpha_i \ \sigma_y \ \sqrt{M} \ L \ I_{yy}} =$$

$$\frac{(0.0145) \ (20 \times 12)^2 \ (8.00) \ \sqrt{30 \times 10^{+6}}}{1.41 \ (33,000) \ \sqrt{\frac{33.3 \times 20 \times 20}{386} + \frac{10.3 \times 4 \times 20}{386} + \frac{26 \times 2 \times 20}{386}} \ \sqrt{13 \times 12} \ \sqrt{36.5}} =$$

$$1.66$$

When this combination of scaled pressure and scaled impulse are plotted in Figure 4-30, the point falls below the quasi-static pressure threshold
which means that the column should survive. A similar calculation should also be made for the columns about its strong axis.

$$\frac{P \ A_1 \ L^2}{\alpha_p \ E \ I} = \frac{(1.42) \ (20 \times 12)^2 \ (13 \times 12)^2}{(2.47) \ (30 \times 10^{+6}) \ (171.0)} = 0.157$$

and

$$\frac{i\ A_1\ h\ \sqrt{E}}{\alpha_i\ \sigma_y\ \sqrt{M\ L\ I}} =$$

$$\frac{(0.0145)\ (20 \times 12)^2\ (9.94)\ \sqrt{30 \times 10^{+6}}}{(1.41)\ (33,000)\ \sqrt{\dfrac{33.3 \times 20 \times 20}{386} + \dfrac{10.3 \times 4 \times 20}{386} + \dfrac{26 \times 2 \times 20}{386}}\ \sqrt{13 \times 12}\ \sqrt{171}} =$$

0.906

As would be anticipated, the column is also stable about its strong axis.

Column Bending Damage

In the weak direction, the W 10 x 39 column in the center of the west wall will be clamped-clamped. Because the blast load travels in the W-E direction, the applied pressure and impulse will be the reflected values. Using Figure 4-26 as a beam bending solution gives for the scaled pressure:

$$\frac{P\ b\ L^2}{\psi_p\ \sigma_y\ Z} = \frac{(3.00)\ (20 \times 12)\ (13 \times 12)^2}{(23.10)\ (33,000)\ (14.0)} = 1.64$$

and for the scaled reflected impulse:

$$\frac{i\ b\ \sqrt{EI}}{\psi_i\ \sqrt{\rho A}\ \sigma_y\ Z} = \frac{(0.030)\ (20 \times 12)\ \sqrt{30 \times 10^{+6}}\ \sqrt{36.5}\ \sqrt{386}}{(0.861)\ \sqrt{\dfrac{39}{12} + \dfrac{3.4 \times 20}{12}}\ (33,000)\ (14.0)} = 3.94$$

This response is well inside all of our asymptotes and far up in the quasi-static loading realm. So, the columns will be severely bent, but the exact magnitude will not be obtained from a bending solution because of the bifurcation or instability associated with plastic bending solutions in the quasi-static loading realm. The amount of deformation will eventually be limited by extensional forces; however, the deformation will be large. This failure mechanism is a very serious one for this structure and indicates that a redesign is needed if we want this structure to survive a blast loading of this magnitude.

Wall Siding Damage

Figure 4-29 for extensional behavior can be used to estimate what happens to siding. The applied pressure and impulse loads will be normally reflected ones. Basically, the scaled pressure and scaled impulse terms in Figure 4-29 are obtained by treating the siding as strips running the full 13 feet from the roof to the foundation of the building. After rearranging the scaled pressure and scaled impulse expressions so that properties are expressed in terms of unit widths, we obtain for scaled pressure:

$$\frac{P\ b\ L\ E^{1/2}}{\sigma_y^{3/2}\ A} = \frac{P_r\ E^{1/2}\ L}{\sigma_y^{3/2}\ (A/b)} = \frac{(3.00)\ \sqrt{30 \times 10^{+6}}\ (13 \times 12)}{(33,000)^{3/2}\left(\frac{0.75}{12}\right)} = 6.84$$

and for the scaled impulse:

$$\frac{i\ b\ E^{1/2}}{\rho^{1/2}\ \sigma_y\ A} = \frac{i\ E^{1/2}\ g^{1/2}}{\left(\frac{\rho g A}{b}\right)^{1/2}\ \sigma_y\left(\frac{A}{b}\right)^{1/2}} = \frac{(0.030)\left(30 \times 10^{+6}\right)^{1/2}(386)^{1/2}}{\left(\frac{3.40}{144}\right)^{1/2}(33,000)\left(\frac{0.75}{12}\right)^{1/2}} = 2.55$$

This combination of scaled pressures and impulses is very close to a μ ratio ($\varepsilon_{max}/\varepsilon_y$) of 4.0. So, the siding will be stretched plastically to a maximum strain of:

$$\varepsilon_{max} = \mu\ \frac{\sigma_y}{E} = 4.0\left(\frac{33,000}{30 \times 10^{+6}}\right) = 4.40 \times 10^{-3}$$

The maximum deformation w_o in the siding can be computed from the insert in Figure 4-29.

$$\frac{\pi^2 w_o^2}{2 L^2} = \frac{\pi^2 w_o^2}{2 (12 \times 13)^2} = \epsilon_{max} = 4.40 \times 10^{-3}$$

or

$$w_o = 4.66 \text{ inches } (118.0 \text{ mm})$$

Summary of Example Problem

The siding will be damaged, but the damage level might be acceptable because it is easily replaced. The roof beams will be deformed, but this damage could be acceptable because the deformation is small. Much more serious is the bending of columns in the wall facing the blast. This damage would probably cause wall collapse.

Other members in this building still remain to be analyzed. In the next structural chapter, you will learn how to analyze this building for sidesway and how to treat multi-story structures.

Many of the principles developed in this chapter have been illustrated in this example problem.

CHAPTER 4

LIST OF SYMBOLS

A	cross-section area of beam
A_1	loaded area for column buckling
b	beam width
c	distance from neutral axis to beam outer fiber
E	Young's modulus

f	Coulomb resisting force
$g(t)$	time function for load
h	beam depth; column height
I	second moment of area (area moment in inertia); total impulse
I_o	impulse asymptote in P*-I diagram
I_{xx}, I_{yy}	second moments of area about specific axes through a cross section
i, i_s, i_r	specific blast impulses
K.E.	kinetic energy
k	linear spring energy
L	length of beam
M	bending moment
M_p	plastic bending moment
M_y	yield bending moment
m, m_1, m_2	lumped masses
P	peak applied pressure
P*	peak applied force
P_m	maximum overpressure
P_o*	force asymptote in P*-I diagram
P_r	peak reflected blast overpressure
P_s	peak side-on overpressure
P_o	ambient air pressure
$p(t)$	time history of pressure
$p*(t)$	time history of force
R	distance from explosive source
S, S_x, S_y	elastic section moduli
S.E.	strain energy
t	time
t_d	pressure pulse duration
t_{d1}, t_{d2}, t_{d3}	times describing complex pressure pulse

t_m	time to maximum response
t_{max}	time at which response of a simple mechanical system is a maximum
t_r	pressure rise time
V	dynamic reaction
Wk	external work
w	beam or plate deflection
w_o	maximum beam or plate deflection
X, Y	half spans of plate
x	displacement
x_{max}	maximum displacement of a simple mechanical system
x_o	initial displacement
$\dot{x}$	velocity
$\ddot{x}$	acceleration
Z, Z_{xx}, Z_{yy}	plastic section moduli
$\alpha, \beta, \gamma, \delta$	ratios describing complex pulse amplitudes and durations
α_i, α_p	coefficients in beam response solutions
δ	deflection
$\varepsilon, \varepsilon_m, \varepsilon_{max}, \varepsilon_{xx}, \varepsilon_{xy}, \varepsilon_{yy}$	strains
μ	ductility ratio
ν	Poisson's ratio
ρ	density
$\sigma, \sigma_{max}, \sigma_y, \sigma_{xx}, \sigma_{yy}$	stresses
ϕ_i, ϕ_p	coefficients in plate response solution
ψ	a function of ωT
$\psi_p, \psi_i, \psi_\varepsilon \; \psi_{w_o}$	dimensionless coefficients in beam response solutions
$\omega, \omega_1, \omega_2, \omega_L, \omega_H$	circular natural frequencies

CHAPTER 5

NUMERICAL METHODS FOR STRUCTURAL ANALYSIS

The application of numerical methods for structural analysis is introduced by solving for the motions of simple one- and two-degree-of-freedom
systems.   These solutions, obtained by direct numerical integration of the
equations of motion, are easily understood, and a good "feel" for the behavior of structures under dynamic loads is obtained by examining the time history of the motions of the structure.   Both elastic and plastic behavior are
treated and procedures for deriving one-degree-of-freedom equivalent systems
to represent the behavior of distributed systems such as beams, plates and
frames are presented.   Numerical methods, once developed for simple systems,
are then extended to treat the elastic-plastic response of more complex
structural arrangements.

ONE-DEGREE-OF-FREEDOM SYSTEMS

Equation of Motion

Consider the spring, mass, and dashpot of Figure 5-1a.   The equation
of motion is obtained from Newton's first law, which equates the sum of the
external forces acting on the body to its inertia.   Taking only the forces
in the direction of motion, the differential equation is

$$F(t) - kx - c\dot{x} = m\ddot{x} \qquad (5-1)$$

where $\dot{x}$ and $\ddot{x}$ represent the velocity and acceleration of the mass, respectively, and damping, which is proportional to velocity, has been assumed.

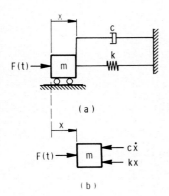

Figure 5-1. (a) One-Degree-Of-Freedom Spring-Mass System
              (b) Free Body Diagram

Although shown offset, the viscous force, $c\dot{x}$, also acts through the center

of mass or the mass is restrained against rotations. The solution of Equa-

tion (5-1) gives the motion of the mass of Figure 5-1. This simple equa-

tion could be solved readily in closed form, but here it will be integrated

numerically to obtain the displacements.

Numerical Integration

Many methods have been proposed for the time integration of Equation

(5-1). The procedure applied to this problem will be the average accelera-

tion method. This is also known as the Timoshenko method [Timoshenko (1928)].

In this method, the velocity and displacement at any time t are written as

$$\dot{x}_t = \dot{x}_{t-\Delta t} + \frac{1}{2}\left(\ddot{x}_t + \ddot{x}_{t-\Delta t}\right)\Delta t \tag{5-2a}$$

$$x_t = x_{t-\Delta t} + \frac{1}{2}\left(\dot{x}_t + \dot{x}_{t-\Delta t}\right)\Delta t \tag{5-2b}$$

Substituting Equation (5-2a) into Equation (5-2b), we have

$$x_t = x_{t-\Delta t} + \dot{x}_{t-\Delta t}\Delta T + \frac{1}{4}\left(\ddot{x}_t + \ddot{x}_{t-\Delta t}\right)\Delta t^2 \tag{5-3}$$

If Equations (5-2a) and (5-3) are substituted into the equation of motion [Equation (5-1)] written for time t, an expression for acceleration at time t in terms of displacements and velocities at time t - Δt is obtained.

$$\ddot{x}_t = \frac{1}{\left(m + \frac{1}{2} C\Delta t + \frac{1}{4} k \Delta t^2 \right)} \Bigg[F(t) - k$$

$$\cdot \left(x_{t - \Delta t} + \dot{x}_{t - \Delta t} \Delta t + \frac{1}{4} \ddot{x}_{t - \Delta t} \Delta t^2 \right) \qquad (5\text{-}4)$$

$$- c \left(\dot{x}_{t - \Delta t} + \frac{1}{2} \ddot{x}_{t - \Delta t} \Delta t \right) \Bigg]$$

Now Equation (5-1) can be integrated using Equations (5-2) and (5-4) if the initial values of displacement and velocity are specified and F(t) is known. External forces which depend upon motions of the structure could also be treated, but, for structures which are loaded by air blast waves, the loading and structural response can be regarded as decoupled with neg-ligible error. The integrating procedure is as follows:

Step 1. At t = 0: Compute $\ddot{x}_{t = 0}$ using Equation (5-1) and speci-fied values for $x_{t = 0}$, $\dot{x}_{t = 0}$, and F(0).

Step 2. Increment time: t = t + Δt.

Step 3. At t = t + Δt, compute: $\ddot{x}_t$ using Equation (5-4), $\dot{x}_t$ using Equation (5-2a), and x_t using Equation (5-2b).

Step 4. Repeat Steps 2 and 3 until some specified time or until the maximum displacement, stresses or strains are reached.

Note that if $x_{t = 0}$ and $\dot{x}_{t = 0}$ are zero, then the structure (the mass M in this case) will not move so long as F(t) = 0; however, when F(t) > 0, Equation (5-4) will give an acceleration and motion will start. Thus, this system of equations is self-starting and requires no special starting provisions.

Many other schemes for numerical integration are available. Some
of the well known ones are the Newmark-Beta method, the Huboult method,
the Wilson-θ, and the linear acceleration method. These methods, including the average acceleration method used here, have been studied and compared by many investigators [McNamara (1975), Tillerson (1975), and Stricklin and Haisler (1974)]. No one method is ideal for every problem. The
average acceleration method was chosen primarily for its simplicity, successful history of use by the authors, and because it has been shown to be
unconditionally stable for linear systems [McNamara (1975)]. It corresponds identically to the Newmark method with $\gamma = 1/2$ and $\beta = 1/4$.

<u>Numerical Solution</u>

Consider the example of Figure 5-2. Damping has been taken as zero,
and a triangular force time history has been assumed.

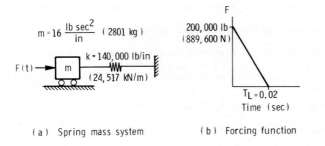

(a) Spring mass system   (b) Forcing function

Figure 5-2. One-Degree-Of-Freedom Elastic System

As a starting point, compute the period of the system.

$$T_N = 2\pi \sqrt{\frac{m}{k}} = 2\pi \sqrt{\frac{16}{140,000}} = 0.0672 \text{ sec}$$

To integrate the equation of motion, an integration time step less than or
equal to about one-tenth of the fundamental period is adequate for a one-
degree-of-freedom system. Using the average acceleration method, stability
is assured; however, the size of Δt affects the accuracy of the solution.

For this case, Δt was chosen as

$$\Delta t = 0.0035 \text{ sec}$$

Substituting Δt, plus the values of the problem parameters, into Equations (5-1) and (5-4), the following equations for calculating the acceleration of the mass, M, are obtained

$$@ \ t = 0, \ \ddot{x} = \frac{200,000 - 140,000 \ X_o}{16}$$

$$(5-5)$$

$$= 12,500 - 8,570 \ X_o$$

$$0 < t < 0.02, \ \ddot{x} =$$

$$\frac{200,000 - 10 \times 10^6 \ t - 140,000 \left[x + 0.0035 \ \dot{x} + \frac{(0.0035)^2}{4} \ \ddot{x} \right]_{t - \Delta t}}{16 + \frac{1}{4} (140,000) (0.0035)^2} \quad (5-6)$$

$$= \frac{200,000 - 10 \times 10^6 \ t - 140,000 \left[x + 0.0035 \ \dot{x} + 3.0625 \times 10^{-6} \ \ddot{x} \right]_{t - \Delta t}}{16.429}$$

where X_o denotes the initial displacement and the bracketed term in Equation (5-6) contains values of displacement, velocity, and acceleration at the last step, $t - \Delta t$. Once X_o has been set, the integration procedure is started by evaluating Equations (5-5), (5-6), (5-2a), and (5-2b) in that order and proceeding as outlined in the last section.

Table 5-1 gives the numerical integration for this example, and displacements for two cycles are plotted in Figure 5-3. Only 11 entries were required to evaluate the response for one cycle. Without damping, the oscillation will continue indefinitely at an amplitude of 1.20 inches (30.5 mm) since the load is no longer acting on the structure.

Table 5-1. Numerical Solution for Displacements, One-Degree-Of-Freedom Spring-Mass System of Figure 5-2*

Column 1	Column 2	Column 3	Column 4	Column 5	Column 6	Column 7
Time	$(200{,}000 - 10 \times 10^6\ t)$	$(x + 0.0035\ \dot{x} + 3.0625 \times 10^{-6}\ \ddot{x})_{t - \Delta t}$	Column 2 - Column 3	Column 4 ÷ 16.429	[Eq. (4-2a)]	[Eq. (4-2b)]
0.0000	200000.00	0.00	200000.00	12500.00	0.0000	0.0000
0.0035	165000.00	5359.37	159640.63	9717.15	38.8800	0.0680
0.0070	130000.00	32743.04	97256.96	5919.92	66.2449	0.2520
0.0105	95000.00	70279.37	24720.63	1504.72	79.2380	0.5066
0.0140	60000.00	110396.29	- 50396.29	- 3067.57	76.5030	0.7792
0.0175	25000.00	145252.34	-120252.34	- 7319.63	58.3254	1.0151
0.0210	0.00	167555.22	-167555.22	-10198.90	27.6680	1.1656
0.0245	0.00	172366.99	-172366.99	-10491.79	- 8.5407	1.1991
0.0280	0.00	159185.34	-159185.34	- 9689.44	- 43.8578	1.1074
0.0315	0.00	129386.30	-129386.30	- 7875.60	- 74.5967	0.9001
0.0350	0.00	86080.60	- 86080.60	- 5239.63	- 97.5483	0.5988
0.0385	0.00	33788.94	- 33788.94	- 2056.70	-110.3169	0.2351
0.0420	0.00	- 22029.96	22029.96	1340.94	-111.5695	-0.1533
0.0455	0.00	- 75549.14	75549.14	4598.59	-101.1753	-0.5256
0.0490	0.00	-121181.74	121181.74	7376.20	- 80.2194	-0.8430
0.0525	0.00	-154164.16	154164.16	9383.80	- 50.8894	-1.0724
0.0560	0.00	-171053.35	171053.35	10411.83	- 16.2470	-1.1899
0.0595	0.00	-170086.25	170086.25	10352.96	20.0914	-1.1832
0.0630	0.00	-151363.83	151363.83	9213.35	54.3324	-1.0530
0.0665	0.00	-116840.50	116840.50	7111.95	82.9017	-0.8128
0.0700	0.00	- 70120.18	70120.18	4268.14	102.8168	-0.4878
0.0735	0.00	- 16080.00	16080.00	978.77	111.9989	-0.1119
0.0770	0.00	39638.78	- 39638.78	- 2412.77	109.4894	0.2757
0.0805	0.00	91219.65	- 91219.65	- 5552.44	95.5503	0.6346
0.0840	0.00	133278.09	-133278.09	- 8112.49	71.6367	0.9271
0.0875	0.00	161423.61	-161423.61	- 9825.68	40.2449	1.1229
0.0910	0.00	172718.09	-172718.09	-10513.16	4.6519	1.2015
0.0945	0.00	165982.50	-165982.50	-10103.17	- 31.4267	1.1546
0.0980	0.00	141919.97	-141919.97	- 8638.51	- 64.2246	0.9873
0.1015	0.00	103042.39	-103042.39	- 6272.08	- 90.3181	0.7168
0.1050	0.00	53408.19	- 53408.19	- 3250.90	-106.9833	0.3715
0.1085	0.00	1801.29	- 1801.29	109.64	-112.4805	-0.0125
0.1120	0.00	- 56822.74	56822.74	3458.74	-106.2359	-0.3953
0.1155	0.00	-105912.45	105912.45	6446.77	- 88.9012	-0.7368
0.1190	0.00	-143945.95	143945.95	8761.83	- 62.2862	-1.0014
0.1225	0.00	-166952.90	166952.90	10162.24	- 29.1690	-1.1614
0.1260	0.00	-172531.61	172531.61	10501.81	6.9930	-1.2002
0.1295	0.00	-160099.71	160099.71	9745.09	42.4251	-1.1137
0.1330	0.00	-130954.98	130954.98	7971.09	73.4284	-0.9110

*Units are English; displacements in Column 7 are in inches.

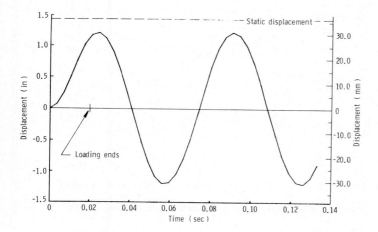

Figure 5-3.   Displacements for the One-Degree-Of-Freedom System of Figure 5-2

The static solution for the peak force is shown as the dotted line
in the figure.   For these conditions, the deflection, based on a static
solution and the peak value of the loading, would have been overestimated
by 19 percent.   This behavior is characteristic of blast loaded structures
because the duration of the loading is usually short relative to the fundamental response period of the structure.

Effect of Load Duration

At this point, it is useful to consider the effect on the response
of the spring mass system produced by varying the duration of the loading
while holding the peak applied force constant.   It is convenient to express the duration of the loading as a function of the period of the system, T_L/T_N.   For the results of Figure 5-3, we have

$$\frac{T_L}{T_N} = \frac{0.02}{0.0672} \approx 0.30$$

Five more cases were computed for the spring-mass system of Figure 5-2

372

with the following T_L/T_N ratios:

$$T_L/T_N$$

0.1
0.5
1.0
5.0
10.0

The results are given in Figure 5-4. Note that when the duration becomes long relative to the period of the system, the loading approaches a step load and the deflection approaches twice the static value. This is a well known result and corroborates the model and integration procedure. Two other observations can be made from these results. Increasing the duration of the loading always increased the deflection of the mass. This is always true for a triangular loading which has its maximum value at t = 0. Likewise, decreasing the load duration decreased the response. This indicates that a very large load applied to a mass for a very short period of time produces little motion. These generalizations are easy to apply to a one-degree-of-freedom system, but are more difficult to apply to struc-

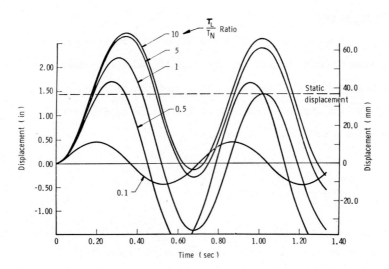

Figure 5-4. Effect of Load Duration on Response of a One-Degree-Of-Freedom System

tures which have many components of different strength, mass, and frequency. For example, a blast wave which hardly moves a large building might break all windows and deform light wall coverings. Thus, each component of the building must be examined for the blast loading.

Impulsive Loading

The concept of an impulsive load was introduced in Chapter 4. It was explained that an impulsive loading could be applied as an initial velocity. To determine when an initial velocity is a good approximation to a short duration loading, consider again the equation of the spring-mass system of Figure 5-1. Rewrite and rearrange the differential equation, Equation (5-1), as

$$\ddot{x} = \frac{F(t)}{m} - \frac{k}{m} x - \frac{c}{m} \dot{x} \qquad (5-7)$$

If the initial velocity and displacement are zero, at early time the mass has moved very little, and so kx/m and $c\dot{x}/m$ are approximately zero. Integrating over the load duration, T_L, gives

$$\dot{x} = \int_0^{T_L} \ddot{x}\ dt = \frac{1}{m} \int_0^{T_L} F(t)\ dt \qquad (5-8)$$

which is recognized as the impulse-momentum equation. If the load duration is short so that the loading is over before the mass has displaced significantly, then the loading can be replaced by an initial velocity computed by Equation (5-8) without appreciable error.

Again, the numerical example of Figure 5-2 will be used to evaluate the effect produced by replacing the forcing functions with an initial velocity as computed from Equation (5-8). Response of the spring-mass system was recomputed for the four cases with the smallest ratio of T_L/T_N.

These were

$$T_L/T_N$$

0.1
0.3
0.5
1.0

The initial velocity is simply the area under the force-time curve divided by the mass. For the triangular pulse of Figure 5-2b, this initial velocity is given in Table 5-2.

Displacements are determined by the numerical integration procedure of Table 5-1. For this integration, the initial displacement was zero, the initial velocity was taken from Table 5-2, and the forcing function F(t) was zero.

These new displacements are plotted in Figure 5-5. By comparing with the displacements of Figure 5-4, obtained by numerical integration using the actual F(t), the accuracy of the initial velocity approximation can be determined. Notice that for T_L/T_N = 0.10, the initial velocity

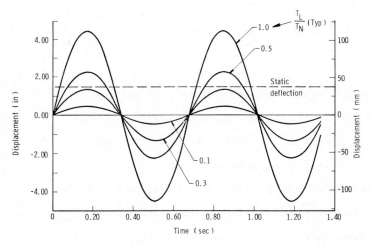

Figure 5-5. Effect of Replacing F(t) by an Initial Velocity Computed by Equation (5-8)

Table 5-2. Initial Velocity Computed
From Equation (5-8)

T_L/T_N	$\dot{x} = \dfrac{1}{2m} F\,T_L$, in/sec (m/sec)
0.1	42 (1.07)
0.3	125 (3.18)
0.5	210 (5.33)
1.0	420 (10.67)

approximation is very accurate, whereas at large values of T_L/T_N, the initial velocity approximation greatly overestimates the loading. The errors in peak displacement produced by replacing F(T) with an initial velocity computed by Equation (5-8) are:

Displacement, inch (mm)

T_L/T_N	F(t)	$\dot{x}_o$	% Error
0.1	0.45 (11.4)	0.45 (11.4)	0%
0.3	1.20 (30.5)	1.34 (34.0)	12%
0.5	1.69 (42.8)	2.24 (56.9)	32%
1.0	2.20 (55.9)	4.48 (118.8)	104%

Three observations can be made from these comparisons:

- For values of T_L/T_N less than 0.1, the loading can be treated as an impulse and replaced by an initial velocity given by Equation (5-8).
- Replacing F(t) by an initial velocity computed by Equation (5-8) always gives an upper bound on the response.
- A true impulsive load will always produce a lower response than the peak value of the load applied statically.

These observations for a single-degree-of-freedom system also hold true for multiple-degree-of-freedom systems. For the first condition above, the loading is said to be impulsive with respect to the structure (it is always structure dependent), or in the impulsive loading realm. Note also that for T_L/T_N values up to 0.30, a reasonable and conservative estimate is obtained.

Damping

Damping has been omitted in calculations performed previously and is almost always ignored when computing the response of structures to blast loading. The principal reasons that damping is ignored for this type of analysis are:

- Only one cycle of response of the structure is of interest.
- In one cycle, the attenuation of the responses produced by structural damping is small.
- Ignoring damping is a conservative approach.
- Damping values for structures are seldom known.
- The energy dissipated through plastic deformation is much greater than that dissipated by normal structural damping.

To demonstrate the effect of damping in a single-degree-of-freedom system, results of Figure 5-3 will be recalculated with different damping values. To define damping in Equation (5-1), the parameter c has units of force divided by velocity so that the product $c\dot{x}$ produces a force which opposes motion of the mass. Damping is seldom defined in these units, but rather as some percentage of critical damping. For free vibration, this is the damping value that will permit the mass to return to its equilibrium position when displaced, but prevent further oscillation. For critical damping, c is

$$c = c_{CR} = 2 \sqrt{km} \qquad (5-9)$$

For convenience, damping will be defined as some percentage of critical damping, so that Equation (5-1) becomes

$$m\ddot{x} - \beta \, c_{CR} \, \dot{x} - kx = F(t) \qquad (5-10)$$

It is generally accepted that for buildings, damping values fall between 2 percent and 10 percent of critical damping.

For the example problem of Figure 5-2, the response was recalculated with β = 5 percent and 10 percent. These results are shown in Figure 5-6. In the response to blast loading, only the first peak and subsequent rebound are of interest since the load is not cyclical. For the first peak and rebound, the attenuation is:

	Displacement, inch (mm)		% Attenuation	
Damping	First Peak	Rebound	First Peak	Rebound
0% c_R	1.20 (30.5)	-1.20 (-30.5)	0.0	0
5% c_{CR}	1.11 (28.2)	-0.95 (-24.1)	7.5	21
10% c_{CR}	1.03 (26.2)	-0.76 (-19.3)	14.2	37

Even at 10 percent c_R, which can be considered as the upper limit of damping for steel framed buildings, the peak response is attenuated by only 14 percent. Thus, most analysts will neglect damping in the response of blast loaded buildings and err slightly on the side of conservatism.

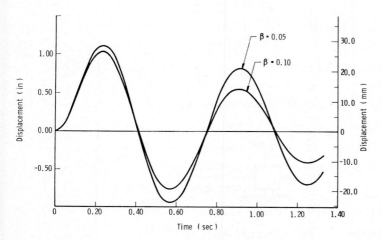

Figure 5-6. Effect of Damping on Displacements of the One-Degree-Of-Freedom System of Figure 5-2

Frames

To show that simple one-degree-of-freedom systems can be used to describe the gross behavior of structural components, a typical building frame
will be analyzed to determine its response to a dynamic load. Consider the
frame of Figure 5-7. Initially, the girder is treated as rigid so that the
frame stiffness can be determined easily and so that the effect of girder
flexibility can be demonstrated by a later example. Also, the usual situation in blast loaded buildings is a loading distributed over the face of a
building (over the column on the loaded end of the frame in our example).
Here again, in this example, the loading is confined to the plane of the
girder so that the column itself does not respond locally to the loading.
The combined beam bending and lateral sway of a frame is treated in another
section of this chapter.

Stiffness of the frame for lateral sway is twice the stiffness of
each column. This can be found from the deflection for a simple cantilever
as shown in Figure 5-7c. The deflection, δ, is equal to twice that of a

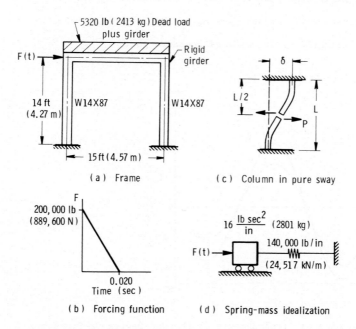

(a) Frame   (c) Column in pure sway

(b) Forcing function   (d) Spring-mass idealization

Figure 5-7. Simple Frame With Rigid Girder

cantilever of length L/2. From elementary strength of materials, we find,

$$\delta = 2 \; \frac{P \; (L/2)^3}{3EI} = \frac{PL^3}{12EI}$$

which gives, for the stiffness of each column,

$$k = \frac{P}{\delta} = \frac{12EI}{L^3}$$

The total frame stiffness is

$$k = k_1 + k_2 = \frac{12E \left(I_1 + I_2 \right)}{L^3} \tag{5-11}$$

In the <u>Manual of Steel Construction</u> (1980), the area and principal section moment of inertia of the WF14X87 beam-column are given as

$$A = 25.6 \; in^2 \; (0.0165 \; m^2)$$

$$I = 967 \; in^4 \; (0.0004025 \; m^4)$$

Combining these values with the frame height [14 ft (4.27 m) to the center of the girder] and the modulus of elasticity [30 x 10^6 psi (207,000 MPa)] gives a frame stiffness of

$$k = 139,720 \; lb/in \approx 140,000 \; lb/in \; (84,500 \; kN/m) \tag{5-12}$$

To compute the total mass, 1/3 of the column mass plus the mass of the girder is added to the dead load. This gives a mass of:

$$m = [0.283 \ (2 \times 25.6) \ (15 \times 12)/3 + 5320]/386 =$$

$$15.88 \ \text{lb-sec}^2/\text{in} \simeq 16 \ \text{lb-sec}^2/\text{in} \ (2801 \ \text{kg})$$

Adding only one-third of the mass of the column to the mass of the girder and dead load is an approximate way of treating the column mass. It neglects part of the mass and, thus, is slightly conservative. Other choices that the analyst can make are to take 37 percent of the beam mass, which is the diagonal term of a F.E. consistent mass matrix [Przemieniecki (1968)] or one-half of the beam mass, which is consistent with a lumped mass matrix formulation [Cook (1974)].

The equivalent one-degree-of-freedom spring-mass system which approximates the frame in sway is shown in Figure 5-7d. This system is identical to that used in the example of Figure 5-2, so the response is the same as that given in Figure 5-3.

Girder Flexibility

To add girder flexibility to the frame of the previous example (Figure 5-7), a W16X88 girder was substituted for the rigid one. As before, joints are rigid and the combined weight of the girder and the 4000 lb (1814 kg) dead load equals 5320 lb (2413 kg).

With a flexible girder, the frame is now indeterminate. In general, there are three redundant reactions in the frame (six reactions, but three can be found from the equations of static equilibrium of the frame). If the frame is symmetric, the number of redundancies can be reduced by arguments of symmetric reactions. We will solve for the case where the girder and each column can be different. Because matrix methods will be used in later examples, they will be introduced here where the sizes of the matrices are small.

In this example, the effect of the dead load on deflections in the girder will be ignored, as will axial deformation in the columns and girder. This approach simplifies the problem and, in general, is an acceptable one for blast loaded frame structures. These restrictions will be removed in a later section of this chapter entitled Multi-Degree-Of-Freedom Systems.

A direct, rather than a general, approach will be used to derive the system equations. For the frame of Figure 5-8, the degrees-of-freedom will be restricted to three, a lateral displacement and rotations at the ends of the girder. These three displacements are numbered and applied to each frame member in Figure 5-9. Notice that displacement 1 is applied to the tops of members 1 and 3 and omitted in member 2. Displacement 1 is a rigid body displacement of the girder and affects the inertia properties of the frame, but not the lateral stiffness of the girder that is of interest here.

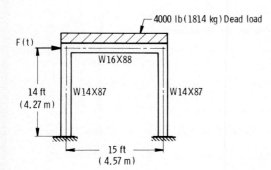

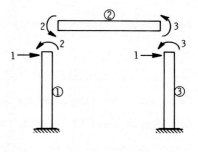

Figure 5-8. Frame With Flexible Girder   Figure 5-9. Displacements in the Frame

A stiffness matrix for each member is obtained by giving each displacement in the member a unit value while holding all others fixed. This is illustrated in Figure 5-10 which gives the stiffness matrix for a two-dimensional beam in bending only (no axial deflections). Numbered arrows give the positive directions of the forces and displacements. Each column in the stiffness matrix is comprised of the forces required to hold the beam in the displaced position shown directly above the column. For col-

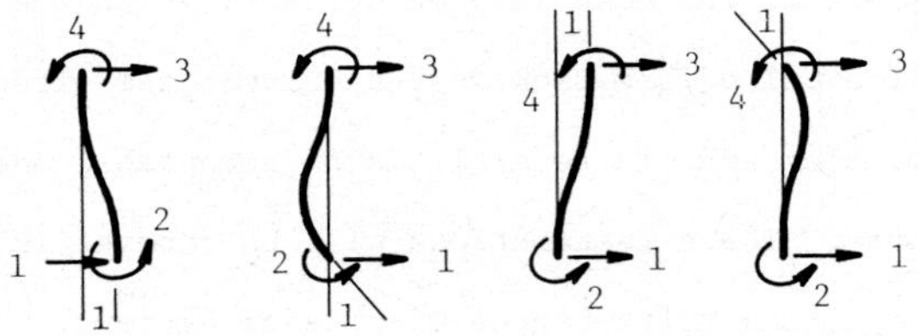

$$
\begin{bmatrix} F_1 \\ F_2 \\ F_3 \\ F_4 \end{bmatrix} = \begin{bmatrix} \dfrac{12EI}{L^3} & \dfrac{-6EI}{L^2} & \dfrac{-12EI}{L^3} & \dfrac{-6EI}{L^2} \\[2ex] \dfrac{-6EI}{L^2} & \dfrac{4EI}{L} & \dfrac{6EI}{L^2} & \dfrac{2EI}{L} \\[2ex] \dfrac{-12EI}{L^3} & \dfrac{6EI}{L^2} & \dfrac{12EI}{L^3} & \dfrac{6EI}{L^2} \\[2ex] \dfrac{-6EI}{L^2} & \dfrac{2EI}{L} & \dfrac{6EI}{L^2} & \dfrac{4EI}{L} \end{bmatrix} \begin{bmatrix} \delta_1 \\ \delta_2 \\ \delta_3 \\ \delta_4 \end{bmatrix}
$$

Figure 5-10. Stiffness Matrix for 2-D Beam in Bending

umn one, the displaced position corresponds to a unit displacement of δ_1 (with all other displacements zero); for column two, it corresponds to a unit displacement of δ_2, etc. Because one end of the beam is "fixed," the forces in each column are easily derived by considering the deflections of a cantilever beam produced by a lateral load and moment applied at the displaced end. Equilibrium equations then give the reactions at the fixed end.

To derive the first column in the stiffness matrix, consider the cantilever beam of Figure 5-11 with a unit displacement at position 1. From elementary strength of materials [for example see Den Hartog (1949)], the displacements of the free end are:

$$
\delta_1 = 1 = \frac{F_1 L^3}{3EI} + \frac{F_2 L^2}{2EI} \tag{5-13}
$$

$$\delta_2 = 0 = \frac{F_1 L^3}{2EI} + \frac{F_2 L}{EI} \qquad (5\text{-}14)$$

Solving Equations (5-13) and (5-14), relationships between δ_1 and F_1 and δ_2 and F_2 can be found. These are:

$$F_1 = \frac{12EI}{L^3} \delta_1 \qquad (5\text{-}15)$$

$$F_2 = -\frac{6EI}{L^2} \delta_1 \qquad (5\text{-}16)$$

The two conditions of equilibrium now give

$$\Sigma F_x = 0: \quad F_3 = -F_1 = -\frac{12EI}{L^3} \delta_1 \qquad (5\text{-}17)$$

$$\Sigma M_z = 0: \quad F_4 = -F_1 L - F_2 = -\frac{12EI}{L^3} \delta_1 L + \frac{6EI}{L^2} \delta_1$$

$$(5\text{-}18)$$

$$F_4 = -\frac{6EI}{L^2} \delta_1$$

Dividing Equations (5-15) through (5-18) by δ_1 yields the stiffness terms

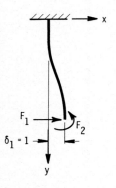

Figure 5-11. Displaced Cantilever Beam

in column 1 of Figure 5-10. The stiffness matrix can also be derived from energy methods and variational techniques, but this approach requires the selection of appropriate interpolation functions. Explanation of these methods are given by Zenkiewicz (1971), Cook (1974), and Gallagher (1975).

For the displacements chosen, the force-displacement equations can be written directly by choosing the appropriate terms from Figure 5-10. For example, member one has displacements in directions 1 and 2 of Figure 5-9 which correspond to displacement directions 3 and 4 of Figure 5-10. Therefore, the appropriate terms are found at the intersection of rows 3 and 4 with columns 3 and 4. Thus, the force deflection equation for all members can be found from Figure 5-10 and they are:

Member 1:
$$\begin{bmatrix} F_1 \\ \\ F_2 \end{bmatrix} = \begin{bmatrix} \dfrac{12EI_1}{L_1^3} & \dfrac{6EI_1}{L_1^2} \\ \\ \dfrac{6EI_1}{L_1^2} & \dfrac{4EI_1}{L_1} \end{bmatrix} \begin{bmatrix} \delta_1 \\ \\ \delta_2 \end{bmatrix} \tag{5-19}$$

Member 2:
$$\begin{bmatrix} F_2 \\ \\ F_3 \end{bmatrix} = \begin{bmatrix} \dfrac{4EI_2}{L_2} & \dfrac{2EI_2}{L_2} \\ \\ \dfrac{2EI_2}{L_2} & \dfrac{4EI_2}{L_2} \end{bmatrix} \begin{bmatrix} \delta_2 \\ \\ \delta_3 \end{bmatrix} \tag{5-20}$$

Member 3:
$$\begin{bmatrix} F_1 \\ \\ F_3 \end{bmatrix} = \begin{bmatrix} \dfrac{12EI_3}{L_3^3} & \dfrac{6EI_3}{L_3^2} \\ \\ \dfrac{6EI_3}{L_3^2} & \dfrac{4EI_3}{L_3} \end{bmatrix} \begin{bmatrix} \delta_1 \\ \\ \delta_3 \end{bmatrix} \tag{5-21}$$

Force-deflection equations for the frame are obtained by applying the equations of equilibrium at the joints. The sizes of the matrices are expanded to include the three displacements and forces, and stiffness contributions (internal forces) at common nodes and directions are summed. With moment continuity at the joints, this gives:

$$
\begin{bmatrix} F_1 \\ -- \\ F_2 \\ F_3 \end{bmatrix}
=
\begin{bmatrix}
\left(\dfrac{12EI_1}{L_1^3}+\dfrac{12EI_3}{L_3^3}\right) & \dfrac{6EI_1}{L_1^2} & \dfrac{6EI_3}{L_3^2} \\
\dfrac{6EI_1}{L_1^2} & \left(\dfrac{4EI_1}{L_1}+\dfrac{4EI_2}{L_2}\right) & \dfrac{2EI_2}{L_2} \\
\dfrac{6EI_3}{L_3^2} & \dfrac{2EI_2}{L_2} & \left(\dfrac{4EI_2}{L_2}+\dfrac{4EI_3}{L_3}\right)
\end{bmatrix}
\begin{bmatrix} \delta_1 \\ -- \\ \delta_2 \\ \delta_3 \end{bmatrix}
\tag{5-22}
$$

The first of Equations (5-22) gives the relationship between a lateral force applied in direction 1 and the three displacements. To obtain the desired relationship between the lateral force and displacement in direction 1, i.e., the lateral spring constant for the frame, the rotations, δ_2 and δ_3 must be expressed in terms of the lateral displacement, δ_1. The last two of Equations (5-22) yield the rotations as a function of δ_1. Noting that the applied moments, F_2 and F_3, are zero for this case, we have [by expanding Equation (5-22)]:

$$
F_1 = \left(\frac{12EI_1}{L_1^3}+\frac{12EI_3}{L_3^3}\right)\delta_1 + \frac{6EI_1}{L_1^2}\delta_2 + \frac{6EI_3}{L_3^2}\delta_3
\tag{5-23a}
$$

$$
0 = \frac{6EI_1}{L_1^2}\delta_1 + \left(\frac{4EI_1}{L_1}+\frac{4EI_2}{L_2}\right)\delta_2 + \frac{2EI_2}{L_2}\delta_3
\tag{5-23b}
$$

$$0 = \frac{6EI_3}{L_3{}^2}\,\delta_1 + \frac{2EI_2}{L_2}\,\delta_2 + \left(\frac{4EI_2}{L_2} + \frac{4EI_3}{L_3}\right)\delta_3 \qquad (5\text{-}23\text{c})$$

Solving for δ_2 and δ_3 in terms of δ_1 and substituting into Equation (5-23a), the desired spring constant, k, is obtained

$$F_1 = k\,\delta_1 \qquad (5\text{-}24)$$

where

$$k = \left(\frac{12EI_1}{L_1{}^3} + \frac{12EI_3}{L_3{}^3}\right) - \frac{6EI_1}{L_1{}^2}$$

$$\cdot\left[\frac{\dfrac{6EI_3}{L_3{}^2} - \left(\dfrac{4EI_2}{L_2} + \dfrac{4EI_3}{L_3}\right)\left(\dfrac{L_2}{2EI_2}\right)\left(\dfrac{6EI_1}{L_1{}^2}\right)}{\dfrac{2EI_2}{L_2} - \left(\dfrac{4EI_2}{L_2} + \dfrac{4EI_3}{L_3}\right)\left(\dfrac{L_2}{2EI_2}\right)\left(\dfrac{4EI_1}{L_1} + \dfrac{4EI_2}{L_2}\right)}\right] - \left(\dfrac{L_2}{2EI_2}\right)\left(\dfrac{6EI_3}{L_3{}^2}\right) \qquad (5\text{-}25)$$

$$\cdot\left[\frac{6EI_1}{L_1{}^2} - \left(\dfrac{4EI_1}{L_1} + \dfrac{4EI_2}{L_2}\right)\frac{\dfrac{6EI_3}{L_3{}^2} - \left(\dfrac{4EI_2}{L_2} + \dfrac{4EI_3}{L_3}\right)\left(\dfrac{L_2}{2EI_2}\right)\left(\dfrac{6EI_1}{L_1{}^2}\right)}{\dfrac{2EI_2}{L_2} - \left(\dfrac{4EI_2}{L_2} + \dfrac{4EI_3}{L_3}\right)\left(\dfrac{L_2}{2EI_2}\right)\left(\dfrac{4EI_1}{L_1} + \dfrac{4EI_2}{L_2}\right)}\right]$$

Notice that the first term of Equation (5-25) is the lateral stiffness for a rigid girder. The second and third terms account for girder flexibility.

This same result could have been obtained by matrix operations. First, Equation (5-22) is partitioned as shown by the dotted lines in the matrices. Written symbolically, this produces

$$
\begin{bmatrix} F_\alpha \\ -- \\ F_\beta \end{bmatrix} = \begin{bmatrix} k_{\alpha\alpha} & \vdots & k_{\alpha\beta} \\ ----+---- \\ k_{\beta\alpha} & \vdots & k_{\beta\beta} \end{bmatrix} \begin{bmatrix} \delta_\alpha \\ -- \\ \delta_\beta \end{bmatrix} \tag{5-26}
$$

where

$$
\{ F_\alpha \} = F_1, \{ F_\beta \} = \begin{bmatrix} F_2 \\ F_3 \end{bmatrix}, \{ \delta_\alpha \} = \delta_1 \text{ and } \{ \delta_\beta \} = \begin{bmatrix} \delta_2 \\ \delta_3 \end{bmatrix}
$$

The braces represent column vectors. For $F_\beta = 0$, as before, the second set of equations in (5-26) gives

$$
\{0\} = \begin{bmatrix} k_{\beta\alpha} \end{bmatrix} \{ \delta_\alpha \} + \begin{bmatrix} k_{\beta\beta} \end{bmatrix} \{ \delta_\beta \}
$$

or

$$
\{ \delta_\beta \} = - \begin{bmatrix} k_{\beta\beta} \end{bmatrix}^{-1} \begin{bmatrix} k_{\beta\alpha} \end{bmatrix} \{ \delta_\alpha \} \tag{5-27}
$$

Substituting for $\{ \delta_\beta \}$ in the first of Equation (5-26), we find

$$
\{ F_\alpha \} = \begin{bmatrix} k_{\alpha\alpha} \end{bmatrix} \{ \delta_\alpha \} - \begin{bmatrix} k_{\alpha\beta} \end{bmatrix} \begin{bmatrix} k_{\beta\beta} \end{bmatrix}^{-1} \begin{bmatrix} k_{\beta\alpha} \end{bmatrix} \{ \delta_\alpha \}
$$

$$
\tag{5-28}
$$

$$
= \left(\begin{bmatrix} k_{\alpha\alpha} \end{bmatrix} - \begin{bmatrix} k_{\alpha\beta} \end{bmatrix} \begin{bmatrix} k_{\beta\beta} \end{bmatrix}^{-1} \begin{bmatrix} k_{\beta\alpha} \end{bmatrix} \right) \{ \delta_\alpha \}
$$

Since $\left\{F_\alpha\right\} = F_1$ and $\left\{\delta_\alpha\right\} = \delta_1$, the lateral stiffness, k, is

$$\left[k\right] = \left[k_{\alpha\alpha}\right] - \left[k_{\alpha\beta}\right]\left[k_{\beta\beta}\right]^{-1}\left[k_{\beta\alpha}\right] \qquad (5\text{-}29)$$

The matrix operations in Equation (5-29) yield Equation (5-25) obtained previously.

Now, we can examine the effect produced by the girder flexibility. Except for the flexible girder, member 2, the frame is identical to the previous example. Properties of the WF16X88 girder are

$$A = 25.9 \text{ in}^2 \ (0.0167 \text{ m}^2)$$

$$I = 1220 \text{ in}^4 \ (0.0005078 \text{ m}^4)$$

Substituting these values plus those for the columns of Figure 5-8 into Equation (5-25), the lateral stiffness with the flexible girder is

$$k = 107,000 \text{ lb/in} \ (18,700 \text{ kN/m}) \qquad (5\text{-}30)$$

This value is about 23 percent lower than that for a rigid girder given by Equation (5-12).

As noted previously, the mass of the frame is unchanged from the last example. Evaluating the response of the frame for the forcing function of Figure 5-7b, the displacements of Figure 5-12 are obtained. Lateral displacements of the frame with and without girder flexibility are given. As expected, including girder flexibility in the analysis increased displacements and the response period of the frame.

It is instructive to examine the maximum bending moments in the columns for the two conditions. With a rigid girder, the moments are computed from the lateral displacement for a member in pure sway (Figure 5-7c). For each column, i, the end moments are given by

$$M_{i_{max}} = k_i \ x_{max} \ \frac{H}{2} = \frac{1}{2} \ H \ k_i \ x_{max} \qquad (5\text{-}31)$$

With equal columns, one-half of the total stiffness [Equation (5-12)] can be used. The maximum displacement is determined from the curve labeled "rigid girder" in Figure 5-12, and the frame height is given in Figure 5-7.

  With flexible girders, the rotations at the girder-column intersection must be known. These are given in terms of the lateral displacement by Equation (5-27). To solve for the maximum moments, the rotations are evaluated from the lateral displacements at each step in the solution. Knowing the lateral displacement and rotations, the forces and moments acting on each member are determined from the member Equations (5-19) through (5-21). These equations give the moments directly except at the base of the columns. At the base of column 1 (Figure 5-9), for example, the moment is simply

$$M = F_2 + F_1 \ H \qquad (5\text{-}32)$$

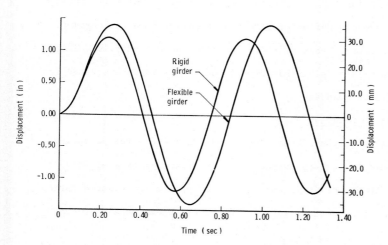

Figure 5-12. Lateral Frame Displacements Showing the Effect of Girder Flexibility

where F_2 and F_1 are the moment and lateral force, respectively, at the top

of the column, and H is the column height.

Following the procedure outlined above for the frame with a flexi-

ble girder and using Equation (5-31) for a rigid girder, the maximum mo-

ments in the column were obtained. These, along with the maximum lateral

displacements, are given in Table 5-3.

Table 5-3. Maximum Moments in the Column

	Rigid Girder	Flexible Girder
Maximum lateral displacement	1.207 in. (30.66 mm)	1.406 in. (35.71 mm)
Maximum moment at top of column	7,082,970 in-lb (800,234 N·m)	5,536,310 in-lb (625,442 N·m)
Maximum moment at base of columns	7,082,970 in-lb (800,234 N·m)	7,103,526 in-lb (802,556 N·m)

Notice that, although displacements increased when girder flexibil-

ity was included in the problem, maximum moments in the two frames are al-

most equal; so, for this example, the assumption of rigid girders is a

reasonable one for sizing the columns for lateral loads. This will gener-

ally be true for frames in which the girder is stronger than the columns.

Also, when elastic-plastic behavior is included in the analysis, the influ-

ence of girder flexibility is further diminished. This is demonstrated in

another section of this chapter.

Elastic-Plastic Behavior

The numerical solution of the elastic one-degree-of-freedom spring-

mass system of Figure 5-2 can be extended easily to accommodate elastic-

plastic behavior. To do so, it is convenient to replace the elastic spring

restoring force, kx, by a restoring force, R(x), which can be any general

function of displacement. Rate effects, proportional to $\dot{x}$ could also be

included, but will not be treated here.

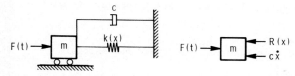

(a) Spring-mass system in which   (b) Free-body diagram
the stiffness varies with
displacement

Figure 5-13. Nonlinear Spring

From the free-body diagram of Figure 5-13b, the equation of motion
is written as:

$$F(t) - R(x) - c\dot{x} = m\ddot{x} \qquad (5-33)$$

As before, we estimate the velocity and displacement at time t using the
average acceleration method. Substituting these expressions given by
Equations (5-2a) and (5-3) into Equation (5-33), the following equation
for the acceleration at time t is obtained.

$$\ddot{x}_t = \frac{1}{m + \frac{1}{2}c\Delta t}\left[F(t) - R(x_t) - c\left(\dot{x}_{t-\Delta t} + \frac{\Delta t}{2}\ddot{x}_{t-\Delta t}\right)\right] \qquad (5-34)$$

Whereas, Equation (5-4) for a linear elastic system gave the acceleration
at time t in terms of displacement, velocities, and accelerations at $t - \Delta t$,
Equation (5-34) with the nonlinear term, $R(x_t)$, depends upon the displacement
at time t. Thus, instead of the direct integration approach used for the
elastic system, we now choose a predictor-corrector method in which the displacement at x_t is estimated (predicted) and then corrected. This procedure
of predicting and correcting can be continued until a convergence or error
criterion is met, but with small time steps, a single iteration is usually
sufficient. Constant acceleration over the time step is chosen for the prediction. A correction is then made using the predicted displacement and

velocities at the end of the time step. The step by step procedure is as follows:

Step 1. Compute $\ddot{x}$ at t = 0 from Equation (5-33) and the initial conditions.

Step 2. Increment time by Δt.

Step 3. Set $\ddot{x}_t = \ddot{x}_{t-\Delta t}$ (t - Δt = 0, initially).

Step 4. Compute $\dot{x}_t$ and x_t from Equations (5-2).

Step 5. Compute $R(x_t)$.

Step 6. Compute $\ddot{x}_t$ from Equation (5-34).

Step 7. (a) If predictor pass, return to Step 4; (b) if corrector pass, set $x_{t-\Delta t} = x_t$, $\dot{x}_{t-\Delta t} = \dot{x}_t$, $\ddot{x}_{t-\Delta t} = \ddot{x}_t$ and go to Step 2.

To demonstrate the procedure and to illustrate the effect of plasticity on the response, the example of Figure 5-2 is repeated with the bilinear spring force as shown in Figure 5-14. The initial elastic portion of the curve is unchanged, and the yield force in the spring is reduced to 75 percent of the maximum force developed in the elastic spring. This insures that yielding will occur, and the effect of plasticity can be observed. The numerical integration is given in Table 5-4 to illustrate the predictor-corrector method, and the computed displacements are shown in Figure 5-15. Maximum amplitude of the elastic system (from Figure 5-3) is denoted by the dashed line.

Figure 5-15 shows that under equal dynamic loads, the strong spring is stressed just to yield (elastic solution) and the weaker spring yields, as shown by the displacements. The significant point is that the peak deformation of the weaker spring was very nearly the same as for the stronger

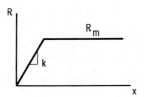

Figure 5-14. Nonlinear Resistance Function

Table 5-4. Numerical Integration for Elastic-Plastic Behavior

Column 1	Column 2	Column 3	Column 4	Column 5	Column 6	Column 7	Column 8	Column 9	Column 10
			$\dot{x}_t =$		$x_t =$	$F(t) =$			
		$\frac{\Delta T}{2}(\ddot{x}_t +$	$x_{t-\Delta T} +$	$\frac{\Delta T}{2}(\dot{x}_t +$	$x_{t-\Delta T} +$	$200{,}000 -$			$\ddot{x}_t =$
Time	Pass	$\ddot{x}_{t-\Delta t})$	Column 3	$\dot{x}_{t-\Delta T})$	Column 5	$10 \times 10^6\,t$	R	$F(t) - R$	Column 9 : 16
0.000	P	0.000	0.00	0.000	0.000	200000.00	0.00	200000.00	12500.000
0.004	P	50.000	50.00	0.100	0.100	160000.00	14000.00	146000.00	9125.000
	C	43.250	43.25	0.086	0.086	160000.00	14000.00	146000.00	9125.000
0.008	P	36.973	80.22	0.247	0.086	160000.00	12110.00	147890.00	9243.125
	C	27.651	70.90	0.228	0.315	120000.00	46682.30	73317.70	4582.356
0.012	P	18.982	89.88	0.322	0.636	80000.00	44072.27	75927.73	4745.483
	C	8.354	79.26	0.300	0.615	80000.00	89091.75	- 9091.75	- 568.234
0.016	P	- 1.529	77.73	0.314	0.929	40000.00	86116.07	- 6116.07	- 382.254
	C	-11.515	67.74	0.294	0.909	40000.00	126000.00	- 86000.00	- 5375.000
0.020	P	-21.500	46.24	0.228	1.137	0.00	126000.00	-126000.00	- 7875.000
	C	-26.500	41.24	0.218	1.127	0.00	126000.00	-126000.00	- 7875.000
0.024	P	-31.500	9.74	0.102	1.229	0.00	126000.00	-126000.00	- 7875.000
	C	-31.500	9.74	0.102	1.229	0.00	126000.00	-126000.00	- 7875.000
0.028	P	-31.500	-21.76	-0.024	1.205	0.00	126000.00	-126000.00	- 7875.000
	C	-31.500	-21.76	-0.024	1.205	0.00	126000.00	-126000.00	- 7875.000
0.032	P	-31.500	-53.26	-0.150	1.055	0.00	101629.87	-101629.87	- 6351.867
	C	-28.45	-50.21	-0.144	1.061	0.00	102482.82	-102482.82	- 6405.176
0.036	P	-25.621	-75.83	-0.252	0.809	0.00	67189.87	- 67189.87	- 4199.367
	C	-21.209	-71.42	-0.243	0.818	0.00	68425.12	- 68425.12	- 4276.570
0.040	P	-17.106	-88.53	-0.320	0.498	0.00	23639.11	- 23639.11	- 1477.445
	C	-11.508	-82.93	-0.309	0.509	0.00	25206.62	- 25206.62	- 1575.414
0.044	P	- 6.302	-89.23	-0.344	0.165	0.00	- 22998.58	22998.58	1437.411
	C	- 0.276	-83.21	-0.332	0.177	0.00	- 21311.40	21311.40	1331.962
0.048	P	5.328	-77.88	-0.322	-0.145	0.00	- 66414.91	66414.91	4150.932
	C	10.966	-72.24	-0.311	-0.134	0.00	- 64836.28	64836.28	4052.268
0.052	P	16.209	-56.03	-0.257	-0.391	0.00	-100752.21	100752.21	6297.013
	C	20.699	-51.54	-0.248	-0.382	0.00	- 99495.15	99495.15	6218.447
0.056	P	24.874	-26.67	-0.156	-0.538	0.00	-121393.76	121393.76	7587.110
	C	27.611	-23.93	-0.151	-0.533	0.00	-120627.31	120627.31	7539.287
0.060	P	30.157	6.23	-0.035	-0.568	0.00	-125584.44	125584.44	7849.028
	C	30.776	6.85	-0.034	-0.567	0.00	-125410.94	125410.94	7838.184
0.064	P	31.353	38.20	0.090	-0.477	0.00	-112798.40	112798.40	7049.900
	C	29.776	36.62	0.087	-0.480	0.00	-113239.84	113239.84	7077.490
0.068	P	28.310	64.93	0.203	-0.277	0.00	- 84804.62	84804.62	5300.289
	C	24.756	61.38	0.196	-0.284	0.00	- 85799.85	85799.85	5362.491
0.072	P	21.450	82.83	0.288	0.005	0.00	- 45422.32	45422.32	2838.895
	C	16.403	77.78	0.278	-0.006	0.00	- 46835.53	46835.53	2927.221
0.076	P	11.709	89.49	0.335	0.329	0.00	0.05	- 0.05	- 0.003
	C	5.854	83.63	0.323	0.317	0.00	- 1639.20	1639.20	102.450
0.080	P	0.410	84.04	0.335	0.653	0.00	45311.13	- 45311.13	- 2831.945
	C	- 5.459	78.18	0.324	0.641	0.00	43667.86	- 43667.86	- 2729.242
0.084	P	-10.917	67.26	0.291	0.932	0.00	84389.66	- 84389.66	- 5274.354
	C	-16.007	62.17	0.281	0.922	0.00	82964.39	- 82964.39	- 5185.275
0.088	P	-20.741	41.43	0.207	1.129	0.00	111971.40	-111971.40	- 6998.312
	C	-24.367	37.80	0.200	1.122	0.00	110956.16	-110956.16	- 6934.760
0.092	P	-27.739	10.06	0.096	1.217	0.00	124358.24	-124358.24	- 7772.398
	C	-29.414	8.39	0.092	1.214	0.00	123889.17	-123889.17	- 7743.073
0.096	P	-30.972	-22.58	-0.028	1.186	0.00	119913.93	-119913.93	- 7494.620
	C	-30.475	-22.09	-0.027	1.187	0.00	120053.06	-120053.06	- 7503.316
0.100	P	-30.013	-52.10	-0.148	1.038	0.00	99280.13	- 99280.13	- 6205.008
	C	-27.417	-49.50	-0.143	1.043	0.00	100007.19	-100007.19	- 6250.449
0.104	P	-25.002	-74.51	-0.248	0.795	0.00	65284.15	- 65284.15	- 4080.259
	C	-20.661	-70.17	-0.239	0.804	0.00	66499.45	- 66499.45	- 4156.216
0.108	P	-16.625	-86.79	-0.314	0.490	0.00	22551.56	- 22551.56	- 1409.473
	C	-11.131	-81.30	-0.303	0.501	0.00	24089.74	- 24089.74	- 1505.609
0.112	P	- 6.022	-87.32	-0.337	0.164	0.00	- 23123.05	23123.05	1445.190
	C	- 0.121	-81.42	-0.325	0.176	0.00	- 21470.60	21470.60	1341.912
0.116	P	5.368	-76.05	-0.315	-0.139	0.00	- 65561.83	65561.83	4097.614
	C	10.879	-70.54	-0.304	-0.128	0.00	- 64018.63	64018.63	4001.165
0.120	P	16.005	-54.53	-0.250	-0.378	0.00	- 99039.23	99039.23	6189.952
	C	20.382	-50.16	-0.241	-0.370	0.00	- 97813.51	97813.51	6113.344
0.124	P	24.453	-25.70	-0.152	-0.521	0.00	-119054.41	119054.41	7440.901
	C	27.108	-23.05	-0.146	-0.516	0.00	-118310.98	118310.98	7394.436
0.128	P	29.578	6.53	-0.033	-0.549	0.00	-122936.31	122936.31	7683.519
	C	30.156	7.11	-0.032	-0.548	0.00	-122774.42	122774.42	7673.401
0.132	P	30.694	37.80	0.090	-0.458	0.00	-110200.00	110200.00	6887.500
	C	29.122	36.23	0.087	-0.461	0.00	-110640.10	110640.10	6915.006

* Units are English; displacements in Column 6 are in inches.

P = predictor pass

C = corrector pass

spring (elastic solution). This occurs because only a slight increase in deformation is required in the weaker system to equal the strain energy in the stronger system.

The ductility ratio, μ, is a common criterion for evaluating plas-

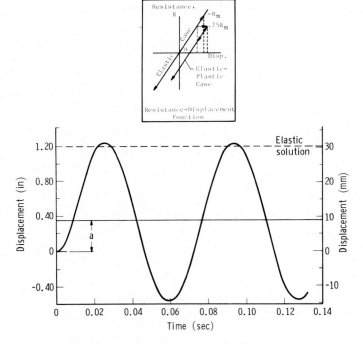

Figure 5-15. Displacements of One-Degree-Of-Freedom System of
Figure 5-2 With a Bi-Linear Resistance Function

tic behavior in structures. It is the total deformation divided by the
deformation at which yielding occurs. For the weak spring, this ratio is

$$\mu = 1.35$$

A ductility ratio of three generally is acceptable for reusable structures,
if other criteria such as hinge rotations and lateral sway are met (see
Chapter 8). Thus, it is apparent that a substantial reduction in strength
still produces an acceptable design for peak overloads when even modest
plasticity is permitted.

We have shown that numerical methods can be used to solve easily for
displacements of linear and nonlinear one-degree-of-freedom systems to a
general dynamic load. Furthermore, some fairly complex structures, such

as frames, can be modeled as a one-degree-of-freedom system, depending upon
the behavior being studied. Following this approach, Biggs (1964) and others
[Newmark (1950), Norris, et al. (1959), and U. S. Army Corps of Engineers
Manual EM 1110-345-415 (1975)] have developed one-degree-of-freedom approximations for the elastic-plastic behavior of beams and plates to dynamic loads.

One-Degree-Of-Freedom Equivalent Systems

To derive one-degree-of-freedom "equivalent" systems for what are normally regarded as multi-degree-of-freedom structural elements requires that
some assumptions be made about the behavior of the structure. Such assumptions were inherent in the one-degree-of-freedom models used to analyze the
frame structures of Figures 5-7 and 5-8. For example, the force-displacement relations of the columns were based upon static solutions. Inertia
effects in the columns would have altered local column displacements and
thus, the force-displacement relationships. Inertia effects in the columns
were handled in an approximate manner by adding 1/3 of the column's weight
to the dead load and girder. In a similar manner, deformation patterns
must be assumed for other elements in order to characterize the behavior of
the total element in terms of a single displacement at a single point.

As an example, consider the simply supported beam of Figure 5-16. If
the displacement along the beam can be described in terms of a single variable, for example, the center displacement, w_o, then the motions are reduced
to a single-degree-of-freedom. This is accomplished by assuming a deformation pattern for the beam which is some function of w_o. Various choices can
be made, e.g., the fundamental mode shape, the static deformed shape for the
load distribution of the dynamic loading, or simply some approximate shape
which resembles the fundamental mode or static deformed shape and matches
the appropriate boundary conditions. For elastic behavior, Biggs (1964)
chooses the static deformed shape of the structure and for plastic behavior,
plastic hinges at locations of the maximum moments. For the simply support-

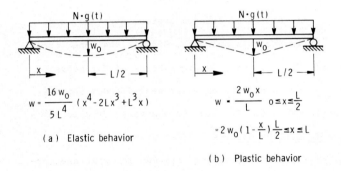

$$w = \frac{16\,w_0}{5\,L^4}\,(x^4 - 2Lx^3 + L^3 x)$$

(a) Elastic behavior

$$w = \frac{2\,w_0\,x}{L} \quad 0 \le x \le \frac{L}{2}$$

$$= 2\,w_0\,(1 - \frac{x}{L})\,\frac{L}{2} \le x \le L$$

(b) Plastic behavior

Figure 5-16. Deformation Patterns for Simple Beam

ed beam, uniformly loaded by the distributed load $N \cdot g(t)$ [$g(t)$ is the time function for the load], the static deformed shape for elastic behavior is given in Figure 5-16a. Figure 5-16b shows the hinge mechanism for plastic behavior.

Reducing displacements to a single-degree-of-freedom does not guarantee similarity between the response of a one-degree-of-freedom system and the structural element. One way that similarity can be obtained is to equate the energies of the distributed and single-degree-of-freedom systems. At any time, the internal strain energy and the system kinetic energy must equal the external work as shown by Equation (5-35).

$$W = U + KE \tag{5-35}$$

where W is the external work, U is the internal strain energy, and KE is the kinetic energy. If the kinematic similarity (equal displacements and velocities in this case) is maintained between the distributed and single-degree-of-freedom systems, then equating energies of the two systems will assure similarity of the computed behavior.

1. Elastic Behavior

As a first example, we chose the fundamental mode of the simply supported beam as the deformed shape. This choice will permit a check on frequency with the exact solution to demonstrate kinematic equivalence of the distributed and equivalent systems. The deformed shape for the fundamental mode is

$$w(x) = w_o \sin \frac{\pi x}{L} \tag{5-36}$$

Evaluating W, U, and KE for this deformation pattern gives

$$W = \int_o^L N \cdot g(t) \ w(x) \ dx = \frac{2}{\pi} N \cdot g(t) \ w_o L \tag{5-37a}$$

$$U = \frac{EI}{2} \int_o^L \left(\frac{d^2 w(x)}{dx^2} \right)^2 dx = \frac{\pi^4 EI w_o^2}{4L^3} \tag{5-37b}$$

$$KE = \frac{1}{2} \int_o^L [\dot{w}(x)]^2 \ \rho A \ dx = \frac{1}{4} \rho A L \ \dot{w}_o^2 \tag{5-37c}$$

where E is the material modulus, I is the beam section moment of inertia, A is the beam cross-sectional area, ρ is the material density, and w_o is the beam center displacement and is some function of time. These same energies, evaluated for the single-degree-of-freedom system of Figure 5-17 are

$$W = F_e \cdot g(t) \ w_o \tag{5-38a}$$

$$U = \frac{1}{2} k_e w_o^2 \tag{5-38b}$$

$$KE = \frac{1}{2} m_e \dot{w}_o^2 \tag{5-38c}$$

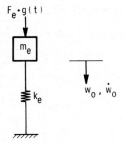

Figure 5-17. Equivalent One-Degree-Of-Freedom System

Equating Equations (5-37) and (5-38) and recognizing that $\rho AL = m$ (total beam mass) and $NL = F$ (total beam static load), the following relationships are obtained:

$$K_L = \frac{F_e}{F} = 0.6366 \qquad (5-39a)$$

$$k_e = \frac{\pi^4 \, EI}{2L^3} = 0.634 \, k \qquad (5-39b)$$

$$K_m = \frac{m_e}{m} = 0.500 \qquad (5-39c)$$

where k is the spring constant for the simply supported beam ($N \cdot L$ divided by static center deflection), K_L is the load factor, and K_m is the mass factor. Notice that the equivalent spring constant is very nearly equal to the static spring constant multiplied by the load factor.

Because these relationships were developed on the basis of kinematic equivalence of the displacement w_o, it is instructive to compare the frequency given by the equivalent mass and spring constants with the exact solution for a simply supported beam. The exact solution for the fundamental mode is given by Den Hartog (1956) as

$$\omega = \pi^2 \sqrt{\frac{EI}{mL^3}} \tag{5-40}$$

where m is the total beam mass which is equal to ρAL. For the equivalent system,

$$\omega = \sqrt{\frac{k_e}{m_e}} = \sqrt{\frac{\pi^4 \, EI}{2L^3 \, (0.5m)}} = \pi^2 \sqrt{\frac{EI}{mL^3}} \tag{5-41}$$

As expected, the assumption of the fundamental mode for the deformed shape gives a frequency which is exactly equal to the beam's fundamental frequency.

Now, K_L, K_m, and k_e will be determined for the static deformed shape of the beam for comparison with the results of Biggs (1964). The static deformed shape is given in Figure 5-16, and for this case, the energies of the distributed system are:

$$W = N \cdot g(t) \, \frac{16 \, w_o L}{25} \tag{5-42a}$$

$$U = \frac{3072 \, EI \, w_o^2}{125L^3} \tag{5-42b}$$

$$KE = \frac{1984}{7875} \, \rho \, AL \, \dot{w}_o^2 \tag{5-42c}$$

Equating to Equation (5-38), as before, we obtain

$$K_L = \frac{F_e}{F} = 0.64 \tag{5-43a}$$

400

$$k_e = \frac{6144}{125L^3} EI = 0.64 \ k \tag{5-43b}$$

$$K_m = \frac{m_e}{m} = 0.5026 \approx 0.50 \tag{5-43c}$$

These values of the mass and load factors match those given by Biggs (1964) and others and shown here in Table 5-5. Note also that the spring rate of the equivalent is exactly equal to the load factor times the spring constant for the distributed system. If we use these values of stiffness and mass to compute the frequency of the equivalent system, we find

$$\omega = 9.889 \sqrt{\frac{EI}{mL^3}} \tag{5-44}$$

which is within 0.2 percent of the fundamental mode of the beam [Equation (5-39)]. Because the static deformed shape is made up of many normal modes, in addition to the first, some difference in frequency is to be expected. If the beam truly deforms dynamically with the static deformed shape, then it would respond with the equivalent system frequency.

Several observations can be made as a result of these calculations for elastic behavior:

- A rational method is available for deriving an equivalent one-degree-of-freedom system for a distributed structure.
- Kinematic equivalency is maintained if the assumed deformation pattern exactly matches the behavior of the distributed system.
- The values of K_L, K_M, and k_e match those given by Biggs (1964) and others for the static-deformed shape.
- For the static-deformed shape, the equivalent stiffness, k_e, can be computed from the static stiffness of the distributed system and the load factor.

Table 5-5. Transformation Factors for Beams and One-Way Slabs

Simply Supported

Loading Diagram	Strain Range	Load Factor K_L	Mass Factor K_M		Load-Mass Factor K_{LM}		Maximum Resistance R_m	Spring Constant k	Dynamic Reaction V
			Concentrated Mass*	Uniform Mass	Concentrated Mass*	Uniform Mass			
$F = pL$	Elastic	0.64	....	0.50	....	0.78	$\frac{8M_p}{L}$	$\frac{384.0\ EI}{5L^3}$	$0.390\ R + 0.110\ F$
	Plastic	0.50	....	0.33	....	0.66	$\frac{8M_p}{L}$	0	$0.380\ R_m + 0.120\ F$
F	Elastic	1.00	1.00	0.49	1.00	0.49	$\frac{4M_p}{L}$	$\frac{48.0\ EI}{L^3}$	$0.780\ R - 0.280\ F$
	Plastic	1.00	1.00	0.33	1.00	0.33	$\frac{4M_p}{L}$	0	$0.750\ R_m - 0.250\ F$
$\frac{F}{2}\ \frac{F}{2}$	Elastic	0.87	0.76	0.52	0.87	0.60	$\frac{6M_p}{L}$	$\frac{56.4\ EI}{L^3}$	$0.525\ R - 0.025\ F$
	Plastic	1.00	1.00	0.56	1.00	0.56	$\frac{6M_p}{L}$	0	$0.520\ R_m - 0.020\ F$

*Equal parts of the concentrated mass are lumped at each concentrated load.

Source: U. S. Army Corps of Engineers (1975).

- The fundamental mode shape and the static-deformed shape give very similar results for a uniform loading. Using the static-deformed shapes should give the best results for other than uniform loading because the deflected shape will not match the fundamental mode so closely.

2. Plastic Behavior

Exactly the same procedure is followed for deriving the equivalent system for plastic behavior. Using the hinged shape of Figure 5-16b, the system energies are

$$W = \int_0^L N \cdot g(t) \; w(x) \; dx = \frac{1}{2} N \cdot g(t) \; L \; w_o \tag{5-45a}$$

$$U = \int_0^\theta M_p \; d\theta = M_p \left(2 \; \frac{w_o}{L/2} \right) = \frac{4w_o}{L} \; M_p \tag{5-45b}$$

$$KE = \frac{1}{2} \int_0^L \rho A \; [\dot{w}(x)]^2 \; dx = \frac{1}{6} \rho \; AL \; \dot{w}_o^2 \tag{5-45c}$$

For the equivalent system, only the strain energy is changed. For fully plastic behavior, it becomes

$$U = R_e w_o \tag{5-46}$$

where R_e is the resistance of the equivalent system. Equating energies of the two systems, load and mass factors, and the equivalent resistance for plastic behavior, are obtained.

$$K_L = \frac{F_e}{F} = 0.50 \tag{5-47a}$$

$$R_e = \frac{4M_p}{L} = 0.50\ R_m \qquad\qquad (5\text{-}47b)$$

$$K_m = \frac{m_e}{m} = \frac{1}{3} \qquad\qquad (5\text{-}47c)$$

where R_m is the maximum static resistance for the uniform beam (total distributed load for the development of a plastic hinge at the beam center). The
resistance function for elastic-plastic behavior of the equivalent system is
bilinear as shown in Figure 5-18.

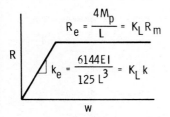

R

$$R_e = \frac{4M_p}{L} = K_L R_m$$

$$k_e = \frac{6144EI}{125\ L^3} = K_L k$$

W

Figure 5-18. Resistance of the Equivalent System

Load and mass factors and resistance functions have been developed by Biggs (1964) and others for many different structural elements and
different loading conditions for each element. Transformation factors have
been developed for:

- simply supported beams
- clamped beams
- proposed cantilevers
- one- and two-way slabs with edge supports
- two-way slabs with interior supports

The resistance functions are not always as simple as that derived for a simply supported beam. A clamped beam, for example, has an elastic-plastic
range intermediate to the elastic and fully plastic ranges. This range is
produced when hinges form at the fixed ends and before the central hinge is

404

formed. Tables of additional transformation factors for clamped beams and other structural components are given in Appendix A.

3. <u>Displacements of the Equivalent System</u>

Displacements of these equivalent systems can be computed using the numerical methods of a previous section in this chapter entitled "Numerical Integration." The only difference in this solution and those described earlier for a one-degree-of-freedom system is that the mass, load and resistance all change when yielding occurs. If the procedure of changing the load, mass, and resistance when yielding occurs is followed, then the response of each system will differ because, in general, changes in the mass and load factors, when yielding occurs, are different for each component. Thus, the approach taken by Biggs was to compute the response of one-degree-of-freedom systems for bilinear resistance functions as in Figure 5-18 and for constant mass and load. The resistance, mass, and load factors are then chosen to match the type of behavior expected. If the behavior is expected to be elastic, then values for the elastic range are used. If substantial plasticity is expected, then values for plastic behavior are used. Averages can be used if modest yielding is expected.

Graphical solutions for a triangular loading are presented in Figure 5-19. The maximum response, y_m, and the time to maximum response, t_m, are given as functions of:

- the resistance to load ratio, R_m/F
- the ratio of the load duration to the fundamental period of the system, T_L/T_N
- the deflection of the system at yield, y_{el}

These curves and similar ones for different loading functions can be used to solve for deflections in any system for which F_1, R_m, and y_{el} can be determined. For elastic behavior of the simply supported beam of Figure 5-16a, the values for entry into Figure 5-19 would be

$$\frac{R_e}{F_e} = \frac{K_L R_m}{K_L F} = \frac{R_m}{F} = \frac{8M_p/L}{NL}$$

$$y_{el} = \frac{R_e}{k_e} = \frac{K_L R_m}{K_L k} = \frac{R_m}{k} = \frac{8M_p/L}{384EI/5L^3}$$

$$T = 2\pi\sqrt{\frac{m_e}{k_e}} = 2\pi\sqrt{\frac{K_L m}{K_L k}} = 2\pi\sqrt{\frac{0.5m}{0.64k}}$$

plus whatever load duration T_L is specified for the applied loading. Notice that F denotes the total force applied to the real system and F_e denotes the force applied to the equivalent system. Graphical solutions for other types of loads are given in Appendix B.

4. Load-Mass Factors and Shear Reactions

Two other parameters in Table 5-5 which need clarification are the load-mass factor, K_{Lm}, and the dynamic reaction, V. The load-mass factor is simply the mass factor divided by the load factor. To see why a separate factor is defined, consider the equation of motion of the one-degree-of-freedom equivalent system.

$$m_e \ddot{x} + k_e x = F_e(t)$$

which can also be written as:

$$K_m m\ddot{x} + K_L kx = K_L F(t) \qquad (5-48)$$

Dividing Equation (5-48) by K_L gives

$$K_{Lm} m\ddot{x} + kx = F(t) \qquad (5-49)$$

406

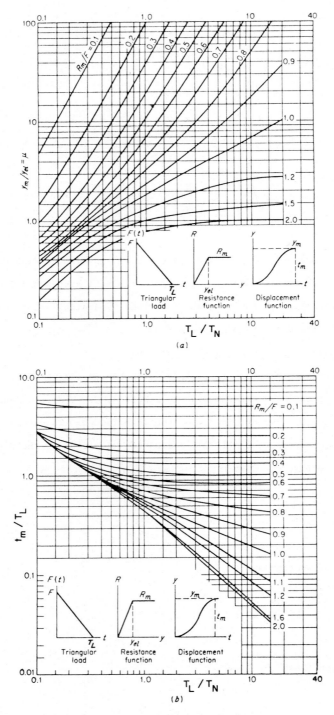

Figure 5-19. Maximum Response of Elastic-Plastic One-Degree-Of-Freedom
System for a Triangular Load
[U. S. Army Corps of Engineers (1975)]

where $K_{Lm} = K_m/K_L$. Thus, equivalent motions can be obtained by simply adjusting the mass of the system. The period of the equivalent system can also be computed with just the load-mass factor.

$$T = 2\pi\sqrt{\frac{m_e}{k_e}} = 2\pi\sqrt{\frac{K_m m}{K_L k}} = 2\pi\sqrt{\frac{K_{Lm} m}{k}} \qquad (5\text{-}50)$$

Regarding dynamic shear reactions in the equivalent systems, Biggs (1964) states,

> It is important to recognize that the dynamic reactions
> of the real structural element have no direct counter-
> part in the equivalent one-degree system. In other
> words, the reaction of the equivalent system, i.e.,
> the spring force, is not the same as the real reac-
> tion. This is true because the simplified system was
> deliberately selected so as to have the same dynamic
> deflection as the real element, rather than the same
> force or stress characteristics.

To determine the dynamic reaction, V, of Table 5-5, a free-body diagram of the element, including inertia effects, must be used. For the uniformly loaded, simply supported beam of Figure 5-16, such a free-body diagram is given in Figure 5-20. Inertia forces are distributed according to the assumed deformation pattern and are opposed to the motion. We consider only the period of time, t_m, up to the maximum response.

From a free-body diagram of one-half of the beam, Figure 5-20b, an expression for V(t) is obtained by taking moments at x = a, the line of action of the resulting inertia forces.

408

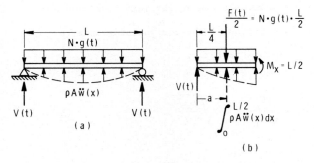

Figure 5-20.   Simply Supported Beam in Equilibrium Under
Dynamic Loading

$$V(a) - M_{x = L/2} - \frac{F(t)}{2} \left(a - \frac{L}{4} \right) = 0 \tag{5-51}$$

For elastic behavior, locations of the resultant of the inertia forces and
the mid-span moment are determined from the static deformed shape.   They
are

$$a = \frac{61}{192} L$$

$$M_{x = L/2} = \frac{48EI}{5L^2} w_o$$

Substituting into Equation (5-51), the dynamic reaction at the beam and _for_
elastic behavior is

$$V(t) = \frac{30.216EI}{L^3} w_o + 0.1066 \, F(t) \tag{5-52}$$

where $F(t)$ is the total dynamic load on the beam [$N \cdot g(t) \cdot L$] and w_o is understood to be a function of time.   If yielding occurs before the maximum load
occurs, then the beam plastic moment, M_p, replaces $M_{x = L/2}$ in Equation
(5-51) and V becomes

$$V(t) = \frac{192M_p}{61L} = 0.1066 \; F(t) \qquad \text{(after yielding)} \qquad (5\text{-}53)$$

Further, M_p can be expressed in terms of the maximum resistance R_m to obtain the form in Table 5-5.

$$V(t) = \frac{24}{61} R_m + 0.1066 \; F(t) \simeq 0.39 \; R_m + 0.11 \; F(t) \qquad (5\text{-}54)$$

Equation (5-54) was developed for the static deformed shape of the beam which applies up to yielding. If the linear deformation pattern of Figure 5-16b were used, slightly different results would be obtained. Notice that in Equation (5-52), the maximum reaction may occur when w_o is a maximum, when $F(t)$ is a maximum, or somewhere in between; however, for Equations (5-53) or (5-54), the maximum reaction occurs when $F(t)$ is a maximum because the beam has yielded before $F(t)_{max}$ occurs. Although Equation (5-52) should be used for elastic behavior and Equation (5-54) for plastic behavior, Equation (5-54) will always give the highest value.

These reactions have been computed for the assumption that the beam responds in its static deformed shape. For strong, short duration blast loading, this will not be true, and high curvature may occur near the supports, producing very high shear forces in the element. Thus, the equations for shear reactions given in Table 5-5 for simply supported beams or similar equations for other elements should not be used for overpressures on structural elements produced by close-in explosive sources.

TWO-DEGREE-OF-FREEDOM SYSTEMS

Spring-Mass

Two-degree-of-freedom systems can be used to represent somewhat more complex structures than the one-degree-of-freedom systems studied in the previous section, and their solutions can be obtained readily using the numerical methods already described. For example, consider the two-degree-of-freedom system of Figure 5-21a, which is acted upon by time dependent forces $F_1(t)$ and $F_2(t)$. From the free-body diagram in part b, the equations of motion of the masses are:

<u>Mass 1:</u>

$$\Sigma \ F_x = m_1 \ddot{x}_1$$

$$F_1(t) + k_2 \ (x_2 - x_1) - k_1 x_1 - c_1 \dot{x}_1 + c_2 \ (\dot{x}_2 - \dot{x}_1) = m_1 \ddot{x}_1 \qquad (5-55)$$

<u>Mass 2:</u>

$$\Sigma \ F_x = m_2 \ddot{x}_2$$

$$F_2(t) - k_2 \ (x_2 - x_1) - c_2 \ (\dot{x}_2 - \dot{x}_1) = m_2 \ddot{x}_2 \qquad (5-56)$$

1. <u>Numerical Integration</u>

As for the one-degree-of-freedom system of Figure 5-1, the average acceleration method will be used to obtain expressions for the acceleration, velocity and displacement which can be integrated in time to find

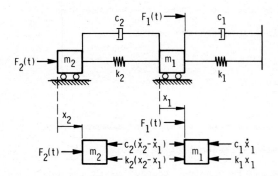

Figure 5-21. Two-Degree-Of-Freedom Spring-Mass System

the response of the system; however, if the equations are placed in the same form as Equation (5-4), i.e., if the accelerations at time t are written in terms of displacements and velocities at time $t - \Delta t$, then two equations are obtained which must be solved simultaneously for $\ddot{x}_{1_t}$ and $\ddot{x}_{2_t}$. It is easier for two-degrees-of-freedom to adopt the predictor-corrector method used for the plasticity solution. In this method, current values (i.e., at $t = t$ rather than at $t = t - \Delta t$) of x and $\dot{x}$ are retained in the expressions for $\ddot{x}$, and, to solve for the motion, current values are estimated (predicted) and then corrected. Following this approach, the equations are:

$$\left(\dot{x}_t \right)_i = \left(\dot{x}_{t - \Delta t} \right)_i + \frac{1}{2} \left(\ddot{x}_t + \ddot{x}_{t - \Delta t} \right)_i \Delta t \qquad i = 1, 2 \qquad (5\text{-}57a)$$

$$\left(x_t \right)_i = \left(x_{t - \Delta t} \right)_i + \frac{1}{2} \left(\dot{x}_t + \dot{x}_{t - \Delta t} \right)_i \Delta t \qquad i = 1, 2 \qquad (5\text{-}57b)$$

$$\left(\ddot{x}_1 \right)_t = \frac{1}{m_1} \left[F_1(t) - \left(k_1 + k_2 \right) x_1 + k_2 x_2 - \left(c_1 + c_2 \right) \dot{x}_1 + c_2 \dot{x}_2 \right]_t \qquad (5\text{-}58a)$$

$$\left(\ddot{x}_2 \right)_t = \frac{1}{m_2} \left[F_2(t) - k_2 x_2 + k_2 x_1 - c_2 \dot{x}_2 + c_2 \dot{x}_1 \right]_t \qquad (5\text{-}58b)$$

Numerical integration follows closely that given in a previous section of this chapter for elastic-plastic behavior, i.e.,

Step 1. Compute $\ddot{x}_1$ and $\ddot{x}_2$ at time $= 0$ from Equations (5-58) and the initial conditions.

Step 2. Increment time by Δt.

Step 3. Set $\left(\ddot{x}_i \right)_t = \left(\ddot{x}_i \right)_{t - \Delta t}$ for $i = 1, 2$ ($t - \Delta t = 0$, initial-ly).

Step 4. Compute $\left(\dot{x}_t \right)_i$ and $\left(x_t \right)_i$ for $i = 1, 2$ from Equations (5-57).

412

Step 5. Compute $\left(\ddot{x}_1\right)_t$ and $\left(\ddot{x}_2\right)_t$ from Equations (5-58) using values of x_t and $\dot{x}_t$ from Step 4.

Step 6. (a) If predictor pass, return to Step 4; (b) if corrector pass, set

$$\left.\begin{aligned}
\left(x_i\right)_{t-\Delta t} &= \left(x_i\right)_t \\[1em]
\left(\dot{x}_i\right)_{t-\Delta t} &= \left(\dot{x}_i\right)_t \\[1em]
\left(\ddot{x}_i\right)_{t-\Delta t} &= \left(\ddot{x}_i\right)_t
\end{aligned}\right\} \text{for } i = 1, 2$$

and go to Step 2.

2. Numerical Solution

To demonstrate the numerical procedure, consider the system of Figure 5-22. It is the same as the one-degree-of-freedom system of Figure 5-2 except another similar system has been added between it and the ground. $F_1(t)$ is zero and damping is ignored.

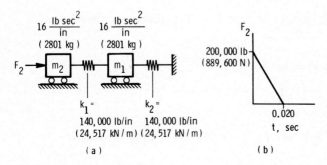

Figure 5-22. Two-Degree-Of-Freedom Example Problem

Substituting the values of Figure 5-22 into Equations (5-58), expressions for the acceleration become

$$\left(\ddot{x}_1 \right)_t = -8,750 \left(2 \; x_1 - x_2 \right)_t \quad \text{for all } t \tag{5-59a}$$

$$\left(\ddot{x}_2 \right)_t = 12,500 - 625,000 \; t - 8,570 \left(x_2 - x_1 \right)_t , \quad t \le T_L \tag{5-59b}$$

Using these two equations plus Equations (5-57) and following the step by step procedure outlined above, motions of the two-degree-of-freedom system, given in Table 5-6, are obtained. Computed displacements, including the relative displacement between m_1 and m_2, are plotted in Figure 5-23. More cycles of the motion have been plotted than were calculated in Table 5-6.

By comparing these results with the one-degree-of-freedom system of Figure 5-3, two effects produced by coupling between the two systems are apparent. One effect is that the frequencies of the responding system have changed. The coupled system has both a higher and lower frequency than each mass treated individually. The other effect is that the deflection in the

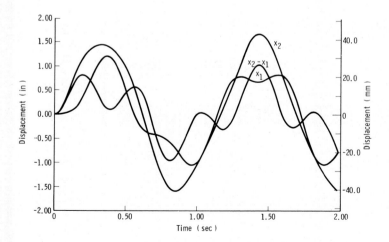

Figure 5-23. Response of Two-Degree-Of-Freedom System

Table 5-6. Predictor-Corrector Solution for Two-Degree-Of-Freedom System of Figure 5-22*

Column 1	Column 2	Column 3	Column 4	Column 5	Column 6	Column 7	Column 8	Column 9	Column 10	Column 11	Column 12	Column 13	Column 14	Column 15	Column 16
Time	Pass	$.002(\ddot{x}_t + \ddot{x}_{t-\Delta T})_1$	$\dot{x}_t = \dot{x}_{1t-\Delta T} +$ Col 3	$.002(\dot{x}_t + \dot{x}_{t-\Delta T})_1$	$x_{1t} = x_{t-\Delta T} +$ Col 5	$F_1(t)/M_1$	$(2x_1 - x_2)_t$	$-8750\,x_{1t}$	$.002(\ddot{x}_t + \ddot{x}_{t-\Delta T})_2$	$\dot{x}_{2t} = \dot{x}_{2t-T} +$ Col 10	$.002(\dot{x}_t + \dot{x}_{t-\Delta T})_2$	$x_{2t} = x_{2t-\Delta T} +$ Col 12	$F_2(t)/M_2$	$(x_2 - x_1)_t$	$\ddot{x}_{2t} =$ Col 14 $- 8750\,x_{15}$
0.000	P	0.000**	0.000**	0.000	0.0000**	0.00	0.000	0.000	0.000	0.000	0.000	0.0000**	12500.000	0.000	12500.000
0.004	P	0.000	0.000	0.000	0.0000	0.00	-0.100	875.000	50.000	50.000	0.100	0.1000	10000.000	0.100	9125.000
0.004	C	1.750	1.750	0.003	0.0035	0.00	-0.079	695.625	43.250	43.250	0.086	0.0865	10000.000	0.083	9273.750
0.008	P	2.782	4.532	0.013	0.0161	0.00	-0.302	2638.650	37.095	80.345	0.247	0.3337	7500.000	0.318	4720.781
0.008	C	6.669	8.419	0.020	0.0238	0.00	-0.268	2343.284	27.989	71.239	0.229	0.3155	7500.000	0.292	4948.141
0.012	P	9.373	17.792	0.052	0.0763	0.00	-0.488	4265.663	19.793	91.032	0.325	0.6400	5000.000	0.564	67.083
0.012	C	13.218	21.636	0.060	0.0839	0.00	-0.453	3960.260	10.030	81.269	0.305	0.6205	5000.000	0.537	305.203
0.016	P	15.841	37.477	0.118	0.2022	0.00	-0.544	4757.069	1.221	82.490	0.328	0.9480	2500.000	0.746	-4026.100
0.016	C	17.435	39.071	0.121	0.2054	0.00	-0.520	4549.697	-7.442	73.827	0.310	0.9307	2500.000	0.725	-3846.616
0.020	P	18.199	57.270	0.193	0.3980	0.00	-0.399	3492.469	-15.386	58.441	0.265	1.1952	0.000	0.797	-6975.356
0.020	C	16.084	55.155	0.180	0.3938	0.00	-0.395	3456.969	-21.644	52.183	0.252	1.1827	0.000	0.789	-6902.853
0.024	P	13.828	68.983	0.248	0.6421	0.00	-0.052	455.346	-27.611	24.572	0.154	1.3362	0.000	0.694	-6073.657
0.024	C	7.825	62.980	0.236	0.6301	0.00	-0.079	694.481	-25.953	26.230	0.157	1.3395	0.000	0.709	-6207.736
0.028	P	2.778	65.758	0.257	0.8876	0.00	0.380	-3327.816	-24.831	1.399	0.055	1.3948	0.000	0.507	-4438.355
0.028	C	-2.267	57.713	0.241	0.8785	0.00	0.341	-2984.327	-21.292	4.938	0.062	1.4019	0.000	0.530	-4641.063
0.032	P	-11.937	41.776	0.207	1.0785	0.00	0.772	-6758.483	-18.564	-13.626	-0.017	1.3845	0.000	0.306	-2677.973
0.032	C	-19.486	38.228	0.192	1.0634	0.00	0.734	-6425.584	-14.638	-9.700	-0.010	1.3924	0.000	0.329	-2878.777
0.036	P	-25.702	12.525	0.102	1.1649	0.00	0.999	-8742.943	-11.515	-21.215	-0.062	1.3305	0.000	0.166	-1449.599
0.036	C	-30.337	7.891	0.092	1.1556	0.00	0.975	-8530.707	-8.657	-18.357	-0.056	1.3362	0.000	0.181	-1580.728
0.040	P	-34.123	-26.232	-0.037	1.1189	0.00	0.988	-8641.878	-6.323	-24.680	-0.086	1.2502	0.000	0.131	-1148.582
0.040	C	-34.345	-26.454	-0.037	1.1185	0.00	0.985	-8618.971	-5.459	-23.816	-0.084	1.2519	0.000	0.133	-1167.598
0.044	P	-34.476	-60.930	-0.175	0.9437	0.00	0.740	-6475.756	-4.670	-28.486	-0.105	1.1473	0.000	0.204	-1781.580
0.044	C	-30.189	-56.644	-0.166	0.9523	0.00	0.760	-6647.270	-5.898	-29.714	-0.107	1.1448	0.000	0.193	-1685.078
0.048	P	-26.589	-83.233	-0.280	0.6725	0.00	0.333	-2909.498	-6.740	-36.454	-0.132	1.0125	0.000	0.340	-2975.005
0.048	C	-19.114	-75.757	-0.265	0.6875	0.00	0.368	-3216.289	-9.320	-39.034	-0.137	1.0074	0.000	0.320	-2799.035
0.052	P	-12.865	-88.623	-0.329	0.3587	0.00	-0.111	974.911	-11.196	-50.230	-0.179	0.8288	0.000	0.470	-4113.585
0.052	C	-4.483	-80.240	-0.312	0.3755	0.00	-0.073	635.518	-13.825	-52.859	-0.184	0.8236	0.000	0.448	-3920.884
0.056	P	2.542	-77.698	-0.316	0.0596	0.00	-0.462	4038.845	-15.684	-68.543	-0.245	0.5782	0.000	0.521	-4560.290
0.056	C	9.349	-70.891	-0.302	0.0732	0.00	-0.432	3778.233	-16.962	-69.821	-0.243	0.5808	0.000	0.504	-4410.795
0.060	P	15.113	-55.779	-0.253	-0.1801	0.00	-0.624	5458.633	-17.643	-87.464	-0.315	0.2636	0.000	0.444	-3882.470
0.060	C	18.474	-52.418	-0.247	-0.1734	0.00	-0.613	5359.777	-16.587	-86.408	-0.312	0.2657	0.000	0.439	-3842.427
0.064	P	21.439	-30.979	-0.167	-0.3402	0.00	-0.570	4984.857	-15.370	-101.778	-0.376	-0.1107	0.000	0.230	-2008.871
0.064	C	20.689	-31.728	-0.168	-0.3417	0.00	-0.580	5075.304	-11.703	-98.111	-0.369	-0.1034	0.000	0.238	-2085.396
0.068	P	16.161	-15.567	-0.086	-0.4280	0.00	-0.343	3005.419	-8.342	-106.453	-0.409	-0.5126	0.000	-0.085	739.714
0.068	C	12.997	-2.570	-0.095	-0.4363	0.00	-0.371	3249.490	-2.691	-100.802	-0.398	-0.5013	0.000	-0.065	568.390
0.072	P	7.289	-8.278	-0.036	-0.4840	0.00	-0.045	395.227	2.274	-98.528	-0.399	-0.8873	0.000	-0.427	3739.755
0.072	C	2.824	-5.454	-0.027	-0.4726	0.00	-0.081	706.002	8.616	-92.186	-0.386	-0.9000	0.000	-0.403	3528.869
0.076	P	-2.175	-10.453	-0.037	-0.5114	0.00	0.205	-1793.329	14.115	-78.071	-0.341	-1.2279	0.000	-0.716	6268.516
0.076	C	-6.090	-16.543	-0.054	-0.5214	0.00	0.174	-1522.488	19.595	-72.591	-0.330	-1.2169	0.000	-0.695	6085.152
0.080	P	-8.431	-18.884	-0.059	-0.5754	0.00	0.308	-2692.848	24.341	-48.250	-0.242	-1.4586	0.000	-0.883	7727.936
0.080	C	-10.214	-29.097	-0.096	-0.5801	0.00	0.292	-2553.426	27.626	-44.965	-0.235	-1.4521	0.000	-0.872	7629.476
0.084	P	-8.935	-27.819	-0.093	-0.6761	0.00	0.219	-1914.288	30.518	-14.447	-0.119	-1.5709	0.000	-0.895	7830.002
0.084	C	—	—	—	-0.6735	0.00	0.223	-1952.010	30.919	-14.046	-0.118	-1.5701	0.000	-0.897	7845.353

---

* Units are English; displacements in Columns 6 and 13 are in inches.

† For the "P" pass, $\ddot{x}_t = \ddot{x}_{t-\Delta T}$

** Initial conditions

second spring is less when coupled to a flexible support (system 1) than
when it is rigidly supported. The peak relative displacement $(x_2 - x_1)$ is
about 1.00 inch, whereas the deflection of the rigidly supported mass given
in Figure 5-3 is about 1.21 inches. This is generally true for blast loaded
components because the loading time is short relative to the structural frequencies. If the loading duration is longer, some "tuning" between the load
duration and the system period may occur, and this generalization would not
hold. It is instructive to investigate the effect of load duration and the
relative stiffness of the two coupled systems upon the maximum displacements
which occur.

Coupling in Two-Degree-Of-Freedom Systems

Often, in the analysis of buildings, structural elements must be analyzed which are supported by flexible structures, or conversely, structural
elements must be analyzed which have the loading transferred to them by the
structure which they support. These conditions arise for wall panels supported by girts, girts loaded by wall panels and supported by columns, roof
joists supported by roof girders, etc. It is important to understand the
effect of coupling which occurs between such systems, particularly if one-
degree-of-freedom solutions are being used for their analysis.

By changing the stiffness of k_1, in the example of Figure 5-22, the
effect of support flexibility upon the response of a structural element can
be shown. Likewise, the effect of load transfer through flexible structures
to the support can be examined. Both of these effects were studied using
the one- and two-degree-of-freedom models studied as examples in this chapter. Some results are given in Figure 5-24 as a function of the relative
periods of the support structure (system 1) and the supported structure
(system 2). A schematic diagram shows the systems analyzed and the results
compared.

To show the effect of support flexibility on the response of the supported structure, the displacements of mass m_2 were computed with both rigid
and flexible support. Results plotted are the maximum relative displacements

416

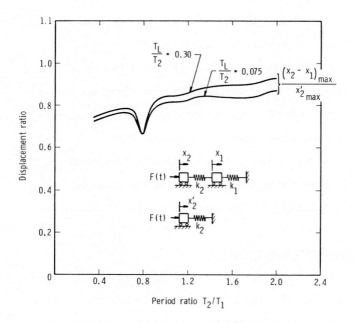

Figure 5-24(a).   Effect of Support Stiffness on
Maximum Displacements

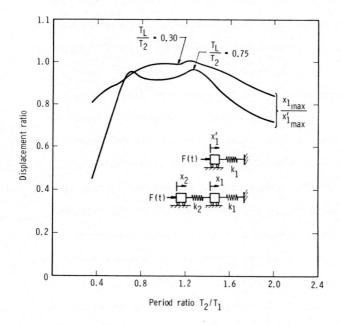

Figure 5-24(b).   Load Transfer Through Flexible Structure

$(x_2 - x_1)$ for the two-degree-of-freedom system divided by the maximum displacement of m_2 with rigid support. Calculations were made with two different forcing functions. One is identical to that of Figure 5-2b, and the other has equal force magnitude but a duration, T_L, of 0.005 sec. Differences in the load duration are expressed as the ratio of the load duration to the period of system 2. The two ratios are

$$T_L/T_2 = 0.02/0.067 \approx 0.30$$

$$T_L/T_2 = 0.005/0.067 \approx 0.075$$

Except for the dip at $T_2/T_1 \approx 0.80$, the motions of mass 2 relative to mass 1 increase almost monotonically with increasing values of T_2/T_1, i.e., with increasing values of k_1. If the results were extended to larger T_2/T_1 ratios (stiffer support), the relative response of the coupled system would approach results for the rigid support and the ratio

$$\frac{\left(x_2 - x_1\right)_{max}}{x_{2_{max}}}$$

would approach 1. At $T_2/T_1 = 0.80$, the frequencies of the coupled system are an even multiple (3 to 1) and the phasing of the displacements may explain the dip in relative displacement. For the two loading conditions considered, the load duration does not have much influence. Usually, the support structure will be stiffer than the structure being supported, $T_2/T_1 > 1.0$, and so the error produced by analyzing for a rigid support should be less than 20 percent; however, before any firm conclusions are drawn, the results should be expanded to include larger T_2/T_1 ratios and different loading functions.

The effect of load transfer through flexible structure is also shown in Figure 5-24 for the two loading conditions. Here the load duration has a more pronounced effect. Maximum load transfer to the substructure occurs over a range of frequencies which span $T_2/T_1 = 1$. In this region, there is little conservatism in assuming full load transfer. Outside this band of frequencies, the attenuation of the load on the substructure can be pronounced, especially for short duration loading. Longer duration loading transfers more load to the substructure over a wider range of relative frequencies. For the loading cases considered, assuming full transfer is conservative; however, for longer duration loading, response ratios

$$\frac{x_{1_{max}}}{x'_{1_{max}}}$$

may exceed one.

Even though these results are not conclusive, they can provide some guidance if the loading and frequency ratios match those in Figure 5-24. Generally speaking, it will be conservative to neglect flexibility in the support and to assume full load transfer to the supporting member.

Two-Story Frame with Rigid Girders

As for the single-story frame analyzed earlier in this chapter, models of two-story frames can be developed very simply under the assumption that the girders are rigid. Furthermore, for single-story frames, the assumption of rigid girders reduced peak displacements only slightly and produced good estimates of design moments in the columns. Similar comparisons will be made for two-story frames.

Consider the two-story frame of Figure 5-25. Members are the same as for the one-story frame of Figure 5-7, but a second story, identical to the first, has been added. The stiffness for each floor is given by Equation (5-11) and is the same as that computed for the single-story frame, or

419

$$k_1 = k_2 = 139,720 \ \text{lb/in} \approx 140,000 \ \text{lb/in} \ (24,500 \ \text{kN/m})$$

The mass of the second floor is the same as for the one-story frame, but the first floor has additional mass produced by the second-story columns. As before, 1/3 of the mass of the column is added to the floor weight. This ne-

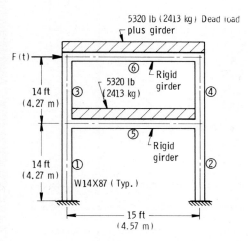

Figure 5-25. Two-Story Frame With Rigid Girders

glects 1/3 of the column, but is felt to be a realistic way of distributing the column weight, and the error introduced should produce conservative results. For the two-story frame, the masses are

$$m_1 = m_2 + \frac{1}{3} \rho \ (A_3 + A_4) \ (14' \ x \ 12) = 18 \ \text{lb-sec}^2/\text{in} \quad (3151 \ \text{kg})$$

$$m_2 = 15.88 \approx 16.0 \ \text{lb-sec}^2/\text{in} \quad (2801 \ \text{kg})$$

Response of the frame will be computed for the distributed overpressure of Figure 5-26a and for a frame spacing of 15 ft in a building. The pressure acts over an area which is 14 ft x 15 ft for the first floor load

and 7 ft x 15 ft for the roof load. If coupling between the frame and the
structural elements, which transfers the loading to the frame, is ignored,
the forces in Figure 5-26b are obtained.

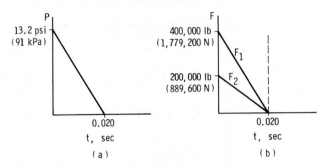

Figure 5-26. Forcing Function for Two-Story Frame

Motions for the frame are determined using the equations of motion
and predictor-corrector solution outlined earlier in this chapter. Displacements, computed for zero initial velocity and displacement and no damping,
are given in Figure 5-27. Because the top floor is exactly the same and has
the same loading as the one-story frame of Figure 5-7, it is instructive to

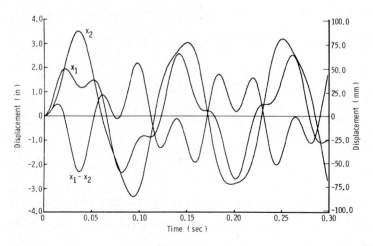

Figure 5-27. Displacements of Two-Story Frame (Figure 5-26) With Rigid Girders

compare the results for the two cases. Referring to Figure 5-3 (which gives displacements for the frame of Figure 5-7) and Figure 5-27, it is obvious that larger displacements were obtained in this example; however, if the relative displacements between floors 1 and 2 are compared to the one-story frame displacements, they are seen to be much larger (2.30 in. for the two-story frame; 1.20 in. for the one-story frame). If the first floor had not been loaded, a smaller deflection would have been predicted from the results of Figure 5-24 for a value of $T_1/T_2 \simeq 1$; however, with the loading on the first floor, Figure 5-24 does not apply. In general, it is not safe to design the second floor as though it is rigidly attached to the ground.

Moments in the frame can be computed readily from the lateral displacements and the value for k for each column. In this case, the columns are equal so that the moments, which are also equal at each end of the columns, are

Lower Columns:

$$M_1 = \frac{1}{2} k_1 x_{i_{max}} \cdot \frac{H_1}{2} = \frac{k_1 x_{i_{max}} H_1}{4} = \frac{(140{,}000)\ (2.597)\ (14')\ (12)}{4}$$

$$= 15.270 \times 10^6 \text{ in. lb} \quad (1.725 \times 10^6 \text{ N·m})$$

Upper Columns:

$$M_2 = \frac{1}{2} k_2 \left(x_2 - x_1 \right)_{max} \frac{H_2}{2} = \frac{1}{4} k_2 \left(x_2 - x_1 \right)_{max} H_2$$

$$= \frac{1}{4} (140{,}000)\ (2.229)\ (14')\ (12)$$

$$= 13.518 \times 10^6 \text{ in. lb} \quad (1.527 \times 10^6 \text{ N·m})$$

where k_1 and k_2 are the total floor stiffnesses and H_1 and H_2 are the heights between floors. For these W14X87 columns, the yield moment is

$$M_y = Z_p \, \sigma_y$$

The Steel Construction Manual (1980) gives 151 in^3 (0.00247 m^3) for the section modulus, Z_p. Taking the yield stress as 36,000 psi (248.21 MPa), the yield moment of each column is found to be

$$M_y = (151)(36,000) = 5.436 \times 10^6 \text{ in. lb} \qquad (0.6033 \times 10^6 \text{ N} \cdot \text{m})$$

Thus, yielding would have occurred had plasticity been included in the analysis. These moments will be compared with those calculated with flexible girders in the following section. Also, results for this same frame, which are based on elastic-plastic behavior, are given in a later section on multi-degree-of-freedom systems.

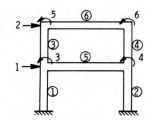

(a) Displacement and force directions

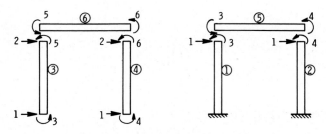

(b) Force and displacement for each member

Figure 5-28. Displacement and Force Directions as a Two-Story Frame

Two-Story Frame with Flexible Girders

As for the one-story frames of Figure 5-9, to include girder flexibility in the computation of the frame lateral stiffness requires that rotational degrees-of-freedom be introduced in the frame at member intersections. Displacement and force directions are shown in Figure 5-28a and member dis-

Member 1:

$$
\begin{bmatrix} F_1 \\ \\ F_3 \end{bmatrix} = \begin{bmatrix} \dfrac{12EI_1}{L_1^3} & \dfrac{6EI_1}{L_1^2} \\ \\ \dfrac{6EI_1}{L_1^2} & \dfrac{4EI_1}{L_1} \end{bmatrix} \begin{bmatrix} \delta_1 \\ \\ \delta_3 \end{bmatrix} \tag{5-60a}
$$

Member 2:

$$
\begin{bmatrix} F_1 \\ \\ F_4 \end{bmatrix} = \begin{bmatrix} \dfrac{12EI_2}{L_2^3} & \dfrac{6EI_2}{L_2^2} \\ \\ \dfrac{6EI_2}{L_2^2} & \dfrac{4EI_2}{L_2^2} \end{bmatrix} \begin{bmatrix} \delta_1 \\ \\ \delta_4 \end{bmatrix} \tag{5-60b}
$$

Member 3:

$$
\begin{bmatrix} F_1 \\ \\ F_2 \\ \\ F_3 \\ \\ F_5 \end{bmatrix} = \begin{bmatrix} \dfrac{12EI_3}{L_3^3} & \dfrac{-12EI_3}{L_3^3} & \dfrac{-6EI_3}{L_3^2} & \dfrac{-6EI_3}{L_3^2} \\ \\ \dfrac{-12EI_3}{L_3^3} & \dfrac{12EI_3}{L_3^3} & \dfrac{6EI_3}{L_3^2} & \dfrac{6EI_3}{L_3^2} \\ \\ \dfrac{-6EI_3}{L_3^2} & \dfrac{6EI_3}{L_3^2} & \dfrac{4EI_3}{L_3} & \dfrac{2EI_3}{L_3} \\ \\ \dfrac{-6EI_3}{L_3^2} & \dfrac{6EI_3}{L_3^2} & \dfrac{2EI_3}{L_3} & \dfrac{4EI_3}{L_3} \end{bmatrix} \begin{bmatrix} \delta_1 \\ \\ \delta_2 \\ \\ \delta_3 \\ \\ \delta_5 \end{bmatrix} \tag{5-60c}
$$

424

Member 4:

$$
\begin{bmatrix} F_1 \\ \\ F_2 \\ \\ F_4 \\ \\ F_6 \end{bmatrix} =
\begin{bmatrix}
\dfrac{12EI_4}{L_4^3} & \dfrac{-12EI_4}{L_4^3} & \dfrac{-6EI_4}{L_4^2} & \dfrac{-6EI_4}{L_4^2} \\[14pt]
\dfrac{-12EI_4}{L_4^3} & \dfrac{12EI_4}{L_4^3} & \dfrac{6EI_4}{L_4^2} & \dfrac{6EI_4}{L_4^2} \\[14pt]
\dfrac{-6EI_4}{L_4^2} & \dfrac{6EI_4}{L_4^2} & \dfrac{4EI_4}{L_4} & \dfrac{2EI_4}{L_4} \\[14pt]
\dfrac{-6EI_4}{L_4^2} & \dfrac{6EI_4}{L_4^2} & \dfrac{2EI_4}{L_4} & \dfrac{4EI_4}{L_4}
\end{bmatrix}
\begin{bmatrix} \delta_1 \\ \\ \delta_2 \\ \\ \delta_4 \\ \\ \delta_6 \end{bmatrix}
\tag{5-60d}
$$

Member 5:

$$
\begin{bmatrix} F_3 \\ \\ F_4 \end{bmatrix} =
\begin{bmatrix}
\dfrac{4EI_5}{L_5} & \dfrac{2EI_5}{L_5} \\[14pt]
\dfrac{2EI_5}{L_5} & \dfrac{4EI_5}{L_5}
\end{bmatrix}
\begin{bmatrix} \delta_3 \\ \\ \delta_4 \end{bmatrix}
\tag{5-60e}
$$

Member 6:

$$
\begin{bmatrix} F_5 \\ \\ F_6 \end{bmatrix} =
\begin{bmatrix}
\dfrac{4EI_6}{L_6} & \dfrac{2EI_6}{L_6} \\[14pt]
\dfrac{2EI_6}{L_6} & \dfrac{4EI_6}{L_6}
\end{bmatrix}
\begin{bmatrix} \delta_5 \\ \\ \delta_6 \end{bmatrix}
\tag{5-60f}
$$

placements in Figure 5-28b. Applying unit displacements in each direction while restraining all others, force-displacement relations for each element are obtained as demonstrated previously in Figures 5-10 and 5-11.

Expanding the matrices to accommodate all forces and displacements and adding stiffness contributions at common displacements, a system of equations for the frame is obtained.

$$
\begin{Bmatrix} F_1 \\ F_2 \\ --- \\ F_3 \\ F_4 \\ F_5 \\ F_6 \end{Bmatrix} = 2E
\left[
\begin{array}{cc|cccc}
6\left(\dfrac{I_1}{L_1^{\,3}} + \dfrac{I_2}{L_2^{\,3}} + \dfrac{I_3}{L_3^{\,3}} + \dfrac{I_4}{L_4^{\,3}}\right) & & & & \text{(SYMMETRIC)} & \\[2.5ex]
-6\left(\dfrac{I_3}{L_3^{\,3}} + \dfrac{I_4}{L_4^{\,3}}\right) & 6\left(\dfrac{I_3}{L_3^{\,3}} + \dfrac{I_4}{L_4^{\,3}}\right) & & & & \\[2.5ex]
\hline
3\left(\dfrac{I_1}{L_1^{\,2}} - \dfrac{I_3}{L_3^{\,2}}\right) & \dfrac{3I_3}{L_3^{\,2}} & 2\left(\dfrac{I_1}{L_2} + \dfrac{I_3}{L_3} + \dfrac{I_5}{L_5}\right) & & \text{(SYMMETRIC)} & \\[2.5ex]
3\left(\dfrac{I_2}{L_2^{\,2}} - \dfrac{I_4}{L_4^{\,2}}\right) & \dfrac{3I_4}{L_4^{\,2}} & \dfrac{I_5}{L_5} & 2\left(\dfrac{I_2}{L_2} + \dfrac{I_4}{L_4} + \dfrac{I_5}{L_5}\right) & & \\[2.5ex]
-\dfrac{3I_3}{L_3^{\,2}} & \dfrac{3I_3}{L_3^{\,2}} & \dfrac{I_3}{L_3} & 0 & 2\left(\dfrac{I_3}{L_3} + \dfrac{I_6}{L_6}\right) & \\[2.5ex]
-\dfrac{3I_4}{L_4^{\,2}} & \dfrac{3I_4}{L_4^{\,2}} & 0 & \dfrac{I_6}{L_6} & 2\left(\dfrac{I_4}{L_4} + \dfrac{I_6}{L_6}\right)
\end{array}
\right]
\begin{Bmatrix} \delta_1 \\ \delta_2 \\ --- \\ \delta_3 \\ \delta_4 \\ \delta_5 \\ \delta_6 \end{Bmatrix}
$$

Partitioning the matrix, equations relating lateral forces and moments to the displacements are obtained,

$$
\begin{bmatrix} F_\alpha \\ -- \\ F_\beta \end{bmatrix} =
\begin{bmatrix} k_{\alpha\alpha} & | & k_{\alpha\beta} \\ ---- & + & ----- \\ k_{\beta\alpha} & | & k_{\beta\beta} \end{bmatrix}
\begin{bmatrix} \delta_\alpha \\ -- \\ \delta_\beta \end{bmatrix}
\tag{5-62}
$$

where

$$
\{\delta_\alpha\} = \begin{bmatrix} \delta_1 \\ \delta_2 \end{bmatrix} \qquad \{\delta_\beta\} = \begin{bmatrix} \delta_3 \\ \delta_4 \\ \delta_5 \\ \delta_6 \end{bmatrix}
$$

and likewise, for the forces $\{F_\alpha\}$ and $\{F_\beta\}$. Note that $\{\delta_\beta\}$ represents the

426

rotations in the frame and $\left\{ F_\beta \right\}$ represents the applied moments. If applied moments are known (in this case, they are zero), then the rotations $\left\{ \delta_\beta \right\}$ can be found in terms of $\left\{ \delta_\alpha \right\}$. These were given previously by Equations (5-27). Likewise, the lateral stiffness of the frame [K] is found from Equation (5-29).

Equations for the lateral motions of the frame are essentially the same as described in the section on two-degree-of-freedom spring-mass systems, where the k terms are computed from Equation (5-29) and the masses are identical to those for the frame with rigid girders. Lateral displacements, computed from the equation of motion can then be used to compute the rotations from Equation (5-27). Knowing the lateral displacements and rotations, the moments in the frame members can be computed from the member equations. This procedure was outlined for the one-story frame of Figure 5-9.

Calculated lateral displacements for the two-story frame of Figure 5-25, but with a W16X88 beam substituted for the rigid girder, are given in Figure 5-29. As expected, displacements are increased over those computed for rigid girders. Maximum moments and displacements in the frame for

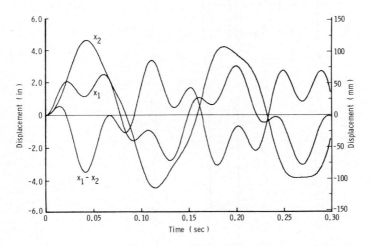

Figure 5-29. Displacements of Two-Story Frame With Flexible Girders

flexible and rigid girders are given in Table 5-7. The yield moment in
the members computed from the plastic section modulus and a yield stress
of 36,000 psi (248.21 MPa) are given for comparison. It is apparent that
yielding in this frame would have occurred, had it been permitted, but
the amount of plastic deformation cannot be judged from these results.

Table 5-7. Comparison of Bending Moments and Displacements in
the Two-Story Frame

	Rigid Girders	Flexible Girders
Maximum Displacement of Floor 1	2.60 in. (66.0 mm)	3.05 in. (77.3 mm)
Maximum Displacement of Floor 2	3.51 in. (89.1 mm)	4.64 in. (117.9 mm)
Maximum Relative Displacement	2.30 in. (58.4 mm)	3.48 in. (88.4 mm)
Maximum Moment in First Floor Columns	15.270×10^6 in. lb* $(1.725 \times 10^6$ N·m)	15.678×10^6 in. lb* $(1.771 \times 10^6$ N·m)
Maximum Moment in Second Floor Columns	13.518×10^6 in. lb* $(1.527 \times 10^6$ N·m)	11.488×10^6 in. lb* $(1.298 \times 10^6$ N·m)

*Yield Moments = 5.436×10^6 in. lb $(0.6083 \times 10^6$ N·m).

Relative to the moments computed for the frame with rigid girders,
moments in the columns with flexible girders are lower for the second floor
but slightly higher for the first floor. This is true even though maximum
displacements have increased considerably, and is in general agreement with
the observations made for the one-story frame. Thus, evaluating columns
under the assumption of rigid girders is a simplifying assumption which
will give reasonable, but not necessarily conservative, results. The effects of including plasticity and axial degrees-of-freedom in these examples
are shown in a later section.

428

MULTI-DEGREE-OF-FREEDOM SYSTEMS

Equation of Motions

Force-deformation relationships were developed in the section for
two-story frames with multi-degrees-of-freedom. The approach taken was to
derive the lateral stiffness of each floor so that only the floor inertia
and floor lateral displacements were retained in the equation of motion.
Rotational degrees-of-freedom were expressed in terms of the lateral displacements and the matrices were "condensed."

This approach can be extended in several ways to treat larger problems. One way is to simply add more floors and columns and treat only the
inertia effects of the floors as before. Another approach is to subdivide
the members and include local inertia effects in the columns and girders.
The second approach has the disadvantage of greatly increasing the number
of degrees-of-freedom in the problem. This introduces higher frequencies
into the system of equations which make numerical integration more difficult and lengthy because smaller integration time steps are required. Many
programs are available which utilize this latter approach. They include
many general purpose programs as well as some developed specifically for
frame analysis. Some of these will be described later in this section.
Here, the approach of modeling the building response only in terms of the
lateral motions of its floors will be outlined in more general terms and
the method of accounting for plasticity in the frame structure will be described.

In Equation (5-26), the static force-deflection relationships for
frame structures were given in partitioned form as

$$\begin{bmatrix} F_\alpha \\ -- \\ F_\beta \end{bmatrix} = \begin{bmatrix} k_{\alpha\alpha} & \vdots & k_{\alpha\beta} \\ ---- & + & ---- \\ k_{\beta\alpha} & \vdots & k_{\beta\beta} \end{bmatrix} \begin{bmatrix} \delta_\alpha \\ -- \\ \delta_\beta \end{bmatrix}$$

(5-26)
repeated

where δ_α represents the lateral displacements of the floors, δ_β represents the rotations at the column-girder intersection, F_α represents the lateral time-dependent loads applied to the floors, and F_β represents the external time-dependent moments applied at the column-girder intersections. For blast loaded structures, these forces are time-dependent and inertia effects must be included. If damping is neglected, then Newton's First Law gives

$$
\begin{bmatrix} F_\alpha \\ -- \\ F_\beta \end{bmatrix}
\begin{bmatrix} k_{\alpha\alpha} & | & k_{\alpha\beta} \\ ----+---- \\ k_{\beta\alpha} & | & k_{\beta\beta} \end{bmatrix}
\begin{bmatrix} \delta_\alpha \\ -- \\ \delta_\beta \end{bmatrix}
=
\begin{bmatrix} m_{\alpha\alpha} & | & m_{\alpha\beta} \\ ----+---- \\ m_{\beta\alpha} & | & m_{\beta\beta} \end{bmatrix}
\begin{bmatrix} \ddot{\delta}_\alpha \\ -- \\ \ddot{\delta}_\beta \end{bmatrix}
\tag{5-63}
$$

Because only the lateral masses of the floors are being included in the dynamics of the building, for this problem $m_{\alpha\beta} = m_{\beta\alpha} = m_{\beta\beta} = 0$ and the equations of motion are

$$
\begin{bmatrix} m_{\alpha\alpha} \end{bmatrix} \{ \ddot{\delta}_\alpha \} + \begin{bmatrix} k_{\alpha\alpha} \end{bmatrix} \{ \delta_\alpha \} + \begin{bmatrix} k_{\alpha\beta} \end{bmatrix} \{ \delta_\beta \} = \{ F_\alpha \}
\tag{5-64}
$$

If external forces are limited to transverse forces at the floors, then

$$
\{ F_\beta \} = \{ 0 \}
$$

and the rotations $\{ \delta_\beta \}$ are given by Equation (5-27)

$$
\{ \delta_\beta \} = - \begin{bmatrix} k_{\beta\beta} \end{bmatrix}^{-1} \begin{bmatrix} k_{\beta\alpha} \end{bmatrix} \{ \delta_\alpha \}
\tag{5-27}
$$

(5-27) repeated

Substituting into Equation (5-64) gives

$$
\begin{bmatrix} m_{\alpha\alpha} \end{bmatrix} \{ \ddot{\delta}_\alpha \} + \left(\begin{bmatrix} k_{\alpha\alpha} \end{bmatrix} - \begin{bmatrix} k_{\alpha\beta} \end{bmatrix} \begin{bmatrix} k_{\beta\beta} \end{bmatrix}^{-1} \begin{bmatrix} k_{\beta\alpha} \end{bmatrix} \right) \{ \delta_\alpha \} = \{ F_\alpha(t) \}
\tag{5-65}
$$

The original stiffness matrix is derived from the member equations as explained earlier in this chapter. There, the element equations were formulated directly in terms of the "global" displacements. The advantage of formulating the equation directly in terms of the global displacements is that boundary degrees-of-freedom have already been eliminated and transformation from element to global coordinates is not necessary. This approach is satisfactory for moderate size problems, particularly when the arrangement of the members is regular and parallel to global axes. For large complicated structural arrangements, it is easier to formulate member equations in local member coordinates and to then transform the stiffness equations to global coordinates. These methods are clearly explained by many authors, e.g., see Cook (1974) and Gallagher (1975), and will not be repeated here.

The linear Equation (5-65) is solved numerically using the average acceleration method discussed earlier in this chapter. In matrix form, Equations (5-2a), (5-2b) and (5-4) become

$$\left\{\dot{X}_t\right\} = \left\{\dot{X}_{t-\Delta t}\right\} + \frac{\Delta t}{2}\left\{\ddot{X}_t + \ddot{X}_{t-\Delta t}\right\} \tag{5-66a}$$

$$\left\{X_t\right\} = \left\{X_{t-\Delta t}\right\} + \frac{\Delta t}{2}\left\{\dot{X}_t + \dot{X}_{t-\Delta t}\right\} \tag{5-66b}$$

$$\left\{\ddot{X}_{\alpha_t}\right\} = \left(\left[m_{\alpha\alpha}\right] + \frac{\Delta t^2}{4}[K]\right)^{-1}\left[\left\{F_\alpha(t)\right\} - [K]\left(\left\{X_{\alpha_{t-\Delta t}}\right\} + \right.\right.$$

$$\tag{5-67}$$

$$\left.\left.\Delta T\left\{X_{\alpha_{t-\Delta t}}\right\} + \frac{\Delta T^2}{2}\left\{\ddot{X}_{\alpha_{t-\Delta t}}\right\}\right)\right]$$

where $[K] = \left[k_{\alpha\alpha}\right] - \left[k_{\alpha\beta}\right]\left[k_{\beta\beta}\right]^{-1}\left[k_{\beta\alpha}\right]$ and damping has been omitted. The step-by-step solution of these equations is straightforward and has already been described.

Plastic Behavior

A method for including plasticity in frames has been suggested by
Majid (1972) and Yamada, et al. (1968). It is based upon the formation of
plastic hinges at the ends of the members and computes the rotations of these
hinges during plastic deformation. Yamada, et al. (1968) have compared results
obtained using this approach for a statically loaded frame to solutions obtained by Neal (1977, first published in 1956) and reported good agreement.
The procedure will be illustrated for the one-story frame with flexible girders analyzed earlier in this chapter.

A schematic of the frame showing the degrees-of-freedom and plastic
hinge rotations is given in Figure 5-30. Two-degrees-of-freedom have been
added at the base of the columns. Although elastic rotations are zero at
these locations, plastic hinges can occur and so the degrees-of-freedom
were added to develop the stiffness coefficients and permit direct compu-

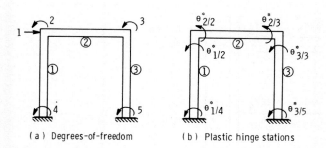

(a) Degrees-of-freedom   (b) Plastic hinge stations

Figure 5-30.   Degrees-Of-Freedom and Plastic Hinges in
One-Story Frame

tation of the moments at the base of the columns. Element equations for
member 2 [Equation (5-20)] are unchanged from the previous example. Equations for members 1 and 3 are changed to reflect the additional degrees-
of-freedom. Referring to the original equations [Equations (5-19) through
(5-21)], the new equations with the additional degrees-of-freedom are easily written as

$$
\text{Member 1:} \quad
\begin{bmatrix} F_1 \\ F_2 \\ F_4 \end{bmatrix}
=
\begin{bmatrix}
\dfrac{12EI_1}{L_1^3} & \dfrac{6EI_1}{L_1^2} & \dfrac{6EI_1}{L_1^2} \\[2ex]
\dfrac{6EI_1}{L_1^2} & \dfrac{4EI_1}{L_1} & \dfrac{2EI_1}{L_1} \\[2ex]
\dfrac{6EI_1}{L_1^2} & \dfrac{2EI_1}{L_1} & \dfrac{4EI_1}{L_1}
\end{bmatrix}
\begin{bmatrix} \delta_1 \\ \delta_2 \\ \delta_4 \end{bmatrix}
\qquad (5\text{-}68a)
$$

$$
\text{Member 2:} \quad
\begin{bmatrix} F_2 \\ F_3 \end{bmatrix}
=
\begin{bmatrix}
\dfrac{4EI_2}{L_2} & \dfrac{2EI_2}{L_2} \\[2ex]
\dfrac{2EI_2}{L_2} & \dfrac{4EI_2}{L_2}
\end{bmatrix}
\begin{bmatrix} \delta_2 \\ \delta_3 \end{bmatrix}
\qquad (5\text{-}68b)
$$

$$
\text{Member 3:} \quad
\begin{bmatrix} F_1 \\ F_3 \\ F_5 \end{bmatrix}
=
\begin{bmatrix}
\dfrac{12EI_3}{L_3^3} & \dfrac{6EI_3}{L_3^2} & \dfrac{6EI_3}{L_3^2} \\[2ex]
\dfrac{6EI_3}{L_3^2} & \dfrac{4EI_3}{L_3} & \dfrac{2EI_3}{L_3} \\[2ex]
\dfrac{6EI_3}{L_3^2} & \dfrac{2EI_3}{L_3} & \dfrac{4EI_3}{L_3}
\end{bmatrix}
\begin{bmatrix} \delta_1 \\ \delta_3 \\ \delta_5 \end{bmatrix}
\qquad (5\text{-}68c)
$$

Yamada, et al. (1968) modifies the elastic equations above after the formation of a plastic hinge by adding the plastic hinge rotations to the degrees-
of-freedom. For member 1, the element equations with plastic hinges at each
end become

$$
\begin{bmatrix} F_1 \\ \\ F_2 \\ \\ F_4 \end{bmatrix} = \begin{bmatrix} \dfrac{12EI_1}{L_1^{\,3}} & \dfrac{6EI_1}{L_1^{\,2}} \\ \\ \dfrac{6EI_1}{L_1^{\,2}} & \dfrac{4EI_1}{L_1} \\ \\ \dfrac{6EI_1}{L_1^{\,2}} & \dfrac{2EI_1}{L_1^{\,2}} \end{bmatrix} \begin{bmatrix} \delta_1 \\ \\ \delta_2 \end{bmatrix} + \begin{bmatrix} \dfrac{6EI_1}{L_1^{\,2}} \\ \\ \dfrac{2EI_1}{L_1} \\ \\ \dfrac{4EI_1}{L_1} \end{bmatrix} \begin{bmatrix} \delta_4 \end{bmatrix} +
$$

$$
\begin{bmatrix} \dfrac{6EI_1}{L_1^{\,2}} \\ \\ \dfrac{4EI_1}{L_1} \\ \\ \dfrac{2EI_1}{L_1} \end{bmatrix} \theta^{*}_{1/2} + \begin{bmatrix} \dfrac{6EI_1}{L_1^{\,2}} \\ \\ \dfrac{2EI_1}{L_1} \\ \\ \dfrac{4EI_1}{L_1} \end{bmatrix} \theta^{*}_{1/4}
$$

(5-69)

where $\theta^{*}_{1/2}$ is the plastic hinge rotation corresponding to degree-of-freedom 2 of member 1 and $\theta^{*}_{1/4}$ corresponds to degree-of-freedom 4 of member 1. The term multiplied by δ_4 will be zero because the elastic rotation, δ_4, is always zero. Note that the column matrices for the hinge rotations are comprised of the corresponding terms for the elastic rotations, δ_2 and δ_4. Note also that plastic hinges can occur in each end of each member so that there are six possible plastic hinges as shown in Figure 5-30. When plastic hinges form, the moments at the hinge rotations are replaced by the plastic moments, M_p. Thus, when yielding occurs simultaneously at both ends, the moment equations become

$$
\begin{bmatrix} F_2 \\ \\ F_4 \end{bmatrix} = \begin{bmatrix} M_{P1/2} \\ \\ M_{P1/4} \end{bmatrix} = \begin{bmatrix} \dfrac{6EI_1}{L_1^2} & \dfrac{4EI_1}{L_1} \\ \\ \dfrac{6EI_1}{L_1^2} & \dfrac{2EI_1}{L_1} \end{bmatrix} \begin{bmatrix} \delta_1 \\ \\ \delta_2 \end{bmatrix} + \begin{bmatrix} \dfrac{4EI_1}{L_1} & \dfrac{2EI_1}{L_1} \\ \\ \dfrac{2EI_1}{L_1} & \dfrac{4EI_1}{L_1} \end{bmatrix} \begin{bmatrix} \theta^*_{1/2} \\ \\ \theta^*_{1/4} \end{bmatrix} \qquad (5\text{-}70)
$$

By simultaneous solution of Equation (5-70), the hinge rotations are obtained.

$$
\begin{bmatrix} \theta^*_{1/2} \\ \\ \theta^*_{1/4} \end{bmatrix} = \begin{bmatrix} \dfrac{4EI_1}{L_1} & \dfrac{2EI_1}{L_1} \\ \\ \dfrac{2EI_1}{L_1} & \dfrac{4EI_1}{L_1} \end{bmatrix}^{-1} \left(\begin{bmatrix} M_{P1/2} \\ \\ M_{P1/4} \end{bmatrix} - \begin{bmatrix} \dfrac{6EI_1}{L_1^2} & \dfrac{4EI_1}{L_1} \\ \\ \dfrac{6EI_1}{L_1^2} & \dfrac{2EI_1}{L_1} \end{bmatrix} \begin{bmatrix} \delta_1 \\ \\ \delta_2 \end{bmatrix} \right) \qquad (5\text{-}71a)
$$

If yielding occurs at one end only, a simultaneous solution of Equation (5-70) is not required. The hinge rotation at the nonyielding end is a constant and the appropriate equation from Equation (5-70) gives the hinge rotation at the end where yielding occurs. Thus, for yielding only at degree-of-freedom 2, the plastic hinge rotation is obtained from the first of Equation (5-70). This gives

$$
\theta^*_{1/2} = \frac{L_1}{4EI_1} \left(M_{P1/2} - \begin{bmatrix} \dfrac{6EI_1}{L_1^2} & \dfrac{4EI_1}{L_1} \end{bmatrix} \begin{bmatrix} \delta_1 \\ \delta_2 \end{bmatrix} - \frac{2EI_1}{L_1} \theta^*_{1/4} \right) \qquad (5\text{-}71b)
$$

Likewise, for yielding only at degree-of-freedom 4, the plastic hinge rotation is

$$
\theta^*_{1/4} = \frac{L_1}{4EI_1} \left(M_{P1/4} - \begin{bmatrix} \dfrac{6EI_1}{L_1^2} & \dfrac{2EI_1}{L_1} \end{bmatrix} \begin{bmatrix} \delta_1 \\ \delta_2 \end{bmatrix} - \frac{2EI_1}{L_1} \theta^*_{1/2} \right) \qquad (5\text{-}71b)
$$

435

Equations (5-71) further reduce to:

<u>Both Ends Yielding:</u>

$$\theta^*_{1/2} = \frac{L_1}{6EI_1}\left(2M_{P_{1/2}} - M_{P_{1/4}}\right) - \frac{\delta_1}{L_1} - \delta_2 \qquad (5\text{-}72a)$$

$$\theta^*_{1/4} = \frac{L_1}{6EI_1}\left(2M_{P_{1/4}} - M_{P_{1/2}}\right) - \frac{\delta_1}{L_1} \qquad (5\text{-}72b)$$

<u>Yielding Only At Degree-Of-Freedom 2:</u>

$$\theta^*_{1/2} = \frac{L_1 M_{P_{1/2}}}{4EI_1} - \frac{3}{2L_1}\delta_1 - \delta_2 - \frac{1}{2}\theta^*_{1/4} \qquad (5\text{-}72c)$$

<u>Yielding Only At Degree-Of-Freedom 4:</u>

$$\theta^*_{1/4} = \frac{L_1 M_{P_{1/4}}}{4EI_1} - \frac{3}{2L_1}\delta_1 - \frac{1}{2}\left(\delta_2 + \theta^*_{1/2}\right) \qquad (5\text{-}72d)$$

It is clear from these equations that δ_2 is the total rotation at degree-of-freedom 2 and that the plastic hinge rotation, $\theta^*_{1/2}$, takes the opposite sign to δ_2. This is a consequence of adding the plastic hinge rotations to Equation (5-69) instead of substracting them. Thus, the quantity $\left(\delta_2 + \theta^*_{1/2}\right)$ in Equation (5-72d) gives the elastic rotation of member 1 at degree-of-freedom 2.

In a similar manner, equations for the plastic hinge rotation are derived for members 2 and 3. These are:

436

<u>Member 2:</u>

$$
\begin{bmatrix} F_2 \\ \\ F_3 \end{bmatrix} = \begin{bmatrix} \dfrac{4EI_2}{L_2} & \dfrac{2EI_2}{L_2} \\ \\ \dfrac{2EI_2}{L_2} & \dfrac{4EI_2}{L_2} \end{bmatrix} \begin{bmatrix} \delta_2 \\ \\ \delta_3 \end{bmatrix} + \begin{bmatrix} \dfrac{4EI_2}{L_2} \\ \\ \dfrac{2EI_2}{L_2} \end{bmatrix} \theta^*_{2/2} + \begin{bmatrix} \dfrac{2EI_2}{L_2} \\ \\ \dfrac{4EI_2}{L_2} \end{bmatrix} \theta^*_{2/3} \qquad (5\text{-}73)
$$

<u>Both Ends Yielding:</u>

$$
\theta^*_{2/2} = \frac{L_2}{6EI_2}\left(2M_{P2/2} - M_{P2/3}\right) - \delta_2 \qquad (5\text{-}74a)
$$

$$
\theta^*_{2/3} = \frac{L_2}{6EI_2}\left(2M_{P2/3} - M_{P2/2}\right) - \delta_3 \qquad (5\text{-}74b)
$$

<u>Yielding Only At Degree-Of-Freedom 2:</u>

$$
\theta^*_{2/2} = \frac{L_2\, M_{P2/2}}{4EI_2} - \frac{1}{2}\left(\delta_3 + \theta^*_{2/3}\right) - \delta_2 \qquad (5\text{-}74c)
$$

<u>Yielding Only At Degree-Of-Freedom 3:</u>

$$
\theta^*_{2/3} = \frac{L_2\, M_{P2/3}}{4EI_2} - \frac{1}{2}\left(\delta_2 + \theta^*_{2/2}\right) - \delta_3 \qquad (5\text{-}74d)
$$

Member 3:

$$
\begin{bmatrix} F_1 \\ \\ F_3 \\ \\ F_5 \end{bmatrix} = \begin{bmatrix} \dfrac{12EI_3}{L_3^3} & \dfrac{6EI_3}{L_3^2} \\ \\ \dfrac{6EI_3}{L_3^2} & \dfrac{4EI_3}{L_3} \\ \\ \dfrac{6EI_3}{L_3^2} & \dfrac{2EI_3}{L_3} \end{bmatrix} \begin{bmatrix} \delta_1 \\ \\ \delta_3 \end{bmatrix} + \begin{bmatrix} \dfrac{6EI_3}{L_3^2} \\ \\ \dfrac{4EI_3}{L_3} \\ \\ \dfrac{2EI_3}{L_3} \end{bmatrix} \theta^*_{3/3} + \begin{bmatrix} \dfrac{6EI_3}{L_3^2} \\ \\ \dfrac{2EI_3}{L_3} \\ \\ \dfrac{4EI_3}{L_3} \end{bmatrix} \theta^*_{3/5}
\tag{5-75}
$$

Both Ends Yielding:

$$
\theta^*_{3/3} = \frac{L_3}{6EI_3}\left(2M_{P3/3} - M_{P3/5}\right) - \frac{\delta_1}{L_3} - \delta_3
\tag{5-76a}
$$

$$
\theta^*_{3/5} = \frac{L_3}{6EI_3}\left(2M_{P3/5} - M_{P3/3}\right) - \frac{\delta_1}{L_3}
\tag{5-76b}
$$

Yielding Only At Degree-Of-Freedom 3:

$$
\theta^*_{3/3} = \frac{L_3 M_{P3/3}}{4EI_3} - \frac{3}{2L_3}\delta_1 - \delta_3 - \frac{1}{2}\theta^*_{3/5}
\tag{5-76c}
$$

Yielding Only At Degree-Of-Freedom 5:

$$
\theta^*_{3/5} = \frac{L_3 M_{P3/5}}{4EI_3} - \frac{3}{2L_3}\delta_1 - \frac{1}{2}\left(\delta_3 + \theta^*_{3/3}\right)
\tag{5-76d}
$$

438

To obtain the system of equations, member matrices are expanded to
include all forces and all degrees-of-freedom and internal forces at common
degrees-of-freedom add. Each plastic hinge rotation is independent of all
others so their contributions are summed separately. The system of equations thus obtained is:

$$
\begin{bmatrix} F_1 \\ -- \\ F_2 \\ F_3 \\ -- \\ F_4 \\ F_5 \end{bmatrix}
=
\begin{bmatrix}
\left(\dfrac{12EI_1}{L_1^3}+\dfrac{12EI_3}{L_3^3}\right) & \dfrac{6EI_1}{L_1^2} & \dfrac{6EI_3}{L_3^2} \\
\dfrac{6EI_1}{L_1^2} & \left(\dfrac{4EI_1}{L_1}+\dfrac{4EI_2}{L_2}\right) & \dfrac{2EI_2}{L_2} \\
\dfrac{6EI_3}{L_3^2} & \dfrac{2EI_2}{L_2} & \left(\dfrac{4EI_3}{L_3}+\dfrac{4EI_2}{L_2}\right) \\
\dfrac{6EI_1}{L_1^2} & \dfrac{2EI_1}{L_1} & 0 \\
\dfrac{6EI_3}{L_3^2} & 0 & \dfrac{2EI_3}{L_3}
\end{bmatrix}
\begin{bmatrix} \delta_1 \\ -- \\ \delta_2 \\ \delta_3 \\ -- \end{bmatrix}
$$

(5-77)

$$ + \sum_{i=1}^{6} \{C\}_i \, \theta^*_i $$

In this equation the plastic hinge rotations and their coefficient matrices
have been numbered consecutively. The order can be arbitrary as long as it
is consistent and carried over to the member equations. For example, $\{C\}_1$
might be

$$\{C\}_1 = \begin{bmatrix} \dfrac{-6EI_1}{L_1^2} \\[2em] \dfrac{4EI_1}{L_1} \\[2em] 0 \\[2em] \dfrac{2EI_1}{L_1} \\[2em] 0 \end{bmatrix} \quad \text{then } \theta_1^* = \theta_{1/2}^*$$

Optionally, the coefficient matrices and the plastic hinge rotations can be expressed in matrix form as

$$\sum_{i=1}^{6} \{C\}_i\, \theta_i^* = \begin{bmatrix} \dfrac{6EI_1}{L_1^2} & \dfrac{6EI_1}{L_1^2} & 0 & 0 & \dfrac{6EI_3}{L_3^2} & \dfrac{6EI_3}{L_3^2} \\[1.5em] \hline \dfrac{4EI_1}{L_1} & \dfrac{2EI_1}{L_1} & \dfrac{4EI_2}{L_2} & \dfrac{2EI_2}{L_2} & 0 & 0 \\[1.5em] 0 & 0 & \dfrac{2EI_2}{L_2} & \dfrac{4EI_2}{L_2} & \dfrac{4EI_3}{L_3} & \dfrac{2EI_3}{L_3} \\[1.5em] \hline \dfrac{2EI_1}{L_1} & \dfrac{4EI_1}{L_1} & 0 & 0 & 0 & 0 \\[1.5em] 0 & 0 & 0 & 0 & \dfrac{2EI_3}{L_3} & \dfrac{4EI_3}{L_3} \end{bmatrix} \begin{bmatrix} \theta_{1/2}^* \\[1.5em] \theta_{1/4}^* \\[1.5em] \theta_{2/2}^* \\[1.5em] \theta_{2/3}^* \\[1.5em] \theta_{3/3}^* \\[1.5em] \theta_{3/5}^* \end{bmatrix} \qquad (5\text{-}78)$$

To find the force-displacement relationship for lateral displacement only, partition the matrices as shown in Equations (5-77) and (5-78).

$$
\begin{bmatrix} F_\alpha \\ -- \\ F_\beta \\ -- \\ F_\gamma \end{bmatrix} = \begin{bmatrix} k_{\alpha\alpha} & \vdots & k_{\alpha\beta} \\ ----&--&---- \\ k_{\beta\alpha} & \vdots & k_{\beta\beta} \\ ----&--&---- \\ k_{\gamma\alpha} & \vdots & k_{\gamma\beta} \end{bmatrix} \begin{bmatrix} \delta_\alpha \\ -- \\ \delta_\beta \end{bmatrix} + \begin{bmatrix} C_\alpha \\ -- \\ C_\beta \\ -- \\ C_\gamma \end{bmatrix} \{\theta^*\}
\qquad (5\text{-}79)
$$

In this system of equations F_α represents the lateral force applied to the top of the frame, F_β represents the external moments applied at the top of the columns and F_γ represents the reactions developed at the base of the columns. Note that $\{\delta_\gamma\}$ is not included because deflections are not permitted at the base of the columns in this example.

For the lateral forces $\{F_\alpha\}$, we have:

$$
\{F_\alpha\} = \begin{bmatrix} k_{\alpha\alpha} \end{bmatrix} \{\delta_\alpha\} + \begin{bmatrix} k_{\alpha\beta} \end{bmatrix} \{\delta_\beta\} + \begin{bmatrix} C_\alpha \end{bmatrix} \{\theta^*\}
\qquad (5\text{-}80a)
$$

and for the moments $\{F_\beta\}$:

$$
\{F_\beta\} = \begin{bmatrix} k_{\beta\alpha} \end{bmatrix} \{\delta_\beta\} + \begin{bmatrix} k_{\beta\beta} \end{bmatrix} \{\delta_\beta\} + \begin{bmatrix} C_\beta \end{bmatrix} \{\theta^*\}
\qquad (5\text{-}80b)
$$

As before, the applied forces (moments), $\{F_\beta\}$, are zero so that the elastic rotations, obtained from Equation (5-80b), are

$$
\{\delta_\beta\} = -\begin{bmatrix} k_{\beta\beta} \end{bmatrix}^{-1} \left(\begin{bmatrix} k_{\beta\alpha} \end{bmatrix} \{\delta_\alpha\} + \begin{bmatrix} C_\beta \end{bmatrix} \{\theta^*\} \right)
\qquad (5\text{-}81)
$$

Substituting Equation (5-81) into Equation (5-80a), the force-displacement relationships in terms of the lateral displacement $\{\delta_\alpha\}$ and the plastic hinge rotations are

$$\{F_\alpha\} = \left([k_{\alpha\alpha}] - [k_{\alpha\beta}][k_{\beta\beta}]^{-1}[k_{\beta\alpha}]\right)\{\delta_\alpha\}$$

(5-82)

$$+ \left([C_\alpha] - [k_{\alpha\beta}][k_{\beta\beta}]^{-1}[C_\beta]\right)\{\theta^*\}$$

The equation of motion for frame lateral displacement can now be written as

$$[m_\alpha]\{\ddot{\delta}_\alpha\} + \left([k_{\alpha\alpha}] - [k_{\alpha\beta}][k_{\beta\beta}]^{-1}[k_{\beta\alpha}]\right)\{\delta_\alpha\}$$

(5-83)

$$= \{F_\alpha\} - \left([C_\alpha] - [k_{\alpha\beta}][k_{\beta\beta}]^{-1}[C_\beta]\right)\{\theta^*\}$$

Where, for the single story frame, $\{\delta_\alpha\} = \delta_1$, $[m_\alpha] = m$, the mass concentrated at the roof, and $\{F_\alpha\} = F$, the lateral force on the roof. Note that in the absence of plastic hinge rotations ($\theta^* = 0$), Equation (5-83) reduces to the elastic equations of motion [Equation (5-65)] and that the k-matrices and C-matrices are not altered during the solution process. Changes in stiffness, associated with the formation of plastic hinges, are accounted for by the reduction in external loading.

The general solution procedure for Equation (5-83) is as follows:

Step 1. Compute $\{\delta_\alpha\}$ by numerical integration of Equation (5-83) with $\{\theta^*\} = 0$, initially.

Step 2. Compute rotations at the column-girder intersections from Equation (5-81).

Step 3. Compute the moments in the members from the element equations [Equations (5-69), (5-73), and (5-75)].

Step 4. Check for the formation of a plastic hinge, i.e., do the beam moments exceed the allowable plastic moments?

Step 5. If plastic hinges occur, compute their value by the appropriate member equation [Equations (5-72), (5-74), or (5-76)].

Step 6. Substitute the hinge rotations into Equation (5-83) and
continue the integration.

A predictor-corrector solution, such as described earlier in this
chapter, is suggested for this nonlinear solution.

Although somewhat tedious, this approach can obviously be applied
to more general structures than the one-story frame used in the example.
This has been done by one of the authors for arbitrarily shaped two-dimensional framework. The program was used to compute the response of one-
and two-story frames (which were examined in earlier sections) to show the
changes in response produced by different requirements in the frame models.
Results of this study are given in the next section.

Modeling of Frames

Using the same procedures for treating plastic hinges which were
demonstrated for the one-story frame in the preceding section, a more
general program was written for modeling two-dimensional framework. Features of the program, most of which will be demonstrated in later examples,
are:

- models elastic-plastic behavior of arbitrarily shaped two-dimensional frames

- includes axial and bending degrees-of-freedom with provisions for
  eliminating axial degrees-of-freedom

- applied transient forces can vary in time and space

- includes provisions for computing static deflections to static
  loads

- transient solution can be started from static displaced position
  and the effect of the static loads can be included in the transient solution

- includes provisions for treating the effects of dynamic and static axial loads on the plastic bending moments in the members

The effects of axial loads on the plastic bending moments in beams were not addressed in earlier sections of this chapter because axial degrees-of-freedom were omitted. Here, axial degrees-of-freedom can be retained and this effect included. Formulas which relate the fully plastic bending moments in beams to axial loads are given by Neal (1977). These formulas are straightforward and will not be repeated. As will be shown in the examples to follow, the effects produced by considering axial loads are small for one- and two-story frames. Although not demonstrated here, the effects of axial loads can be important for tall structures.

In duplicating results for the examples previously considered, some slight differences can be expected. These will be produced principally by the way in which the mass of the structure is modeled. In earlier examples of one- and two-story frames, one-third of the column mass was added at the floors of the structure. In the multi-degree-of-freedom program used here, one-half of the mass is added at the nodes so that for models with many nodes, none of the mass of the structure is omitted. Also, rotational masses, omitted in the simple models, are included. Differences in the response produced by these effects will not be large in these examples.

1. Single Story Frames

The single story frames of Figures 5-7 and 5-8 were reexamined using the finite element computer program described above. Conditions identical to those analyzed earlier were repeated and new cases were analyzed to show the effects produced by various modeling refinements. The loading on the frame, from which the force-time history of Figure 5-7b was derived, is the pressure pulse given in Figure 5-26a. This pressure of 13.2 psi (91 kPa) acts over a width (between frames) of 15 ft (4.57 m) and produces a distributed vertical load of approximately 2380 lb/in (416.8 kN/m). Further, the loading corresponds to an explosion of 1000 lb (454 kg) TNT at a distance of 110 ft (33.53 m) from the blastward side of the frame. This information was necessary for computing distributed loads on the blastward column and for computing sweeping loads on the top of the frame, both of which were considered in some of the cases.

444

Table 5-8. Modeling Comparison for the Single Story Frame

Sketch	Case No.	No. Element	Material Behavior	Flexible Girder	Axial Degree-Of-Freedom	Static Defls	Wks	$N_y = f(A_d)$	ΔT, sec	Maximum Sway, in. (mm)	Time of Maximum Sway, y, sec	Location Of Plastic Hinges, v#	Maximum Hinge Rotation θ#, degrees	Residual Maximum Hinge Rotation, degrees
	1	3	E						0.0002	1.1838 (30.069)	0.0239			
	2	3	E	•					0.0002	1.3808 (35.072)	0.0265			
	3	3	E-P	•					0.0002	1.2200 (30.988)	0.0253	⊖:1',2; ⊙:3,4	0.1155°	0.0998°
	4	3	E-P	•	•				0.0002	1.4111 (35.842)	0.0275	⊖:1',4	0.1556°	0.0814°
	5	3	E-P	•	•				0.0001	1.4217 (36.111)	0.0278	⊖:1',4	0.1596°	0.0841°
	6	3	E-P	•		•	•		0.0001	1.4279 (36.269)	0.0279	⊖:1',4	0.1597°	0.0808°

Table 5-8. Modeling Comparison for the Single Story Frame (Continued)

Sketch	Case No.	No. Element	Material Behavior	Flexible Girder	Axial Degree-Of-Freedom	Static Defls	Wks	$N_y = F'(\lambda_d)$	T, sec	Maximum Sway, in. (mm)	Time of Maximum Sway, sec	Location of Plastic Hinges, no	Maximum Hinge Rotation θ, degrees	Residual Maximum Hinge Rotation, degrees
	7	3	E-P	•	•	•		•	0.0001	1.4300 (36.322)	0.0279		0.1659°	0.0864°
	8	6	E-P	•					0.0001	1.4534 (36.916)	0.0275		0.1968°	0.1381°
	9	10	E-P	•					0.000025	1.4506 (36.845)	0.0277		0.2264°	0.1429°
	10	10	E-P	•					0.00001	1.7794 (45.197)	0.0305		0.4607°	0.3376°
	11	10	E-P	•					0.00002	1.8134	0.0315		0.4471°	0.3252°
	12	10	E-P	•		•	•	•	0.00002	1.79377	0.0309		0.5309°	0.3762°

Maximum Hinge is Denoted by +.

Results from these calculations are given in Table 5-8. The model configuration, location of the applied loads and load magnitude are given in the sketches. In all cases except for the sweeping load, the load is represented by a triangular pulse which rises from zero to its maximum value in 0.0001 sec and decays to zero in 0.02 sec. The sweeping load is calculated using side-on overpressures which decay from 5.2 psi (35.8 kPa) at the blastward side of the frame to 4.2 psi (28.96 kPa) at the leeward side and traverse the frame in 0.0095 sec. Rise times for the sweeping load depend upon the sweep rate across the elements in the top of the frame. Decay times are nominally 0.02 sec. The next three columns in Table 5-8 identify the case number, the number of elements in the model and the material behavior. Elastic behavior is denoted by E and elastic-plastic behavior by E-P. A dot (•) in any one of the next five columns indicates that this particular effect has been included in the model. These columns are:

- Flexible girder - rotations and yielding of the girder are permitted.

- Axial degrees-of-freedom - axial extensions and compression of the frame members are permitted. These degrees-of-freedom produce higher frequencies in the responses.

- Static deflections - deflections produced by the static (dead) load are calculated and used as the initial displacements at the beginning of the transient solution.

- WKs (work of the static forces) - static forces are added to the dynamic forces during the transient solution and any work that they produce as the structure deflects is accounted for.

- $MY = f\ (A_d)$ - the fully plastic bending moments in the beams and columns are expressed as a function of the dynamic axial forces in these members.

Subsequent columns give the integration time step, the maximum sway of the frame, locations of plastic hinges, the maximum hinge rotation which occurs in the frame and, finally, the maximum residual plastic hinge rotation.

Note that the locations of the plastic hinges are keyed to the element and degree-of-freedom numbers in the sketch and that the location of the maximum hinge rotation is denoted by the symbol (≠). The maximum residual rotation is always at the same location as the maximum rotation.

The first two cases in Table 5-8 repeat calculations performed earlier for the elastic frame, and can be compared to the results in Figure 5-12 and Table 5-3. Note that slightly smaller deflections (sway) were obtained with the F.E. program. This difference of less than two percent is attributed to slightly greater mass in the finite element model as noted previously. All other cases in Table 5-8 include elastic-plastic behavior (as indicated by the formation of plastic hinges), and these cases can be compared to study the effect produced by increasing levels of sophistication in the model. The effects studied are:

- girder flexibility
- axial degrees-of-freedom
- static loads
- reduction in the fully plastic bending moments produced by dynamic axial loads
- modeling the girder and columns with more than one element
- distributed loading on the blastward column
- distributed loading on the blastward column and a sweeping load on the roof

Observations from these studies are given in following paragraphs. These observations should not be construed as being correct for all situations because of the limited number of configurations investigated; however, they can provide useful insight into the proper modeling of frames.

a. Girder Flexibility

The effect of girder flexibility is shown by comparing Cases 3 and 4. With a rigid girder, the column is stiffer which reduces frame sway, and also equal plastic hinge rotations occur at the top and bottom of the columns. Allowing the girder to flex eliminates the plastic hinge at the top of the column and increases the rotation of the hinge at its base. The increase in the maximum rotation is 35 percent. For larger loads and greater deflections, the plastic hinge would reoccur at the top of the column and the percent difference in the hinge would be much smaller than 35 percent.

b. Axial Degrees-Of-Freedom

The effects produced by including axial degrees-of-freedom in the response (but without accounting for its effect on the fully plastic bending moments) are shown in Cases 4 and 5. These effects are small indeed with the maximum sway increasing by only 0.8 percent and the plastic hinge rotation by only 2.6 percent. Thus, it is only necessary to include axial degrees-of-freedom in such models if the effect of the axial loads on the plastic bending moments is to be considered as in Case 7.

c. Static Loads

The effects of static loads are included by computing the static deflections as starting displacements for the transient solution and by applying both the static and dynamic loads during the transient solution. Comparing Cases 5 and 6, it is seen that including the static forces in the analysis had negligible effects. If there were larger dead loads, for example in the center of the girder, or if the columns were inclined to the vertical, static loads would have been more noticeable. Frame geometry and the amount of dead load should always be considered before omitting the effect of the static loads.

d. Reduction in Plastic Moments Produced by Axial Loads

Cases 6 and 7 show the effect produced by treating the allowable fully plastic bending moments in the frame members as functions of the axial loads. Both static and dynamic forces affect the moments. Frame sway increased by 0.15 percent and the plastic hinge rotation increased by 3.9 percent. Although not shown in Table 5-8, the allowable plastic bending moment in the columns decreased 1.0 percent (maximum) as a result of the axial loading. A more pronounced effect can be inferred from Cases 11 and 12. Here, the combined effect of static loads and axial forces caused yielding to occur in the girder and increased the maximum plastic hinge rotation by 19 percent. The maximum sway was reduced very slightly. It should be noted that, although the percentage change was significant, the magnitude of the change was only 0.084 degrees.

    e. Modeling Frame Members With More Than One Element

        The effects produced by using more than one element to model the girder and columns is demonstrated by Cases 7, 8 and 9. Increasing the number of elements redistributes the mass of the frame and also causes more mass to respond to the loading (mass is now assigned to the lower column nodes which were assigned to ground). Also, with several nodes along the column and girder, they now have local responses to the applied loads. The results show a 1.8 percent increase in lateral sway and a 36.5 percent increase in the maximum plastic hinge rotations. The total increase in the plastic hinge was 0.061 degrees. Local response of the blastward column, produced by frame accelerations, appears to be the primary reason for the increase in the maximum hinge rotation.

    f. Distributed Loading on the Blastward Column

        In Cases 1 through 9, the loading on the frame was a constant value obtained by distributing one-half of the total load on the blastward face to the roof and one-half to the floor. Distributing the load over all column nodes, as in this example, increases the total load on the frame and increases the response of the blastward column. Comparing Cases 9 and 10 shows that the higher loading increased frame sway by 22.7 percent and the maximum plastic hinge rotation by 103.5 percent. The maximum hinge rotation is occurring at the base of the blastward column and the sharp increase in its value is associated with column response coupled to the frame sway. It is clear from this comparison that local column response is important in its design and that a single load applied to the top of a one-story frame should be greater than one-half of the to-

450

tal loading on the blastward side in order to produce realistic frame response.

g. Sweeping Loads on the Roof

Cases 10 and 11 demonstrate the effect produced by including the sweeping roof load. For this example, the roof loads produced by the side-on overpressures have a magnitude of about one-third that of the loads produced by the reflected pressures on the blastward column. Frame sway increases slightly and the maximum hinge rotation, which occurs at the base of the blastward column, decreases slightly. The decrease in the plastic hinge rotation is produced by the negative moment on the top of the column (or reduction in positive moment) associated with bending of the girder. As discussed earlier (see "Reduction in Plastic Moment Produced by Axial Loads"), the girder yielded in Case 12. This was produced by the reduction in moment produced by the axial loads. Yielding was slight in the girder and is of little consequence. Thus, the effect of the roof loads is small in this example, and generally, this will be true. Girders are usually stronger than columns and the applied loads are smaller because of the difference in magnitude between the side-on and reflected overpressures.

2. Two-Story Frame

The two-story frames of Figures 5-25 and 5-28 were also reexamined to study the effects of plasticity and axial loads on the frame responses. Results from these calculations are given in Table 5-9. The first 10 columns in the table are the same as in Table 5-8, and have already been explained. The next six columns give the results obtained for the cases run. They give maximum sway (lateral displacements) for the second floor and the roof, and the maximum and residual plastic hinge rotations at the base of the first story and second story columns.

Cases 1 and 2 of Table 5-9 are a repeat of earlier calculations, but are performed with the finite element program. These results compare with those given in Table 5-7 and in Figures 5-27 and 5-29. Notice that for the rigid girder, Case 1, the results in Table 5-8 are very close to those

Table 5-9. Modeling Comparison for the Two Story Frame

Sketch	Case No.	No. Element	Material Behavior	Flexible Girder	Axial Degree-of-Freedom	Static Detls	Wks	$M_f = F(\lambda_d)$	ΔT	Maximum Sway, in. (mm)		Plastic Hinge Rotations at Base of Columns			
										Second Story	Roof	Maximum First Story	Residual First Story	Maximum Second Story	Residual Second Story
	1	6	E						0.0001	2.52 (64.0)	3.63 (92.9)				
	2	6	E	•					0.0001	2.07 (52.7)	4.67 (118.7)				
	3	6	E-P	•					0.0001	3.41 (86.6)	4.98 (126.5)	+0.78363°	+0.78363°	0.57198°	0.39716°
	4	6	E-P	•					0.0001	3.37 (85.6)	5.59 (142.0)	+0.79161°	+0.76209°	0.50668°	0.49452°
	5	6	E-P	•	•	•	•	•	0.0001	3.50 (89.0)	5.68 (144.3)	+0.84305°	+0.78954°	0.53389°	0.52076°

in Table 5-7. In Case 2, for a flexible girder, the roof sway is almost the same, but the sway of the second floor is 32 percent lower than that calculated previously. Just as for the one-story frame, differences in the results can be attributed principally to the differences in the way the column masses were distributed in the finite element program and in the simplified analyses. Differences in the masses shift the frequencies of the system. For Case 2, this slight shift caused the peak in the sway displacements of the floors to be exactly out of phase as shown in Figure 5-31. This phasing attenuates the response of the second floor and the relative motions between the floors, as can be seen by comparing Figure 5-31 with Figure 5-29. Also, a close look at Figure 5-29 shows that this phenomenon was close to occurring in the earlier calculations. A small shift in frequencies can also be important when the responding frequencies of the system are near a characteristic time of the loading or near resonances with a repeated loading.

The effect of plasticity on the response of the frame is shown by comparing Cases 1 and 2 with Cases 3 and 4. Yielding in the frame has "smoothed out" the response so that the difference between modeling the frame with a rigid or flexible girder is less pronounced. Notice that the maximum plastic hinge rotations are about the same in Cases 3 and 4 and that the differences in the sway displacements between the two cases are smaller when plasticity occurs. Figure 5-32 gives graphical results for Case 4. Comparing Figures 5-31 and 5-32, the "smoothing" or damping of the response produced by plasticity is clearly seen.

The last case, Case 5, demonstrates the effects produced by including the work of the static forces and the reduction in plastic bending moments produced by axial forces. The effects are measurable, but small, for this case and can usually be ignored safely. As noted earlier, static loads and dynamic axial loads can be important for certain problems, such as for multi-story buildings or for buildings with high roof loading, either from static or dynamic loads.

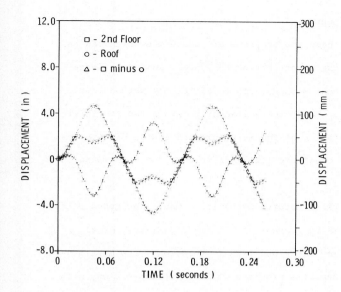

Figure 5-31.   Results for Case 2, Table 5-9

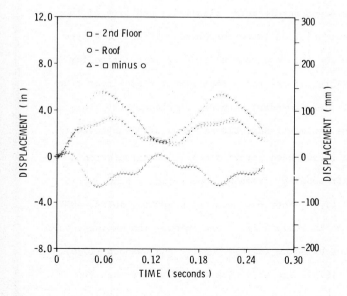

Figure 5-32.   Results for Case 4, Table 5-9

454

Available Computer Programs

Many general purpose computer programs are available which can be
used to calculate the response of structures to dynamic loading. These
programs utilize finite element methods, finite difference methods, or
some combination of the two. Most finite element programs use finite elements for spatial representation of the structure and finite differences
in time. Finite difference codes use finite differences in both space and
time.

A very large number of structural mechanics computer programs exist,
as evidenced in the summaries by Fenves, et al. (1973), Pilkey, et al.
(1974), and Perrone and Pilkey (1977 and 1978). Here, only the most general and readily available computer programs will be discussed along with
a few programs that appear to be particularly well suited to blast-resistant design.

Fourteen widely used finite element computer programs that include
provisions for static and dynamic structural behavior are listed in Table
5-10. Of these, the first seven permit metal plasticity and five of these
permit a nonlinear transient solution to be performed. This is the type
of analysis that is needed for the most general type of calculation for
blast loaded structures. Three of the codes that permit a nonlinear transient solution also permit time-dependent boundary displacements. This
feature is required for the nonlinear analysis of structures subjected to
ground shock. Thus, these three codes permit a nonlinear calculation to be
performed for combined blast loading, fragment impact, and ground shock.

Additional features of the four most general programs, ADINA, ANSYS,
MARC, and NASTRAN, are given in Table 5-11. More information on these four
codes is found in references by Bathe (1976), DeSalvo and Swanson (1979),
MARC Analysis Corporation (1975), and McCormick (1976). Note that two of
the four programs include substructuring, which can be used to separate
portions of the structure that remain elastic from the nonlinear part, and

Table 5-10. Comparison of Capabilities of General Purpose Computer Programs

Capabilities			ADINA	ANSYS	ASKA III-1, 2	MARC	NASTRAN	NEPSAP	NONSAP	STARDYNE	SUPERB	SAP IV	NISA	DANUTA	SDRC	STRUDL II
Static			X	X	X	X	X	X	X	X	X	X	X	X	X	X
Dynamic			X	X	X	X	X	X	X	X	X	X	X	X	X	X
Nonlinear Transient			X	X	O	X	X^2	X	X^3	O	O	O	X	O	X	O
Elements	1-D		X	X	X	X	X	X	X	X	X	X	X	X	X	X
	2-D		X	X	X	X	X	X	X	X	X	X	X	X	X	X
	3-D		X	X	X	X	X	X	X	X	X	X	X	X	X	O
Shells	Shells of Revolution		O	X	X	X	X	X	X	X	X	X	X	X	X	O
	Arbitrary		O	X	X	X	X	X	X	X	X	X	X	X	X	X
Thermal Loadings			X	X	X	X	X	X	X	X	X	X	X	X	X	X
Creep			X	X	X	Creep & Relaxation	O	X	O	O	O	O	O	O	O	O
Time Dependent Boundary Displacement			O	X	O	X	X	O	O	X	O	O	O	O	O	O
Temperature Dependent Material Properties			X	X	X	X	O	X	O	O	O	O	O	O	X	O
Geometric Nonlinearities			X	X	X	X	X	X	X	O	O	O	O	O	X	X
Large Strains			X	X	O	X	O	X	X	O	O	O	O	O	X	X
Material Model	Metal Plasticity		X	X	X	X	X^2	X	X	O	O	O	O	O	O	O
	Soils/Rocks		O	O	O	X	O	O	O	O	O	O	O	O	O	O
Material Symmetry	Isotropic		X	X	X	X	X	X	X	X	X	X	X	X	X	X
	Anisotropic		X	X	X	X	X	X	X	O	O	X	O	O	O	X
Crack (3-D)			O	X	O	X	O	O	O	O	O	O	O	O	O	O
Geometry Plottings			O	X	O	X	X	O	O	X	X	O	X	X	X	O

1 Commercial Version (Proprietary)
X = Yes
O = No
2 1-D Elements Only
3 2-D Elements Only

Table 5-11. Additional Features of the Most Suitable Programs for Elastic-Plastic Behavior

Feature Of Program	ANSYS	ADINA	MARC	NEPSAP
Element Library	General	General	General	General
Material Treatment	General	Moderately General	General	General
Plasticity Theory	Isotropic Hardening, Kinematic Hardening, Ramberg-Osgood or Any Other Power-Law Representation	Isotropic Hardening, Kinematic Hardening	Isotropic Hardening, Kinematic Hardening	Isotropic Hardening, Kinematic Hardening
Method of Solving Linear Equations	Wave Front	Partitioning	Gausian Elimination	Skyline
Ease of Use	Easy	Easy	Moderate	Moderate
Documentation	Extensive	Extensive	Extensive	Limited
Restart	X	X	X	X
Substructuring	X	X	O	O
Multipoint Constraints (Tying Modes)	X	O	X	X
Pre- and Post-Processors	X	X	X	X
Equilibrium Checks	O	X	X	X
Method of 3-D Crack Analysis	Singularity Element Formulation	O	Singularity Element Formulation	O
Automatic Mesh Generators	X	O	X	X
Node Numbering	Does Not Remember Nodes Internally		Requires Pre-Processor	
Cost of Run	Average	Average	Expensive	Average
Proprietary	X	X	X	X

X = Available or Yes

O = Not Available or No

three permit constraint equations, which can be used to eliminate unwanted
degrees-of-freedom. Still, the model of a building constructed with these
programs will, in general, be complex and expensive to use. Thus, this
approach is recommended only for small problems or for critical structures
which warrant detailed and expensive analyses.

Four other programs, which offer unique capabilities for nonlinear
analysis, are included in this review. These programs were developed specifically for nonlinear behavior, but they are not as well known and widely
distributed as the codes above. The four codes are:

- AGGIE I
- DEPROSS
- DYNFA
- PETROS 4

Of these four codes, AGGIE I is the most general. The other codes were developed for the response of specific types of structures to transient loads.

1. AGGIE I [Haisler (1977 and 1978)]

This finite element code is an extension to the SAP and NONSAP
codes included in Table 5-10. It includes provisions for a nonlinear transient solution with all elements. Nonlinearities include material and
large displacements. Element types include a three-dimensional truss, two-
dimensional isoparametric solid, and a three-dimensional isoparametric solid. The three-dimensional isoparametric solid can represent a thick shell.
Time-dependent displacements are not now permitted.

2. DEPROSS [Wu and Witman (1972)]

These programs calculate the elastic-plastic response of impulsively loaded, simple structures. The structures are represented by discrete masses connected by massless lengths. Extensional deformation is
distributed in the lengths, and bending deformation is concentrated at the
joints (mass points). The beam cross sections are further idealized by a
number of flanges separated by material that carries only shear and no ax-

458

ial stresses. The equations of motion of the mass points are cast in finite difference form, and the time history of the response is found by a stepwise integration process. Material behavior is inelastic strain hardening, and strain rate effects can be included. Brief descriptions of the three programs are given below.

a. <u>DEPROSS 1</u>

This program calculates the dynamic response of beams and circular rings that are subjected to axisymmetric impulsive loading. Beams can be simply supported or clamped and rings clamped or free. The program requires that the cross sections be rectangular and uniform.

b. <u>DEPROSS 2</u>

The dynamic response of unbonded, concentric circular rings is calculated by this program. As for DEPROSS 1, the ring section must be rectangular and uniform, and the impulse loading must be axisymmetric. The concentric rings may consist of different materials and have different thicknesses, but must be the same width. In addition, the two rings must be initially concentric, but not necessarily in contact.

c. <u>DEPROSS 3</u>

This program is similar to DEPROSS 1 but applies to circular plates and spherical shells. Plates can be simply supported and clamped. Shells must be clamped. Again, the thickness of the plates or shells must be uniform, and the impulse loading must be axisymmetric.

3. <u>DYNFA [Stea, et al. (1977)]</u>

This program was designed specifically for the analysis of frame structures subjected to blast loading. The program is based upon standard matrix methods of structural analysis and a lumped parameter representation of the frame. Numerical integration by the linear acceleration method is used to solve for frame displacements. Inelastic behavior of the frame members is included by the formation of plastic hinges at the nodes whenever the combined axial load and bending moment capacity of a member is exceeded. The recommended modeling procedure is to include nodes at the quarter points of a beam or column member, resulting in five nodes and four elements per member. Although the program is limited to two-dimensional frames, the use of four elements per member can result in a fairly large number of elements and degrees-of-freedom, even for fairly simple frames; however, such a representation is necessary in order to study the combined effect of local beam and column response (to local loading) and gross frame motions. In addition to metal plasticity, nonlinear effects which are accounted for in the program include the P-Δ effect produced by large sway of axially loaded members. This effect is accounted for by the addition of a shear couple to the loading. Documentation for this program includes a procedure for the preliminary sizing of members in single story frames based upon the formation of plastic hinges (mechanism analysis) and general guidelines for the design of frame structures to resist blast loading.

4. PETROS 4 [Pirotin, et al. (1976)]

The PETROS 4 program was developed to predict the arbitrarily large deflection, elastic-plastic transient responses of arbitrarily shaped shells. Shells can be thin, multilayered and have variable thickness; they can have temperature distributions and undergo various types of deformation in response to arbitrary initial velocity distributions, transient loads, and temperature histories. Strain-hardening and strain-rate sensitive material behaviors are taken into account.

The program is based upon a finite-difference solution to the governing shell equations. Displacement boundary conditions can be very general and include time-dependent translations. Applied forces can vary arbitrarily in both space and time. Up to nine different types of shells can be analyzed by PETROS 4, including shells with variable thickness, multimaterials, multilayers (with hard or soft bonding), and shells whose thicknesses vary with time.

New computer codes are written and published almost daily, so the analyst must stay abreast of new developments. It is recommended that computer programs which have a history of successful use by the engineering community be used whenever possible. Test cases created by the user should also be run as a further check on the accuracy of the results.

CHAPTER 5

LIST OF SYMBOLS

A	cross-sectional area of beam
b	beam width
E	Young's modulus
F	applied force
F_e	equivalent force
F_1, F_2, F_3, etc.	forces
g(t)	time function for load
H	beam depth; column height
I	second moment of area (area moment in inertia); total impulse
i	specific blast impulse
KE	kinetic energy
K_L	load factor
K_{LM}	load-mass factor
K_M	mass factor
k	linear spring constant
k_e	equivalent spring constant
L	length of beam
L_1, L_2, L_3, etc.	lengths of beam members in a frame
ℓ	beam length
M	bending moment

M_p	plastic bending moment
M_y	yield bending moment
m, m_1, m_2	lumped masses
m_e	equivalent mass
$p(t)$	time history of pressure
R	resisting force
R_e	resistance of equivalent system
R_m	maximum static resistance for uniform beam
S	elastic section modulus
T_L	duration of load
T_N	natural period of system
t	time
t_d	pressure pulse duration
t_m	time to maximum response
t_r	pressure rise time
U	strain energy
V	dynamic reaction
WK	external work
w	beam or plate deflection
$\dot{w}$	transverse velocity of beam element
w_o	maximum beam or plate deflection
x	displacement
$\dot{x}$	velocity
$\ddot{x}$	acceleration
x_o	initial displacement
Y_{el}	deflection at yield
Y_m	maximum response
Z_p	plastic section modulus
α, β	general subscripts in matrix formulation for force-displacement relations
β	percent of critical damping

Δt	time increment
δ	deflection
δ_{ST}	deflections under static load
θ^*	plastic hinge rotations
μ	ductility ratio
ρ	density
σ, σ_{max}	stresses
ω	circular natural frequencies

CHAPTER 6

FRAGMENTATION AND MISSILE EFFECTS

GENERAL DISCUSSION AND DEFINITIONS

By the definition of an explosion given in Chapter 2, a blast wave must develop as a consequence of any explosion even though the wave may not be strong enough to cause damage. Most discussions, testing, accident investigation, and explosive accident prediction, therefore, involves studies of blast waves and their effects. But often, significant damage in accidental explosions is caused by the impact of fragments or objects which were generated during the explosions and hurled against targets (or "receivers") at high speed. In this chapter, we first define some of the parlance of fragment and missile generation and impact effects. We then discuss the generation of fragments and missiles by various types of explosions, how to predict fragment trajectories and impact conditions, and the state of knowledge of impact effects for classes of fragments and missiles occurring in accidental explosions.

Primary Fragments or Missiles

The term "primary fragment" denotes a fragment from a casing or container for an explosive source. If the source is a true high explosive, the container or casing usually ruptures into a very large number of small primary fragments which can be projected at velocities up to several thousand meters per second by the explosion [Zaker (1975b)]. For bomb and shell casings, typical masses of damaging fragments recovered in field tests are about 1 gram [Zaker (1975b)]. These primary fragments, although irregular, are usually of "chunky" geometry, i.e., all linear dimensions are of the same order of magnitude.

At the other extreme are containers which burst or split while containing a high pressure gas or vapor, and form only one or two missiles [Pittman (1976)]. A number of intermediate cases are also possible, with containers burst by runaway internal reactions which are not detonations, or with brittle containers failed by internal pressure [Esparza and Baker (1977a) and Pittman (1972a and 1972b)]. The velocities for this class of primary fragment are usually much lower than fragments from high explosive detonations, ranging only up to hundreds of meters per second. Also, the fragments (or more properly, missiles) are often quite massive -- up to kilograms -- and of elongated or flat geometry.

Secondary Fragments or Missiles

Containers or casings which fragment or burst during accidental explosions are not the only sources of fragments and missiles. The blast waves from severe explosions can interact with objects located near the explosion source, tear them loose from their moorings if they are fastened down, and accelerate them to velocities which can cause impact damage. The objects could be small tools, materials such as pipes and lumber, parts of buildings and other structures disrupted by the explosion, large pieces of equipment such as autos or portable generators, etc. The usual terms for these potentially damaging objects are "secondary missiles" or "secondary fragments." Still another term in vogue with NASA is "appurtenances." We will use the three terms as synonyms.

Drag-type and Lifting-type Fragments

Once fragments or missiles have been formed and accelerated by the explosion, they will fly through the air until they impact some target (receiver), or the ground. The forces acting on the fragments and affecting their trajectories are inertia, gravitation, and fluid dynamic forces. The fluid dynamic forces are determined by the instantaneous velocity of the fragment through the air, the air density, and the shape and orientation of the fragment at each instant in time. Generally, fragments are quite irregular in shape and may be tumbling, so completely accurate description of the fluid dynamic forces during flight is difficult, if not impossible. In the trajectory analysis for fragment flight, one usually resorts to some simplified description of the fluid dynamic forces, and uses the concepts from aerodynamics of division of these forces into components called drag (along the trajectory or normal to the gravity vector) and lift (normal to the trajectory or opposing gravity). Then, the force components are given at any instant by

$$F_L = C_L (1/2) \rho V^2 A_L \qquad (6\text{-}1)$$

and

$$F_D = C_D (1/2) \rho V^2 A_D \qquad (6\text{-}2)$$

where C_L and C_D are lift and drag coefficients determined empirically as a function of shape and orientation with respect to the velocity vector, and the magnitude of the velocity V. In the equations, ρ is air density and $q = (1/2) \rho V^2$ is termed the dynamic pressure. If a fragment is of chunky shape, so that $C_D \gg C_L$ for any flight orientation, it is called a drag-type fragment. If, on the other hand, $C_L \geq C_D$ for some flight orientation, the fragment is called a lifting-type fragment. We will discuss later methods of trajectory predictions for both types of fragments.

Terms Relating to Fragment and Missile Impact Effects

Trajectory analyses and/or test results can give predictions of fragment ranges, masses, impact velocities, and even the probability of striking a given target [see Zaker (1975a and 1975b) and Baker, Kulesz, et al. (1975)]. But, their impact (or terminal ballistic) effects on various targets or receivers help to determine the degree of missile hazard. We will define here a number of terms related to missile impact effects.

An impacting fragment can cause damage to a multitude of types of receivers by striking and either penetrating or rebounding without penetrating. The term penetration usually means that the fragment or missile disrupts or displaces some of the target material during impact, but does not pass through the target. The missile may or may not remain lodged in the target. On the other hand, if the missile passes entirely through, the target is said to have been perforated.

Impacting missiles may damage a target by simple momentum transfer, and various wave transmission effects, either in conjunction with penetration or perforation, or in their absence. Special terms employed for some of these effects are spalling or scabbing, indicating the process by which impact-induced compression waves in solids cause failures in tension after wave reflection from a free surface. The process is quite well described by Rinehart (1975). In brittle materials such as concrete or plaster, spalling can occur for missile impacts at relatively low velocities, less than 100 meters per second [Baker, Hokanson, et al. (1976)].

GENERATION OF FRAGMENTS AND MISSILES BY EXPLOSIONS

Primary Fragments

A primary fragment, as mentioned above, denotes a fragment from a casing or container for an explosive source. The explosive source can be a high explosive, high pressure gas or vapor, or chemical which, in a runaway reaction, generates a pressure high enough to burst the containment vessel. We noted earlier that fragments from bomb and shell casings containing high explosives are usually small (one gram or less), "chunky" in shape, and have initial velocities of several thousand meters per second. The size and shape of fragments from vessels containing high pressure gas or vapor or hazardous materials which can generate a gas from a runaway reaction can vary from small to large "chunky" fragments, disc-shaped fragments, or even large portions of the vessel. The size and shape of the fragments will depend greatly on the metallographic history of the vessel, its physical condition (such as dents, grooves, bends, or internal cracks or flaws), and the condition of joints, most notably weld joints. Tests with flight weight, thin-walled, propellant tanks for space vehicles [Willoughby, et al., Volume II (1968b)] revealed that the fragments from these vessels were both "chunky" and "lifting" fragments. A "lifting" fragment is one which has a diameter several times greater than its thickness. Sometimes these fragments are referred to as "disc" shaped fragments and they tend to have a significant coefficient of lift which allows them to sail through the air.

Much work has been done by the Department of Defense in determining velocities of pieces of bomb and shell casings for military purposes. Since this work is not directly applicable to industrial accidents, it will not be covered here. Velocities of fragments from spherical vessels bursting from high internal pressures have been analytically determined by the works of Grodzovskii and Kukanov (1965) and Taylor and Price (1971), and further expanded to a larger number of fragments by Bessey (1974) and to cylindrical geometries by Bessey and Kulesz (1976). In these analyses, the energy of the confined gas is partitioned between the kinetic energy of the fragment, the energy of the gas escaping between the cracks between the fragments as they are formed, and the energy of the expansion of the internal gas. The amount of energy expended to burst the containment vessel is considered small compared to the total amount of energy available.

Equal Fragments

A schematic depicting the essential characteristics of the Taylor and Price solution for a sphere bursting in half is shown in Figure 6-1. Before accelerating into an exterior vacuum, the sphere has internal volume V_{oo} and contains a perfect gas of adiabatic exponent (ratio of specific heats) γ and gas constant R with initial pressure p_{oo} and temperature θ_{oo} (Figure 6-1a). At a time $t = 0$, rupture occurs along a perimeter Π, and the two fragments are propelled in opposite directions due to forces ap-

(a)

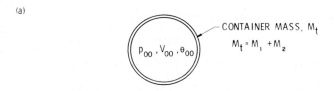

(b)

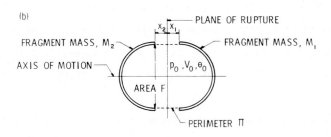

(c)

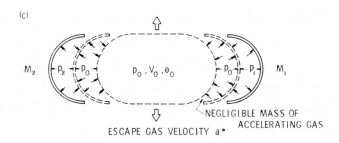

Figure 6-1. Parameters for Sphere Bursting Into Two Halves

plied against the area F which is perpendicular to the axis of motion of
the fragments (Figure 6-1b). The mass of the two fragments M_1 and M_2 is
considered large relative to the mass of the remaining gas at elevated pressure (Figure 6-1c). Using the equations of motion of the fragments, the
equation of state, and one-dimensional flow relationships, one can obtain
the velocities of the fragments. Detailed velocity calculations are given
in Appendix C.

A similar approach can be taken for determining the velocity of
pieces of a sphere bursting into a large number of fragments (Figure 6-2a),
a cylinder bursting into two halves perpendicular to its axis of symmetry
(Figure 6-2b) and a cylinder bursting into a large number of strip fragments

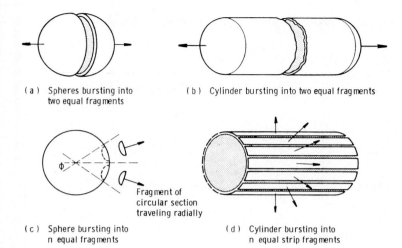

(a) Spheres bursting into two equal fragments

(b) Cylinder bursting into two equal fragments

(c) Sphere bursting into n equal fragments

Fragment of circular section traveling radially

(d) Cylinder bursting into n equal strip fragments

Figure 6-2. Assumed Fragmentation Patterns
    [Baker, Kulesz, et al. (1978)]

(Figure 6-2c). These cases have been formulated by Bessey (1974) and Bessey and Kulesz (1976) and involved changes in the mass of the fragments, the area F over which the forces act and the perimeter of the rupture Π. After making the appropriate changes and substitutions, a set of equations similar to those given in Appendix C can be derived and solved simultaneously using the Runge-Kutta method of numerical iteration.

A comparison is made in Baker, Kulesz, et al. (1975) between the computer code predictions for the velocity of fragments from spheres bursting into a large number of pieces and some experimental data. D. W. Boyer, et al. (1958) had measured fragment velocities from bursting glass spheres of various dimensions where the contained gas was air or helium (He). Table 6-1 lists the data from the Boyer report and compares it with values obtained from a computer code which determined the velocity of fragments from a bursting sphere using the method described above. The results of the computer calculation are within about 10 percent of Boyer's experimental results. Table 6-2 lists data for fragments from bursting titanium alloy spheres filled with nitrogen. The experimental data are from a report by Pittman

Table 6-1. Initial Fragment Velocities, V_i, From Bursting Glass Spheres

$(\rho = 2.6 \text{ Mg/m}^3)$

Sphere Characteristics				Initial Fragment Velocities	
	Wall			V_i (Boyer)[*]	V_i (Code)
Radius	Thickness		Pressure		
mm	mm	Type	Pa	m/s	m/s
12.7	1.0	Air	2.25×10^6	51.8	51.5
25.4	1.0	Air	2.25×10^6	75.6	80.2
63.5	1.9	Air	1.38×10^6	69.8	78.0
12.7	1.0	He	2.25×10^6	44.2	38.4
25.4	1.0	He	2.25×10^6	79.4	61.6

[*] V_i values were obtained by averaging measured values for similar cases from Table 1 of Boyer (1958) for the 12.7 and 25.4 mm radius spheres and by measurements from Figure 13 of Boyer (1958) for the 63.5 mm radius sphere.

Table 6-2. Initial Fragment Velocities, V_i, From Bursting Titanium Alloy Spheres

$(\rho = 4.46 \text{ Mg/m}^3)$

Sphere Characteristics				Initial Fragment Velocities	
	Wall			V_i (Pittman)[*]	V_i (Code)
Radius	Thickness		Pressure		
mm	mm	Type	Pa	m/s	m/s
117	2.74	N	5.51×10^7	366 ± 15	352
343	9.19	N	5.51×10^7	342 ± 30[1]	339
343	9.19	N	5.51×10^7	426 ± 27[2]	339
343	9.19	N	5.51×10^7	448 ± 30[2]	339

[*] Values taken from Pittman (1972a).

[1] This value was based on velocity measurements using a strobe photographic technique.

[2] These values were based on velocity measurements using breakwire measurement techniques.

(1972a) who measured the velocities of fragments emanating from the spheres with breakwire and strobe photographic technique. The computer code for predicting velocities by the method developed above predicted velocities from the small diameter sphere within 10 percent while velocities predicted for the large diameter spheres are low by 15 percent where breakwire techniques were used to measure fragment velocities (the breakwire technique is less accurate than photography). The computer predictions agree well with the measured data where strobe photography was used to measure the fragment velocities.

The method developed by Taylor and Price (1971) and modified by Baker, et al. (1975) for calculating velocities of fragments from bursting spherical and cylindrical pressure vessels have been used to provide velocities of various fragments which could be plotted in some form of prediction curve. The development of the necessary equations, the numerical iteration method used to solve simultaneously the differential equations and the computer programs can be found in Appendix IV.A and Appendix IV.C of Baker, Kulesz, et al. (1975). The only assumptions included here are those needed to determine fragment velocities.

The basic assumptions are:

- The vessel with gas under pressure bursts into equal fragments. If there are only two fragments, and the vessel is cylindrical, the vessel bursts perpendicular to its axis of symmetry. If there are more than two fragments, and the vessel is cylindrical, strip fragments (end caps are ignored) are formed and expand radially about the axis of symmetry (see Figure 6-2).

- The cylindrical containment vessel has hemispherical end caps. (These are ignored when the vessel bursts into multiple fragments.)

- The thickness of the containment vessel is uniform.

- Vessels have a length-to-diameter (L/D) ratio of 10.0 for cylinders or 1.0 for spheres.

- Contained gases are either hydrogen (H_2), air, argon (Ar), helium (He) or carbon dioxide (CO_2).

To calculate the velocities of the fragments from <u>cylindrical</u> containment vessels bursting into two fragments, the technique used in Taylor and Price (1971) and modified in Baker, Kulesz, et al. (1975) was used to calculate the velocity of the fragments at the point when the exit area from the opening in the cylinder was equal to the combined cross-sectional areas of the two fragments. The computer program was then stopped and final velocity was determined by using a computer program (GASROC) based on ideal gas law isentropic expansion through the open cross section (exit area) of each fragment. Input to this "rocketing" computer code was the initial velocity from the computer code above, the state variables of the gas at the start of the calculations, the mass of the fragment, and geometrical aspects of the fragment.

Figure 6-3 contains plots of the velocity term versus the pressure term for two fragments, ten fragments and one hundred fragments from spherical or cylindrical vessels. Three separate regions have been bounded to

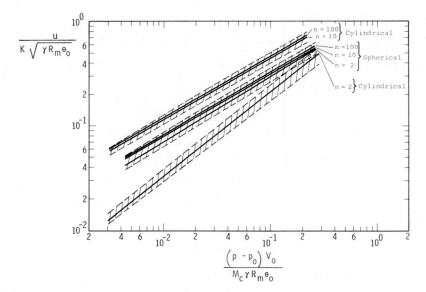

Figure 6-3. Scaled Fragment Velocity Versus Scaled Pressure

account for scatter: 1) cylindrical vessels bursting into multiple fragments, 2) spherical vessels bursting into halves or multiple fragments, and
3) cylindrical vessels bursting into two fragments. Estimates of the initial
velocities of cylinders and spheres can be extracted from the nondimensional terms read directly from the appropriate bounded regions on the graph.
The two nondimensional terms in Figure 6-3 are:

- Nondimensional pressure term

$$= \frac{(p - p_o) V_o}{M_c \gamma R_m \theta_o} = \frac{(p - p_o) V_o}{M_c a_{gas}^2} =$$

$$\frac{(\text{pressure} - \text{atm. pressure}) (\text{Volume})}{(\text{Mass of container}) (\text{Sound speed of the gas})^2}$$

- Nondimensional velocity term

$$= \frac{u}{K\sqrt{\gamma R_m \theta_o}} = \frac{u}{Ka_{gas}} =$$

$$\frac{(\text{velocity})}{(\text{constant}) (\text{sound speed of the gas})}$$

where K equals 1.0 for equal fragments.

The technique for predicting initial fragment velocities for spherical or cylindrical pressure vessels bursting into equal fragments requires
knowledge of the internal pressure P, internal volume V_o, mass of the container M_c, ratio of specific heats γ, ideal gas constant adjusted for the
gas R_m, and the absolute temperature of the gas θ_o at burst. Table 6-3
contains the corresponding γ's and R_m's for the gases for which this analysis is appropriate.

474

Table 6-3. Summary of Ratios of Specific Heat and
Ideal Gas Constants for Different Gases

Ideal Gas Constant R_m

Gas	Ratio of Specific Heats γ	$\left(\dfrac{m^2}{sec^2 \cdot {}^\circ K}\right)$
Hydrogen	1.400	4124.0
Air	1.400	287.0
Argon	1.670	208.1
Helium	1.670	2078.0
Carbon Dioxide	1.225	188.9

In summary, in order to estimate the initial velocity of fragments from pressurized spheres and cylinders which burst into _equal_ fragments, one should use the following procedures:

Step 1. Calculate the nondimensional pressure term

$$\frac{(p - p_o) \, V_o}{M_c \gamma R_m \theta_o}$$

Step 2. Locate the corresponding value of the nondimensional velocity term

$$\frac{u}{K \sqrt{\gamma R_m \theta_o}}$$

and solve for velocity u. (Note: K = 1.0 for equal fragments.)

Note: Axes of Figure 6-3 are nondimensional terms and merely require that one use a self-consistent set of units.

Cylinders With Length-to-Diameter Ratio of 10.0 Bursting Into Two
Unequal Fragments

The Taylor and Price (1971) method modified by Baker, Kulesz, et al.

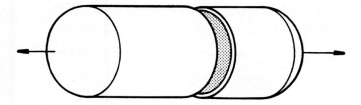

Figure 6-4. Assumed Breakup Into Two Unequal Fragments

(1975) for calculating velocities of fragments from bursting spherical and
cylindrical gas vessels has been expanded to provide initial velocities of
unequal fragments from cylindrical vessels. The development of the necessary equations and the subsequent computer program UNQL are explained in
depth in Appendix D of Baker, Kulesz, et al. (1978).

The basic assumptions are:

- The vessel with gas under pressure breaks into two unequal fragments along a plane perpendicular to the cylindrical axis, and
  the two container fragments are driven in opposite directions
  (see Figure 6-4).

- The containment vessel is cylindrical and has hemispherical end
  caps.

- The thickness of the containment vessel is uniform.

- Vessels have a length-to-diameter (L/D) ratio of 10.0.

- Contained gases are either hydrogen (H_2), air, argon (Ar), helium (He) or carbon dioxide (CO_2).

The ideal gas law isentropic expansion computer code (GASROC) described above was used to calculate velocities after the cylindrical open
area between cylindrical fragments was equal to the combined cross-sectional
areas of the two fragments.

The technique for predicting initial fragment velocities for fragments
from a cylinder (L/D = 10.0) which breaks into two unequal fragments perpendicular to its axis of symmetry is identical to that for equal fragments ex-

cept for the value of the constant K. The value of K depends on the ratio of the fragment mass to the total mass of the cylinder as shown in Figure 6-5. To estimate the initial velocity of a fragment from a pressurized cylinder (L/D = 10.0) which bursts into <u>unequal</u> fragments, one should use the following procedures:

Step 1. Calculate the nondimensional pressure term =

$$\frac{(p - p_o) V_o}{M_c \gamma R_m \theta_o}$$

Step 2. Locate the corresponding value of the nondimensional velocity term

$$\frac{u}{K \sqrt{\gamma R_m \theta_o}}$$

in the region bounded for n = 2 (cylindrical vessels).

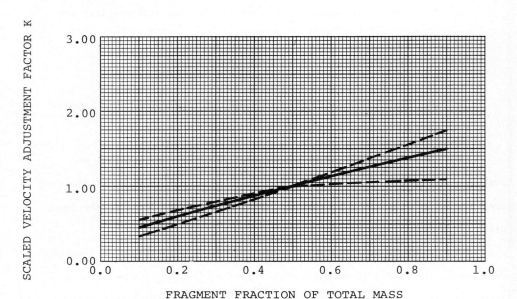

Figure 6-5. Adjustment Factor K for Unequal Mass Fragments

Step 3. Determine the value of K from Figure 6-5.

Step 4. Solve for velocity u.

Note: Axes of Figure 6-3 are nondimensional terms and merely require that one uses a self-consistent set of units.

Secondary Fragments

Secondary fragments, as explained previously, are objects located near an explosive source which are torn loose from their moorings, if they are fastened down, and are accelerated through interaction with the blast wave(s). There are two basic types of secondary fragments: 1) unconstrained and 2) constrained. For the purposes of this discussion, we will consider unconstrained objects first and then constrained objects. Fragments from glass window breakage will also be discussed after discussing constrained objects.

1. Unconstrained Secondary Fragments

To predict velocities to which appurtenances are accelerated by explosions, one must consider the interaction of blast waves with solid objects. Figure 3-8 [Baker, et al. (1974b) and Baker, Kulesz, et al. (1975)] shows schematically, in three stages, the interaction of a blast wave with an irregular object. As the wave strikes the object, a portion is reflected from the front face, and the remainder diffracts around the object. In the diffraction process, the incident wave front closes in behind the object, greatly weakened locally, and a pair of trailing vortices is formed. Rarefaction waves sweep across the front face, attenuating the initial reflected blast pressure. After passage of the front, the body is immersed in a time-varying flow field. Maximum pressure difference during this "drag" phase of loading is the stagnation pressure.

To predict the effect of a blast wave on an appurtenance, it is necessary to examine the net transverse pressure on the object as a function of time. This loading, somewhat idealized, has been shown in Figure 3-9. After time of arrival t_a, the net transverse pressure rises linearly from zero to a maximum peak reflected pressure P_r in time $(T_1 - t_a)$. For an

object with a flat face nearest the approaching blast wave, this time interval is zero. Pressure then falls linearly to drag pressure in time
$(T_2 - T_1)$ and decays more slowly to zero in time $(T_3 - T_2)$

       Once the time history of net transverse pressure loading is
known, the prediction of appurtenance velocity can be made. The basic
assumptions for unconstrained secondary fragments are that the appurtenance behaves as a rigid body, that none of the energy in the blast wave
is absorbed in breaking the appurtenance loose from its moorings or deforming it elastically or plastically, and that gravity effects can be
ignored during this acceleration phase of the motion. The equation of motion of the object is then

$$A \, p(t) = M\ddot{x} \tag{6-3}$$

where A is the area of the object presented to the blast front, $p(t)$ is the
net transverse pressure according to Figure 3-8, and x is the displacement
of the object (dots denote derivatives with respect to time). The object
is assumed to be at rest initially, so that

$$x(0) = 0, \; \dot{x}(0) = 0 \tag{6-4}$$

Equation (6-3) can be integrated directly. With use of the initial conditions, Equation (6-4), this operation yields, for appurtenance velocity,

$$\dot{x}(T_3) = \frac{A}{M} \int_0^{(T_3 - t_a)} p(t) \, dt = \frac{A}{M} i_d \tag{6-5}$$

where i_d is the total drag and diffraction impulse. The integration in
Equation (6-5) can be performed explicitly if the pressure time history
is described by suitable mathematical functions, or performed graphically

or numerically if p(t) cannot be easily written in function form. In either case, Equation (6-5) yields the desired result -- a predicted velocity for an object. The integral in Equation (6-5) is merely the area
under the curve in Figure 3-9.

For shocks of intermediate strength, $P_s/p_o \leq 3.5$ where P_s is
side-on overpressure and p_o is atmospheric pressure, the solution of Equation (6-5) can be found in Baker, Kulesz, et al. (1975). A rather long
equation develops which, in nondimensional form, can be expressed as

$$\bar{V} = \frac{M V a_o}{p_o A (KH + X)} = f\left(\frac{P_s}{p_o}, \frac{C_D i_s a_o}{P_s (KH + X)}\right) \tag{6-6a}$$

$$\bar{V} = f(\bar{P}, \bar{i}) \tag{6-6b}$$

where M is the mass of the object, V is the velocity of the object, a_o is
the velocity of sound in air, p_o is atmospheric pressure, A is mean presented area of the object, K is the constant (4 if appurtenance is on the ground
and 2 if appurtenance is in air), H is the minimum transverse distance of
the mean presented area of object, X is the distance from the front of the
object to location of its largest cross-sectional area, P_s is the peak incident overpressure, C_D is the drag coefficient, and i_s is the incident specific impulse.

A pictorial explanation of the appurtenance parameters A, H, and
X is shown in Figure 6-6. Values for drag C_D can be found in Hoerner (1958)
for various shaped objects. A summary of drag coefficients for objects of
various shapes is given in Table 3-2. For computational purposes, Equation
(6-6) is presented in Figure 6-7 where $\bar{P} = P_s/p_o$, $\bar{i}$ is the nondimensional
term containing i_s, and $\bar{V}$ is the nondimensional term containing V. The figure contains several curves for different values of $\bar{V}$ and is quite useful for
the range of $\bar{P}$ and $\bar{i}$ presented.

480

2. Constrained Secondary Fragments

Westine (1977) has developed an approximate engineering procedure for estimating secondary fragment velocities when constrained objects are exposed to explosive detonations. This objective was accomplished by dividing the problem into two parts -- the first for estimating the specific impulse imparted to flat, cylindrical, and spherical secondary fragments in the vicinity of cylindrical explosive charges, and the second for estimating the velocity of constrained beams of any material and cross-sectional area which could become secondary fragments because of this impulsive load.

To calculate the specific impulse imparted to the target, Westine (1977) conducted a model analysis for a target aligned parallel to a line charge which is larger than the target. After eliminating terms which are

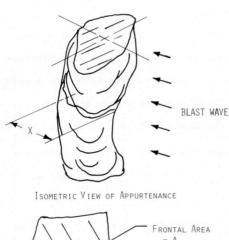

BLAST WAVE

ISOMETRIC VIEW OF APPURTENANCE

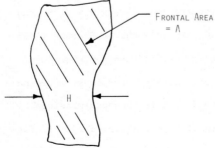

FRONTAL AREA
= A

H

FRONT VIEW OF APPURTENANCE

Figure 6-6. Pictorial Explanation of Appurtenance Variables

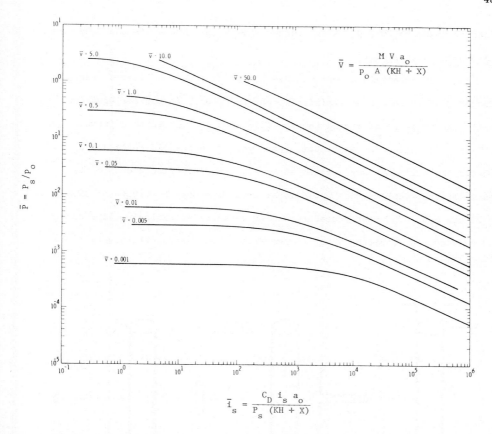

$$\bar{V} = \frac{M V a_o}{P_o A (KH + X)}$$

$$\bar{i}_s = \frac{C_D \, i_s \, a_o}{P_s \, (KH + X)}$$

Figure 6-7. Nondimensional Object Velocity $\bar{V}$ as a Function of
Nondimensional Pressure $\bar{P}$ and Nondimensional Impulse $\bar{i}$

invariant under similar atmospheric conditions such as the density of air,
atmospheric pressure and the speed of sound and assuming that the effect of
length of the target relative to the radius of the explosive is relatively
insignificant, one has

$$\frac{i}{\beta R_e} = \frac{\psi i}{\beta R_e} \left[\frac{R}{R_e}, \frac{R_t}{R_e}, \frac{\ell_e}{R_e} \right] \tag{6-7}$$

where i is the specific acquired impulse (N s/m^2 = Pa · s), β is the nondimensional shape factor, R_e is the radius of the explosive (m), R is the

standoff distance (m), R_t is the target radius (m), and ℓ_e is the length of the explosive line (m).

        An experimental program was conducted at the Ballistic Research Laboratories to determine empirically the functional format for Equation (6-7). In order to determine the specific impulse applied to a target when a blast from a line charge occurs, tests were conducted in which small spheres and cylinders were placed at various standoff distance from the center line of various size cylindrical explosive Comp B charges as shown in Figure 6-8. The test procedure was to detonate an explosive line charge and measure the resulting velocity imparted to unconstrained targets. The

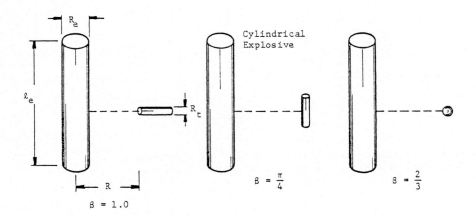

(a) Exposed Flat Face   (b) Exposed Cylindrical   (c) Exposed Spherical
                                   Surface   Surface

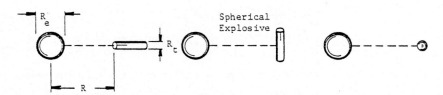

Figure 6-8. Target Orientation for Unconstrained Tests

specific impulse imparted to the target could then be determined from

$$i = \frac{MV}{A\beta} \tag{6-8}$$

where M is the mass of the object, V is the velocity of the object, and A is the projected area of the object. The results of the curve fit to the experimental data are shown in Figure 6-9. The ordinate in this figure has a quantity called R_{eff} in it instead of $R_e(\ell_e/R_e)^{0.333}$. This quantity R_{eff} stands for the effective radius of the equivalent sphere of explosive which could be formed from a cylinder of radius R_e and length ℓ_e. The term R_{eff} is related to R_e and ℓ_e through

$$\text{Cylindrical Charge:} \quad R_{eff} = 0.9086 \left(\frac{\ell_e}{R_e}\right)^{0.333} R_e \tag{6-9a}$$

$$\text{Spherical Charge:} \quad R_{eff} = R_e \tag{6-9b}$$

The existence of two straight line regions for values of R/R_e less than and greater than 5.25 (cylindrical charges) is apparent in Figure 6-9. In the near field where R/R_e is less than 5.25, the slope of the line for cylindrical charges in Figure 6-9 is minus 1.0 which means that $(i/\beta R_{eff})$ times (R/R_e) equals a constant for invariant R_e/R_t. The normally reflected specific impulse close to the line charge is thus caused primarily by momentum of the explosive products. In other words, the impulse close to the charge is caused by adding up the mass times the velocity products of all the particles from the explosive, casing, and engulfed air. Because the specific impulse i is caused by momentum in explosive products, it decays with standoff distance inversely as the surface area of a cylinder enclosing the line source, which equals $2\pi R \ell_e$. Only as standoff distances grow larger do the effects of momentum loss through air drag and gravitational effects re-

484

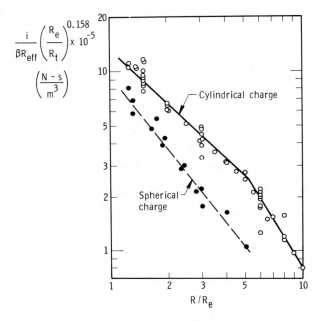

$$\frac{i}{\beta R_{eff}} \left(\frac{R_e}{R_t}\right)^{0.158} \times 10^{-5}$$

$$\left(\frac{N-s}{m^3}\right)$$

Figure 6-9.   Specific Acquired Impulse for Unconstrained
Secondary Fragments

duce this phenomenon sufficiently for shock wave phenomena to become more important and the slope of the line to change, as shown in Figure 6-9.

A similar relationship holds for objects exposed to nearby spherical charges of Comp B as shown by the bottom curve in Figure 6-9 which was determined by directly applying Westine's relationship to the data developed by Kineke (1976) for cylinders exposed end-on to spherical charges. Because the specific impulse i imparted to a target close to the charge is caused by momentum in explosive products, it decays with standoff distance inversely as the surface area of a sphere enclosing the spherical source, which equals $4\pi R^2$. (Standoff distance is the distance of the center of the charge to the nearest face of object.) The area of this sphere of engulfed air is larger than the area of the cylinder of engulfed air described above whenever R > ℓ_e. Therefore, the specific impulse i imparted to objects exposed to spherical charges should decrease more rapidly with distance (for R > ℓ_e) for objects exposed to cylindrical charges oriented as in Figure 6-8. This relationship is demonstrated by the steeper slope of the bottom curve (spherical charge) in Figure 6-8. When R equals R_e, the target is in contact with the charge and specific impulse imparted to the target should approach the same value for spherical and cylindrical charges. If one extends the curves in Figures 6-8 to R/R_e equal to one, one can observe that this relationship holds within the scatter of the data.

Baker (1966) performed a similar analysis in order to predict normally reflected impulses close to spherical explosive charges. He experimentally determined that the scaling law applies for distances corresponding to a mass of engulfed air which is considerably less than (approximately onetenth) the mass of the explosive. For a spherical Comp B explosive source, this would correspond to

$$\frac{R}{R_e} = \left[\frac{(0.1)\ \rho_{expl}}{\rho}\right]^{1/3} = 5.07$$

Examining the bottom curve in Figure 6-9 one will notice a transition in the curve near R/R_e equal to 5.07. However, lack of sufficient data hinders the determination of an accurate experimental transition point. Thus, the curves in Figure 6-9 should not be used at distances beyond those shown by the lines in the figure. For longer distances from the explosive charge, Figure 6-7 can be used.

The straight lines plotted in Figure 6-9 can easily be put in equation form. These predictive equations are:

Cylindrical Charges:

$$\frac{i}{\beta R_{eff}}\left(\frac{R_e}{R_t}\right)^{0.158} =$$

(6-10a)

$$1.320 \times 10^6 \left(\frac{R}{R_e}\right)^{-1} \text{N} \cdot \text{s/m}^3 \text{ for } 0.13 \leq \frac{R}{R_e} \leq 5.25$$

$$\frac{i}{\beta R_{eff}} \left(\frac{R_e}{R_t}\right)^{0.158} = \tag{6-10b}$$

$$4.59 \times 10^6 \left(\frac{R}{R_e}\right)^{-1.75} \text{N} \cdot \text{s/m}^3 \text{ for } 5.25 \leq \frac{R}{R_e} \leq 10.0$$

Spherical Charges:

$$\frac{i}{\beta R_{eff}} \left(\frac{R_e}{R_t}\right)^{0.158} = \tag{6-10c}$$

$$1.078 \times 10^6 \left(\frac{R}{R_e}\right)^{-1.4} \text{N} \cdot \text{s/m}^3 \text{ for } 0.13 \leq \frac{R}{R_e} \leq 5.07$$

If an explosive other than Composition B is used, the value for impulse i obtained from Figure 6-9 and the previous three equations need to be adjusted as follows:

$$i_{expl} = \frac{\left[\left(\Delta H_{Comp \ B}\right)^{1/2} \rho_{Comp \ B}\right]}{\left[\left(\Delta H_{expl}\right)^{1/2} \rho_{expl}\right]} \left(i_{Comp \ B}\right) \tag{6-11}$$

where ΔH is energy (heat of detonation) per unit mass, ρ is density, subscript "Comp B" represents Composition B explosive, subscript "expl" represents explosive being used, and $i_{Comp \ B}$ is the value of i obtained from Figure 6-9 or the previous three equations.

The second half of Westine's analysis which is included in Appendix D consisted of the development of a method to determine the amount of energy consumed in freeing the appurtenance from its moorings. The strain energy U consumed in fracturing a cantilever beam was estimated by assuming a deformed shape and substituting the appropriate mechanics relationships for different modes of response in both ductile and brittle beams. A number of different solutions resulted which had sufficient similarities to permit generalizations after the strain energies were developed. In particular, it was found that for the four different modes of failure hypothesized (ductile bending, brittle bending, ductile shear, and brittle shear), the strain energy at failure equaled the product of toughness, volume and a constant (see Appendix D). For some modes of failure, the constant was a weak function of the cross-sectional shape of the fragment, but this constant varies very little. A second major conclusion was that the toughness appears to be the only mechanical property of importance. All four solutions give the result that this area under the stress-strain curve times the volume of the specimen times a constant equals the strain energy expended in fracturing the specimen.

After curve fitting to experimental data, Westine (1977) concluded that the velocity of the constrained fragment could be described by

$$\frac{\sqrt{\rho_s}\, V}{\sqrt{T}} = -\,0.2369 + 0.3931 \left(\frac{i\,b}{\sqrt{\rho_s T}\, A}\right)\left(\frac{\ell}{b/_2}\right)^{0.3}$$

$$\text{for } \left(\frac{i\,b}{\sqrt{\rho_s T}\, A}\right)\left(\frac{\ell}{b/_2}\right)^{0.3} \ge 0.602 \qquad\qquad (6\text{-}12)$$

$$V = 0 \;\; \text{for } \left(\frac{i\,b}{\sqrt{\rho_s T}\, A}\right)\left(\frac{\ell}{b/_2}\right)^{0.3} \le 0.602$$

where V is fragment velocity, ρ_s is fragment mass density, T is toughness of fragment material, b is loaded width of beam, ℓ is length of target, A is cross-sectional area, and i is specific impulse.

This pair of equations applies for cantilever beams of any material and any cross-sectional area. To estimate the velocity, the specific impulse i imparted to the beam is an estimate from the standoff distance and line charge geometry using the equation in Figure 6-9. Substituting this impulse, beam properties, and beam geometry into Equation (6-12) gives the fragment velocity (see Figure 6-10). If the quantity

$$\left(\frac{i\,b}{\sqrt{\rho_s T\,A}} \right) \left(\frac{\ell}{b/2} \right)^{0.3}$$

is less than 0.602, the fragment will not break free; hence, its velocity is zero.

An equation similar in format to Equation (6-12) but with different coefficients for slope and intercept can also be used for beams with other boundary or support conditions. Although Westine did not have a large quantity of data available to demonstrate this observation, enough data existed on clamped-clamped beams to show that the coefficients -0.6498 instead of -0.2369 and 0.4358 instead of 0.3931 work better for this boundary condition.

3. Fragments from Glass Window Breakage

The breakage of glass windows through interaction with a blast wave can be a major source of penetrating fragments. Because of this, Fletcher, Richmond and Jones (1973 and 1976) conducted blast experiments to obtain information on glass fragments from breaking window panes. From their statistical analysis of the data, they were able to establish functional relationships among several variables. To use their data in conjunction with the human body penetration work of Sperrazza and Kokinakis

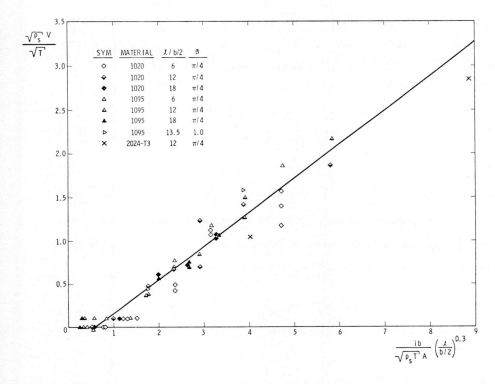

Figure 6-10. Scaled Fragment Velocities for Constrained Cantilever Beams

(1967 and 1974) which are discussed later in this document, it is necessary
to obtain the velocity and area to mass ratio of the glass fragments from
the work of Fletcher, et al. (1971 and 1976). After converting their equations to SI units, the geometric mean frontal area A' of fragments becomes

$$A' = 6.4516 \times 10^{-4} \, e^{\left[2.4 - \sqrt{12.5 + (5.8566 \times 10^{-5} \, P_e)^2} \right]} \qquad (6\text{-}13)$$

for P_e in range 0 to 96.5 kPa, where P_e is effective peak overpressure in
Pascals.

The effective peak overpressure P_e is equivalent to incident peak
overpressure P_s for windows oriented side-on or back-on to the approaching

490

blast wave and reflected peak overpressure P_r for windows oriented face-on to the approaching blast wave. For $\bar{P}_s \leq 3.5$ ($\bar{P}_s = P_s/p_o$ where p_o is atmospheric pressure), $\bar{P}_r$ (= P_r/p_o) can be determined from

$$\bar{P}_r = 2\bar{P}_s + \frac{(\gamma + 1)\, \bar{P}_s^{\,2}}{(\gamma - 1)\, \bar{P}_s + 2\gamma} \tag{6-14}$$

where $\gamma = 1.4$.

Atmospheric pressure p_o can be acquired from Figure 6-11 at sea level and various altitudes above sea level. For $\bar{P}_s > 3.5$, $\bar{P}_r$ can be calculated from Equation (3-5). Effective peak pressure P_e is

$$P_e = \bar{P}_s\, p_o \tag{6-15}$$

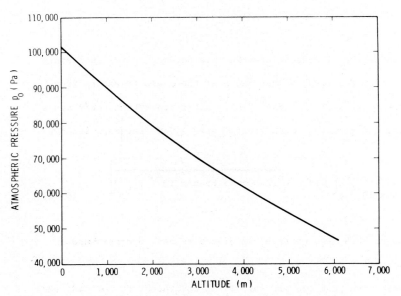

Figure 6-11. Atmospheric Pressure Versus Altitude Above
Sea Level

for windows oriented side-on or back-on the approaching blast wave, or

$$P_e = \bar{P}_r \, P_o \tag{6-16}$$

for windows oriented face-on the approaching blast wave.

If one assumes that all fragments are square, then mass M can be determined from

$$M = \rho y^2 h \tag{6-17}$$

where ρ is the density of glass (2471 kg/m^3), y is the length of a square edge in meters and h is the thickness of the glass in meters.

If all glass fragments travel flat side forward, then

$$A = A' \tag{6-18}$$

If all glass fragments travel edge forward, then

$$A = h \sqrt{A'} \tag{6-19}$$

Thus, for these two cases, the ratio A/M is

$$\frac{A}{M} = \frac{1}{h \, \rho} \tag{6-20}$$

or

$$\frac{A}{M} = \frac{1}{\rho \sqrt{A'}} \tag{6-21}$$

Whichever gives the lower value for A/M should be chosen for safety reasons.

The geometric mean velocity can be acquired from Fletcher, et al. (1973 and 1976). After converting their equations to SI units,

$$V = \left[(0.2539) + (1.826 \times 10^{-4}) \, (h - 7.62 \times 10^{-4})^{-0.928}\right]$$

$$\times \left[0.3443 \, P_e^{0.547}\right] \, m/s \qquad (6-22)$$

for P_e in the range of 690 Pa to 689 kPa and $h \geq 7.62 \times 10^{-4}$ m.

TRAJECTORIES AND IMPACT CONDITIONS

Trajectories

After a fragment has acquired an initial velocity, that is, the fragment is no longer accelerated by an explosion or pressure rupture, two forces act on the fragment during its flight. These are gravitational forces and fluid dynamic forces. Fluid dynamic forces are visually subdivided into drag and lift components. The effect of drag and lift will depend both on the shape of the fragment and its direction of motion with respect to the relative wind. The fluid dynamic force components of drag and lift at any instant can be expressed as

$$F_D = C_D \, (1/2) \, \rho \, V^2 \, A_D \qquad (6-23)$$

and

$$F_L = C_L \, (1/2) \, \rho \, V^2 \, A_L \qquad (6-24)$$

where C_D and C_L are drag and lift coefficients determined empirically as a function of shape and orientation with respect to the velocity vector.

In a simplified trajectory problem [Baker, Kulesz, et al. (1975)]

where the fragment is considered to move in one plane, equations of motion can be written for acceleration in the X and Y directions.

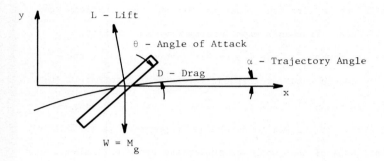

The acceleration in the Y direction is:

$$\ddot{Y} = -g - \frac{A\,C_D\,\rho\,(\dot{X}^2 + \dot{Y}^2)}{2M}\sin\alpha + \frac{A\,C_L\,\rho\,(\dot{X}^2 + \dot{Y}^2)}{2M}\cos\alpha \qquad (6\text{-}25)$$

and for the X direction

$$\ddot{X} = -\frac{A\,C_D\,\rho\,(\dot{X}^2 + \dot{Y}^2)}{2M}\cos\alpha - \frac{A\,C_L\,\rho\,(\dot{X}^2 + \dot{Y}^2)}{2M}\sin\alpha \qquad (6\text{-}26)$$

where A is the area of the fragment, C_D is the drag coefficient, and C_L is the lift coefficient. At t = 0

$$\dot{X} = V_i \cos\alpha_i \qquad (6\text{-}27)$$

$$\dot{Y} = V_i \sin\alpha_i \qquad (6\text{-}28)$$

where V_i is the initial velocity and α_i is the initial trajectory angle.

494

 It is assumed that the fragment is spinning about its Y axis. This motion gives the required stability for flight and allows the fragment to maintain a constant angle with respect to the relative wind. For "chunky" fragments, the coefficient of lift can be set to zero and the equations reduced to the proper form for fragments experiencing drag and no lift.

 The equations shown above can be solved simultaneously using the Runge-Kutta method and can be used for fragment velocities up to Mach 1 for STP conditions, which is around 340 m/s (1100 ft/sec). Baker, Kulesz, et al. (1975) have exercised the program to determine fragment range for a number of conditions. Some of this work was duplicated by Baker, Kulesz, et al. (1978) and put in a more convenient form as shown in Figure 6-12.

 These curves were developed by performing a model analysis to generate dimensionless parameters which describe the general problem, next using the computer code FRISB to determine ranges for selected cases, and then plotting the results to form the curves. It should be noted that, in generating these curves, several initial trajectory angles were used in the analysis to obtain the maximum range for the respective fragments. For ease in understanding the use of these curves, the example which follows

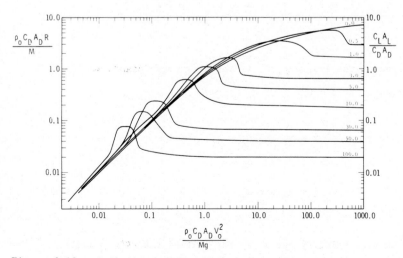

Figure 6-12. Scaled Curves for Fragment Range Prediction

is presented. The procedure for determining fragment range is:

Step 1. Calculate the lift/drag = $C_L A_L / C_D A_D$ for the fragment.

Step 2. Calculate the velocity term = $\rho_o C_D A_D V^2 / Mg$ for the fragment.

Step 3. Select the curve on the graph for the appropriate lift/drag ratio; locate the velocity term on the horizontal axis; find the corresponding range term, $\rho_o C_D A_D R / M$ and determine the range, R.

Note that, for lift to drag ratios $C_L A_L / C_D A_D$ that are not on the curve, a linear interpolation procedure can be used to determine the range from the curve. Interpolation in the steep areas of the curve can cause considerable error and it is recommended that, for these cases, the computer code FRISB be exercised.

In addition to determining fragment range, the above method also gives other impact parameters such as acceleration and velocity at impact and total time of flight. Maximum height attained by the fragment can also be determined if necessary by examining the results of intermediate calculations during the iterative solution generation.

Most of the fragments from explosions are usually "chunky" in shape and have a lift coefficient (C_L) of 0.0. However, in some cases, where one predicts a breakup pattern which involves a large number of plate-like fragments, lift on fragments can become an important consideration. The discussion in Appendix E gives a technique for calculating the normal force coefficients and lift and drag force coefficients of <u>plates</u> having square or circular shapes.

IMPACT EFFECTS

Effects on Structures or Structural Elements

Structures which can be damaged by fragments include frame or masonry residences, light to heavy industrial buildings, office buildings, public buildings, mobile homes, cars, and many others. Damage can be superficial, such as denting of metal panels or breakage of panes of glass. But, massive fragments can cause more extensive damage such as perforation of wooden

roofs, severe crushing of mobile homes or cars, etc. Most of the fragments
will be nonpenetrating and will cause damage by imparting impulsive loads
during impact. The impacts will almost certainly be of short enough duration to be purely impulsive for almost any "target" structure or structural
component. Impact conditions with large fragments which can be certain to
cause significant structural damage can probably also be established by
equating kinetic energy in the fragment to energy absorption capability for
typical roof panels, roof supporting beams, etc.

1. Impact of Fragments Against Thin Metal Targets

The structures that are considered here are metal plates and
sheets. There does not appear to be any effect of the curvature of the
target; therefore, it is reasonable to use data for flat targets and apply
them to any general shape that may be of interest.

The methods which follow [Baker, Kulesz, et al. (1975)] are
based upon an examination of data of fragment and hailstone impact upon
metal sheets and plates [Kangas (1950), McNaughtan and Chisman (1969) and
Recht (1970)]. In these studies, synthetic hailstones (ice spheres) were
fired at target sheets of aluminum alloys, and various shapes of fragments
were fired at steel targets. A model analysis was performed using the

Table 6-4. List of Parameters for Penetration
of Metal Sheets and Plates

a	radius of fragment (assuming spherical shape)
h	thickness of target
V	velocity of fragment
δ	permanent deflection of target at point of impact
ρ_r	density of fragment (projectile)
ρ_t	density of target
σ_t	yield stress of target material

methods described by Baker, Westine and Dodge (1973). The parameters of interest are listed in Table 6-4.

This analysis is concerned with plastic deformation, which makes the parameter σ_t more important than the modulus of elasticity of the target material. Also, the fragment is assumed to be either a rigid body, or a very weak, crushable body, which makes the strength of the fragment an unnecessary parameter. The model analysis and a study of the data result in the nondimensional terms in Table 6-5. When $(\delta h/a^2)$ is plotted versus

$$\left(\frac{\rho_p \, V}{\sqrt{\sigma_t \, \rho_t}} \right),$$

the data follow a straight line with some scatter in the data points (see Figure 6-13). The line intersects the horizontal axis at a positive value of velocity. This is expected because there is a finite fragment velocity below which no permanent target deflection occurs.

For given fragment properties, a given target, and a given normal component of fragment velocity, δ can be obtained. Of course, for very low fragment velocities, there is no permanent deflection.

Table 6-5. Nondimensional Terms for Penetration
of Metal Sheets and Plates

$\left(\dfrac{\rho_p \, V}{\sqrt{\sigma_t \, \rho_t}} \right)$ dimensionless projectile velocity

$\left(\dfrac{\delta h}{a^2} \right)$ dimensionless target deflection

$\left(\dfrac{h}{a} \right)$ dimensionless target thickness

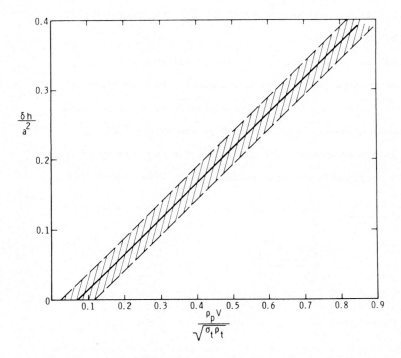

Figure 6-13. Nondimensional Deflection Versus Nondimensional
Velocity for "Chunky," Crushable Fragments

This method was developed for impacts not very close to the edge
of a sheet or plate. For fragment impact near the edge of an unsupported
or simple-supported sheet or plate, the deflection may be twice the deflection that would be otherwise expected.

The V_{50} limit velocity is defined as the velocity at which a
projectile will have a 50 percent chance of penetrating a given target.
Knowing the properties of the projectile (fragment) and the target, V_{50}
can be obtained from Figure 6-14.

In this figure, a is the radius of the fragment (assuming a
spherical shape), h is the thickness of the target, ρ_p is the density of
the fragment (or projectile), ρ_t is the density of the target material, and
σ_t is the yield stress of the target material.

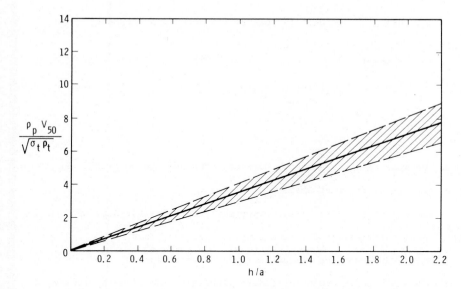

Figure 6-14.   Nondimensional Limit Velocity Versus Nondimensional Thickness
for "Chunky," Non-Deforming Fragments

The solid line in Figure 6-14 gives the relationship between limit velocity and target thickness. As the graph shows, there is uncertainty
in this relation.   For hard fragments which are less likely to deform, a
lower nondimensional limit velocity (more conservative) should be chosen.
For softer fragments, a higher limit velocity can be used.   At this time,
it is not known whether this relationship holds for values of h/a greater
than about 2.2.

This method is good for the impact of a fragment with its velocity normal to the target surface.   For oblique impacts, the normal component of the velocity should be used.   According to one report [McNaughtan
and Chisman (1969)], for oblique impacts, the penetration velocity is minimum at an angle of 30° from the normal direction.   The difference between
the penetration velocities at 0° and at 30° may be as great as 20 percent.
Therefore, if oblique impact is expected, the penetration velocity obtained
by use of Figure 6-14 should be multiplied by 0.8.

This analysis has been formulated for spherical fragments. To apply this to fragments of other shapes, let

$$a = \left(\frac{m}{\rho_p \; \frac{4\pi}{3}} \right)^{1/3} ,$$

where m is the mass of the fragment. More research must be done to determine other effects of fragment shape.

Table 6-6 is a list of the important properties (density and yield stress) of a few selected fragment and target materials.

Baker, Hokanson and Cervantes (1976) have conducted a number of experiments in which solid wooden cylinders with length-to-diameter ratios

Table 6-6. Material Properties

Material	Density ρ kg/m^3	Yield Stress σ Pa
Steel	7850	
1015		$3.46 - 4.49 \times 10^8$
1018		3.66×10^8
1020 (large grained)		4.42×10^8
1020 (sheet)		3.11×10^8
A36		2.49×10^8
Aluminum Alloys (sheet)	2770	
2024-0		8.85×10^7
2024-T3		3.66×10^8
2024-T4		3.66×10^8
6061-T6		2.42×10^8
Titanium Alloy	4520	
6Aℓ4V		1.11×10^9

of 31.1 were impacted end on into thin mild steel targets. Fitting a curve
to the data, they developed the following penetration equation:

$$\frac{\rho_p V_{50}^2}{\sigma_t} = 1.751 \left(\frac{h}{d}\right) \left(\frac{\ell}{d}\right)^{-1} + 144.2 \left(\frac{h}{d}\right)^2 \left(\frac{\ell}{d}\right)^{-1} \qquad (6-29)$$

where V_{50} is the striking velocity for 50 percent perforation, ρ_p is the
density of the projectile, σ_t is the yield strength of the target, h is
the thickness of the target, ℓ is the length of the projectile, and d is
the diameter of the projectile. Limits for Equation (6-29) are:

$$5 \le \ell/d \le 31$$

$$0.05 \le h/d \le 0.10$$

$$0.01 \le \frac{\rho_p V_{50}^2}{\sigma_t} \le 0.05$$

2. Impact of Fragments on Roofing Materials

Nearly any impact of a fragment upon the roof of a building will
cause at least some superficial damage. Damage which only affects the ap-
pearance but which does not interfere with the performance of the roofing
will not be discussed here. Serious damage includes cracking and complete
penetration.

Because of the many kinds of roofing and the scarcity of data of
fragment impact upon roofing materials, the following discussion will be
kept as general as possible, presenting only the lower limits of damage for
groupings of roofing materials, with the understanding that these are not
known very accurately.

An analysis for impact upon metal targets leads one to believe
that the important projectile property is momentum. Until more information

is obtained, it must be assumed that momentum is also important in impact upon roofing materials. [The following discussion is based upon data by Greenfield (1969) in which synthetic hailstones were projected at roofing material targets. The velocities in the tests correspond to the terminal fall velocities of hailstones of the particular sizes used.]

The roofing materials can be separated into three classes: as- phalt shingles, built-up roofs (alternate layers of bitumen and reinforc- ing membranes, often topped with pebbles or crushed stone), and miscellan- eous materials (asbestos cement shingles, slate, cedar shingles, clay tile, and sheet metal). Lower limits of fragment momentum for serious damage to common roofing materials are given in Table 6-7. For oblique impact, the component of the velocity normal to the surface of the roof should be used in the calculation of momentum.

In general, any fragment which strikes a roofing material will probably exceed the momentum required to produce serious damage. This is true because most of the fragments will be large, drag-type fragments, ex- periencing little or no lift which might allow it to "settle" on the roof. To determine the vertical component of the striking velocity V_{y_f} (given the initial vertical component of velocity V_{y_o} of the fragment) for the simple case where $y_o = y_f$ and V_{y_o} is greater than 0, the following equation may be used:

$$V_{y_f} = -\sqrt{Mg/K_y}\ \sin\left[\tan^{-1}\left(V_{y_o}\sqrt{K_y/Mg}\right)\right] \qquad (6\text{-}30)$$

where M is the fragment mass, g is the gravity constant, V_{y_o} is the initial vertical component of velocity, and K_y is equal to $C_D A_D \rho/2$ where C_D is the drag coefficient, A_D is the area presented in the vertical direction, and ρ is the density of air. Impact conditions for other cases may be estimated using numerical approximations to solve the equations of motion.

Table 6-7. Fragment Impact Damage for Roofing Materials

[Greenfield (1969)]

Roofing Material	Minimum Fragment Momentum For Serious Damage (mv) kg m/s	Comments
Asphalt shingles	0.710	Crack shingle
	6.120	Damage deck
Built-up roof	<0.710	Crack tar flood coat
	2.000	Crack surface of conventional built-up roof without top layer of stones
	>4.430	With a 14 kg/m^2 top layer of slag, there was no damage up to 4.43 kg m/s, which was the maximum momentum of the test
Miscellaneous		
0.003 m (1/8") asbestos cement shingles	0.710	
0.006 m (1/4") asbestos cement shingles	1.270	
0.006 m (1/4") green slate	1.270	
0.006 m (1/4") grey slate	0.710	
0.013 m (1/2") cedar shingles	0.710	
0.019 m (3/4") red clay tile	1.270	
Standing seam terne metal	4.430	Plywood deck cracked

Aged shingles may sustain serious damage at lower fragment momentum than that which is given in the table. Also, the tests were conducted at room temperature. The limiting momentum would be greater for shingles at a lower temperature.

3. Impact of Projectiles Against Reinforced Concrete Targets

Several tests by various researchers have been conducted to examine the problem of penetration of concrete targets by low-velocity projectiles. Most of these tests were conducted to examine the possibility of
tornado-borne missiles damaging concrete nuclear reactor containment walls.
Since insufficient data are currently available to form an empirical equation to predict penetration or spall threshold velocity for concrete panels,
some of the data of several researchers will be presented to demonstrate the
current state-of-the-art.

Baker, Hokanson and Cervantes (1976) performed several experimental tests in which model wooden utility poles or model schedule 40 pipe were
impacted normal to the center of concrete test panels. As was the case for
their tests with sheet steel impacts, the targets were sufficiently large to
confine residual deformation to the central portions of the target. Deformation profiles were somewhat irregular, with discontinuities near fracture
planes, but were quite similar in nature to profiles reported by Vassalo
(1975). Although the limited number of tests conducted precluded repeat
shots, the formation of spall craters and fragments appeared to be consistent from test to test. Baker, Hokanson and Cervantes later reviewed available test data obtained both in the United States and abroad, and summarized
the available data. Their summary is included here.

a. Steel Pipe Missiles

There are a total of 66 tests in which steel pipe missiles
were impacted against concrete panels. Nine tests for short pipes (6.06 <
ℓ/d < 9.74) were reported by Vassalo (1975). Stephenson (1976) and Stephenson, et al. (1975) present the results of 14 full-scale tests in which long
pipes (13.8 < ℓ/d < 36.3) were propelled by a rocket sled into concrete
panels. The most complete series of tests is reported by Jankov, et al.
(1976), where the results of 36 quarter scale (4.5 < ℓ/d < 24.3) missile
penetration tests are presented. Unfortunately, about half of these tests
were conducted against panels which had been impacted previously. The re-

maining seven tests were reported by Baker, Hokanson and Cervantes (1976). This series of tests is unique in that the panels are thought to be more representative of actual concrete containment structures than the panels in any of the previously cited reports. This is because the rebars were spaced very close together. The model pipe missile not only could not pass between the rebars, but impacted against at least four wires. Three of the tests were for pipe missiles with a 30° nose angle. The results of these three tests indicate that a considerable amount of the projectile kinetic energy is expended in deforming the nose of the missile, leaving less energy to deform the target. Apparently, blunt-end pipe missiles represent the most severe threat to concrete panels.

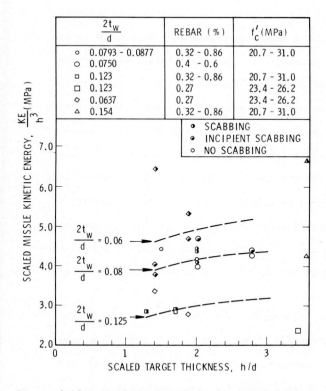

Figure 6-15. Scabbing Threshold for Steel Pipes on
            Reinforced Concrete Panels

Figure 6-15 presents the scabbing threshold for blunt-end steel pipe missiles penetrating reinforced concrete panels. The format of the figure is the same as originally presented by Jankov, et al. (1976); scaled kinetic energy versus scaled target thickness. Note that the vertical axis is dimensional. This axis could have been made nondimensional by dividing KE/h^3 by the strength of the concrete F_c' or by some other material strength. This was not done in Figure 6-15 because the appropriate material strength should not only be the concrete strength, but should be a combination of concrete rebar, and perhaps, missile strength. In Figure 6-15, only the tests which appear to be close to the scabbing threshold are presented. The data appear to group into three sets depending on the wall thickness to diameter ratio $(2t_w/d)$ of the missile. The dashed lines on the figure are used to aid the reader in visualizing the separation of the data into groups. Clearly, not enough data are available to predict, with any degree of confidence, the scabbing threshold of pipe missiles impacting concrete panels. Incidentally, the data from Baker, Hokanson and Cervantes (1976) are not presented because these points were closer to the penetration threshold. However, the indication is that more kinetic energy was required to initiate spalling for these panels $(2t_w/d = 0.140$, rebar percent = 0.485) than would be indicated by the $2t_w/d = 0.125$ line on Figure 6-15. Since the scaled wall thickness of the missile was greater, less kinetic energy should be required to initiate spalling. Evidently, the presence of the larger quantity of rebar is the cause of this apparent anomaly. Further efforts in this area of research should include systematic investigations of the significance of the influence of different amounts of rebar.

Rotz (1975) has presented a simple empirical equation for predicting the thickness of threshold of spalling, T, for steel pipe missiles. We do not include this formula here because it is dimensional and can be shown to give inaccurate predictions for tests conducted at any other scale than the small series on which Rotz based the equation. In

later unpublished work, however, Rotz has modified his equation to make
it dimensionally homogeneous.

      b.   Utility Pole Missiles

         There is a total of 15 tests in which model utility poles
were fired against concrete panels. Nine tests are reported by Vassalo
(1975), Stephenson (1976), Stephenson, et al. (1975), and Jankov, et al.
(1976), while four tests are given by Baker, Hokanson and Cervantes (1976).
The remaining tests are for composite concrete and steel panels reported by
Ting (1975). Spall damage of any level was observed in only three tests.
Two of these tests [Jankov, et al. (1976)] were conducted at velocities
well above the postulated velocity of a tornado-accelerated utility pole.
The other test in which spallation occurred [Baker, Hokanson and Cervantes
(1976)] was the only one in which the projectile did not fail on impact.
Apparently, the utility pole missile is not a threat to the type of con-
crete walls found in nuclear containment structures.

      c.   Rod Missiles

         A total of 75 tests have been conducted in which solid
steel rod projectiles were fired at concrete targets. Twenty-six of these
tests were conducted in the United States by Barber (1973), Vassalo (1975),
Ting (1975), Stephenson (1976), and Jankov, et al. (1976). Barber, Ting
and Stephenson's data are for long ℓ/d (16.8 to 38.7) rods while Jankov,
et al. and Vassalo's data are for relatively short ℓ/d (1.75 to 4.0) rods.
The remaining 49 tests were conducted by the French research establishment.
Gueraud, et al. (1977) presents nine tests for long ℓ/d (15 to 40) rods,
Fiquet and Dacquet (1977) presents 22 and Goldstein and Berriaud (1977)
presents 18 short ℓ/d rod tests. The panels from these three references
had much heavier reinforcing than did the panels of other researchers. In
many cases, five layers of rebar were used, each layer more closely spaced
than the layers found in the American panels. The influence of the heavier rebar, alluded to previously in the section on pipe missiles, is clearly shown in Figure 6-16. The more densely reinforced panels require con-

508

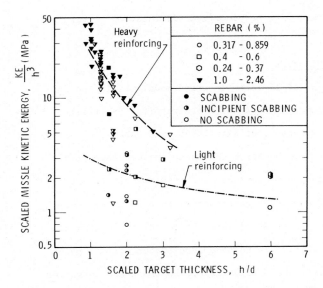

Figure 6-16.   Scabbing Threshold for Solid Rod Missiles
Impacting Reinforced Concrete Panels

siderably more kinetic energy to induce scabbing than do panels which are
lightly reinforced.   The lines on Figure 6-16 are shown to emphasize the
difference between heavily and lightly reinforced panels.

    4.   Spallation

        Westine and Vargas (1978) have developed a model to predict in-
cipient spallation from targets which are struck by fragments whose cross-
sectional width at impact is much less than the lateral dimensions of the
target.   In their analysis, they consider as a worst case a cylindrical
fragment strikes a plate normally (at zero angle of obliquity) as shown in
Figure 6-17.

        The high pressures associated with this impact process send a
stress wave into the material in a fashion similar to that described for
the air blast wave.   The major differences in this impact are that the time
histories of the stress waves are not necessarily triangular.   Waves now go
up the fragment as well as into the target, and some wave dissipation occurs
because the loading is applied locally, rather than uniformly, to a surface.

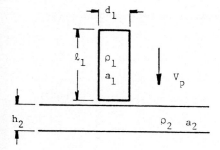

Figure 6-17.   Sketch of a Fragment Impact

Mathematically, the solution to this problem is not an easy one; however, dimensional analysis, physical reasoning, and test data can be applied to develop an empirical solution which designers can use to determine the threshold of spall.

Using dimensional analysis, the threshold of spall can be determined from

$$\frac{P}{\sigma} = f\left(\frac{ia}{Ph_2}, \frac{d_1}{h_2}\right) \tag{6-31}$$

where P is the peak contact pressure, σ is the ultimate strength of the target material, i is the specific impulse imparted to the target, a is the speed of sound in the target material, h is the target thickness, and d_1 is the impact diameter of the fragment.

This relationship essentially states that the peak stress relative to the ultimate strength of the target material is some function of the duration of loading (i/P) relative to the transit time (h/a) for a wave through the material; and, for nonuniform loadings as in Figure 6-17, a function of the relative dimensions of the fragment and the target (d_1/ h_2). The ratio (ia/Ph) is, in fact, the number of transits of the wave through the target material before the fragment comes to rest.

For an impact as in Figure 6-17, the conservation of mass, momentum, and energy equations allow one to determine the peak contact pressure P at impact as

$$P = \frac{\rho_2 a_2 V_p}{\left(1 + \frac{\rho_2 a_2}{\rho_1 a_1}\right)} \tag{6-32}$$

where the subscript 1 denotes the fragment and the subscript 2 denotes the target and all symbols are defined in Figure 6-17. The specific impulse i imparted to the target equals the momentum per unit area in the fragment and is given by Equation (6-33).

$$i = \rho_1 \ell_1 V_p \tag{6-33}$$

Substituting for i and P in Equation (6-31) and cancelling terms gives

$$\left[\frac{\rho_2 a_2 V_p}{\sigma_2 \left(1 + \frac{\rho_2 a_2}{\rho_1 a_1}\right)}\right] = f\left[\frac{d_1}{h_2}, \frac{\rho_1 \ell_1}{\rho_2 h_2}\left(1 + \frac{\rho_2 a_2}{\rho_1 a_1}\right)\right] \tag{6-34}$$

The three parameter space of nondimensional numbers defined by Equation (6-34) can be given a functional format empirically using test data. Basically, the influence of the parameter (d_1/h_2) is to geometrically reduce or dissipate the scaled peak pressure

$$\bar{P} = \left[\frac{\rho_2 a_2 V_p}{\sigma_2 \left(1 + \frac{\rho_2 a_2}{\rho_1 a_1}\right)}\right] \tag{6-35}$$

So, Equation (6-34) can be expressed as

$$\frac{\left(\dfrac{d_1}{h_2}\right)^N \rho_2 a_2 V_p}{\sigma_2 \left(1 + \dfrac{\rho_2 a_2}{\rho_1 a_1}\right)} = f\left[\frac{\rho_1 \ell_1}{\rho_2 h_2}\left(1 + \frac{\rho_2 a_2}{\rho_1 a_1}\right)\right] \qquad (6\text{-}36)$$

Experimental test data were plotted and the coefficient N was found to equal 0.4. Figure 6-18 is a plot based on an unpublished compilation of test data by Hokanson. Three different material combinations with a wide range in impedance mismatches are included in Hokanson's data. The data point, on which Figure 6-18 is based, demonstrates a broad range of target thickness to fragment diameter ratios (h_2/d_1). A factor of 2.0 has been added in the denominator of the abscissa because striker plates give rectangular rather than triangular pulse histories. For rectangular pulses, the quantity (T_{wave}/T_{trans}) equals 1/2 this quantity in a triangu-

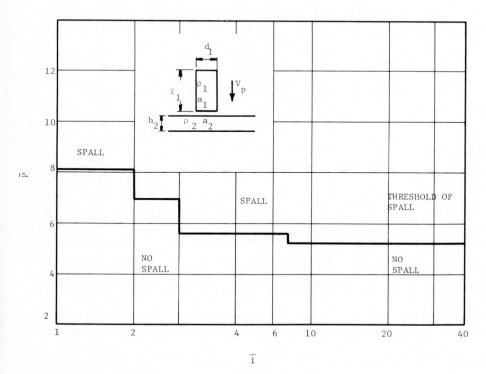

Figure 6-18. Spall Threshold for Fragment Impacts

lar pulse, so the factor 2.0 appears in the denominator to retain the

(T_{wave}/T_{trans}) significance associated with the abscissa.

A solid line which has discrete steps in it separates the spall regime from the no spall regime. Notice that each step occurs at the end of each wave transit. The step is assumed to be caused by subsequent reinforcement with each reflection from the front face of the armor as more energy is added. After about three reflections, the significance of each reflection becomes less important until finally some limiting threshold for an infinite number of reflections is reached at

$$\bar{P}_1 = \bar{P}\left(\frac{d_1}{h_2}\right)^{0.4} = \left[\frac{(d_1/h_2)^{0.4} \; \rho_2 a_2 V_p}{\sigma_2\left(1 + \frac{\rho_2 a_2}{\rho_1 a_1}\right)}\right] = 5.25 \qquad (6\text{-}37)$$

To use the results presented in Figure 6-17, one computes the two dimensionless quantities

$$\bar{P}\left(\frac{d_1}{h_2}\right)^{0.4} \quad \text{and} \quad \left[\frac{\rho_1 \ell_1}{2\rho_2 h_2}\left(1 + \frac{\rho_2 a_2}{\rho_1 a_1}\right)\right]$$

and plots the result. If the computed point falls above the threshold, spall should result, but if it falls below the threshold, no spall should occur.

The result which has been presented is developed for normal (zero angle of obliquity) impacts. Oblique impacts can be computed using detailed computer programs to trace waves; however, normal impacts are a worst case. All oblique impacts result in lower stress levels than from normal impacts.

For a very quick and crude rule-of-thumb estimate of the effectiveness of reinforced concrete panels in resisting penetration by steel fragments, it can be assumed that one inch of mild steel is equivalent to

nine inches of concrete, i.e., if it is known that a one-inch thickness
of mild steel will defeat a particular fragment threat, it can be estimated that nine inches of reasonable quality reinforced concrete will also defeat the fragment. When more realistic estimates of concrete penetration are desired, the above or other suitable methods [Baker, Westine,
et al. (1980)] should be used.

Effects on People

Injury to people due to fragment impact is usually divided into two
categories. One injury category involves penetration and wounding by small
fragments and the other involves blunt trauma by large, nonpenetrating
fragments. Both of these human body fragment injury areas will be examined
in the pages which follow.

1. Penetrating Fragments

Undoubtedly a great deal of research has been conducted to produce classified wound ballistics equations for the military. Although thorough unclassified equations of this type do not exist, some publicly available body penetration data have been accumulated in recent times and some
relatively simple analyses have been performed. More reliable damage criteria will undoubtedly be produced as the state-of-the-art improves.

Sperrazza and Kokinakis (1967) concerned themselves with a ballistic limit velocity V_{50} for animal targets. The V_{50} velocity is the
striking velocity at which one expects half the impacting missiles to per-
forate an object. They found that this velocity depended on the area to
mass ratio, that is

$$V_{50} \propto \frac{A}{M} \qquad (6\text{-}38)$$

where A is cross-sectional area of the projectile along the trajectory,
and M is the mass of the projectile. They fired steel cubes, spheres and
cylinders of various masses up to 0.015 kg into 3-mm thick isolated skin

(human and goat) to establish a ballistic limit. One of their assumptions was that, if the projectile penetrates the skin, its residual velocity would be sufficient enough to cause severe damage. This cautious assumption is appropriate for establishing a certain margin of safety in the calculation. Their conclusions were that, in the range of their data for steel cubes, spheres, and cylinders, V_{50} depended linearly on projectile A/M ratio. Specifically,

$$V_{50} = 1247.1 \ \frac{kg}{m \cdot s} \left(\frac{A}{M}\right) + 22.03 \ m/s \qquad (6-39)$$

for A/M $\leq$ 0.09 m^2/kg, M $\leq$ 0.015 kg where A/M is in m^2/kg, and V_{50} is in m/s.

Kokinakis (1974) fired plastic sabots end-on into 20 percent gelatin that was 1 cm thick. The sabots were fired end-on since this represents the worst case, and 20 percent gelatin was used because this ballistically simulates isolated human skin. The linear relation of V_{50} versus A/M formulated by Sperrazza and Kokinakis (1967) is plotted in Figure 6-19. The average values for these experiments are located on this graph. Circles on the figure represent the initial experiments using steel cubes, spheres and cylinders weighing up to 0.015 kg (0.033 lb_m), and each average value represents as many as 30 data points. The line drawn on the graph is a least-squares fit to these average values. Upward pointed triangles represent the average values for the subsequent experiments with end-on plastic sabots. These average values also lie near the line drawn for the prior study, thus adding a degree of confidence in the analysis.

Unfortunately, other authors have not presented their penetration data in the same form as Sperrazza and Kokinakis. Glasstone (1962) expressed the probability of glass fragments penetrating the abdominal cavity in terms of the mass of the glass fragments. To compare Glasstone's conclusions with that of Sperrazza and Kokinakis, it is necessary to make a few assumptions. The first assumption is that the glass fragment velocity for

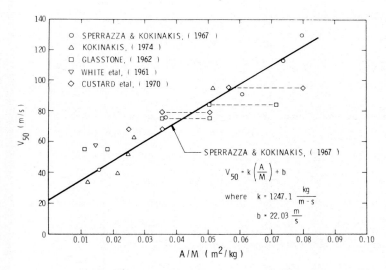

Figure 6-19.   Ballistic Limit (V_{50}) Versus Fragment Area/Mass for Isolated Human and Goat Skin

50 percent probability of penetration of the abdominal cavity is biologically equivalent to the ballistic limit velocity V_{50} for penetrating isolated human skin. Glasstone only specifies the mass of the glass required for penetration and does not give its cross-sectional area, thickness or density. For the purpose of comparing the conclusions of Glasstone with those of Sperrazza and Kokinakis, it was assumed that glass fragments are propelled edge-on, which is probably the worst case, and that they are square with thicknesses of 3.175 mm (1/8 inch) to 6.35 mm (1/4 inch). It was also assumed that the glass fragments have an average density of 2471 kg/m^3 [Fletcher, et al. (1971)]. With these assumptions, it is not difficult to calculate A/M. If the glass fragment has a thickness h, and edge length y, then for volume

$$V = y^2 t \qquad (6\text{-}40)$$

where V is the volume of the fragment, y is the edge length, and h is the thickness.   Thus, the mass of the fragment is

$$M = \rho \, y^2 h \qquad (6\text{-}41)$$

where ρ is the density of the glass.   Rearranging Equation (6-41) gives
the edge length,

$$y = \sqrt{\frac{M}{\rho \, h}} \qquad (6\text{-}42)$$

The area-to-mass ratio A/M, assuming edge-on impact, is

$$\frac{A}{M} = \frac{h \, y}{M} \qquad (6\text{-}43)$$

or from Equations (6-41), (6-42), and (6-43),

$$\frac{A}{M} = \sqrt{\frac{h}{\rho \, M}} \qquad (6\text{-}44)$$

Glasstone's criteria for 50 percent probability of glass fragments penetrating the abdominal cavity are shown in Table 6-8.   This table also contains
the estimates for A/M for glass thicknesses of 3.175 mm (1/8 inch) and 6.35
mm (1/4 inch).   The velocity values and calculated values for A/M which
fall in the range of values used by Sperrazza and Kokinakis are plotted as
squares in Figure 6-19.   The dashed lines indicate a range of A/M values
for thickness values from 2.175 mm (1/8 inch) to 6.35 mm (1/4 inch).   Even
with the crude assumptions mentioned above, the calculated points fall very
near the line drawn on Figure 6-19.

White, et al. (1961) also related skin penetration velocity to
the masses of impacting fragments.   He concluded that slight skin laceration occurred when spherical bullets with mass 0.0087 kg (0.0191 lb$_m$) were
propelled into the body at 57.9 m/s (190 ft/sec).   Assuming that the den-

Table 6-8.   50 Percent Probability of Glass Fragments
Penetrating Abdominal Cavity

[Glasstone (1962)]

Mass of Glass Fragment kg	Impact Velocity m/s	A/M [3.175 mm thick] m²/kg	A/M [6.35 mm thick] m²/kg
0.0001	125	0.1136	0.1603
0.0005	84	0.0507	0.0717
0.0010	75	0.0358	0.0507
0.0100	55	0.0113	0.0160

sity ρ of steel is 7925 kg/m^3, the A/M ratio can be calculated from

$$\frac{A}{M} = \frac{\pi\, r^2}{M} \qquad (6\text{-}45)$$

where r is the radius of the spherical penetrator, or

$$\frac{A}{M} = \frac{\pi}{M} \left(\frac{3M}{4\,\pi\,\rho}\right)^{2/3} \qquad (6\text{-}46)$$

Using Equation (6-46) and the mass and density mentioned above, A/M becomes 0.0148 m^2/kg. The velocity value given above (57.9 m/s) and the calculated value for A/M are plotted on Figure 6-19 as a downward pointed triangle. This point appears to be a little higher than expected, especially since only slight skin laceration is expected at these velocities instead of 50 percent penetration.

Custard and Thayer (1970), like Glasstone, specify velocity as a function of mass only for 50 percent penetration. Making the assumptions that the thickness of the glass can vary from 3.175 mm (1/8 inch) to 6.35 mm (1/4 inch), that the fragments travel edge-on and are square, and that the density of glass is 2471 kg/m^3, A/M was calculated from Equation (6-44). The results are plotted on Figure 6-19 as diamonds and agree fairly well with the conclusions of Sperrazza and Kokinakis. Thus, for values of A/M up to 0.09 m^2/kg and values of M up to 0.015 kg (0.033 lb), the functional relationship expressed in Equation (6-39) and drawn as a solid line in Figure 6-19 is an adequate representation of 50 percent probability of skin penetration by a projectile that can result in serious wounds.

2. Nonpenetrating Fragments

Very limited information for body damage from nonpenetrating objects are contained in Table 6-9. It should be noted that damage is dependent on fragment mass and velocity only. The table also only contains one fragment mass value. One can logically assume that larger masses propelled

Table 6-9. Tentative Criteria for Indirect Blast Effects From
            Nonpenetrating Fragments

[White (1968), Clemedson, et al. (1968), and White (1971)]

Mass	Event	Extent of Damage	Impact Velocity
4.54 kg (10 lb$_m$)	Cerebral Concussion	Mostly "safe"	3.05 m/s (10 ft/sec)
		Threshold	4.57 m/s (15 ft/sec)
	Skull Fracture	Mostly "safe"	3.05 m/s (10 ft/sec)
		Threshold	4.57 m/s (15 ft/sec)
		Near 100%	7.01 m/s (23 ft/sec)

at the same velocities shown in the table will produce more damage than the
4.54 kg (10 lb) mass presented in the table.

      Figures 6-20 and 6-21 contain personnel fragment impact damage
criteria as presented by Ahlers (1969). For fragment weights greater than
4.54 kg, the criteria for threshold head impact injuries are slightly lower

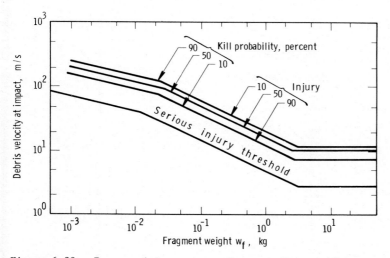

Figure 6-20. Personnel Response to Fragment Impact (Abdomen and Limbs)

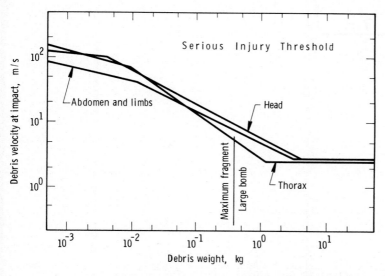

Figure 6-21. Personnel Response to Fragment Impact (Serious Injury Threshold)

520

(more conservative) than those of Table 6-9. The percentage next to a particular curve in Figure 6-20 denotes the percent of people (for a large
sample) that would die if subjected to any of the impact conditions detailed by the curve. The serious injury threshold curves on Figures 6-20 and
6-21 specify the debris velocity and weight combinations below which no
serious injuries are expected to occur.

CHAPTER 6

EXAMPLE PROBLEMS

Cylinder Burst Example (Velocity Calculation)

A cylinder containing helium bursts into 1/4 (M_1) and 3/4 (M_2) of
its total mass M_c. Initial conditions are: M_c = 100 kg, $(P - P_o)$ = 20
MPa, T_o = 300°K, γ = 1.67 (Table 6-3), R_m = 2078 m^2/sec^2 - °K (Table 6-3),
M_1 = 25 kg, M_2 = 75 kg, and V_o = $(\pi/4)$ $(0.2)^2$ (2) = 6.283 x 10^{-2} m^3.

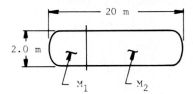

From Figure 6-5,

$$K_1 = 0.67, \quad K_2 = 1.32$$

Using Figure 6-3, calculate the abscissa:

$$\frac{(P_o - P_a) V_o}{M_c \gamma R_m \theta_o} = \frac{(20 \times 10^6)(6.283 \times 10^{-2})}{(100)(1.67)(2078)(300)} = 1.207 \times 10^{-2}$$

Read the ordinate for n = 2, cylindrical:

$$\frac{u}{K \ (\gamma R_m \theta_o)^{1/2}} = 3.5 \times 10^{-2}$$

For Fragment No. 1, K_1 = 0.67 and

$$u_1 = (3.5 \times 10^{-2}) \ (0.67) \ [(1.67) \ (2078) \ (300)]^{1/2} = 23.9 \ m/s$$

For Fragment No. 2, K_2 = 1.32 and

$$u_2 = (3.5 \times 10^{-2}) \ (1.32) \ [(1.67) \ (2078) \ (300)]^{1/2} = 47.1 \ m/s$$

Appurtenance Velocity Example

A 1.0 m^3 cube resting on the ground is struck face-on by a blast wave. Initial conditions are: M = 1000 kg, a_o = 340 m/s, p_o = 100 kPa, A = 1 m^2, K = 4, H = 1 m, X = 0, P_s = 70 kPa, C_D = 1.05 (Table 3-2), and i_s = 3.5 kPa · s (100 ms duration wave).

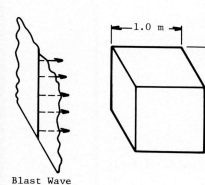

Blast Wave

From Equation (6-6):

$$\bar{P}_s = \frac{P_s}{P_o} = \frac{70 \times 10^3}{100 \times 10^3} = 0.7$$

$$\bar{i}_s = \frac{C_D i_s a_o}{P_s (KH + X)} = \frac{(1.05) (3.5 \times 10^3) (340)}{(70 \times 10^3) [(4) (1) + (0)]} = 4.46$$

Using Figure 6-7, interpolate between $\bar{V} = 1.0$ and $\bar{V} = 5.0$. Estimate that $\bar{V} = 2.5$. From Equation (6-6):

$$V = \frac{\bar{V} p_o A (KH + X)}{M a_o} = \frac{(2.5) (100 \times 10^3) (1) [(4) (1) + (0)]}{(1000) (340)} = 2.9 \text{ m/s}$$

Fragment Maximum Range Example

A steel manhole cover is hurled edge-on at an angle of attack of 10°. Initial conditions are: $V_o = 150$ m/s, $\rho_o = 1.293$ kg/m^3, $M = 200$ kg, $A_1 = (\pi/4) (0.8 \text{ m})^2 = 0.50$ m^2 (Figure E-1), $A_2 = (0.8 \text{ m}) (0.05 \text{ m}) = 0.04$ m^2 (Figure E-1), $C_{L_1} = 0.42$ (Table E-2), $C_{L_2} = -0.36$ (Table E-2), $C_{D_1} = 0.075$ (Table E-2), $C_{D_2} = 2.02$ (Table E-2), and $g = 9.8$ m/s^2.

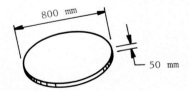

800 mm

50 mm

Use Figure 6-12 and Equations (E-1) and (E-2) (see Table E-2):

Step 1:

$$\frac{C_{L_1} A_1 + C_{L_2} A_2}{C_{D_1} A_1 + C_{D_2} A_2} = \frac{(0.42) (0.50) + (-0.36) (0.040)}{(0.075) (0.50) + (2.02) (0.040)} = 1.65$$

Step 2:

$$\frac{\rho_o \left(C_{D_1} A_1 + C_{D_2} A_2 \right) V_o^2}{Mg} =$$

$$\frac{(1.293) \ [(0.075) \ (0.50) + (2.02) \ (0.040)] \ (150)^2}{(200) \ (9.8)} = 1.76$$

Step 3:

$$\bar{R} = \frac{\rho_o \left(C_{D_1} A_1 + C_{D_2} A_2 \right) R}{M} = 1.25$$

$$R = \frac{\bar{R} M}{\rho_o \left(C_{D_1} A_1 + C_{D_2} A_2 \right)} =$$

$$\frac{(1.25) \ (200)}{(1.293) \ [(0.075) \ (0.50) + (2.02) \ (0.040)]} = 1630 \text{ m}$$

Plate Denting Example

A 2 kg chunk of concrete strikes an aluminum alloy roof panel at 10 m/s. The panel is 1.25 mm thick with $\sigma_t = 300$ MPa, $\rho_t = 2500$ kg/m^3. The concrete density $\rho_p = 2000$ kg/m^3. The effective radius of the fragment is:

$$a = \left(\frac{M}{\rho_p \frac{4\pi}{3}} \right)^{1/3} = \left[\frac{(2) \ (3)}{(2000) \ (4\pi)} \right]^{1/3} = 0.062 \text{ m}$$

Using Figure 6-13, the abscissa is:

$$\frac{\rho_p V}{\sqrt{\sigma_t \rho_t}} = \frac{(2000) \ (10)}{\sqrt{(300 \times 10^6) \ (2500)}} = 0.0231$$

From Figure 6-13, this is <u>below</u> the threshold for damage, so a deflection δ of <u>0</u> is predicted. That is, the plate is not dented.

Plate Perforation Example

A steel ball with radius a of 12 mm strikes the same aluminum alloy plate of the Plate Denting Example at a velocity of 500 m/s. Thus, ρ_t = 2500 kg/m^3, σ_t = 300 MPa, and ρ_p = 8000 kg/m^3. For this example, calculate the thickness h for the limit velocity V_{50} = V (500 m/s). The ordinate of Figure 6-14 is:

$$\frac{\rho_p V_{50}}{\sqrt{\rho_t \, \sigma_t}} = \frac{(8000)\ (500)}{\sqrt{(2500)\ (300 \times 10^6)}} = 4.62$$

From Figure 6-14,

$$\frac{h}{a} = 1.28$$

Thus, h = (1.28) (12) = 15.4 mm of aluminum alloy.

Steel Plate Perforation Example for a Utility Pole Missile

A utility pole with a diameter d of 6.305 m and a length of 4.57 m strikes a steel target with a yield strength σ_t of 2.76 x 10^8 Pa and a thickness h of 25.4 mm. The density ρ_p of the pole is 636 kg/m^3.

What is the threshold velocity V_s for penetration? Using Equation (6-29):

$$\frac{\rho_p V_s^2}{\sigma_t} = 1.751 \left(\frac{h}{d}\right) \left(\frac{\ell}{d}\right)^{-1} + 144.2 \left(\frac{h}{d}\right)^2 \left(\frac{\ell}{d}\right)^{-1}$$

$$= 1.751 \left(\frac{0.0254}{0.305}\right) \left(\frac{4.57}{0.305}\right)^{-1} + 144.2 \left(\frac{0.0254}{0.305}\right)^2 \left(\frac{4.57}{0.305}\right)^{-1}$$

$$= 0.0765$$

$$V_s = \left(\frac{0.0765 \; \sigma_t}{\rho_p} \right)^{1/2}$$

$$V_s = \left[\frac{(0.0765) \; (2.76 \times 10^8)}{(636)} \right]^{1/2} = 188 \; m \; (s)$$

CHAPTER 6

LIST OF SYMBOLS

A, A'	presented area of an object
A_D	drag area
A_L	lift area
a	fragment radius
a, a_2	sound speed in target material
a_1	sound speed in impacting fragment
a_{gas}	speed of sound in a gas
a_o	sound velocity in air
a_o, a_*	gas velocities
b	loaded width of beam
C	mass of gas; a constant
C_D	drag coefficient
C_L	lift coefficient
C_w, C_p	wave speeds
d	projectile diameter
d_1	diameter of impacting fragment
E	Young's modulus
F	an area
F_D	drag force
F_L	lift force
g	acceleration of gravity

g, g_1, g_2	nondimensional displacements
H, $\bar{X}$	dimensions
h	beam depth
h, h_2	thickness of target material
I	total impulse
I_d	drag transverse impulse
I_s	incident specific impulse
I_{st}	total impulse absorbed by structural constraint
i	specific acquired impulse
K	a constant related to relative masses of vessel fragments
k	coefficient of discharge
L	lift
ℓ	beam length; projectile length
ℓ_1	length of impacting fragment
ℓ_e	length of explosive line
M, M_1, M_2, M_t, M_c	masses
M_y	yield moment
N	an exponent
n	number of fragments from a bursting vessel
P	peak contact pressure
P_e	effective peak pressure
P_o, P_1, P_2, P_*	pressures
P_1	modified impact pressure
P_s	peak side-on overpressure
$\bar{P}_s$, $\bar{P}_r$, etc.	barred quantities are nondimensional forms of corresponding dimensional quantities
P_a	atmospheric pressure
q	dynamic pressure
R	gas constant; fragment range; standoff distance
R_e	radius of explosive

R_{eff}	effective radius
R_t	target radius
r	radius of a spherical penetrator
T	toughness
U	strain energy
u	velocity of fragment
V	fragment volume
V, V_i	fragment velocities
V_o	reservoir volume
V_{oo}, V_o	internal volumes in pressure vessel
V_p	fragment impact velocity
V_s	striking velocity
V_{50}	threshold impact velocity for perforation
w_o	maximum deflection
$\dot{X}$, $\dot{Y}$	velocity components
$\ddot{X}$, $\ddot{Y}$	acceleration components
$\dot{x}$	velocity
x, x_1, x_2	displacements
y	beam shape function; dimension of glass fragment; edge length of fragment
Z	plastic section modulus
α, α_1	trajectory angles
α, β	nondimensional groups
β	nondimensional shape factor
γ	ratio of specific heats
δ	permanent deformation
ε_{max}	maximum strain
θ	temperature magnitude
θ_{oo}, θ_o	absolute temperatures
ν	Poisson's ratio
ζ	nondimensional temperature

Π	perimeter of a rupture
π_c	a nondimensional group
ρ, ρ_o, ρ_*	densities
ρ_1, ρ_2	material densities
ρ_p	projectile material density
ρ_r	fragment density
ρ_s	density of structural material
ρ_t	target density
σ	ultimate strength of target material
σ_t	yield stress of target material
σ_y	yield stress
τ	dimensionless time
χ	displacement amplitude
ψ	a function

CHAPTER 7

THERMAL RADIATION EFFECTS

INTRODUCTION

Damaging thermal effects are quite often initiated by accidental explosions, and conversely, fires often cause accidental explosions by "cook-
ing-off" pressure vessels, dangerous chemicals, or explosives. When a
pressure vessel is exposed to a fire, the heat can weaken the strength of
the walls while the heat transfer to the contents of the vessel raises the
interior pressure. This combination of high vessel skin temperatures and
elevated pressures can cause the pressure vessel to fail. Elevated temperatures can cause runaway chemical reactions and can lead to detonation in
explosive material. In fact, the words "fire and explosion," or "explosion
and fire," are often mentioned in the same breath and treated as the same
hazard. In-depth study of fires and their effects is in itself a very
broad discipline, and some topics were discussed in Chapter 1. Here, we
consider the limited topic of thermal radiation effects for those classes
of explosions which produce significant fireballs. Accidental explosions
of most propellants (including solid and liquid propellants), runaway chemical explosions, vessel burst followed by confined or unconfined vapor
cloud explosions, pool fires, detonation of high explosives, and nuclear
blasts all can produce fireballs.

In nuclear weapon explosions, thermal radiation effects can be quite
severe, and can constitute a significant part of the damage caused by such
weapons. Thermal radiation pulse characteristics and thermal radiation effects are discussed at some length in Glasstone and Dolan (1977). Since
1946, because of the importance of the thermal pulse as a nuclear weapon
damage mechanism, most of the methods and data on thermal radiation from

530

explosions are limited to nuclear sources. The reader should be careful to differentiate between the thermal radiation effects from chemical explosions and from nuclear explosions. This disparity will be discussed later, but errors can result in the direct carry over from nuclear weapons technology to fireballs from chemical explosions, detonation of high explosives, etc.

Accidental chemical explosions can produce huge fireballs, as is evident from Figure 7-1, taken from Strehlow and Baker (1976). The fireballs may or may not be accompanied by significant air blast yield. Fireballs as large and long-lasting as the one in Figure 7-1 are both hot

Figure 7-1. Fireball from the Rupture of One Tank Car Originally Containing 120 m^3 of LPG. Crescent City, Illinois, June 21, 1970. Notice the Water Tower on Left and the Train on the Right Side for Scale. Only the Sphere is the Fireball. The Column Leading from the Ground to the Fireball is Dust Sucked up by the Rising Hot Air. [Strehlow and Baker (1976)]

enough and large enough to produce thermal radiation damage, either by igniting combustible material or by direct thermal radiation injury to humans. Immersion in a large fireball has, of course, even a higher probability of causing damage.

A careful review of 81 accidents involving injuries and fatalities in the explosives and propellant industries between 1959 and 1968 [Settles (1968)] indicated that radiant heat was a significant factor. Twenty-three of the 81 accidents involved fire only and another 44 involved both fire and an explosion. Fourteen accidents involved a fireless explosion. Seventy-eight fatalities occurred in these 81 accidents, only one of which was the result of blast overpressures. The one blast fatality which occurred was not the result of blast damage to human tissue. It was caused by air drag carrying this person bodily and slamming him into another object. (This is called tertiary blast injury in Chapter 8.) The other 77 fatalities occurred from either flying fragments or the lethal searing of radiant heat. Although the results of this review must be tempered because of Department of Defense or company proprietary restrictions, it is sufficiently detailed to demonstrate that more attention should be paid to fragmentation and thermal effects.

Unfortunately, the subject of fireball growth and radiant thermal energy causing injury and damage has not been well studied, and only a very few pertinent references are available. The general status of thermal hazards from munitions explosions has been reviewed by Rakaczky (1975), while Gayle and Bransford (1975), High (1968), Bader, et al. (1971), and Hasegawa and Sato (1978) discuss fireballs from liquid propellant and fuel explosions. Lastly, Jarrett (1968) presents British safety distance criteria, part of which are based on thermal radiation effects. Because this subject is incompletely developed, portions of what will be presented involve our conjecture concerning what we believe to be an appropriate (but not yet established) self-consistent approach for interrelating fireball size, fireball temperature, standoff distance, and radiative damage thresholds.

For discussion, this problem is subdivided into three areas. In the first, we discuss fireball growth -- its diameter and temperature time history. The second topic deals with the propagation, by radiation, of thermal energy from a fireball. The third area presents criteria for determining if receivers (e.g., people, materials, propellants, explosives) are damaged by the impinging radiation. All three components (growth of source, propagation of radiant energy, and criteria for damage) must be combined before answers can be obtained.

TRANSIENT FIREBALL GROWTH

Transient fireball growth from an explosion or fire can be studied using similitude or model theory. Provided a list of parameters defining a problem is complete, model theory allows this list to be expressed as a fewer number of nondimensional ratios called pi terms. In this problem involving the transient growth of a fireball, the investigator can start with a list of six parameters and reduce it mathematically to two nondimensional ratios. A solution can then be developed by plotting experimental data to interrelate these two nondimensional ratios.[†]

To determine the transient growth of a fireball, as in a BLEVE, the important physical parameters include the energy release E (which is assumed to be instantaneously released), the diameter D of the resulting fireball as a function of time t, the temperature θ of the fireball (which depends upon the fuel), the heat capacity (ρC_p) of air in the fireball, and the Stefan-Boltzmann constant σ (which accounts for energy radiated to the surroundings). This approach treats conduction and convection processes as

---

[†] Baker, Westine, and Dodge (1973) is an excellent reference for learning more about the use and power of modeling techniques.

being secondary to radiation and the temporary storage of heat within a fireball. The six parameters just discussed are listed in Table 7-1 together with their fundamental units of measure in a force F, length L, time T, and temperature θ system.

Table 7-1. Parameters Defining Fireball Size

Symbol	Parameter	Fundamental Unit of Measure
E	Energy release	FL
D	Diameter of fireball	L
t	Time	T
θ	Temperature of fireball	θ
ρC_p	Heat capacity within the fireball	$F/L^2\theta$
σ	Stefan-Boltzmann constant	$F/LT\theta^4$

Table 7-1 describes a six-parameter space of dimensional quantities. Model theory [Baker, Westine, and Dodge (1973)] allows us to reduce the six-parameter space to a two-parameter space of nondimensional numbers (pi terms). The procedure is an algebraic one which introduces no new assumptions into this solution. Two acceptable nondimensional ratios which are obtained by solving in terms of the exponents associated with σ, ρC_p, E, and θ are presented in functional format as Equation (7-1).

$$\left[\frac{\left(\rho C_p\right)^{1/3}\theta^{1/3}\,D}{E^{1/3}}\right] = \psi\left[\frac{\sigma\,\theta^{10/3}\,t}{\left(\rho C_p\right)^{2/3}E^{1/3}}\right] \qquad (7-1)$$

Equation (7-1) states that a nondimensional or scaled fireball diameter

$$\left[\frac{\left(\rho C_p\right)^{1/3}\theta^{1/3}\,D}{E^{1/3}}\right]$$

is a function of a scaled duration or time

$$\left[\frac{\sigma \theta^{10/3} t}{\left(\rho C_p\right)^{2/3} E^{1/3}} \right]$$

Because only two scaled parameters are involved in Equation (7-1), this equation suggests that results from a single experiment and data from accidents can be made nondimensional and plotted to obtain empirically a single functional relationship. To the best of our knowledge this suggestion has not yet been followed; however, research by various investigators have generated some formulas which certainly have interrelationships close to those suggested by Equation (7-1).

If we continue this analysis by assuming that ρC_p and σ are constants in various accident scenarios, then a simplified dimensional version of Equation (7-1) results.

$$\left[\frac{\theta^{1/3} D}{E^{1/3}} \right] = \psi \left[\frac{\theta^{10/3} t}{E^{1/3}} \right] \tag{7-2}$$

In Equation (7-2) the temperature θ is a function largely of the type of hazard. Propellants have a θ near 2500K; whereas, chemical explosives have a θ nearer 5000K, and flammable gases have a θ around 1350K. A table of temperature values for different hazards could be created if finer breakdowns were required. For most hazardous materials, the fireball reaches a maximum size very rapidly and stays at nearly that size for a long duration until it collapses at a time t*. Equation (7-2), thus, implies if the transient buildup is ignored, that:

$$\frac{\theta^{1/3} D}{E^{1/3}} = A_1 \tag{7-3}$$

and

$$\frac{\theta^{10/3} \, t^*}{E^{1/3}} = A_2 \qquad (7-4)$$

where A_1 and A_2 are constant coefficients. Equation (7-4) can be rewritten in a form for comparison with other work. For a given material, the total energy content is directly related to the total mass of the material, thus, Equation (7-4) can also be written as:

$$\frac{\theta^{10/3} \, t^*}{M^{1/3}} = b \qquad (7-4a)$$

where M is the total mass of "exploding" material and b is a constant.

Equation (7-3) stipulates that equal quantities of explosives and propellants with a factor of two difference in their absolute temperatures will only have a 26 percent difference in the diameter of their fireballs. Equation (7-4), however, states that, because of the strong 10/3 exponent on temperature, cooler propellant burns will last greater than 10 times the fireball duration for an equal weight of explosives. These observations are what is experienced in field tests. Physically, this is reasonable since a hotter object will radiate its energy more quickly than a cooler object.

Rakaczky (1975), in a literature review of explosions in the open, made the following observation. The most generally accepted relation between fireball diameter D (m) and the mass of chemicals in a munition M (kg) is:

$$D = 3.76 \, M^{0.325} \qquad (7-5)$$

The exponent in Equation (7-5) was obtained by least-squares fitting to test data. It is fairly close to the 0.333 exponent theoretically obtained from the model analysis.

Similarly, empirical fits by Rakaczky (1975) for fireball duration t* (sec) yielded:

$$t^* = 0.258 \; M^{0.349} \qquad\qquad (7\text{-}6)$$

Again, this exponent is close to the 0.333 exponent shown in Equation (7-4). No limits of applicability were given on Equations (7-5) and (7-6) so they should be used with caution. They probably apply for fireballs with temperatures of approximately 2500°K. If the coefficients of Equations (7-5) and (7-6) are extended to other materials, one should account for temperature differences, especially in duration, by applying Equations (7-3) and (7-4).

High (1968) also made empirical studies in an attempt to estimate the size and radiation from a Saturn V fireball should an accident occur. His equations for maximum fireball diameter and maximum fireball duration based on tests using liquid rocket propellants are quite similar to Rakaczky's results:

$$D = 3.86 \; M^{0.320} \qquad\qquad (7\text{-}7)$$

$$t^* = 0.299 \; M^{0.320} \qquad\qquad (7\text{-}8)$$

High (1968) also presented plots to demonstrate the validity of his solution. Thus, comparisons are seen in Figures 7-2 and 7-3. The exponent 0.32 in both figures is very close to the 1/3 predicted by theory. If the exponents were changed to 1/3, the coefficients in Equations (7-7) and (7-8) would change slightly.

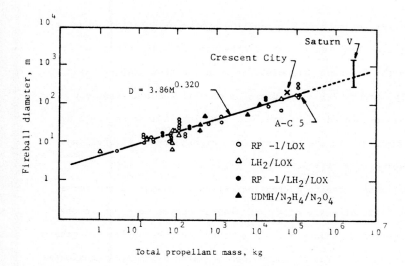

Figure 7-2.   Fireball Diameters for Various Masses and
               Types of Propellants
               [High (1968)]

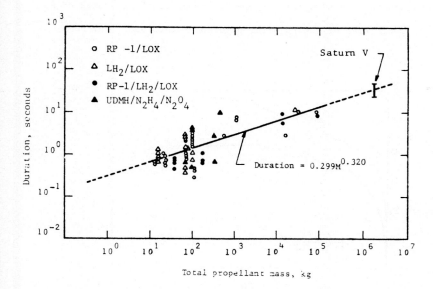

Figure 7-3.   Fireball Duration for Various Masses and
               Types of Propellants
               [High (1968)]

538

Accidents very seldom have a fireball diameter or duration measured, because no one is ever prepared to record observations. The exception to this statement is the Crescent City disaster, where a photographer captured a fireball on film, Figure 7-1. The diameter of this fireball from one tank car BLEVE has been added to Figure 7-2. It plots almost on the line as seen by the symbol x. In the Crescent City disaster, the fireball was 180 meters in diameter for this $6.8 \times 10^{+4}$ kilogram spill of LPG.

High (1968) also presents a transient fireball history curve for a specific accident. His fireball height and radius curves were put in the scaled format suggested by Equation (7-2) so they could be applied to other accidents. The curve for fireball radius can be mathematically approximated with a hyperbola which fits the results excellently and is included as an insert to Figure 7-4.

Often in the area where little research has been performed, some investigators present results which appear to be inconsistent or at odds with another investigator's work. Usually, these "discrepancies" are resolved as a more complete perception of the physical principles involved. For com-

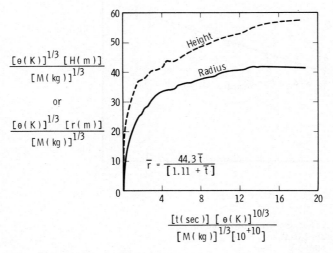

$$\frac{[\theta(K)]^{1/3}\,[H(m)]}{[M(kg)]^{1/3}}$$

or

$$\frac{[\theta(K)]^{1/3}\,[r(m)]}{[M(kg)]^{1/3}}$$

$$\bar{r} = \frac{44.3\,\bar{t}}{[1.11 + \bar{t}]}$$

$$\frac{[t(sec)]\,[\theta(K)]^{10/3}}{[M(kg)]^{1/3}[10^{+10}]}$$

Figure 7-4. Geometric Relationship for Transient Size
of a Fireball

pleteness, an alternate, but accepted and used in some circles, estimate of
fireball duration will be presented.

Other investigators [Bader, et al. (1971) and Hasegawa and Sato (1978)]
obtain similar fits to data on fireball diameters, but very different fits to
data on fireball duration. Bader, et al. (1971) estimate fireball duration by
assuming the fireball is initially hemispherical, but, as buoyant forces begin
to act on the hot gases, the fireball rises and becomes spherically shaped.
When all of the fuel has reacted, the radial growth or expansion is assumed
to stop, and buoyancy then causes the fireball to lift off the ground. Very
soon after the fireball lifts off the ground, the fuel is consumed and the
fireball burns out. In other words, they assume that the propellant burnout
time t_b, liftoff time, and fireball duration coincide. Mathematically,
Bader derives a relationship for duration by equating buoyancy force to the
fluid force resisting motion. The buoyancy force is:

$$F_B = \frac{4}{3} \pi r^3 \rho g \qquad (7\text{-}9)$$

and the fluid resisting force is:

$$F_R = \frac{2}{3} \pi r^3 \rho \left[\frac{2}{r} \left(\frac{dr}{dt}\right)^2 - \frac{d^2 r}{dt^2} \right] \qquad (7\text{-}10)$$

In Equation (7-10), the $\dfrac{d^2 r}{dt^2}$ term is the inertial term and the $\dfrac{2}{r}\left(\dfrac{dr}{dt}\right)^2$ term
is the added mass term created by the gaseous fireball displacing air. The
differential equation which results when Equations (7-9) and (7-10) are
equated is:

$$\frac{d^2 r}{dt^2} - \left(\frac{2}{r}\right)\left(\frac{dr}{dt}\right)^2 + 2g = 0 \qquad (7\text{-}11)$$

The solution to Equation (7-11) is given by:

$$r = \frac{2}{3} \left(\frac{g}{2} \right) t^2 \tag{7-12}$$

Equation (7-12) states that the radius grows as a constant (2/3) times what would be free fall distances, $\frac{g}{2} t^2$. If Equation (7-3) for diameter is substituted into Equation (7-12), then the time t_b when the radius is a maximum, the fuel is consumed, and the fireball burns out is given by:

$$t_b = C \frac{E^{1/6}}{\theta^{1/6}} \tag{7-13}$$

where C is a constant.

The Japanese [Hasegawa and Sato (1978)] compiled experimental test data for small propane, pentane, and octane fireballs. Their least-squares curve fit to test data for time of liftoff gives the relationship:

$$t_b \text{ (sec)} = 1.07 \ M \ (kg)^{0.181} \tag{7-14}$$

for the gases involved, as can be seen in Figure 7-5. The experimental 0.181 is close to the 0.167 exponent predicted by Equation (7-13), and within the scatter in Figure 7-5.

The differences in Figures 7-3 and 7-5 raise the question of which relationship should be used for estimating fireball duration. The answer is that either can be used without causing great inaccuracies. Fireball duration is very difficult to estimate because no sharp abrupt end occurs as the fireball cools down. Differences in definition of duration by various investigators mean that only approximate answers are possible. Table 7-2 compares Equation (7-14) using Figure 7-5 durations to Equation (7-8) using Figure 7-2. Most of the data in these figures fall between a

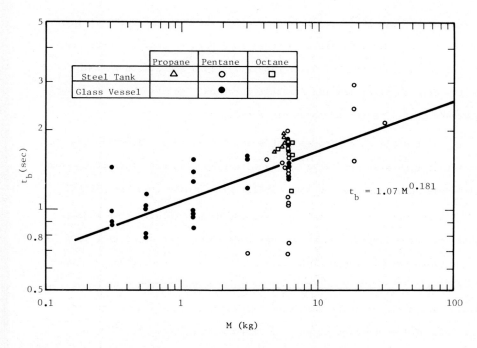

Figure 7-5. Logarithmic Plot of the Duration of Fireball t_b
Versus the Mass of Fuel M
[Hasegawa and Sato (1978)]

Table 7-2. Comparison Between Methods of
Estimating Fireball Durations

Mass M (kg)	Japanese Prediction t_b (sec)	High's Prediction t* (sec)
1.0	1.07	0.30
10.0	1.62	0.627
10^{+2}	2.46	1.31
10^{+3}	3.74	2.74
10^{+4}	5.67	5.72
10^{+5}	8.60	11.94
10^{+6}	13.04	24.95
10^{+7}	19.79	52.13

fuel mass of 0.3 and 10^{+5} kg. In this region, the Japanese estimate larger durations for small quantities of fuel, and High estimates larger durations for large quantities of fuel. Table 7-2 is a comparison between
these two methods of estimating fireball duration. Notice that both give
estimates within a factor of three of each other.

A major difference between High's and the Japanese fireball duration
data is that High's is for large spills with a mass greater than 20 kg, and
the Japanese durations are for small spills with a mass less than 20 kg.

A final comparison of the Japanese and High's solutions to experimentally measured fireball durations was made using unreported test results provided by Mr. Jean Paul Lucotte of SNPE, La Bouche, France. Figure 7-6 is a plot of solid gun propellant mass versus fireball duration.
Both High's and the Japanese solutions are shown in this figure and are
to the left of the observed durations. High's results are for liquid
oxygen, liquid hydrogen and similar liquid propellants which have a 3600°K
fireball temperature rather than the 2500°K temperature which is typical
of gun propellant fireballs. Adjusting High's prediction line for a difference in temperature by using the scaling laws gives the line through
the SNPE test results. A similar adjustment to the Japanese solution only
moves their solution 10 percent which is difficult to see on a log-log
plot. The Japanese solution also moves in the wrong direction because gun
propellants are hotter than propane and similar pressurized flammable
liquids.

Based upon this evidence, we recommend using High's procedure for
large yields, and the Japanese durations only when yields are very small,
less than 10 kg. The effect of flame temperature on duration can be accounted for in Figures 7-2 and 7-5 by using Equations (7-4) or (7-13) to
adjust the duration time appropriately for extremely hot materials such
as an explosive.

One last comment concerning this difference in estimating fireball

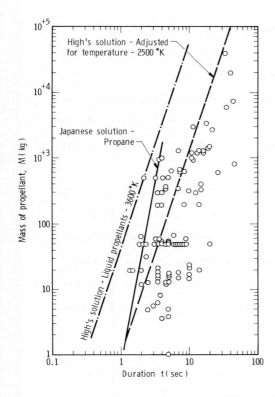

Figure 7-6.   Unreported Fireball Durations for Propellants
            [Provided by French at SNPE]

durations. In our model analysis, conduction and convection were assumed
to be unimportant as heat transfer mechanisms. For small fireballs, and
the growth of small fireballs, this assumption may not be valid. Likewise,
Bader, et al. have made several assumptions in deriving their differential
equations which certainly are not valid in certain regimes. The observation that one approach yields good estimates for very large fireballs, and
the other approach is in agreement for smaller sources, indicates that
these results are the asymptotic or limiting cases of a more general solution. Obviously, this is an area where more research is needed to arrive
at a comprehensive understanding.

One physical effect that has been ignored in the modeling is the
emissivity of the fireball; in other words, how well does the fireball

approximate a true blackbody? For most fireballs of any size, whether they result from an explosive, hydrocarbons, nuclear effects, etc., the fireball is optically thick and the emissivity probably of the order of 0.7 to 1.0. Uncertainties in estimating the emissivity for each source, the relatively small spread in values for practical applications, and the success of the modeling with an "assumed" emissivity of 1.0 warrants the approach which has been taken.

However, there are certain fireballs, hydrogen sources being a very notable example, where the emissivity is very low. For translucent sources such as hydrogen, the fireball results which have been presented are not applicable. So little radiation is generated by hydrogen fireballs that radiant effects outside the fireball are practically nonexistent, although everything inside the fireball could be destroyed by the intense heat.

The "fireball" associated with a pool fire is another class of fireball that can cause thermal radiative damage. Unlike the "BLEVE" fireball, a pool fire is much more of a steady state event. The major problem in a pool fire is predicting the average height of the "fireball." As a first approximation, this is done by equating the energy loss from radiation to the energy in the consumed fuel. Equation (7-15) expresses this relationship:

$$D \, H \, \sigma \, \theta^4 = (\text{Constant}) \, D^2 \, B \qquad\qquad (7\text{-}15)$$

where   H is the average height of the pool fire,

B is the rate at which the fuel is consumed,

D is the diameter of the pool,

σ is Stefan-Boltzmann's constant, and

θ is temperature.

Dividing Equation (7-15) by $D^2 B$ to nondimensionalize it gives:

$$\left(\frac{H}{D}\right)\left(\frac{\sigma\,\theta^4}{B}\right) = \text{Constant} \tag{7-15a}$$

But, Equation (7-15a) ignores convection processes. Convection does occur, and will modify the burning process. Lee and Sears (1959) demonstrate that to study natural convection as would occur in a pool fire, two nondimensional ratios called Prandtl number $\dfrac{C_p\,\mu}{k}$ and Grashof number $\dfrac{D^3\,\rho^2\,g\,\beta\,\theta}{\mu^2}$ need to be included in the analysis. This modification means Equation (7-15) now becomes:

$$\left(\frac{H\,\sigma\,\theta^4}{D\,B}\right) = f\left(\frac{C_p\,\mu}{k},\ \frac{D^3\,\rho^2\,g\,\beta\,\theta}{\mu^2}\right) \tag{7-16}$$

where C_p is specific heat of air,

  μ is viscosity of air,

  k is conductivity of air,

  g is acceleration of gravity, and

  β is coefficient of volumetric expansion.

Lee and Sears (1959) proceed to demonstrate that Prandtl number and Grashof number combine empirically in natural convection studies as indicated by Equation (7-17).

$$\left(\frac{H\,\sigma\,\theta^4}{D\,B}\right) = f\left[\left(\frac{C_p\,\mu}{k}\right)^{1/4} \times \left(\frac{D^3\,\rho^2\,g\,\beta\,\theta}{\mu^2}\right)^{1/8}\right] = \tag{7-17}$$

$$f\left[\frac{C_p^{1/4}\,D^{3/8}\,\rho^{1/4}\,g^{1/8}\,\beta^{1/8}\,\theta^{1/8}}{k^{1/4}}\right]$$

This observation indicates that conductivity and buoyancy effects are important, but natural convection does not depend on viscous effects (μ cancels out of the analysis).

To complete this pool fire analysis, experimental observations made by Hägglund (1977) can be used. Hägglund observed that if the same fuel (fuel oil) was used (that is, if σ, θ, B, C_p, ρ, g, β, and k are held constant), the quantity H/D varies as the inverse of the pool diameter D to the 1/3 power.

$$\frac{H}{D} = \frac{2.6}{D^{1/3}} \qquad (7\text{-}18)$$

Hägglund's observations for fuel oil means that in functional format, Equation (7-17), for any fuel, is given by:

$$\left[\frac{H}{D}\right] \left[\frac{\sigma \left(\rho \ C_p\right)^{2/9} \theta^{37/9} g^{1/9} \beta^{1/9} D^{1/3}}{B \ k^{2/9}}\right] = \text{Constant} \qquad (7\text{-}19)$$

Now, fireball size and duration can be estimated for either a "BLEVE" or a pool fire. The next step is to predict radiation from these fireballs.

PROPAGATION OF THERMAL ENERGY FROM FIREBALLS

The second step in an analysis is to predict the thermal radiant heat flux and, subsequently, the thermal energy per unit area Q at various positions around a fireball. Once again, a model analysis can be used to advantage. If we assume that no atmospheric energy dissipation occurs, the parameters of interest will include: the heat flux q, the diameter of the fireball D, the fireball temperature θ, the standoff distance R, and the Stefan-Boltzmann constant σ which accounts for the radiant energy from the fireball. These parameters, together with their fundamental units of measure, are given in Table 7-3.

A model analysis can, once again, be applied which gives two nondimensional ratios or pi terms, even though there are five parameters and

Table 7-3. Parameters for Transmitting Radiant Energy

Symbol	Parameter	Fundamental Units of Measure
q	Heat flux at some point in space	F/LT
D	Diameter of the fireball	L
θ	Temperature of the fireball	θ
σ	Stefan-Boltzmann constant	$F/LT\theta^4$
R	Standoff distance	L

four fundamental units of measure.[†] This happens because the force F and time T are linearly dependent, or in mathematical terms, because the rank of the matrix used to obtain pi terms is three rather than four. Two pi terms which can be obtained are expressed as Equation (7-20).

$$\left[\frac{q}{\sigma\theta^4}\right] = \psi\left[\frac{R}{D}\right] \tag{7-20}$$

Next, asymptotic relationships can be used to determine a functional format which might be appropriate for Equation (7-15). This is accomplished by deciding how the heat flux q should vary with the standoff distance R when the propagation is in either the near field or the far field. Since σ is a constant, Equation (7-16) can be written as a function which can be inspected to show that the correct limits are approached in both the near and far fields:

$$\left[\frac{q}{\theta^4}\right] = \frac{G\dfrac{D^2}{R^2}}{\left(F + \dfrac{D^2}{R^2}\right)} \tag{7-21}$$

[†]Usually, in similitude analysis, the number of pi terms equals the number of dimensional parameters minus the number of fundamental dimensions.

where G and F are constant coefficients.

When D/R is very large, the thermal flux equation becomes:

$$\left[\frac{q}{\theta^4}\right] = G \qquad \text{(near source)} \tag{7-21a}$$

and whenever D/R is very small, the thermal flux equation becomes:

$$\frac{q\ R^2}{\theta^4\ D^2} = \left(\frac{G}{F}\right) \qquad \text{(distant source)} \tag{7-21b}$$

These limits are appropriate, as near field radiation (D/R large) appears to come from a wall and would be independent of R; whereas, far-field radiation (D/R small) appears to come from a point source and would follow an inverse R^2 law. Thus, Equation (7-21) approaches both limits appropriately. From a limited series of test results by High (1968), for q in $J/m^2 \cdot sec$ and θ in °K, F is equal to 161.7 and G equals 5.26 x 10^{-5}.

The thermal energy per unit area Q can be estimated by integrating Equation (7-21). If fireball size and temperature are treated as constants, q as given by Equation (7-21) can be multiplied by t* as in Equation (7-4a) to predict Q. This procedure yields:

$$\frac{Q}{(bG)\ M^{1/3}\ \theta^{2/3}} = \frac{\dfrac{D^2}{R^2}}{\left(F + \dfrac{D^2}{R^2}\right)} \tag{7-22}$$

High (1968) can also be used to obtain the product (bG). For Q in J/m^2, θ in °K, and M in kg of fuel, this product bG is 2.04 x 10^4 and F still equals 161.7. Table 7-4 was created to compare calculated Q and q to observed results reported by High (1968). Agreement is excellent at a variety of different standoff distances.

Table 7-4. Radiation From Fireball for Saturn V Accident

R(m)	Q_{cal} $(J/m^2) \times 10^{-6}$	Q_{obs} $(J/m^2) \times 10^{-6}$	q_{cal} $(J/m^2 \cdot s) \times 10^{-5}$	q_{obs} $(J/m^2 \cdot s) \times 10^{-5}$
610	3.880	3.880	2.290	2.290
914	1.730	1.730		
1219	0.973	0.970		
1524	0.624	0.624	0.368	0.368
1829	0.433	0.430		

Although Equation (7-21) for q and Equation (7-22) for Q were derived based upon time invariant fireball diameters D, the transient solution would have a very similar format that could be developed if it were
deemed more desirable. For the time variant solution, an empirical curve
fit would be needed for Equation (7-2) and the same similitude analysis
would apply; however, D would now become D(t) and q would be q(t). The
resulting integration to obtain Q from q(t) would be a little more complex,
but the procedure would be the same. The question of how complex a solution is required should await the accumulation of more test data. Fortunately, the same experiments which yield fireball size could also be used
to furnish the G and F transmission coefficients provided thermal flux
gauges were placed at various locations around different fireballs.

If one substitutes Equation (7-3) for fireball size into Equations
(7-21) and (7-22), both free-field radiation solutions become two-parameter
spaces of scaled quantities. Equations (7-21) and (7-22) become:

$$\frac{q}{(G)\ \theta^4} = \frac{1}{\left[1 + \left(\frac{F}{a^2}\right)\frac{R^2}{M^{2/3}}\theta^{2/3}\right]} \qquad (7\text{-}23)$$

and

$$\frac{Q}{(bG)\ M^{1/3}\ \theta^{2/3}} = \frac{1}{\left[1 + \left(\frac{F}{a^2}\right)\frac{R^2\ \theta^{2/3}}{M^{2/3}}\right]} \tag{7-24}$$

Because Equation (7-23) shows that $\frac{q}{\theta^4}$ is a function of $\frac{R\ \theta^{1/3}}{M^{1/3}}$, and Equation (7-24) shows $\frac{Q}{M^{1/3}\ \theta^{2/3}}$ is a function of this same quantity $\frac{R\ \theta^{1/3}}{M^{1/3}}$, Equations (7-23) and (7-24) can be plotted on the same graph.

Figure 7-7 is this plot which can be used for a BLEVE or other violent explosion. Scaled thermal flux q and scaled energy per unit area Q are plotted as functions of scaled distance R.[†] High's results for diameter, Figure 7-2, is the basis for the parameter a in Equations (7-23) and (7-24).

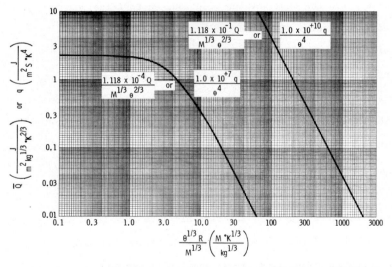

Figure 7-7. Scaled Thermal Flux and Energy for Large Yield Explosions

---

[†]Because fireball thermal emissivity, Stefan-Boltzmann's constant, heat capacity of thermal fireballs, and the energy release per unit mass for various fuels are treated as constants, the equations and graphs are dimension-al rather than nondimensional.

The graph has been folded so all the results can be plotted on the same graph.

Now that q and Q can be determined at various locations, the next step is to determine if receivers will be damaged.

DAMAGE CRITERIA FOR RECEIVERS

A receiver is any object which might be burned. People, explosives, propellants, buildings, and other structures are all examples of receivers. The relationship which will be used as a threshold curve for different amounts of damage is a plot of q, thermal energy flux per unit area, versus Q, thermal energy per unit area.[+] For extremely long duration heat fluxes where the duration is longer than the time required to reach equilibrium, a threshold would be determined solely by the thermal flux rate q. On the other hand, for a very short duration heat flux where the duration is so short that no appreciable energy can be transported away, the threshold would be determined solely by the energy Q. Any values of both q and Q greater than the threshold values would result in damage, and any values of either q or Q less than the threshold would indicate no damage. Figure 7-8 is a qualitative example of a possible q-Q curve for people, explosives, propellants, or materials. Obviously, a q-Q curve could also be presented as a Q versus time t, or q versus t, curve.

The q-Q curve is the thermal counterpart of P-i diagrams for blast waves. Just as the specific impulse i is the time integral of pressure P for short duration loadings, so that heat energy Q is the time integral of the heat flux q for short duration thermal pulses.

One of the most complete threshold criterion for radiation damage presently available is for burning the rods and cones in the eye. This empirically obtained criterion, by Miller and White of Technology, Inc.,

---

[+]For brevity, we will refer to Q as the energy and q as the thermal flux.

552

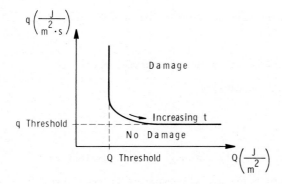

Figure 7-8. Example of q-Q Curve

was obtained by testing on monkeys to interrelate thermal energy Q with

duration t and image size d_i. Figure 7-9 shows their results; the damage

is for partial, and not total blindness because damage only occurs where

the image has been focused. The upper right-hand side of this curve has

little meaning, as the blink time of the eye is on the order of 10^{-2} sec-

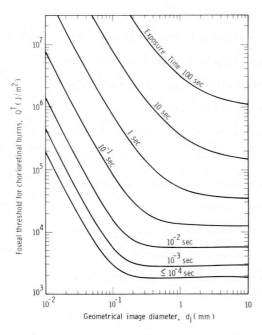

Figure 7-9. Chorioretinal Burn Thresholds for Primates

onds, and a person would react unless he was unaware of possible eye damage.
Notice that for really short durations, less than 10^{-4} sec, and large images,
the criterion is a constant thermal energy of 1.7×10^3 J/m^2. Such a criterion is a time integral of thermal flux q and states that the q-time
history is unimportant for short duration fireballs; only the total energy
being deposited is of importance. For small image areas, more energy Q
can be deposited since some of the energy can be dissipated through conduction into other regions of the eye. For really long durations of 10
to 100 seconds, the criterion in Figure 7-9 is approaching a constant maximum flux q limit which varies dependent upon image size. For long durations, this criterion infers that for a fixed image size, the rate at
which energy is deposited matters whenever the durations are very long.
To use Figure 7-9, the image size can be computed by using geometric pro-
portion optic principles as given by Equation (7-25).

$$\frac{d_i}{\lambda} = \frac{D}{R} \tag{7-25}$$

The parameter λ is the focal length of the eye, approximately 17 mm for
the average person, D is the "diameter" of the fireball, and R is the distance to the fireball.

This very complete curve for loss of eyesight is exactly the type
of q-Q relationship or Q-t relationship that is required to assess thermal
damage (burns) on exposed skin, or for initiating explosive, propellants,
and other hazardous materials.

Another well developed threshold criterion is a q-t curve developed
by Buettner (1950) and used by the Dutch for the initiation of burns on
bare skin. Figure 7-10 is this q-t diagram for separating bearable and unbearable pain domains (a criterion close to second degree burns). Two
lines are drawn because scatter naturally occurs between different individuals. Fifty percent of all observations fall between the two lines seen

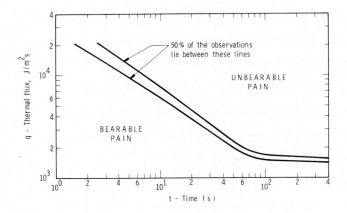

Figure 7-10.   Threshold of Pain From Thermal Radiation on Bare Skin
[Buettner (1950)]

in Figure 7-10.   The experimental criterion used in Figure 7-10 for a per-

son being irradiated is that unbearable pain occurs when a layer 0.1 mm

below the skin's surface exceeds a temperature of 44.8°C.   If this point

is reached, the pain increases sharply, then it declines and later disap-

pears.   This reaction indicates a complete burn in the irradiated skin

area.   An exposure to a thermal flux rate of 1.4×10^3 J/m^{2}s, no matter

how long the exposure is, will not result in pain because an increase in

peripheral blood flow prevents the localized temperature from reaching

44.8°C.   For a given radiation, the painpoint can be reached earlier if

the skin is warmed beforehand, and vice versa.   Notice that for extremely

long durations, Figure 7-10 reaches a constant q criterion, as it should;

and for short durations, the product of q times t (Q) is almost a con-

stant, a result anticipated by our earlier discussion.

We recommend using Figure 7-10 as an exposed skin burn criterion be-

fore using any nuclear weapon data such as that in Glasstone (1977) and

Jarrett (1968).   To extrapolate the nuclear weapon data to predict radia-

tion effects from propellant, explosive, and other chemical reaction fire-

balls is very dangerous, as the wave lengths being transmitted, or black-

body temperatures of the fireballs if one prefers, are orders of magnitude

different. For nuclear weapons, prompt radiation temperatures reach 10^{+7}
K which emit radiated energy in the band of 10^{-2} to 10 nm (0.1 to 100 angstroms).[†] Propellants emit energy at temperatures around 2500K and
explosives at temperatures around 5000K, which is in the area of 200 to
5000 nm (2,000 to 50,000 angstroms).

Several curves for nuclear weapons are available in the literature.
Figure 7-11 is from Jarrett and Figure 7-12 is from Glasstone. Both show
threshold radiant energies for causing burns of various degrees. The ab-
scissae in these figures are explosive yield (at the top of Figure 7-11)

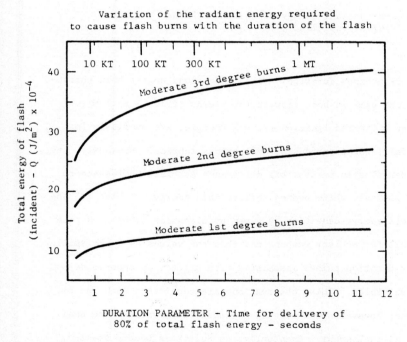

Variation of the radiant energy required
to cause flash burns with the duration of the flash

Figure 7-11.   Nuclear Radiant Energy Exposures for Burns
[Jarrett (1968)]

[†]The much longer duration nuclear fireball radiates at a much lower temperature, approximately 7500K.

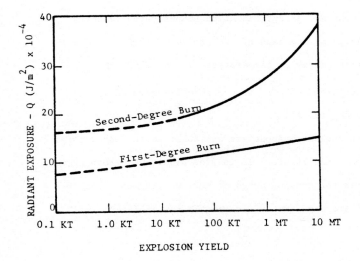

Figure 7-12. Nuclear Radiant Energy Exposure for Burns
[Glasstone (1962)]

which, in turn, determines duration. First and second degree burn thresholds are almost the same in both figures for yields of 10 KT to 1 MT.

Only a few unlimited ignition thermal energies are available for
household materials, forest products, and various fabrics. Obviously, the
color of the material also makes a big difference as black bodies absorb
radiant energy, whereas, white bodies reflect this energy. Tables 7-5 and
7-6 list threshold energy densities as given by Glasstone (1962). Notice
that these are all for nuclear weapons and that the values of Q are smaller
with the shorter duration pulses associated with kiloton as opposed to
megaton explosive yields. These observations are consistent with our projected q-Q curves; however, more testing is needed with yields from small
chemical fires and explosions. Eventually, as durations become smaller,
we would expect threshold energy densities Q to reach a constant value.

In addition to tests on materials, various explosives and propellants should be exposed in hazardous threshold experiments. For more data
on humans, chicken skins are often used as a substitute for human flesh.
Until these thermal threshold criteria are better defined, development of

Table 7-5. Approximate Radiant Exposure for Ignition of
House-Hold Materials and Dry Forest Fuels

Material	Mass Per Unit Area	Ignition Exposure (J/m^2) x 10^{-4}	
	g/m^2	** 20 kilotons	** 10 megatons
Dust mop (oily gray)	---	13	21
Newspaper, shredded	68	8	17
Paper, crepe (green)	34	17	33
Newspaper, single sheet	68	13	25
Newspaper piled flat, surface exposed	---	13	25
Newspapers, weathered, crumpled	34	13	25
Newspaper, crumpled	68	17	33
Cotton waste (oily gray)	---	21	33
Paper, bond typing, new (white)	68	63	126
Paper, Kraft, single sheet (tan)	68	29	59
Matches, paper book, blue heads exposed	---	21	38
Cotton string scrubbing mop, used (gray)	---	25	42
Cellulose sponge, new (pink)	1322	25	42
Cotton string mop, weathered (cream)	---	29	54
Paper bristol board, 3 ply (dark)	339	33	63
Paper bristol board, 3 ply (white)	339	50	105
Kraft paper carton, flat side, used (brown)	543	33	63
Kraft paper carton, corrugated edges exposed, used (brown)	---	50	105
Straw broom (yellow)	---	33	71
Excelsior, Ponderosa pine (light yellow)	2976 g/m^3	21	50
Tampico fiber scrub brush, used (dirty yellow)	---	42	84
Palmetto fiber scrub brush, used (rust)	---	50	105
Twisted paper, auto seat cover, used (multi-color)	440	50	105
Leather, thin (brown)	203	63*	126*
Vinyl plastic auto seat cover	339	67*	113*
Woven straw, old (yellow)	440	67*	138*
Dry rotted wood (punk)	---	17	38
Fine grass	---	21	42
Deciduous leaves	---	25	50
White-pine needles	---	25	59
Coarse grass	---	29	67
Spruce needles	---	33	71
Ponderosa pine needles, brown	---	33	75

* Indicates material was not ignited to sustained burning by the incident thermal
energy indicated.

** Approximately a 4-second duration with a 20 kiloton fireball and a 40-second
duration with a 10 megaton fireball.

predictive procedures for fireball size and the resulting radiant emission
will not provide complete answers. All three areas, fireball growth, radiant transmission, and damage criteria for receivers, must be developed
simultaneously if all parts are to be available for producing complete
answers.

Table 7-6.   Approximate Radiant Exposures for
             Ignition of Fabrics

Material	Mass Per Unit Area	Ignition Exposure $(J/m^2) \times 10^{-4}$	
	g/m^2	20 kilotons	10 megatons
Rayon-acetate taffeta (wine)	102	8	13
Cotton chenille bedspread (light blue)	---	16	33
Doped fabric, aluminized cellulose acetate	---	75	147
Cotton muslin, oiled window shade (green)	271	21	46
Cotton awning canvas (green)	407	21	38
Cotton corduroy (brown)	271	25	46
Rayon twill lining (black)	102	4	8
Cotton venetian blind tape, dirty (white)	---	29	50
Cotton sheeting, unbleached, washed (cream)	102	63	126
Rayon twill lining (beige)	102	33	67
Rayon gabardine (black)	203	13	25
Cotton skirting (tan)	170	29	54
Cotton denim, used (blue)	339	33	54
Cotton and rayon auto seat cover (dark blue)	305	33	54
Acetate shantung (black)	102	38	63
Rayon-acetate drapery (wine)	170	38	67
Rayon marquisette curtain (ivory)	68	38	59
Cotton denim, new washed (blue)	339	38	59
Cotton auto seat upholstery (green, brown white)	339	38	67
Rayon gabardine (gold)	237	38	84
Cotton venetian blind strap (white)	---	67	126
Wool flannel, new washed (black)	237	33	67
Cotton tapestry, tight weave (brown shades)	407	67	126
Wool surface, cotton base, auto seat upholstery (gray)	440	67*	147*
Wool, broadloom rug (gray)	237	67*	147*
Wool pile chair upholstery (wine)	543	67*	147*
Wool pile frieze chair upholstery (light brown)	475	67*	147*
Nylon hosiery (tan)	---	21*	42*
Cotton mattress stuffing (gray)	---	33	67
Burlap, heavy, woven (brown)	610	33	67
Rubberized canvas auto top (gray)	678	67*	117*

*   In these cases, the material was not ignited to sustained burning by the radiant
    exposure indicated.

As we have proceeded through this chapter to assess the thermal radiation damage to a receiver, we first characterized the source whether it
be a BLEVE, a pool fire, or an explosion. Next, an expression was derived
for the transport of the radiation; that is, how the intensity of the radiation decays with distance from the source. Finally, after computing
the energy and energy flux at the receiver, we can determine if receivers
(exposed skin, eye damage, heating of materials, chemicals, explosives,
etc.) are damaged. The following example problem illustrates these principles.

CHAPTER 7

EXAMPLE PROBLEM

Assume that a large tank car with $6.8 \times 10^{+4}$ kilograms of propane
in it "cooks-off" in an accident. Predict: 1) the diameter, duration,
and temperature of the resulting "BLEVE," 2) how far away painful burns
can be expected on exposed skin, and 3) how far away some loss of eyesight
can be expected if someone looks at the accident for 1.0 second. Also,
estimate how all three of the questions would be answered if a detonation
of $6.8 \times 10^{+4}$ kilograms (150,000 lb) of high explosive had occurred instead
of a tank car cook-off.

Equation (7-7) can be used to estimate the diameter of the fireball
for a material with a temperature of approximately 3600°K.

$$D = 3.86 \ M^{0.320} = 3.86 \ (6.80 \times 10^{+4})^{0.320} =$$

$$136.0 \text{ meters (3600°K)}$$

Equation (7-8) can be used to estimate the duration for this same 3600°K
fireball.

$$t* = 0.299 \ M^{0.320} = 0.299 \ (6.80 \times 10^{+4})^{0.320} =$$

$$10.5 \text{ seconds (3600°K)}$$

However, the average temperature of the fireball depends upon the fuel,
which for propane will be approximately 1350°K and not 3600°K.

The scaling law, Equation (7-3), is used to estimate the diameter
of the cooler propane fireball and hotter high explosive fireball. High
explosive has a temperature closer to 5000°K as compared to 3600°K, so

560

for propane the diameter is:

$$D = \frac{136 \text{ meters}}{\left(\frac{1350}{3600}\right)^{1/3}} = 189.0 \text{ meters} \quad \text{(Propane BLEVE)}$$

This estimate is essentially the 185.0 meter diameter fireball observed at Crescent City. For a high explosive, the diameter is:

$$D = \frac{136 \text{ meters}}{\left(\frac{5000}{3600}\right)^{1/3}} = 122.0 \text{ meters} \quad \text{(H.E. Explosion)}$$

Similarly, Equation (7-4) is used to scale the durations for the propane and H.E. fireballs. For the propane duration,

$$t* = \frac{10.5}{\left(\frac{1350}{3600}\right)^{10/3}} = 276.0 \text{ seconds} \quad \text{(Propane BLEVE)}$$

and for the H.E. fireball duration

$$t* = \frac{10.5}{\left(\frac{5000}{3600}\right)^{10/3}} = 3.51 \text{ seconds} \quad \text{(H.E. Explosion)}$$

The H.E. explosion is a smaller and shorter duration fireball that is hotter.

To determine the threshold standoff distance where exposed skin will be burned, Figure 7-10 is used to estimate the threshold radiation flux rate. For the tank car with a 276.0 second BLEVE, Figure 7-10 indicates that q equals approximately $1.4 \times 10^{+3}$ J/m^2s. Next, Figure 7-7 must be used to estimate the standoff distance. To do this, the quantity $\left(\frac{10^{+7} q}{\theta^4}\right)$

is computed. For a propane BLEVE, this quantity equals:

$$\left(\frac{q\ 10^{+7}}{\theta^4}\right)_{BLEVE} = \frac{1.4 \times 10^{+3} \times 10^{+7}}{(1350)^4} = 4.21 \times 10^{-3}$$

Using Figure 7-7, then gives $\frac{R\ \theta^{1/3}}{M^{1/3}}$ of 95, which can be solved for R.

$$\frac{R\ \theta^{1/3}}{M^{1/3}} = \frac{R\ (1350)^{1/3}}{(6.8 \times 10^{+4})^{1/3}} = 95.0$$

or

$$R = 351.0 \text{ meters} \quad \text{(BLEVE Burns)}$$

A similar calculation for high explosive indicates that q equals $1.6 \times 10^{+4}$ J/m²s for a 3.51 second fireball, $\frac{q\ 10^{+7}}{\theta^4}$ equals 2.56×10^{-4}, $\frac{R\ \theta^{1/3}}{M^{1/3}}$ equals 385, and R is 919 meters for the smaller, shorter duration but hotter explosive fireball.

Figure 7-9 must be used to determine if any eye damage occurs. Its use is a trial and error procedure because the image size d_i cannot be determined until the distance R has been calculated. If we assume a fairly large image size for a 1.0 sec exposure, the thermal energy threshold Q^T would equal approximately $4.0 \times 10^{+4}$ J/m². Now the quantity $\left(\frac{1.118 \times 10^{-4}\ Q}{M^{1/3}\ \theta^{2/3}}\right)$ can be computed so Figure 7-7 can be used to determine the scaled distance $\frac{R\ \theta^{1/3}}{M^{1/3}}$. For the propane fireball, this quantity is:

$$\left(\frac{1.118 \times 10^{-4}\ Q}{M^{1/3}\ \theta^{2/3}}\right)_{BLEVE} = \frac{1.118 \times 10^{-4} \times 4.0 \times 10^{+4}}{(6.8 \times 10^{+4})^{1/3}\ (1350)^{2/3}} = 8.97 \times 10^{-4}$$

Figure 7-7 indicates that the scaled threshold distance is:

$$\left(\frac{R\,\theta^{1/3}}{M^{1/3}}\right)_{BLEVE} = \frac{R\,(1350)^{1/3}}{(6.8 \times 10^{+4})^{1/3}} = 207.0$$

or

$$R = 764.0 \text{ meters} \qquad (BLEVE \text{ Burns})$$

The image size and subsequent thermal energy Q^T must be checked to determine if our initial assumed value of Q^T was correct. The image size is computed using Equation (7-25).

$$\frac{d_i}{\lambda} = \frac{d_i}{17.0} = \frac{D}{R} = \frac{189}{764}$$

or

$$d_i = 4.20 \text{ mm}$$

For an image size of 4.2 mm, and an exposure time of 1.0 seconds, the thermal energy Q^T equals $4.0 \times 10^{+4}$ J/m^2 which was our assumed value. This check means that our assumption was correct and no iteration is needed.

For the equal weight high explosive charge, a similar calculation would indicate that Q^T equals $4.0 \times 10^{+4}$ J/m^2, $\frac{1.118 \times 10^{-4}\,Q}{M^{1/3}\,\theta^{2/3}}$ equals 3.75 $\times 10^{-4}$, $\frac{R\,\theta^{1/3}}{M^{1/3}}$ equals 320, and R is 764.0 meters. The image size for the high explosive fireball would be 2.71 mm, which for a 1.0 second duration has a Q^T of $4.0 \times 10^{+4}$ J/m^2 which means this calculation is correct.

For the BLEVE, partial loss of eye sight occurs at a larger stand-off distance than exposed skin burns. This behavior occurs because the lens in the eye focuses thermal energy on a small area of the retina. For the explosive detonation, the threshold for partial loss of eye sight is the same distance as for the BLEVE which shows that for this calculation, partial loss of eye sight is not a sensitive function of temperature.   On the other hand, for the hotter explosive detonation, the standoff distance for the threshold of burns on exposed skins is much larger than the threshold for the cooler BLEVE.

For either a BLEVE or an explosive detonation, the threshold standoff distances for these thermal processes are much larger than thresholds for air blast induced damage.   These observations reinforce the conclusions made by Settle and discussed in the introduction to this chapter that radiant heating is generally a more lethal mechanism than blast.

CHAPTER 7
LIST OF SYMBOLS

$A, A_1, A_2, F, G,$ a, b	constants
B	rate at which fuel is consumed
D	fireball diameter
E	energy release
F_B	buoyancy force
F_R	resisting force
g	acceleration of gravity
H	height of fireball
M	mass of reactive material, fuel, or explosive
Q	thermal energy per unit area
q	heat flux
R	standoff distance
r	fireball radius
t	time

t_b	time for maximum fireball radius
t^*	fireball duration
θ	temperature
ρ	density
ρC_p	heat capacity
σ	Stefan-Boltzmann constant
ψ	a function

CHAPTER 8

DAMAGE CRITERIA

GENERAL

In earlier chapters, we have discussed the characteristics of blast waves generated by accidental explosions and the loads they apply to various "targets"; some of the properties of impacting missiles or fragments; and methods of predicting response and damage to the targets. To complete the assessment of hazards, one must, however, know or assume what character and level of damage is critical, or conversely, what level of damage can be tolerated. In other words, one must establish damage criteria for such diverse "targets" as strong and weak structures, plant facilities, structural elements, vehicles, and people.

DAMAGE CRITERIA FOR BUILDINGS

Criteria Based on Bomb Damage Studies

At the end of World War II, many brick homes in and around London had been damaged from German bombing. This warfare provided many data points on actual structures exposed to different size bombs at various standoff distances. Jarrett's (1968) curve fitted an equation with the format:

$$R = \frac{K \, W^{1/3}}{\left[1 + \left(\frac{7000}{W} \right)^2 \right]^{1/6}} \qquad (8-1)$$

to brick houses with constant levels of damage to relate explosive charge weight W and standoff distance R. The constant K in Equation (8-1) changed with various levels of damage. Equation (8-1) can be converted to a P-i diagram, because side-on overpressure and side-on specific impulse can be calculated from R and W. Figure 8-1 is a side-on pressure versus impulse diagram using Jarrett's curve fit to establish the isodamage contours. The existence of both the quasi-static and impulsive loading realms is apparent in the three different isodamage contours shown in this figure. Levels of damage increase as pressures and impulses increase in Figure 8-1. Actually, this P-i diagram for brick homes can be used for damage criteria for other homes, small office buildings, and light framed industrial buildings because these buildings are structurally similar. Jarrett's Equation (8-1), or Figure 8-1 which is equivalent, is the basis for explosive safe standoff codes in the United Kingdom. The curves are more useful than the equation for establishing damage thresholds for industrial explosions, because the curves do not depend on the concept of "TNT equivalence."

The use of a simple overpressure damage criterion for buildings, such as by Brasie and Simpson (1968), is <u>not</u> recommended. This criterion can cause serious <u>overestimate</u> of structural damage from relatively small explosions where impulsive response predominates. Overpressure is also

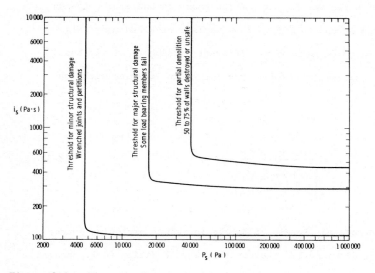

Figure 8-1. Pressure Versus Impulse Diagram for Building Damage

used as a structural damage criterion for blasts from nuclear warheads, which may have durations in seconds for megaton yields [see Glasstone (1962)]. Again, it is usually inappropriate to use damage thresholds from Glasstone for situations where the blast duration is shorter, or even not much longer, than characteristic structural response times.

In the U. S., the Department of Defense Explosives Safety Board is responsible for specifying safe standoff distances in the storage of all military explosives. They presently use criteria of safe standoffs R versus charge weights W given by:

$$\frac{R}{W^{1/3}} = \text{constant} \tag{8-2}$$

where the constant is a function of the class of structure or operation. Table 8-1 is a brief summary of some of the 1980 quantity-distance criteria specified by the Board. We feel that these criteria are inferior to the British criterion, Equation (8-1), because $\frac{R}{W^{1/3}}$ equal to a constant implies constant overpressure criteria (see Table 8-1) which immediately infers response in the quasi-static loading realm. Once again, use of a quasi-static loading realm criterion by government agencies is noted so results will not be misapplied.

Johnson (1967) has proposed a relationship between damage and distance from high explosive detonations of the form

$$\frac{R_{100}}{R_W} = 7.64 \ W^{-0.435} \tag{8-3}$$

where R_{100} is the distance from a 100 lb TNT explosive charge required to achieve some specified level of damage and R_W is distance for equal damage from a TNT charge weight of W pounds. The constants in Equation (8-3) were obtained by a least-squares fit to data for a variety of targets, including

Table 8-1. Some Quantity-Distance Relations from 1980
U. S. Department of Defense Ammunition and
Explosives Safety Standards

Separation Category	Scaled Distance $R/W^{1/3}$, $m/kg^{1/3}$	Side-on Overpressure, k Pa
Barricaded Aboveground Magazine	2.4	190
Barricaded Intraline Operations	3.6	69 - 76
Unbarricaded Aboveground Magazine	4.4	55
Unbarricaded Intraline Operations	7.1	24
Public Traffic Route	9.5	16
Inhabited Building	16 - 20	8.3 - 5.9

aircraft, cantilever beams and wires, 2 1/2-ton trucks, radar antennas, and several sizes of aluminum cylindrical shells. Westine (1972) used Johnson's data, and showed that Equation (8-3) was not consistent with the P-i damage concept, reaching improper asymptotes. He recommended, instead, criteria of the form

$$R = \frac{A\ W^{1/3}}{\left(1 + \frac{B^6}{W} + \frac{C^6}{W^2}\right)^{1/6}} \qquad (8\text{-}4)$$

which can be shown to reach proper asymptotes in the P-i plane. On fitting Johnson's data, Westine showed that Equation (8-4) has significantly less scatter for suitable choices of parameters A, B, and C than Equation (8-3).

Criteria for Onset of Damage or Superficial Damage

The minor damage level shown in Figure 8-1, or significant permanent deformations of more ductile structures, may well be considered completely unacceptable for off-site damage or damage to buildings which normally are occupied by many people. So, other criteria are needed to establish thresholds for damage, or incipient damage levels.

In most buildings, the components which normally fail at the lowest blast loads are windows. Window glass is a brittle material which will shatter when stress reaches the elastic limit. Some types of light siding for industrial buildings are also quite brittle and will shatter on reaching some critical stress. For such materials, criteria for onset of damage are also criteria for catastrophic damage to these building components. Because windows usually have small lateral dimensions between supports, they respond rapidly to blast loads and, therefore, often fall within the quasi-static loading regime for accidental explosions. In this case, use of an overpressure criterion for damage threshold can be appropriate. Mainstone (1971) gives a series of plots for predicting breakage of glass windows by internal gas explosions which are reproduced here as Figures 8-2 and 8-3 for use as quasi-static asymptotes for glass breakage.

To use Figures 8-2 and 8-3, one follows the inserts and the numbered sequence. One reads across from 1, the area of the pane, to 2, the ratio of maximum to minimum face dimension, then up to 3, the "glass factor" (ratio

570

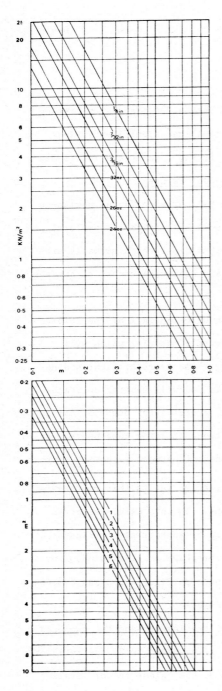

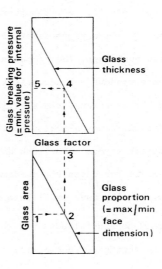

Figure 8-2.   Pressure for Breakage of Sheet Glass Panes
[Mainstone (1971)]

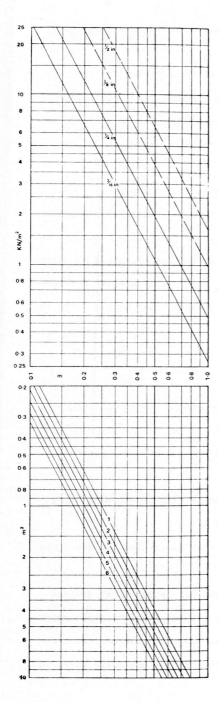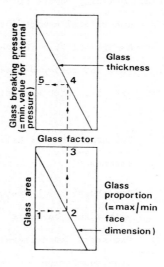

Figure 8-3. Pressures for Breakage of Plate Glass Panes
[Mainstone (1971)]

of area to perimeter) and 4, the nominal thickness, and across again to 5,
the pressure.

Even for ductile structures such as steel-framed or reinforced concrete buildings, one may wish to exclude visible damage and, therefore,
choose a damage criterion of no permanent deformation. Stated differently,
this criterion limits stresses and strains to the elastic regime for the
primary structural material.

Criteria for Significant Permanent Deformation or Damage

The overall building damage curves in Figure 8-1 include several damage levels beyond superficial damage for residential buildings. Another approach to setting damage criteria for residences has been advanced by Wilton
and Gabrielson (1972), who reviewed damage to test dwellings which were exposed to the air blast from high explosives and nuclear air bursts. The test
structures included:

Type I   Two-story, center hall, wood frame house with a full
basement

Type II   Two-story, brick and concrete block, center-hall house
with a full basement

Type III   One-story, wood-frame, ranch-style house on a concrete
slab foundation

Type IV   Two-story, brick apartment house with heavy shear walls
(European-type construction)

Explosive charge yields were all great enough that damage fell in the quasi-
static loading realm.

In assessing levels of damage, Wilton and Gabrielson first divided
each house into certain component groups and evaluated relative replacement
costs of each group, using bid analysis methods common in the construction
industry. The results of this procedure for one type of house are given in
Table 8-2.

Table 8-2. Value of Component Groups for Type I House
            [Wilton and Gabrielson (1972)]

Item	Value (percent of total)
Floor and Ceiling Framing	17.0
Roof Framing and Roof Surface	7.0
Exterior and Interior Wall Framing	16.0
Interior Plaster	11.0
Exterior Sheathing and Siding	8.6
Doors	4.6
Windows	4.8
Foundation and Basement	19.0
Miscellaneous: Stairs, Fireplace, Paint, Trim	12.0
TOTAL	100.0

To perform a damage estimate of a particular house, the quantities of materials damaged (i.e., the number of studs split, the square feet of plaster removed or damaged, the panes of glass broken, etc.) were then computed, and the percentage of each of the elements, shown in Table 8-2, that were damaged or destroyed were determined. By multiplying these percentages by the percent of total value estimate for each of the elements of Table 8-2, a total damage estimate expressed in percent of the total value of the structure was derived. These estimates ranged from about 13 percent to over 80 percent in this particular study, with only the foundation and basement being undamaged for damage percentages over 80 percent.

Residences are almost invariably built with load-bearing walls and of relatively brittle materials. Conversely, many industrial buildings are constructed with steel or reinforced concrete framing which exhibits considerable ductility and can, therefore, undergo rather large plastic deformations without collapse. The concept of allowing some permanent deformation for one-time accidental loads such as explosions is essential to blast resistant design [see Norris, et al. (1959)], and the degree of permanent deformation is often expressed as a <u>ductility</u> <u>ratio</u>

$$\mu = X_m/X_e \tag{8-5}$$

where X_m is maximum deformation and X_e is deformation corresponding to the elastic limit. Healy, et al. (1975), recommended a set of deformation criteria for steel-framed buildings, as follows: (θ_{max} is maximum permanent bend angle at a connection, δ is relative side-sway between stories, H is story height, and L/d is span/depth ratio for a beam element).

(1) Beam elements including purlins, spandrels and girts

   (a)   Reusable structures

       θ_{max} = 1° or μ = 3, whichever governs

   (b)   Non-reusable structures

       θ_{max} = 2° or μ = 6, whichever governs

(2) Frame structures

   (a)   Reusable structures

       For side-sway, maximum δ/H = 1/50

       For individual frame members, θ_{max} = 1°

   NOTE: For σ_y = 36 ksi, θ_{max} should be reduced according to the following relationship for L/d less than 13,

$$\theta_{max} = 0.07\ L/d + 0.09 \tag{8-6}$$

       For the higher yield steels, θ_{max} = 1° governs over the practical range of L/d values.

   (b)   Non-reusable structures

       For side-sway, maximum δ/H = 1/25

       For individual frame members, θ_{max} = 2°

   NOTE: For σ_y = 36 ksi, θ_{max} should be reduced according to the following relationship for L/d less than 13,

$$\theta_{max} = 0.14 \ L/d + 0.18 \hspace{4cm} (8\text{-}7)$$

For the higher yield steels, $\theta_{max} = 2°$ governs over the practical range of L/d values.

(3) Plates

    (a)   Reusable structures

        $\theta_{max} = 2°$ or $\mu = 5$, whichever governs

    (b)   Non-reusable structures

        $\theta_{max} = 4°$ or $\mu = 10$, whichever governs

(4) Cold-formed steel floor and wall panels

    (a)   Reusable structures

        $\theta_{max} = 0.9°$ or $\mu = 1.25$, whichever governs

    (b)   Non-reusable structures

        $\theta_{max} = 1.8°$ or $\mu = 1.75$, whichever governs

(5) Open-web joists

    (a)   Reusable structures

        $\theta_{max} = 1°$ or $\mu = 2$, whichever governs

    (b)   Non-reusable structures

        $\theta_{max} = 2°$ or $\mu = 4$, whichever governs

    NOTE: For joists controlled by maximum end reaction, μ is limited to 1 for both reusable and non-reusable structures.

DAMAGE CRITERIA FOR VEHICLES

Large-scale accidental explosions have often caused extensive damage to vehicles parked nearby or passing by. Usually, the vehicles are cars and trucks. But, in one instance, the massive Texas City explosion in 1947 [Feehery (1977)], the explosion destroyed two light planes which were flying overhead and observing the fire which preceded the explosion.

A number of accidental explosions of relatively small scale have occurred following highway accidents with tanker trucks carrying liquid fuels or combustible chemicals, and with cargo trucks carrying explosives or munitions. Again, other vehicles on the highway or stopped by the accident are damaged by the effects of these explosions.

The damaging effects to vehicles are the direct effects of blast such as window breakage and crushing of the vehicle shell, the secondary blast effect of toppling or overturning of the vehicle, and impact damage from debris hurled by the explosion. Because many accidental explosions also cause large fireballs, vehicles can also be ignited and destroyed by fire.

The military has developed rather detailed damage criteria for military vehicles such as trucks and aircraft. These criteria are not, however, suitable for assessing damage from industrial explosions because they define the ability of the vehicles to function or fail to function in a battle scenario. (A truck can be extensively damaged by blast, for example, and still be capable of running and carrying cargo.) We have not seen any attempt to set damage criteria for vehicles involved in industrial explosions, although Glasstone and Dolan (1977) list criteria for transportation equipment subjected to nuclear air blast (see Table 8-3). Our inclination would be to develop a procedure somewhat similar to that used by Wilton and Gabrielson (1972) for assessing house damage. That is, vehicle damage would be described as a percent of the cost needed for repair or replacement, based on commercial cost estimation practices in the automotive repair business.

Trucks, buses, mobile homes, missiles on the launch pad, and various other objects can be damaged because they overturn when enveloped by a blast wave from an accidental explosion. In Baker, Kulesz, et al. (1975), criteria were developed and prediction graphs prepared to calculate the thresholds for toppling of such "targets." These graphs are presented here.

To determine if a target overturns, we use two different graphs. The first graph, Figure 8-4, allows us to calculate the total average spe-

Table 8-3. Damage Criteria for Land Transportation Equipment
[Glasstone and Dolan (1977)]

Description of Equipment	Damage	Nature of Damage
Motor Equipment (cars and trucks)	Severe	Gross distortion of frame, large displacements, outside appurtenances (doors and hoods) torn off, need rebuilding before use.
	Moderate	Turned over and displaced, badly dented, frames sprung, need major repairs.
	Light	Glass broken, dents in body, possibly turned over, immediately usable.
Railroad Rolling Stock (box, flat, tank, and gondola cars)	Severe	Car blown from track and badly smashed, extensive distortion, some parts usable.
	Moderate	Doors demolished, body damaged, frame distorted, could possibly roll to repair shop.
	Light	Some door and body damage, car can continue in use.
Railroad Locomotives (diesel or steam)	Severe	Overturned, parts blown off, sprung and twisted, major overhaul required.
	Moderate	Probably overturned, can be towed to repair shop after being righted, need major repairs.
	Light	Glass breakage and minor damage to parts, immediately usable.
Construction Equipment (bulldozers and graders)	Severe	Extensive distortion of frame and crushing of sheet metal, extensive damage to caterpillar tracks and wheels.
	Moderate	Some frame distortion, overturning, track and wheel damage.
	Light	Slight damage to cabs and housing, glass breakage.

cific impulse i_t imparted to the target. The second graph, Figure 8-5,

allows us to calculate the average specific impulse i_θ that is the thresh-

old of overturning. If i_t imparted to the target exceeds i_θ, the target

should overturn; however, if i_t is less than i_θ, the loading is insuffi-

cient to overturn the target. Both Figures 8-4 and 8-5 are nondimensional,

so any self-consistent set of units can be used. The scaled total impulse

imparted to a target $\left(\dfrac{a_o \; i_t}{p_o \; H} \right)$ in Figure 8-4 is a function of the scaled free-

field pressure P_s/p_o and the scaled free-field impulse $\left(\dfrac{a_o \; C_D \; i_s}{p_o \; H} \right)$ where p_o

is ambient atmospheric pressure, a_o is ambient sound velocity, P_s is side-

on free-field overpressure, i_s is free-field side-on impulse, H is the

smaller of either the target height or target width, and C_D is an air drag

coefficient. For typical trucks, buses and other vehicles, H is the total

height h of the vehicle. For missiles or other tall narrow objects, H is

the diameter of the missile. The air drag coefficient C_D varies between

1.2 for streamlined cylindrical bodies to 1.8 for long rectangular shapes.

In Figure 8-5 the scaled threshold impulse for just overturning an object

is presented as a function of scaled target height $\left(\dfrac{h}{b} \right)$ and scaled c.g. lo-

cation $\left(\dfrac{h_{cg}}{h} \right)$, where h is the total height of the target, h_{cg} is the height

of the c.g. location, $h_{b\ell}$ is the height for the center of pressure, A is

the presented target area, b is the vehicle track width or depth of tar-

get base, g is the acceleration of gravity, m is the total mass of the

target, and i_θ is the threshold impulse. The analysis assumes that the

target is not initially tilted, the c.g. horizontal location is at $\dfrac{b}{2}$

and the mass is uniformly distributed throughout the target. Use of these

graphs to obtain answers is best presented through an illustrative example

which appears as an Example Problem in this chapter.

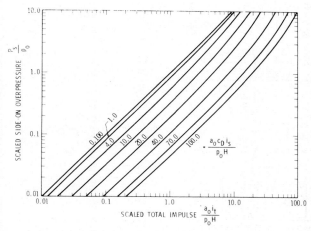

Figure 8-4.   Specific Impulse Imparted to a Target Which Might Overturn

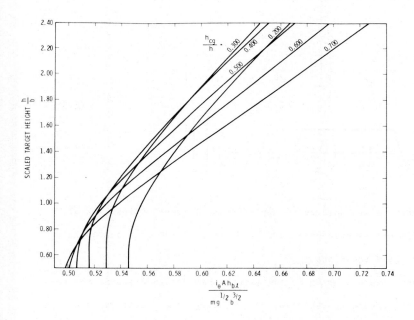

Figure 8-5.   Impulse for Threshold of Overturning

The equation shown graphically in Figure 8-4 is:

$$\frac{a_o i_t}{P_o H} = \frac{1.47 \left(\dfrac{P_s}{P_o}\right)\left(\dfrac{a_o C_D i_s}{P_o H}\right)}{\left[7.0 + \left(\dfrac{P_s}{P_o}\right)\right]} +$$

$$\frac{\left[1.0 + \dfrac{3\left(\dfrac{P_s}{P_o}\right)}{\left[7.0 + \left(\dfrac{P_s}{P_o}\right)\right]}\right]\left(\dfrac{P_s}{P_o}\right)}{\left[1.0 + 0.857 \left(\dfrac{P_s}{P_o}\right)\right]^{1/2}}$$

$$(8\text{-}8)$$

The equation shown graphically in Figure 8-5 is:

$$\frac{i_\theta A h_{b\ell}}{m \, g^{1/2} \, b^{3/2}} = \left\{ \left[\frac{2}{3} + \frac{h^2}{6b^2} + \frac{2h^2}{b^2}\left(\frac{h_{cg}^2}{h^2}\right)\right] \times \right.$$

$$(8\text{-}9)$$

$$\left. \left[\sqrt{\frac{1}{4} + \left(\frac{h^2}{b^2}\right)\left(\frac{h_{cg}^2}{h^2}\right)} - \left(\frac{h}{b}\right)\left(\frac{h_{cg}}{h}\right)\right]\right\}^{1/2}$$

Both equations are derived in Appendix 3A of Baker, Kulesz, et al. (1975). In addition, experimental test data are used there to demonstrate the validity of these equations.

In summary, there seem to be few established damage criteria for accidental explosion damage to vehicles, and military damage criteria are inappropriate.

DAMAGE CRITERIA FOR PEOPLE

Introduction

In the United States, the standards established by the Department of Defense and the Department of Energy for ammunition and explosives safety both consider, in a limited way, the effects of blast and impacts on people. Different levels of accidental exposure of personnel are allowed in the safety standards of these two government departments, depending on the known or expected probability that an accidental explosion will occur. The Department of Defense 1980 standards relate probable personnel injury to the quantity-distance separation categories (shown previously in Table 8-1). These are given in Table 8-4. Note that correlations are only specified for blast overpressure, and impulse does not enter these standards.

The United States Department of Energy specifies three classes of protection for personnel in high explosives storage, handling and processing activities (ERDA Appendix 6301, Facilities General Design Criteria, March 1977). The section of this safety standard is reproduced here as Table 8-5. Note again that only blast overpressure is specified for each of the three safety classes.

The criteria for personnel safety just described differ from criteria in NASA safety workbooks [see Baker, Kulesz, et al. (1975)]. There, the effects of blast overpressure on people are considered in much more depth. The presentation which follows comes essentially from the NASA safety document.

Literature concerning the harmful effects of blast on humans has been published as early as 1768 by Zhar [Burenin (1974)]. However, knowledge of the mechanisms of blast damage to humans was extremely incomplete until World War I, when the physics of explosions were better understood. Since that time, numerous authors have contributed considerable time and effort in the study of blast damage mechanisms and blast pathology. Each accident situation has its own unique environment with trees, buildings,

Table 8-4. Expected Effects to Personnel at Various Overpressure
Exposure Levels

U. S. Department of Defense Ammunition and Explosives Safety
Standards (Source: DOD 5154.45, June 23, 1980)

Overpressure, P_s, kPa (psi)	Q-D Separation Category [R, m for W in kg, (ft for W in lb)]	Personnel Injury
190 (27)	Barricaded Aboveground Magazine $2.4\ W^{1/3}$ $(6\ W^{1/3})$	Occupants of unstrengthened building will be killed by direct action of blast, by being struck by building debris, or by impact against hard surfaces.
69-76, (10-11)	Barricaded Intraline $3.6\ W^{1/3}$ $(9\ W^{1/3})$	Occupants of unstrengthened building are expected to suffer severe injuries or death from direct blast, building collapse, or translation.
55 (8)	Unbarricaded Aboveground Magazine $4.4\ W^{1/3}$ $(11\ W^{1/3})$	Occupants of unstrengthened building will be killed or incur serious injuries by blast to eardrums and lungs, by being blown down or struck by fragments and building debris.
24 (3.5)	Unbarricaded Intraline $7.1\ W^{1/3}$ $(18\ W^{1/3})$	Personnel injuries of a serious nature or possible death are likely from fragments, debris, firebrands or other objects. There is a 10% chance of eardrum rupture.
16 (2.3)	Public Traffic Route $9.5\ W^{1/3}$ $(24\ W^{1/3})$	Personnel may suffer temporary hearing loss or injury from secondary blast effects such as building debris and the tertiary effect of displacement. No fatalities or serious injuries from direct blast are expected.
8.3-5.9, (1.2-0.85)	Inhabited Building $16\ W^{1/3} - 20\ W^{1/3}$, $(40\ W^{1/3} - 50\ W^{1/3})$	Personnel are provided a high degree of protection from death or serious injury with injuries that do occur being principally caused by glass breakage and building debris.

Table 8-5. U. S. Department of Energy Safety Protection Levels
for High Explosives Bays

(Source: ERDA Appendix 6301, Facilities General Design Criteria,
Handbook, March 1977)

In the planning of HE activities to be performed and in the design of HE bays to satisfy these activity requirements, a basic tenet
shall be to limit HE activity hazards exposure to a minimum number of
personnel. Additionally, each bay housing an HE activity must have levels of protection based on the hazard class determined for the activity.
The levels of protection may be accomplished by equipment design, structural design, and/or the provision of operational shields.

The levels of protection required for each hazard class are as
follows:

(a) Class III. Bays for Class III (low accident potential) activities shall provide protection from explosion propagation from bay
to bay within buildings and between buildings which are located at in-
traline or magazine distance. Minimum separation distances may be reduced when HE bays are designed to totally contain the effects of an
accident (blast pressures and missiles).

(b) Class II. Bays for Class II (moderate accident potential)
activities shall, in addition to complying with the requirements for
Class III bays, include design to prevent fatalities and severe injury
to personnel in all occupied areas other than the bay of occurrence.
Prevention of fatalities and severe injuries is satisfied where personnel in occupied areas other than the bay of occurrence will not be exposed to:

(1) Overpressure greater than 15 psi Maximal Effective Pressure.

(2) Structural collapse of the building in the event of an explosion in the HE bay.

For the purpose of this Class II category, access ramps and plant
roads are not considered occupied areas.

(c) Class I. Bays for Class I (high accident potential) activities shall, in addition to complying with the requirements for Class II
bays, provide protection to prevent serious injuries to all personnel,
including personnel performing the activity, personnel in other occupied
areas, and all transient personnel.

Prevention of serious injuries is satisfied where personnel will
not be exposed to:

(1) Overpressures greater than 5 psi Maximal Effective Pressure.

(2) Structural collapse of the building.

(3) Missiles.

This protection may be achieved by controlling debris through
suppression, containment, etc., or by establishing an exclusion area
with positive access control.

hills and various other topographical conditions which may dissipate the energy of the blast wave or reflect it and amplify its effect on an individual. Because of these different variational factors involved in an explosion-human body receiver situation, only a simplified and limited set of blast damage criteria will be included in here. The human body "receiver" will be assumed to be standing in the free-field on flat and level ground when contacted by the blast wave. Excluding certain reflected wave situations, this is the most hazardous body exposure condition. Air blast effects will also be subdivided into two major categories: direct (primary) blast effects and indirect blast effects [White (1968)].

Direct Blast Effects

Direct or primary blast effects are associated with changes in environmental pressure due to the occurrence of the air blast. Mammals are sensitive to the incident, reflected and dynamic overpressures, the rate of rise to peak overpressure after arrival of the blast wave, and the duration of the blast wave [White (1968)]. Specific impulse of the blast wave also plays a major role [White, et al. (1971) and Richmond, et al. (1968)]. Other parameters which determine the extent of blast injury are the ambient atmospheric pressure, the size and type of animal, and possibly age. Parts of the body where there are the greatest differences in density of adjacent tissues are the most susceptible to primary blast damage [White (1968); Damon, et al. (1970); and Bowen, et al. (1968)]. Thus, the air-containing tissues of the lungs are more susceptible to primary blast than any other vital organ [Damon, et al. (1974)]. The ear, although not a vital organ, is the most sensitive. This organ responds to energy levels as low as 10^{-12} watts/m^2 or pressures approximately 2×10^{-5} Pa (2.1×10^{-9} psi). This small force causes an excursion of the eardrum a distance less than the diameter of a single hydrogen molecule [Hirsch, F. G. (1968)].

Pulmonary injuries directly or indirectly cause many of the pathophysiological effects of blast injury [Damon, et al. (1973)]. Injuries include pulmonary hemorrhage and edema [White (1968) and Damon, et al. (1973)], rupture of the lungs [Burenin (1974)], air-embolic insult to the heart and central nervous system [White (1968)], loss of respiratory reserve [White (1968)] and multiple fibrotic foci, or fine scars, of the lungs [Damon, et al. (1970)]. Other harmful effects are rupture of the eardrums and damage to the middle ear, damage to the larynx, trachea, abdominal cavity, spinal meninges and radicles of the spinal nerves and various other portions of the body [Burenin (1974)].

Indirect Blast Effects

Indirect blast effects can be subdivided into three major categories [White (1968)]: secondary effects, tertiary effects and miscellaneous effects.

Secondary effects involve impact by missiles from the exploding device itself or from objects located in the nearby environment which are accelerated after interaction with the blast wave (appurtenances). Characteristics which affect the extent of damage done to a human due to impingement of fragments include the mass, velocity, shape, density, cross-sectional area and angle of impact [White (1968)]. Pathophysiological effects include skin laceration, penetration of vital organs, blunt trauma, and skull and bone fracture [Clemedson, et al. (1968)].

Tertiary effects involve whole-body displacement and subsequent decelerative impact [White (1968)]. In this case, the blast pressures and winds pick up and translate the body. Damage can occur during the accelerating phase or during decelerative impact [Hirsch, A. E. (1968)]. The extent of injury due to decelerative impact is by far the more significant [Glasstone (1962)], and is determined by the velocity change at impact, the time and distance over which deceleration occurs, the type of surface impacted and the area of the body involved [White (1968)]. When the human body is exposed to such acceleration or decelerative impact, the head is the most vulnerable to mechanical injury as well as the best protected area [von Gierke (1971b)]. In addition to injury to the head, vital internal organs can be damaged and bones can be broken as a result of decelerative impact. The impact velocity required to produce a certain percentage of skull fractures is usually less than the impact velocity required to produce the same percentage total body (randomly oriented) impact lethality [White, et al. (1961); Clemedson, et al. (1968); and White (1968)].

Miscellaneous blast effects such as dust and thermal damage are usually considered to be insignificant for conventional blast materials [White (1968)]. However, Settles (1968) discusses the results from 81 industrial accidents involving propellants or explosives that had occurred from 1959-1968 in which there were 78 fatalities and 103 persons injured. Of these 81 accidents, 44 involved both fire and explosion, 23 involved fire only, and only 14 involved a detonating reaction. Settles (1968) carefully distinguishes between an explosion and a detonation. Rupture of a pressurized vessel, for example, is an explosion, but a detonation involves propagation of a shock through a detonable material. In all these accidents, only one of the 78 fatalities was caused by blast, and this was in one of the 14 detonations in which a person was thrown against another object (death by translation). All of the other deaths were from fragments or fire or a combination of the two. In the 14 accidents involving a detonation, 34 fatalities occurred all together. In all of the 78 fatalities, 19 people died exclusively from fire, 58 people died from a combination of fire and fragments, and only one person died from whole body translation due to interaction with the blast wave. Unfortunately, little information is currently available in the open literature concerning damage criteria for people exposed to thermal radiation.

Lung Damage Due to Air Blast Exposure

Portions of the body where there are great differences in density of adjacent tissues are the most susceptible to blast injury [White (1968);

Bowen, et al. (1968); and Damon, et al. (1974)]. The lungs, which contain numerous air sacs, or alveoli, are less dense than surrounding tissues and are, therefore, very sensitive to blast injury. Because of their relatively low density, the air sacs of the lungs are compressed by the implosion of the abdominal and chest walls and upward motion of the diaphragm after interaction with the blast wave. Within tolerable limits of pressure magnitude and rate of increase in external pressure, the body is able to compensate for the rise in external pressure by body wall motion and increases in internal pressure. However, when the inward motion of the body wall is excessively rapid and severe, there is marked distortion of the thoracic organs, including the lungs, causing hemorrhage into airways and shearing and extension of the lungs along and around the relatively "stiff" major bronchi and pulmonary arteries [White, et al. (1971)]. Depending upon the severity and extensiveness of mass hemorrhage and arterial air embolism, death can ensue within a short period.

Investigators have taken two basic approaches in studying the reaction of the body to external forces. Von Gierke (1967, 1971a, 1971c, and 1973), Kaleps and von Gierke (1971), Carmichael and von Gierke (1973) and Fletcher (1971) have studied the possibility of producing biodynamic models which simulate the reaction-response characteristics of the human body.

Von Gierke's model is basically mechanical involving springs, masses and damping mechanisms, while Fletcher's model is fluid-mechanical involving springs, masses, damping mechanisms and gases. Other authors, including many individuals at the Lovelace Foundation for Medical Education and Research in Albuquerque, New Mexico, have analyzed the results of experiments involving laboratory animals and have extrapolated their results, using certain basic assumptions, to the human animal. Since the information acquired by this latter group directly lends itself to the formulation of lethality (or survivability) curves for humans subjected to primary blast damage, their results, with slight modifications, will be used extensively.

Bowen, et al. (1968) and White, et al. (1971) have developed pressure versus duration lethality curves for humans which are especially amenable to this document. Some of the major factors which determine the extent of damage from the blast wave are the characteristics of the blast wave, ambient atmospheric pressure, and the type of animal target, including its mass and geometric orientation relative to the blast wave and nearby objects [White, et al. (1971)]. Although Richmond, et al. (1968), and later White, et al. (1971), both from the Lovelace Foundation, discuss the tendency of the lethality curves to approach iso-pressure lines for "long" duration blast waves, their lethality curves demonstrate dependence on pressure and duration alone. Since impulse, or more properly, specific impulse, is dependent on pressure as well as duration, pressure-impulse lethality or survivability curves appear to be more appropriate. The tendency for pressure-impulse lethality curves to approach asymptotic limits is also very aesthetically appealing from a mathematical point of view.

Also, since both pressure and specific impulse at a specified distance from most explosions can be calculated directly using methods described in this document, it is especially appropriate that pressure-impulse lethality (or survivability) curves be developed.

The human target orientation positions which require the lowest incident pressure-impulse combinations for a specified amount of damage to the human body are standing or lying very near a flat reflecting surface with the incident blast wave approaching the wall at a normal angle of incidence (see Figure 8-6). However, the complexities involved with the shape and type of reflecting surface, the incident angle of the blast wave, and the proximity of the human body target to the reflecting surface are much too involved for this document. Also, the fact that there may not be a suitable reflecting surface near an individual exposed to a blast wave precludes the use of lethality (survivability) curves based on reflection from proximate surfaces. The next most sensitive

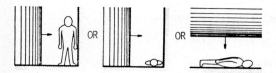

Figure 8-6. Thorax Near a Reflecting Surface Which is
Perpendicular to Blast Winds, Subject Facing
Any Direction

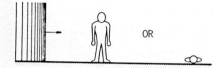

Figure 8-7. Long Axis of Body Perpendicular to Blast
Winds, Subject Facing Any Direction

human body orientation is exposure to the blast wave in the free-field
with the long axis of the body perpendicular to the blast winds with the
subject facing in any direction (see Figure 8-7). This position is a
likely body exposure orientation and will be assumed in this document to
be the position of victims exposed to primary blast damage from explosions.

Researchers at Lovelace [Bowen, et al. (1968) and White, et al.
(1971)] produced scaling laws for pressure and duration effects on animals. It is, therefore, necessary to develop a consistent scaling law
for impulse. Simplifying Lovelace's scaling laws in such a manner that
only the human species or large animals are considered, one is able to
arrive at the following relationships or scaling laws:

(1) The effect of incident overpressure is dependent on the ambient atmospheric pressure. That is,

$$\bar{P}_s = \frac{P_s}{P_o} \qquad (8\text{-}10)$$

590

where $\bar{P}_s$ is scaled incident peak overpressure

P_s is peak incident overpressure

P_o is ambient atmospheric pressure

(2) The effect of blast wave positive duration is dependent on ambient atmospheric pressure and the mass of the human target. That is,

$$\bar{T} = \frac{T \, P_o^{1/2}}{m^{1/3}}$$ (8-11)

where $\bar{T}$ is scaled positive duration

T is positive duration

m is mass of human body

(3) Impulse i_s can be approximated by

$$i_s = \left(\frac{1}{2}\right) P_s T$$ (8-12)

Equation (8-12) assumes a triangular wave shape and is conservative, from an injury standpoint, for "long" duration blast waves which approach square wave shapes because it underestimates the specific impulse required for a certain percent lethality. It is also a close approximation for "short" duration blast waves which characteristically have a short rise time to peak overpressure and an exponential decay to ambient pressure, the total wave shape being nearly triangular. Applying the blast scaling developed at the Lovelace Foundation for peak overpressure and positive duration to the conservative estimate for specific impulse determined by Equation (8-12) above, one can arrive at a scaling law for specific impulse:

$$\bar{i}_s = \frac{1}{2} \bar{P}_s \bar{T}$$ (8-13)

where $\bar{i}_s$ is scaled specific impulse. From Equations (8-10), (8-11), and (8-13),

$$\bar{i}_s = \frac{1}{2} \frac{P_s T}{P_o^{1/2} m^{1/3}} \qquad (8-14)$$

or from Equation (8-12)

$$\bar{i}_s = \frac{i_s}{P_o^{1/2} m^{1/3}} \qquad (8-15)$$

Thus, as indicated by Equation (8-15), scaled specific impulse $\bar{i}_s$ is dependent on ambient atmospheric pressure and the mass of the human target.

As mentioned earlier, the air blast damage survivability curves constructed by researchers at the Lovelace Foundation [Bowen, et al. (1968) and White, et al. (1971)] are based on incident overpressure and duration. It was, therefore, necessary for our applications to modify the survival curves for man applicable to free-stream situations where the long axis of the body is perpendicular to the direction of propagation of the blast wave (see Figure 8-7) so that the axes of the graph would be scaled incident overpressure and scaled specific impulse. To do this, it was necessary to determine the pressure and duration combinations which produced each survivability curve, calculate scaled incident overpressure and scaled specific impulse using Equations (8-10) and (8-14) above, and reconstruct the survivability curves accordingly. These reconstructed curves are shown in Figure 8-8. It should be noted that these curves represent percent survivability, and higher scaled pressure and scaled impulse combinations allow fewer survivors. Presenting the curves in this fashion is advantageous since they apply to all altitudes with different atmospheric pressures and all masses (or sizes) of human bodies. Once one determines the incident

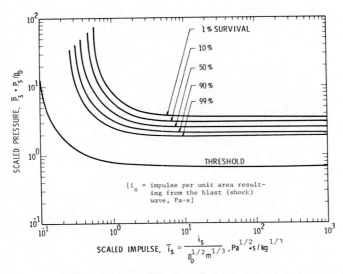

Figure 8-8. Survival Curves for Lung Damage to Man

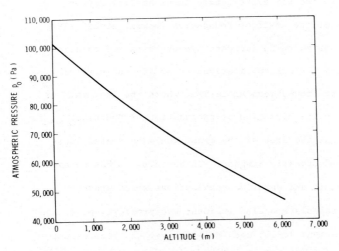

Figure 8-9. Atmospheric Pressure Versus Altitude

overpressure and specific impulse for an explosion, they can be scaled us-

ing Equations (8-10) and (8-15). The proper ambient atmospheric pressure

to use for the scaling can be acquired from Figure 8-9, which shows how

atmospheric pressure decreases with increasing altitude above sea level

[Champion, et al. (1962)]. The value for mass used in the scaling is de-

termined by the demographic composition of the particular area under investigation. It is recommended that 5 kg (11 lb_m) be used for babies,

25 kg (55 lb_m) for small children, 55 kg (121 lb_m) for adult women, and

70 kg (154 lb_m) for adult males. It should be noticed that the smallest

bodies in this case are the most susceptible to injury. More information

on the development of the survivability curves can be found in Appendix

III.B of Baker, et al. (1975).

In summary, the method for assessing lung damage is as follows:

(1) Determine peak incident overpressure P (or P_s) and specific

impulse i (or i_s) at an appropriate distance from the explosion source.

(2) Determine ambient atmospheric pressure (Figure 8-9).

(3) Calculate scaled incident overpressure $\bar{P}_s$ from Equation

(8-10):

$$\bar{P}_s = P_s/p_o$$

(4) Decide on the mass (in kg) of the lightest human to be exposed at this location.

(5) Calculate scaled specific impulse $\bar{i}_s$ from Equation (8-15):

$$\bar{i}_s = \frac{i_s}{p_o^{1/2} \, m^{1/3}}$$

(6) Locate P_s and i_s on Figure 8-8 and determine if these values
are in an acceptable risk area.

Ear Damage Due to Air Blast Exposure

The ear, a sensitive organ system which converts sound waves into

nerve impulses, responds to a band of frequencies ranging from 20 Hz to

20,000 Hz. This remarkable organ can respond to energy levels as low as

10^{-12} watt/m^2, which causes the eardrum to deflect less than the diameter of a single hydrogen molecule [Hirsch, F. G. (1968)]. Not being able to respond faithfully to pulses having periods less than 0.3 millisecond, it attempts to do so by making a single large excursion [Hirsch, F. G. (1968)]. It is this motion which can cause injury to the ear.

The human ear is divided into the external, middle, and inner ear. The external ear amplifies the overpressure of the sound wave by approximately 20 percent and detects the location of the source of sound [Hirsch, F. G. (1968)]. Rupture of the eardrum, or tympanic membrane, which separates the external ear from the middle ear, has captured most of the attention of clinicians although it is not the most severe type of ear injury. The eardrum and ossicles of the middle ear transfer acoustical energy from the external ear to the inner ear where mechanical energy is finally converted into the electrical energy of the nerve impulse. The middle ear is an impedance matching device as well as an amplification stage. The middle ear contains two dampers: the stapedus muscle and associated ligaments which limit the vibration of the stapes when subjected to intense signals, and the tensor tympani muscle and its adjoining ligaments which limit the vibration of the eardrum. The first damper is the most important. These dampers have a reflex time of approximately 0.005 to 0.01 second [Hirsch, F. G. (1968)], which is longer than "fast" rising air blasts. The manner in which the malleus and incus are linked allows far more resistance to inward displacement than to outward displacement. However, if the eardrum ruptures after inward displacement during positive phase of loading of the blast wave, the malleus and incus are less likely to displace as far outward during the negative phase of loading of the blast wave as they would if the eardrum remained intact. In this case, eardrum rupture could be beneficial. The maximum overpressure and its rise time, however, control the characteristics of the negative phase and is, therefore, of prime importance [Hirsch, F. G. (1968)]. Rupture of the eardrum, thus, becomes a good measure of serious ear damage.

Unfortunately, the state-of-the-art for predicting eardrum rupture is not as well developed as that for predicting lung damage from blast waves. A direct relationship, however, has been established between the percentage of ruptured eardrums and maximum overpressure. Examining data from Vadala (1930), Henry (1945), and through personal contacts, Hirsch, F. G. (1968) constructed a graph similar to that shown in Figure 8-10 and concluded that 50 percent of exposed eardrums rupture at an overpressure of 103 kPa (15 psi). White (1968) supports this conclusion for "fast" rising overpressures with durations of 0.003 second to 0.4 second occurring at ambient atmospheric pressure of 101 kPa (14.7 psi). Hirsch, F. G. (1968) also concluded that threshold eardrum rupture for "fast" rising overpressures is 34.5 kPa (5 psi), which is also supported by White (1968) for the range of duration and at the atmospheric pressure mentioned above.

At lower overpressures than those required to rupture eardrums, a temporary loss of hearing can occur. Ross, et al. (1967), have produced a graph of peak overpressure versus duration for temporary threshold shift (TTS). Below the limits of the graphs, a majority (75 percent at least) of those exposed are not likely to suffer excessive hearing loss. According to Ross, et al. (1967), their curves should be lowered 10 dB to protect

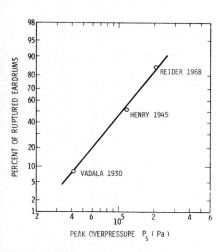

Figure 8-10. Percent Eardrum Rupture Versus Overpressure

596

90 percent of those exposed, lowered 5 dB to allow for a normal angle of
incidence of the blast wave, and increased 10 dB to allow for occasional
impulses. In sum, to assure protection to 90 percent of those exposed and
to allow for normal incidence to the ear (the worst exposure case) of an
occasional air blast, their curves should be lowered 5 dB.

Limits for eardrum rupture and temporary threshold shift, as presented above, are dependent on peak incident overpressure and duration.
Since specific impulse is dependent upon the duration of the blast wave
and since both peak incident overpressure and specific impulse at a specified distance from an explosion can be calculated using methods in this
document, it is especially appropriate that pressure-impulse ear damage
curves be developed from the pressure-duration curves, assuming a triangular shape for the blast wave allows for simple calculations which are conservative from an injury standpoint.

The ear damage criteria presented in Figure 8-11 were developed from
the criteria for eardrum rupture developed by Hirsch, F. G. (1968) and
White (1968) and from the criteria for temporary threshold shift developed
by Ross, et al. (1967). Equation (8-12) was used to calculate specific im-

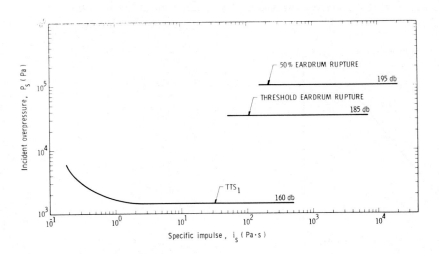

Figure 8-11. Human Ear Damage Curves for Blast Waves Arriving at
Normal Angle of Incidence

pulse, and temporary threshold shift represents the case where 90 percent
of those exposed to a blast wave advancing at normal angle of incidence to
the ear are not likely to suffer an excessive degree of hearing loss. The
threshold for eardrum rupture curve is the location below which no ruptured
ears are expected to occur and the 50 percent eardrum rupture curve is the
location at which 50 percent of ears exposed are expected to rupture.

The method for assessing ear damage is as follows:

(1) Determine peak incident overpressure P (or P_s) and specific
impulse i (or i_s) at an appropriate distance from the particular explosion under consideration.

(2) Locate P_s and i_s on Figure 8-11 and determine if these values
are in an acceptable risk area.[+]

Head and Body Impact Damage Due to Whole Body Displacement

During whole body displacement, blast overpressures and impulses
interact with the body in such a manner that it is essentially picked up
and translated. Tertiary blast damage involves this whole body displacement and subsequent decelerative impact [White (1968)]. Bodily damage
can occur during the accelerating phase or during decelerative impact
[Hirsch, A. E. (1968)]. The extent of injury due to decelerative impact
is the more significant [Glasstone (1962)], however, and is determined by
the velocity change at impact, the time and distance over which deceleration occurs, the type of surface impacted, and the area of the body involved [White (1968)].

Although the head is the most vulnerable portion of the body to
mechanical injury during decelerative impact, it is also the best protected [von Gierke (1971b)]. Because of the delicate nature of the head, many
may feel that translation damage criteria should be based on skull fracture
or concussion. However, since body impact position is likely to be randomly

---

[+]Regions below a particular curve in the figure denote less damage risk
than that which is represented by the curve.

oriented after translation, others may feel that this factor should be taken into account in determining expected amounts of impact damage. In an effort to satisfy proponents of each point of view, both types of impact, essentially head foremost and random body impact orientation, will be considered.

Because of the many parameters involved in decelerative impact, a few assumptions will be made. First of all, translation damage will be assumed to occur during decelerative impact with a hard surface, the most damaging case [Glasstone (1962)]. Another assumption is that, since impact onto only hard surfaces is being considered, translation damage will depend only on impact velocity. That is, impacting only one type of surface precludes the need for considering change in velocity of the body during impact. This assumption, however, is not entirely valid when one considers that the compressibility of various portions of the body can vary considerably.

White (1968, 1971) and Clemedson, et.al. (1968), agree that the tentative criteria for tertiary damage (decelerative impact) to the head should be those presented in Table 8-6. White's (1971) recently revised criteria for tertiary damage due to total body impact are summarized in Table 8-7. It is beneficial to note that the mostly "safe" velocity criteria for each type of impact condition are identical.

Baker, et al. (1975) have developed a method for predicting the blast incident overpressure and specific impulse combinations which will translate human bodies and propel them at the critical velocities presented in Tables 8-6 and 8-7. Borrowing from Hoerner's (1958) conclusions that the human body is similar in aerodynamic shape to a cylinder with length-to-diameter ratio between 4 and 7, and that, in the standing position, man's drag coefficient is between 1.0 and 1.3, Baker, et al. (1975) used an average length-to-diameter ratio of 5.5 and a drag coefficient of 1.3 in their computer program for calculating translation velocity attained

Table 8-6. Criteria for Tertiary Damage
(Decelerative Impact) to the Head

[White (1968, 1971) and Clemedson, et al. (1968)]

Skull Fracture Tolerance	Related Impact Velocity m/s
Mostly "safe"	3.05
Threshold	3.96
50 percent	5.49
Near 100 percent	7.01

Table 8-7. Criteria for Tertiary Damage
Involving Total Body Impact

[White (1971)]

Total Body Impact Tolerance	Related Impact Velocity m/s
Mostly "safe"	3.05
Lethality threshold	6.40
Lethality 50 percent	16.46
Lethality near 100 percent	42.06

by people exposed to blast waves. A drag coefficient of 1.3 was used because this produced a higher velocity for any pressure-impulse blast wave
combination than the lower drag coefficients, thus allowing for a certain
margin of safety in their calculations. The average density of a human was
assumed to be approximately the same as that of water and four body masses
were considered. These masses were 5 kg (11 lb_m) for babies, 25 kg (55 lb_m)
for children, 55 kg (121 lb_m) for women, and 70 kg (154 lb_m) for men. Since
atmospheric pressure and the speed of sound, which were used in the human
body translation velocity calculations, vary with altitude, Baker, et al.

600

(1975) chose sea level, 2000 m (6560 ft), 4000 m (13,100 ft), and 6000 m

(19,700 ft) altitudes when they excercised their computer program.   While

trying to consolidate their results, Baker, et al. (1975) discovered that

for constant density, speed of sound, and atmospheric pressure, the trans-

lation velocity of a human body exposed to a blast wave is a function of the

incident overpressure and the ratio of incident specific impulse over the

mass of the human body to the one-third power.   In functional format,

$$V = f \left(P_s, \frac{i_s}{m^{1/3}} \right)$$ (8-16)

Their results are shown in the figures which follow.   Figure 8-12 contains

the pressure-scaled impulse combinations required to produce the velocities

for various expected percentages of <u>skull</u> <u>fracture</u> (see Table 8-6) at sea

level, while Figure 8-13 contains the pressure-scaled impulse combinations

required to produce the velocities for various expected percentages of

<u>lethality</u> from whole body impact (see Table 8-7) at sea level.   Curves for

other altitudes differ only slightly from the sea level curves.   The pro-

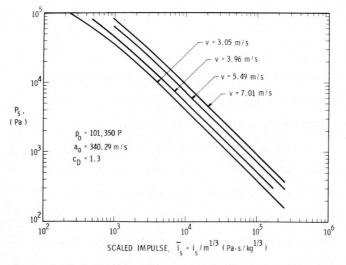

Figure 8-12.   Skull Fracture, 0 m Altitude

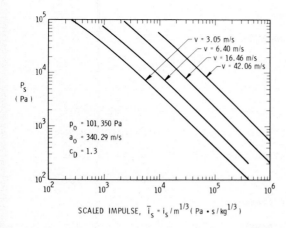

SCALED IMPULSE, $\bar{i}_s = i_s/m^{1/3}$ (Pa • s / kg$^{1/3}$)

Figure 8-13. Lethality From Whole Body Translation,
0 m Altitude

cedure for determining the amount of potential tertiary blast (whole body
displacement) damage is as follows:

(1) Determine peak incident overpressure P (or P_s) and specific
impulse i (or i_s) at an appropriate distance from the particular explosion
situation under consideration.

(2) Determine the lightest representative mass of an exposed human
at this location and calculate $i_s/m^{1/3}$.

(3) Determine the atmospheric pressure or altitude at the blast location and locate the position of the pressure (P_s)-scaled impulse ($i_s/m^{1/3}$)
combination on the appropriate graph of Figure 8-12 for percent skull fracture or Figure 8-13 for percent lethality from whole body impact. Determine
if this pressure-scaled impulse combination is in an acceptable risk area.[+]

[+]Regions below a particular curve in the figures denote less damage risk
 than that which is represented by the curve.

602

CHAPTER 8

EXAMPLE PROBLEMS

Window Glass Breakage

Consider a 1 m square pane of plate glass which is 1/4-inch thick.
Enter Figure 8-3 at 1 m^2 and cross to curve labeled 1 (maximum to minimum
face dimension = 1). This gives a glass factor of 0.25. Go up to glass
thickness curve labeled 1/4-inch and read pressure. This gives 7.5 kN/m^2 =
7.5 kPa as the quasi-static breaking pressure.

Overturning of a Truck by Air Blast

Presume that a 2-1/2 ton truck is located such that the free-field
pressure P_s equals 310 kPa (43.5 psi) and the free-field impulse i_s equals
1210 Pa·s (0.170 psi-sec). The vehicle has a height of 2.93 m (9.61 ft), a
c.g. height of 1.37 m (4.49 ft), a vehicle track width of 1.77 m (5.81 ft),
a presented area from the side of 14.8 m^2 (159 ft^2) and a mass of 5430 kg
(1.19 x 10^4 lb). If we assume that the drag coefficient equals 1.8, that
atmospheric pressure is 101 kPa (14.7 psi) and that the velocity of sound
in air is 329 m/sec (1079 ft/sec), then the scaled free-field pressure
(P_s/p_o) is 3.06 and the scaled free-field impulse $\left(\frac{a_o\, C_D\, i_s}{p_o\, H}\right)$ is 2.41. Entering Figure 8-4 gives a scaled applied impulse ($a_o i_t/p_o H$) of 3.93. Mul-
tiplying the scaled impulse by p_o and by H and dividing it by a_o then gives
the applied specific impulse i_t of 3547 Pa·s (0.498 psi-sec). This numerical value for i_t must be compared to i_θ obtained from Figure 8-5. First,
the scaled target height $\left(\frac{h}{b}\right)$ and scaled c.g. location $\left(\frac{h_{cg}}{h}\right)$ must be calculated. They equal 1.65 and 0.468, respectively. Entering Figure 8-5
yields a scaled critical threshold impulse $\left(\frac{i_\theta\, A\, h_{b\ell}}{m^{1/2}\, b^{3/2}}\right)$ of 0.585. Multiplying the scaled impulse by m, $g^{1/2}$, and $b^{3/2}$ plus dividing by A and $h_{b\ell}$
(which is assumed to equal half the total height of the vehicle) yields
the critical threshold impulse i_θ of 1081 Pa·s (0.113 psi-sec). Because
the applied impulse i_t is greater than the critical threshold impulse i_θ

(3547 Pa·s or 0.498 psi-sec), we predict that the vehicle should overturn. This illustrative example describes an actual experiment included in test data from Appendix 3A of Baker, Kulesz, et al. (1975). The vehicle did overturn, as predicted.

Primary Blast Injury

Consider a 75 kg man at 0 m altitude subjected to a blast wave with an overpressure of 7×10^4 Pa and a duration of 10 ms. What is his chance of survival?

From Figure 8-9 an altitude of 0 m corresponds to an atmospheric overpressure of 1.01×10^5 Pa. Converting 7×10^4 Pa overpressure to scaled overpressure gives $\dfrac{7 \times 10^4}{1.01 \times 10^5} = 6.93 \times 10^{-1} \ \bar{P}_s$. With a scaled impulse of

$$\frac{\left(7 \times 10^4\right)^{1/2} \left(10 \times 10^{-3}\right)}{75^{1/3}} = 6.27 \times 10^{-1} \frac{Pa^{1/2} \cdot s}{kg^{1/3}}$$

from Figure 8-8 we find that the overpressure is below threshold and, therefore, there is no lung damage to man.

Ear Injury

A human is subjected to a blast wave with overpressure of 10^4 Pa and impulse of 100 Pa·s. What are the effects on his ears?

From Figure 8-11, we see that 10^4 Pa is below the threshold of 3.5×10^4 Pa and, therefore, no eardrum rupture occurs.

Tertiary Injury

For the blast loading in the primary blast injury problem, what are the chances of tertiary injury, either skull fracture or whole body translation?

From Figure 8-12, with an overpressure of 7×10^4 Pa and a calculated scaled impulse of

$$\frac{(7 \times 10^4)\,(10 \times 10^{-3})}{75^{1/3}} = 1.66 \times 10^2 \frac{\text{Pa} \cdot \text{s}}{\text{kg}^{1/3}}$$

we find there would be no skull fracture at 0 m altitude. From Figure
8-13, the same pressure and impulse combination gives no injury from whole
body translation.

CHAPTER 8

LIST OF SYMBOLS

A	presented target area; a constant
a_o	sound velocity in air
B, C, K	constants
b	vehicle track width or target base width
C_D	drag coefficient
d	depth of slender member
g	acceleration of gravity
H	target height or width
h	total vehicle height
$h_{b\ell}$	height of center of pressure
h_{cg}	height of center of gravity
i, i_s, i_θ, i_t	blast wave specific impulse
L	length of slender member
P, P_s	blast wave overpressure
$\bar{P}, \bar{T}$, etc.	scaled parameters
P_o	ambient air pressure
R	standoff distance

T	blast wave duration
W	explosive charge weight
X_e	maximum elastic deformation
X_m	maximum deformation
δ	sidesway displacement
θ_{max}	maximum allowable angular displacement
μ	ductility ratio

CHAPTER 9

EXPLOSION EVALUATION PROCEDURES AND DESIGN FOR
BLAST AND IMPACT RESISTANCE

GENERAL

The material and methods presented in earlier chapters cover various aspects of explosions and their effects. This material can be used to help solve two general classes of problems. The first class is the assessment or evaluation of the magnitude and effects of an accidental explosion which has either already occurred or may occur. The second class is the design of plant sites and certain structures in such sites to attenuate or mitigate accidental explosion effects. General methods of approach to these two general classes of problems are discussed in this chapter, together with several rather comprehensive example problems which will illustrate the combined use of several of the many specific prediction methods presented in earlier chapters.

EXPLOSION EVALUATION PROCEDURES

Techniques for Explosion Accident Investigation and Evaluation

All too often, explosion accidents are very poorly investigated and reported. The first concerns of plant managers and others at the site are to minimize further damage, reach and treat casualties, and then to repair damage and resume normal operations as quickly as possible. Much can be learned from detailed study of such accidents, to both evaluate the effects of each particular accident and to help determine how to mitigate or present similar accidents. (We have learned much from studying the case histories of accidents presented in an earlier chapter.) We have considerable experience in explosion accident investigation, and have developed some general procedures which may help you if you have such accidents. These procedures or techniques cover the field investigation of the accident, and the later

evaluation and reporting. They are summarized in Table 9-1. You may not be

able to follow all the steps outlined in these suggested procedures, but you

should try.

Examples of some thorough and well-reported accident investigations

Table 9-1. Techniques for Explosion Accident
Investigation and Evaluation

Field Investigation

• If you are not experienced in such investigations, contact experienced
  investigators immediately. The investigating team should include at
  least:

    (1) an explosion dynamics expert,
    (2) a structural expert, and
    (3) a plant operations expert.

• Leave debris undisturbed, if possible.

• Take many photographs, directed by an experienced investigator. Catalog location and direction of all photos. Include overall pictures and
  aerial views, if possible.

• Make direct field measurements of degree of damage to lightly damaged
  structures, with clear identification of location and orientation of
  these structures. In order of usefulness (best first), these are:

    (1) damage to ductile structures or elements --- steel beams, plates,
        poles, etc.

    (2) damage to wooden structural members --- roof joists, studs, etc.

    (3) glass breakage. Pane dimensions and thickness, and type of glass
        should be recorded.

• Observe and document complete or massive structural destruction.

• Obtain scaled plans or maps of the area, and plans of major damaged
  structures.

• Make as complete a "missile map" as possible. Chart locations of fragments, weigh, measure or describe shapes, and describe materials.

• Record eyewitness accounts.

• Record thermal effects.

• If possible, obtain meteorological data for the nearest weather station
  at the closest time to the accident.

608

Table 9-1.   Techniques for Explosion Accident
Investigation and Evaluation (Continued)

Analysis and Reporting

• Determine the most probable accident scenario, including:

(1) sequence of events prior to the explosion
(2) ignition source location and type
(3) sequence of events during the explosion

• Use as many "damage indicators" for blast damage as possible, and use
the P-i concept to establish blast yield versus distance.

• Correlate blast yield with a known or estimated inventory of energetic
material which could have been released.

• Establish iso-damage contours.

• Determine if blast focusing could contribute to blast damage.

• Correlate and/or report missile map data.

• Correlate and/or report thermal effects.

• Report all results and conclusions as completely as possible.

are Burgess and Zabetakis (1973), Tucker (1976), and National Transportation

Safety Board (1976). A number of the other such reports listed in our bib-

liography are, on the other hand, rather deficient in their depth of cover-

age. We cannot cite some excellent reports because they are kept confiden-

tial by the company or agency who did or sponsored the investigations.

Simplified Blast Analysis

Many aspects of estimating or predicting the blast waves from acci-

dental explosions, and their effects on structures and humans, are covered

in earlier chapters. Although the application of these methods is eased by

a working knowledge of blast physics and/or structural dynamics, we feel

such knowledge is not necessary to conduct at least a simplified blast anal-

ysis. The steps are summarized in Table 9-2, and will be shown in detail in

one of the comprehensive example problems later in this chapter.

Table 9-2. Simplified Blast Analysis

• Define type or types of explosive source (e.g., high explosive, bursting pressure vessel, vapor cloud, explosive dust, etc.).

• Define most probable source location or locations.

• Define maximum energy release and release rate.

• Predict free-field and normally reflected blast wave properties using charts in Chapter 2.

• Use P-i diagrams to predict effects using charts in Chapters 4 and 8.

Simplified Fragment Analysis

Chapter 6 contains much information on generation, flight and impact of fragments. Example problems in that chapter illustrate the use of various prediction curves and equations relating to fragment or missile hazards, but no rationale for fragment analysis is given. We outline in Table 9-3 the steps we feel an investigator should follow in conducting a fragment analysis.

We know of few systematic studies of fragment hazards conducted using the steps outlined in Table 9-3. Usually, explosion hazards evaluations concentrate on blast effects, with very cursory study of potential fragment effects.

Simplified Fireball Analysis

Chapter 7 includes our knowledge on our present capability to estimate fireball characteristics and effects for chemical explosions. Again,

Table 9-3. Simplified Fragment Analysis

- Define worst-case explosion source.

- Determine if missile hazard can be significant.

  (1) Can primary fragments be generated?
  (2) Can secondary fragments be generated?

- Predict fragment velocities and masses.

  (1) For velocities, use figures and equations in Chapter 6.

  (2) For masses, assume some breakup pattern. For ductile vessels, we suggest using N = 2 to N = 10. For high strength vessels, we suggest using N = 100. For brittle materials such as glass or unreinforced masonry, N can be assumed to be very large. Secondary fragments other than glass shards can be assumed to retain their original form.

- Predict maximum fragment ranges for fragments from figures in Chapter 6.

- If a spectrum of masses and velocities can be defined, use maximum range graph to develop a "missile map." A criterion of a specified number of fragments impacting per unit ground area may be an acceptable hazard.

- Predict limit of glass breakage from blast using curves in Chapter 8.

- Predict specific impact effects from figures and tables in Chapter 6.

- Establish safety distances. These distances can be such that:

  (1) no fragments impact at the target,
  (2) a specified fragment number impact per unit surface area, or
  (3) they define the limit for glass damage to humans.

some specific example problems on these effects appear in the chapter, but no procedure for overall estimation appears. Table 9-4 gives such an outline.

Only for nuclear explosions and very large scale accidents with liquid propellants have fireball effects been seriously considered or reported [High (1968) and Glasstone and Dolan (1977)]. Predictions using methods from Chapter 6 appear in one of the comprehensive problems later in this chapter.

Table 9-4. Simplified Fireball Hazards Analysis

- Determine maximum amount of flammable chemical which can be released by an accident.

- Use figures in Chapter 7 to predict fireball size.

- Use figures in Chapter 7 to predict fireball durations.

- Use equations in Chapter 7, with values from text, to obtain an estimate of Q in J/m^2.

- Divide Q by duration to obtain q in $J/m^2 \cdot s$.

- Use figures in Chapter 7 to obtain estimates of effects on humans.

- Use tables in Chapter 7 to obtain ignition thresholds for materials. (These will only apply for quite long duration fireballs.)

- Establish damage radii for effects on humans and materials.

DESIGN PHILOSOPHY AND GUIDELINES

General

In locations where there is potential hazard from accidental explosions, the principles and data given in previous chapters can be used to incorporate significant blast and impact resistance in design of new construction, to evaluate the blast resistance of existing structures, and to choose limits to inventories of potentially explosive materials to minimize accidental explosion hazards. In this section, we give some general design

guidelines or principles for new construction, and follow with specific examples for the two principal kinds of industrial building construction,
etc., steel frames and reinforced concrete frames.

In designing structures to resist external blast or impact loading,
the primary goal should be to protect the inhabitants of the structure. To
do this, one first hopes to design to prevent catastrophic failure of the
entire structure or large portions of the structure. Then, one should try
to minimize the effects on the building inhabitants of flying debris from
the explosion itself or from light or frangible parts of the building. Finally, the effects of blast waves transmitted into the building through
openings should be minimized. Typical structures to be protected would be
office buildings and control rooms in chemical plants, which normally house
a number of people. An important goal for control rooms can be survival of
process control equipment, to allow shutdown of parts of the plant still
under control after the explosion [Balemans and van de Putte (1977)].

The design goals can be somewhat different for plant process buildings. Often, these buildings house only equipment with no or very few people present, and are arranged for remote control of the processes. Here,
the primary design goal may be to minimize damage to the equipment, and injury to people in the structure can be given a lower priority.

Some general design considerations and guidelines for blast-resistant industrial structures have evolved from our staff experience, and experience by others. Some such guidelines follow, largely taken from Baker,
Westine, et al. (1977) and from Forbes (1976).

The design philosophy advanced by Forbes (1976) follows:

- A blast-resistant building should be capable of withstanding an
  external plant explosion of realistic magnitude in order to protect personnel, instruments, and equipment it houses from the
  damaging effects of the blast.

- Structural damage is tolerable if it is not detrimental to the safe operation of the facility during and after the accident.

- In the event of an explosion of intensity (damage potential) in excess of the design values, the structure should "fail" by excessive deformation without any significant loss of its load-carrying capacity, thereby providing an adequate margin of safety against catastrophic collapse.

- The building is expected to resist a major explosion only <u>once</u> in its life. It should, however, have the capacity to support safely the post-explosion conventional design loads with some minor repairs.

We emphasize that, for the structure to fail by excessive deformation, the material must fail in a ductile manner and critical joints must fully develop the strength of the material joined. Progressive failure in the case of an overload is necessary because the building will most often be designed for "realistic" or probable accidents and not for the worst-case accident which conceivably can occur. Progressive collapse offers some assurance against a catastrophe in the event of the worst-case accident.

In addition to the general philosophy suggested by Forbes (1976), which seems reasonable for most industrial plants, we recommend the following guidelines.

- <u>Choose Building Shape and Orientation to Minimize Blast Loading</u>
  The blast-resistant quality of a building is influenced by its shape and its orientation relative to potential explosive sources. If possible, orient buildings with the narrow dimension facing the explosive source. This will minimize the total loading on the structure. Also, avoid interior (reentrant) corners on the blastward side of the building. Overpressures can be amplified by multiple reflections in these corners. It is also desirable to place door openings and windows in the building op-

posite the blastward face. In some cases, the architecture of the building can be changed to improve its blast resistance. Flat walls facing the explosive source see the full reflected pressure. Curved walls, such as a cylinder, receive lower reflected pressures, and the resultant load on the building is less.

- Keep the Building Exterior "Clean"

  Avoid architectural details which may result in unnecessary hazards in an accidental explosion. For example, parapets, copings, signs, and falsework can become dangerous missiles under strong blast loading. Minimize windows and doors in the building and orient them away from explosive sources as noted above.

- Give Attention to Interior Design

  Avoid objects which can fall on people when the building moves or tie the objects securely to the building structure. Special attention should be given to the construction of false ceilings, the mounting of overhead lighting fixtures, and the routing and mounting of air conditioning ducts, all of which can come adrift and fall on building inhabitants.

- Use Good Design and Construction Practices

  Structures in blast-resistant buildings may be strained well beyond their yield points. It is imperative that attention be given to good joint design so that critical joints will develop the full strength of the base material. Also, ductility of the base material must be maintained at all operational temperatures; that is, the NIL ductility temperature of the material must be below the minimum operating temperature of the structure. This is particularly important in welded joints where material properties can change in the heat-affected zone. Reinforcing of the concrete must be designed to assure that ductile rather than brittle failures occur.

615

Buildings and plant structures can be designed in many different ways
and fabricated from different materials. Suggestions for satisfactory types
of construction for blast resistant structures are given in the following
sections.

Materials of Construction

The most important feature of blast-resistant construction is the
ability of the structural elements to absorb large amounts of energy in a
ductile manner without catastrophic failure in the structure as a whole.
Construction materials in blast-protective structures must, therefore, have
ductility as well as strength. Reinforced concrete and structural steel
are the materials most often used in this type of construction. Aluminum
has been used in some special applications but is not addressed here. Windows, doors, roofing and siding must also be chosen for their blast and
fragment resistance.

Reinforced concrete is an excellent material for blast-resistant construction. Its mass provides inertia to help resist the transient blast
loads and a reinforced concrete building possesses continuity and lateral
strength. In addition to its inherent structural strength, reinforced concrete is effective in providing protection against fire and flying debris
which usually accompanies an explosion. The amount, placement and quality
of the steel used in reinforced concrete construction must be chosen properly to assure ductile behavior. Two problems which arise in the use of reinforced concrete, and which need additional research, are reinforcement
for shear resistance and the elimination or control of concrete spall when
overpressures are high. The most reliable guide for shear reinforcement is
provided by Structures to Resist the Effects of Accidental Explosions (1969).
This reference recommends lacing for all structures which are impulsively
loaded (high overpressure region) or which have hinge rotations greater than
two degrees. Stirrups are recommended for other cases. Additional information on dynamic shear are provided by Stea, et al. (1981). Steel fibers

have been used successfully to control spallation in reinforced slabs, but this approach is expensive and not well documented. When it is necessary to control spall fragments, spall plates are recommended.

Structural steel is a ductile, high strength material which is especially suitable for framing in blast-resistant buildings. Structural and intermediate grades, with assured ductility, are preferable. High strength steel with marginal ductility and low grade steels without a well defined NIL ductility temperature should not be used. Structural steels appropriate for blast-resistant structures are designated in Part 2 of the Manual of Steel Construction (1980) and in Plastic Design in Steel (1971).

Windows must also be designed for blast resistance and, perhaps, for fragment resistance. For low overpressures, common window glass can be used if it is sized properly, but missiles or higher pressures than anticipated can produce very dangerous glass fragments. We recommend that windows in blast-resistant structures be glazed with a shock-resistant plastic, such as polycarbonate resin, or with a laminated glass which is designed properly for blast resistance. Commercial suppliers of these materials provide design data for shock loading. Window frames must also be checked for their blast resistance and may require special design. Some commercial windows are held in place by pliable sealant material which may be insufficient in the case of blast loading.

Exterior doors must also be checked for their blast resistance or commercially available blast rated doors must be used. As for windows, the support frames must be able to react the loads from the door. Unless special latches are used, doors should be swung so that the blast loading seats the door against its frame.

Conventional roofing and siding materials can be used for blast-resistant building if the components are chosen properly to withstand the blast loads. Corrugated metal decking supporting a poured concrete slab is good for roofing. The concrete adds mass to resist the impulsive loads and prevents buckling of the corrugations. In sizing the metal decking, the strength of the concrete is usually ignored. Overload produces stretching and progressive collapse of the roof. Open web joists are commonly used to support the roof deck, but care must be taken to avoid instabilities and sudden collapse of these members. Proper lateral bracing is essential. More compact sections, which are less subject to instability, are preferred.

As for the roof, the use of corrugated siding on the exterior of the building is acceptable if it is sized properly to resist the blast loading; however, the corrugated siding is subject to buckling and sudden loss of bending strength in an overload condition. For this reason, we suggest that additional ties to the supporting structure be provided for the siding so that some in-plane forces can be developed under large deformations. This will prevent the siding from being blown away and creating a fragment hazard if an overload occurs.

Types of Construction

Some acceptable types of construction for blast-resistant structures have been mentioned in previous sections. Here, we will elaborate on acceptable construction and note types of construction which are unacceptable or undesirable.

Acceptable construction is any method or arrangement which will resist the design blast loads and fail progressively in a ductile manner in the event of an overload. Rigid-frame, shear wall, monolithic shell, stiffened shell, cast-in-place and precast reinforced concrete construction are all acceptable. Joint design is important in all blast-resistant structures, but it demands special attention in precast concrete. Post-tensioned steel cables have been used successfully by a commercial firm in the United Kingdom to join precast reinforced concrete assemblies for blast-resistant buildings.

Cast-in-place reinforced concrete and steel frame buildings are the most common types of construction, particularly for conventional use buildings which will be loaded by external blast. For internal blast, steel shell structures are very efficient. When fragment hazards are combined with blast loads, double-walled steel shell buildings (usually filled between walls with a material to stop or slow the fragments) or reinforced concrete can be used. Penetrations through shell-type structures require special designs to react the high in-plane loads which are developed in these structures.

Certain types of construction commonly used in ordinary buildings are not recommended for blast-resistant structures. The principal basis for evaluating such construction is its mode of failure if severe overloading occurs. Brittle construction is not suitable for blast-resistant structures. Besides being vulnerable to catastrophically sudden failure under blast overload, it provides a source of debris which can cause major damage when hurled by the blast wind. Nonreinforced concrete, brick, timber, masonry and corrugated plastic and asbestos are examples of this type of construction. These normally should not be used in the exterior shell of blast-resistant structures. If in an otherwise ductile structure, brittle behavior of some elements cannot be avoided, as is the case for axially loaded reinforced concrete columns or for shear walls, the margin of safety for these elements should be increased; that is, their capacity should be downgraded. Additional comments on the use of shear walls and bracing are given in a later section of this chapter.

Foundation Design

For a structure to exhibit any measure of blast resistance, its frame and foundation must be capable of sustaining the large lateral loading. This requirement is similar to that for earthquake-resistant design. In general, structures which are earthquake-resistant are also, to some degree, blast resistant.

The design of foundations of blast-resistant structures introduces problems which are as yet unanswered in civil engineering. This is because there has been very little work done on dynamically loaded footings and other foundations, and current practices are limited to design for static loading. Use of maximum blast overpressures to estimate static loads will probably result in grossly overdesigned foundations. There is considerable evidence from explosion accident investigations that foundations rarely fail, even for strong structures which have essentially been reduced to rubble by air blast. We can only give tentative guidelines for foundation design at present, and some suggestions for dynamic testing and analysis of footings. These suggestions appear in Appendix F.

Containment Structures

A concept which has been used in the nuclear power industry for some time (i.e., surrounding a potentially hazardous operation with a structure designed to contain completely the effects of the worst conceivable accident) is now starting to gain favor in the chemical industry. Although complete containment structures can be expensive and may also impede normal operations, they can provide added safety, particularly for small pilot plant or batch process operations with high potential for runaway chemical reaction or explosion.

Two reported applications of complete containment, both using steel shell structures to contain blast and arrest fragments, are given by Penninger and Okazaki (1980) and Afzal and Chiccarine (1979). We strongly endorse this concept, but caution against the design of such containment structures by engineers who are not thoroughly versed in explosion effects and dynamic structural design.

DESIGN METHODOLOGY

General

The purpose of this section is to provide general guidance in the application of analysis methods for the design of structures to resist the effects of dynamic loads. The methodology presented here covers types of

loading, and structural response to these loads, which are not specifically addressed elsewhere in this text. These loads include fragment impact,
impact by earth ejecta and ground motions. Analytical methods described
in Chapter 5 are suitable for calculating the structural response to these
loads, but these calculations were not treated in the examples. References will be cited as required to supplement the material in the text.

In general terms, the steps that are taken when designing a building for blast resistance are:

- Establish siting, general building layout, and the design criteria.

- Size members for static loads.

- Determine potential sources of accidental explosions and establish their location and magnitude.

- Calculate the blast loads on the building.

- Define the fragments for which the building must be designed.

- Calculate loading from fragment impact.

- Calculate the loading from earth ejecta if cratering throws
  earth against the building.

- Define the ground shock at the building site.

- Perform a preliminary analysis to size the structure for the
  blast loads.

- Perform an analysis considering all loads using either simplified or numerical methods.

- Redesign or resize the structure as required to meet the design criteria.

Design of a building usually involves interaction between several disci-
plines. An architect engineer (AE) is usually responsible for the siting
and general layout of the building. The AE may also size structural members of the building to resist conventional loads. Conventional loads
will include dead weight, equipment operating loads, and loads associated

with natural phenomena such as tornado, fire, and earthquake. Safety and process engineers may define the location and energy of potential explosions that will produce overpressures, fragments, and ground shock for which the building must be designed.

Using Chapter 3 of this book, the designer can, knowing the location and energy in an explosion, predict the blast loading [p(t) or P and i] for which the building must be designed. Estimates of ground shock and fragments that may strike the building can be made using information provided by Baker, et al. (1980).

Once the building has been designed for conventional loads and the blast loads on the building have been established, the analyst is ready to evaluate the strength of the building for these loads. Usually, some allowance can be taken for the fact that dynamic loading will cause the building to respond with a strain rate that is high enough to increase the material yield strength. A good summary of material properties for dynamic analysis is given by Baker, et al. (1980). Analytical methods for structural design have already been developed and/or described in Chapters 4 and 5. This section provides guidance for the choice of a method and its application and for the analysis and sizing of structural members in the building.

Analysis methods described here have been divided into two general categories: simplified methods and numerical methods. In this discussion, the definitions will be:

- Simplified Methods: Refers to single or two-degree-of-freedom approximations to the dynamic behavior of structural elements as described in Chapters 4 and 5. Most often the solution is presented in graphical or in closed form (as in Chapter 4 and Appendix B), but numerical integration is used to obtain solutions in some cases.

- Numerical Methods: Refers to sophisticated multi-degree-of-freedom solutions for the response of structural elements or of complex structural systems. A digital computer is required for these solutions, and the cost can be high if plasticity is included and a dynamic solution is obtained. These methods are described in Chapter 5 under the section entitled "Multi-Degree-Of-Freedom Methods."

Elements of Dynamic Response Analysis

Three basic elements of dynamic response analysis are:

- Description of the loading on the structure,
- Determination of the material properties under dynamic loading, and
- Calculation of the response of structures to the applied loads.

Chapters 4 and 5 explain, in detail, how structures behave when excited by transient blast loading and describe analytical methods for calculating the structural response. This section gives brief summaries of the types of loads on structures produced by explosions and of material properties for rate of strain characteristics of those which occur in blast loaded structures.

1. Transient and Quasi-Static Loads

By their very nature, accidental explosions produce transient loads on buildings. To design for blast resistance, the effect of these transient loads on the response of the building must be determined; however, what seems to be very brief to the observer in real time can be quite long in terms of the vibration periods of a structure so that the loads are classified as short (impulsive), or long (quasi-static), relative to the structural vibration periods. The intermediate region between impulsive and quasi-static, where periods are about the same as loading times, is called the dynamic realm (see Chapter 4).

The types of loads considered herein that can be produced on a building by an explosion are:

- overpressures produced by the blast wave
- a long term gas pressure if the explosion is confined
- ground shock
- fragment impact
- impact from soil if cratering occurs

The most important load in design is usually the overpressure produced by the blast wave. Ground shock is usually of secondary importance. The increase in long term pressure occurs only when there is confinement, and the creation of fragments or soil ejecta will depend upon characteristics of the explosion source (see Chapter 6) and the position of the explosion relative to the ground and adjacent structure.

Each of these loads can be classified as impulsive, dynamic, or quasi-static relative to a specific structural component. If the period of the fundamental vibration of the component is very short, all of the loads may be quasi-static (long duration relative to the structural period). If the period of vibration is long (low frequency), then all of the loads may be impulsive for that component. Even so, for most building structures, impact loads produced by fragments and earth ejecta are idealized as an impulse or initial velocity and the gas pressure increase is usually quasi-static. It is easy to see that the same load can be classified differently for different components in a building.

The way a load affects the structure depends not only upon its duration but upon its rise time and general shape as well. Examples of the effect of different load pulse shapes for an elastic oscillator are given in Chapters 4 and 5 and in the graphs of Appendix B. Figure 5-19, for example, gives the dynamic load factor (DLF) and the time (t_m) at which the maximum response occurs for a triangular force with zero rise time. Appendix B gives solutions for other pulse shapes. Notice that the maximum DLF is always a function of the ratio between some characteristic time of the loading (t_d or t_r) and the fundamental period (T) of the one-degree-of-freedom oscillator. $(DLF)_{max}$ is the ratio of the maximum response of the structure (maximum deflection, stress, etc.) relative to the maximum response produced by a static load of the same magnitude. Notice also that all of the forces give about the same peak value of DLF (approximately 2.0). For rectangular and triangular force pulses, this occurs when the duration is long (both have zero rise times), and for a ramp function to a constant load, it occurs when the rise time is short. These cases all equal or approach the DLF for a step function to a constant load. It is well known that DLF = 2.0 for a step function with $t_d/T > 5.0$ [Biggs (1964)].

Loading from a blast wave from high explosive (HE) charges can be closely approximated by the triangular pulse with zero rise time. If the explosion is produced by vapor or dust clouds, or bursting pressure vessels, very different forcing functions can occur as explained in Chapter 4. Curve shapes for quasi-static pressure loadings are closely approximated by the ramp function to a constant value or, conservatively, by using long duration rectangular or triangular force pulses with zero rise times.

When the forces on the building are either impulsive or quasi-static, simplifications in the analysis are possible. Impulsive loads can be replaced by an initial velocity imparted to the structure. As explained in Chapter 5, this is done by equating the total impulse to the change in momentum of the structure. From the initial velocity, the initial kinetic energy is obtained. Equating the initial kinetic energy to the strain energy absorbed by the structure during deformation (elastic or plastic) will give the impulsive asymptote for the structure. For many structural components, equations can be derived to give this asymptote (see Chapter 4).

When the forces are quasi-static, then, for elastic behavior, a static analysis can be used. This is strictly true only for very slowly applied loads; however, for suddenly applied constant loads (step function),

the results of the static analysis, multiplied by two, are also valid. If
the material deforms plastically, the quasi-static asymptote can be obtained
by equating the work done by the load as the structure deforms to the strain
energy of deformation. As for impulsive loads, simple formulas that give
the quasi-static asymptote can be derived for many structural components.
This was demonstrated in Chapter 4, which also gives graphical solutions
based on the pressure-impulse (P-i) concept for the complete loading realm.

Ground shock produced by an explosion is difficult to define.
For linear systems, it can be represented by a response spectrum. Approximate methods of defining such a spectrum are discussed by Baker, Westine,
et al. (1980), Nicholls, et al. (1971), and Biggs (1964). In nonlinear response problems, ground shock is most conveniently represented as a displacement-time history that is applied to the base of the building.

2. Material Behavior

Under dynamic loads associated with explosions, strain rates in
materials will be in the range of 1 to 100 s^{-1}. These rates are high enough
to increase the yield strength of some structural materials above static
values. A good summary of data on strain rate effects in structural steels,
concrete, and other materials are given by Baker, Westine, et al. (1980).

When using numerical methods, continuous variations in strain
rate that occur in the materials can be included in the analysis. In such
cases, the approximate formulas that relate yield stress to strain rate are
useful; however, when using simplified methods, a simpler way of accounting
for strain rate is required. Norris, et al. (1959) suggests that some average value be used which is based upon the time required to reach the yield
stress. Following this approach, the authors cite average strain rates of
0.02 to 0.2 s^{-1}, and, for ASTM A7 structural steel, the authors recommend
an increase in the yield stress from about 265 MPa to 291 MPa (38,000 psi
to 41,600 psi). A similar increase is probably warranted for other structural steels, but a choice for any material can be made if the strain rate

data are available and if 0.2 s^{-1} is accepted as a reasonable average for structural response to blast loads.

Numerical results reported by Cox, et al. (1978) showed that strain rate is very important for rings that are subjected to uniform dynamic internal pressures. Including the effect of strain rate reduces peak strains to about one-third; however, for beams in the same structure, strain rates had very little effect on calculated peak strains. Peak strains in the beam were only slightly above yield, and in bending only the surface experiences the maximum strain rates. For rings, the rate of straining is uniform over the total cross section of the component; high strain rates significantly increase the total yield loads and, thus, significantly reduce the total strain.

Strain rate effects should be included in numerical methods if it can be done conveniently, particularly when membrane action (significant stretching of the material at the neutral axis) is predominant in the response. For simplified methods, some increase in the yield stress above static values is also warranted. Of course, ignoring strain rate entirely leads to conservative results.

Material damping (the hysteresis which occurs in the stress-strain diagram when a material is subjected to cyclic loads) is small and can be ignored in blast related response problems. Only the first few cycles of response are of interest and the effect of material damping will be insignificant. Overall structural damping will be higher, but it too will be small compared to the damping effects of structural plasticity. Thus, for transient response problems involving material plasticity, ignoring damping will lead to a slightly conservative result.

Analysis and Design Procedures

There are many steps in the procedure which must be followed to design buildings for blast resistance. The steps can be loosely divided into five major areas:

- design requirements

- structural configuration

- preliminary sizing for dynamic loads

- dynamic analysis

- design iterations

The important points to consider in each of these five areas are described in the following sections; however, the emphasis of this chapter is on the analysis of structures for blast loading and not the establishment of design requirements, the layout and siting of a building, or the design for static loads. Results of these tasks are provided as input to the analyst. Of course, interaction should take place between all parties involved in the building design. Interaction early in the design stage may alter the building location, orientation, and structural configuration in ways that improve its blast resistance.

Flow charts are provided in Appendix F that show the steps in the design/analysis procedure. These steps cover all important areas in the design process, but they are not keyed specifically to the five major areas listed above. The flow charts are substantially more detailed in some of the areas than in others. An explanation of each block in the charts is given.

1. Design Requirements

Design requirements for the structure are usually established by the safety engineer, architectural engineer, or by some procuring agency. Often very extensive facility design requirements are established. As a minimum, the design requirements should include:

- A complete description of the accidental explosion for which the structure is to be designed. This description must be in sufficient detail so that blast wave overpressures, fragment impacts, earth ejecta, and ground shock that affect the building can be determined. Alternately, the design loads (pressures, impulse, fragment impacts, etc.) can just be specified for the analyst.

- The level of protection and serviceability of the building after the design accident must be specified. This information will allow the analyst to determine whether or not

  - glass breakage is permitted
  - siding or sheathing must remain in place
  - fragment penetration is permitted
  - the primary structure is to be reusable after the "design" accident

2. Structural Configuration

For many buildings, the location, orientation, and basic structural configuration will be set by the operational functions that it must perform. These will be dictated principally by architects and production personnel. In certain cases, such as for containment vessels, the primary function is blast containment, and here the design configuration may be dictated by the designer/analyst. Regardless of whether this information is provided by others or is determined by the analyst, it should include:

- building location and orientation
- identification of surrounding terrain and neighboring structures
- type of structure, i.e., above ground, buried, steel panel and frame, steel shell, reinforced concrete
- general building layout
- details of primary and secondary structural members designed for static loads

3. Preliminary Sizing for Dynamic Loads

Before detailed analyses are performed on the structure, it is usually advantageous to establish preliminary sizes of the main structural members for the dynamic loads. For frame structures, this can be done by performing a two-dimensional mechanism analysis, by using equivalent one-degree-of-freedom analyses, or by using the design charts and formulas in Chapter 4.

A mechanism analysis requires that the dynamic load factor be estimated for the loads acting on the building. Usually, only the air blast loads (the primary loads) are considered because dynamic load factors are difficult to establish for ground shock or fragment impact. Once the equivalent load has been determined, different collapse mechanisms are checked to find the one that governs the design of the frame. In the analysis, work of the external loads (product of force and distance) is equated to strain energy (products of the plastic moments and hinge rotations). The procedures for mechanism analysis, such as given by Stea, et al. (1977), provide guidance for the choice of a mechanism that will produce an economical design. A mechanism analysis will establish the minimum required plastic yield moment in the frame members from which the member dimensions can be established.

Mechanism analyses for dynamic loads are usually limited to single-story frames. The reason is that an equivalent static loading is difficult to establish for two-story buildings. Mechanism analyses for two-story frames under static loads are described by Neal (1977).

After preliminary analyses (as suggested above) have been performed, numerical methods can be used to check and refine the structural members. If numerical methods are used, it is recommended that the members' sizes first be determined as accurately as possible using the simplified methods. This includes the use of a mechanism analysis, one-degree-of-freedom dynamic analyses, and even small multi-degree-of-freedom numerical models. The purpose is to reduce the design iterations (number of analyses) required with the more sophisticated numerical model. These analyses can be very costly, and the number of cases run should be minimized.

Before a nonlinear transient analysis is performed, it is suggested that an estimate of the cost for such a calculation be made. A good source of cost information is the code developer. The developer is usually aware of similar problems which have been solved with the code (although nonlinear transient solutions are not commonplace) and he knows the code operations which are required for a solution. Thus, after the analyst has selected a code to be used for the calculations and has formulated the problem in specific terms, he should contact the code developer for data on which to base the estimate. It is also possible that the developer can put the analyst in touch with users who have solved similar problems, even though the developer can seldom release information directly. We further suggest that a small test problem, with only a few degrees-of-freedom, but which uses all of the code features which will be required in the calculations, first be solved to check the cost-estimating procedure and confirm the estimate.

4. Dynamic Analysis

Both simplified methods and more sophisticated numerical methods for performing nonlinear dynamic analyses were described in Chapters 4 and 5. With numerical methods, the analyst can usually treat all loads and even secondary effects readily. With simplified methods, secondary effects must be approximated or ignored, and it is often difficult to treat multiple loads simultaneously.

Some questions to be considered when analyzing buildings for blast loads are:

- How are dead loads treated?
- What is the effect of combined axial and bending loads in beam-columns?
- Is it good practice to use shear walls and bracing in blast-resistant design?

The answers to these questions depend, to a large extent, upon the type of analysis performed; however, some general guidelines can be given.

a. Dead Loads

The stresses produced by dead loads are often ignored in blast-resistant design. There are two main reasons for neglecting their effects. One is that the dead loads are usually small relative to the loads produced by an explosion. Thus, if the dead loads do reduce the strength of the structure, the percentage reduction will be small. The other reason is that masses associated with dead loads are often free to move when the building shakes. This makes the true effect of the dead loads very difficult to evaluate. It is emphasized that the mass associated with any structure or equipment that is attached to the building should be included in the mass of the building when performing dynamic analyses.

If dead loads are important in the blast-resistant design, and they can be for multi-story buildings or buildings designed for low overpressures, then their effect should be included. For simplified analyses, the common way of treating dead loads is to account for their effect on the bending capacity of the structural members. This is done by reducing the allowable plastic moments because of the initial static moments and axial loads. Formulas which give the perfectly plastic moment as a function of the axial load are given by Neal (1977). If numerical methods are being used, then the dead loads should be included as added masses or concentrated forces. If masses, representing equipment, etc., are free to move relative to the building, then they can be approximated by fixed (or moving) forces and their mass ignored in the response calculations.

b. Effect of Axial Loads

As discussed above, static axial loads will reduce the allowable plastic moments in beam-columns. Because the dead loads are usually small, this effect will be small also, but it is sometimes used as a convenient way of including the effect of dead loads in simplified analyses. In most nonlinear numerical analyses, the effect of axial loads is included directly because a stress criterion is used to predict the onset of yielding.

Dynamic axial loads will also occur in the members. This effect is ignored in simplified analyses. Because the axial response of beam-columns usually occurs at much higher frequencies than the lateral or bending response, neglecting the effect of dynamic axial loads is justified. In most nonlinear numerical methods, the effect of dynamic axial forces on the members is automatically included unless this mode of behavior is eliminated (by node coupling, etc.). In programs which are developed specifically for frames, plasticity is often accounted for by the formation of plastic hinges (see Chapter 5). In these programs, including the effect of axial loads on the plastic moments is usually optional.

c. Bracing and Shear Walls

Shear walls and bracing, when used properly, can result in efficient structures for resisting blast loads; however, there are two principal objections to these components that keep them from being used frequently. One objection is that they can be subjected to lateral loads (braces are often combined with a wall panel) that may cause premature failure or at least reduce the "in-plane" strength. Another objection is that when these components fail, the failure tends to be sudden and catastrophic, not gradual and progressive. If these components are used, the designer must take care to assure that these components do not fail.

When designing with shear walls or cross bracing, the designer/analyst must recognize that these components are very stiff relative to most other components in the building for loads applied in their plane. Thus, all lateral loads on the building, which act parallel to the shear wall or plane of bracing, tend to be reacted by these components. This concentrates the loads on the component and upon connections to the remainder of the structure. The designer must assure that these components and the attachments can react the total lateral loads with a high margin of safety and also take steps to protect them from blast loads normal to their surface.

The designer/analyst should also be aware that buildings with shear walls or bracing are difficult to treat with simplified methods. The main difficulty is in the calculation of vibration periods. Formulas are not available for estimating frequencies in the plane of these components, and numerical methods are required for accurate calculation. Thus, numerical methods are needed for good design calculations. If simplified methods must be used, then a conservative approach would be to design the braced panels or shear walls for twice the total load on the building or bay as appropriate; however, because of the uncertainty in load transfer through the structure, a multi-degree-of-freedom analysis is recommended.

5. Design Iteration

In order to assure that the design requirements are met, the designer/analyst must set criteria or design allowables for the structure. These criteria may include maximum allowable stress, strain, deformation, or joint rotation. Recommended criteria for blast-resistant design are given in Chapter 8 and by Stea, et al. (1977). The criteria are seldom met on the first attempt, and so design iterations are required. Iterations should be performed using simplified methods such as a mechanism analysis or one-degree-of-freedom equivalent systems. Design iterations with complex numerical methods should be avoided, if possible. Guidance for changing the resistance of the structure must be obtained by examining the results of previous response calculations.

CHAPTER 9

EXAMPLE PROBLEMS

DESIGN OF A REINFORCED CONCRETE WALL

This example is part of a larger problem to design a high explosive processing facility at a munitions plant. A plan view of the building is shown in Figure 9-1. The interior of the building is open except for the

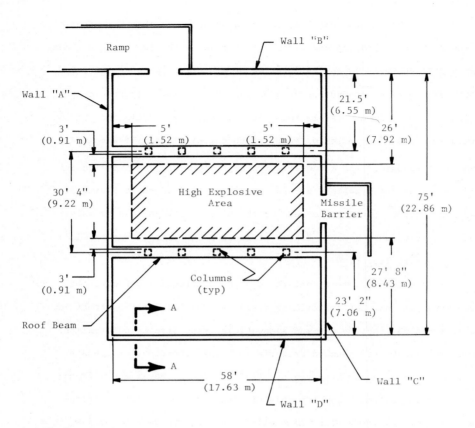

Figure 9-1. High Explosive Wing

square columns which support the roof beams. A 325 lb (147.4 kg) charge of
TNT can be located anywhere within the high explosive (HE) area at a height
above the floor from contact to 7 ft (2.13 m). Openings in the facility
are closed by heavy doors, but these doors cannot withstand the detonation
of the 325 lb (147.4 kg) charges and so they will be blown open in an accidental explosion. The problem addressed here is the design of Wall "D."
No fragment loads will be considered.

A cross section through Wall "D," Section A-A of Figure 9-1, is giv-
en in Figure 9-2. The wall is 15 ft (4.57 m) high and 58 ft (17.68 m) long
(inside dimensions). The wall is long relative to its width and so it can
be analyzed as a strip from floor to ceiling. Lacing, which will be re-

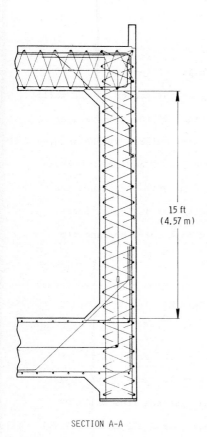

15 ft
(4.57 m)

SECTION A-A

Figure 9-2. Cross-Section Through Wall "D"

quired for this building, is shown in the cross section. The facility is

not designed to be reusable after the worst-case accident. Damage criteria

for reinforced concrete structures are given in <u>Structures to Resist the</u>

<u>Effects of Accidental Explosions</u> (1969). For non-reusability, hinge rota-

tions in the range of 2 to 12 degrees are permitted. In this design, we

chose to limit rotation to 6 degrees or less.

As a first step in the design, we determine the worst-case loading

on the wall from an accidental explosion. The wall will be loaded by blast

waves and by the build-up of quasi-static pressure in the closed volume.

636

Blast Wave Loading

Loading from the blast waves is influenced by the charge location.
In the center of the building between two columns and closest to Wall "D"
is one possibility which would produce high loads in the center of the wall;
however, adjacent to Wall "A" will give a reflection off the side wall as
well as the floor and produce even higher loads, particularly at the intersection of Walls "A" and "D." The pressure produced will not drop significantly for a distance of 15 to 20 ft (4.57 to 6.10 m) from the corner and
so it will produce the worst-case blast wave loading for design purposes.

Using a charge reflection factor of 2 from both the floor and the
wall gives an effective charge weight of

$$W_{eq} = 2 \times 2 \times 325 = 1300 \text{ lb (590 kg) TNT}$$

Because of the asymmetrical charge location and also the presence of the
columns to break up wave reflections, no multiple reflections of the blast
wave will be treated. The charge standoff is the distance from Wall "D"
to the edge of the HE area plus the charge radius. For a 325 lb (147.4 kg)
spherical charge, the radius is 1 ft (0.3 m), which gives a standoff distance R of

$$R = 27 \text{ ft 8 in.} + 1 \text{ ft} = 28 \text{ ft 8 in. (8.73 m)}$$

and a scaled standoff distance of

$$\frac{R}{W^{1/3}} = \frac{8.73}{(590)^{1/3}} = 1.04 \text{ m/kg}^{1/3}$$

For this scaled standoff distance, Figure 2-46 gives the reflected pressure
and reflected impulse on Wall "D" of $P_r = 4,137$ kPa (600 psi), $i_r/W^{1/3} =$
0.057, and $i_r = (8.38)(0.56) = 4.7$ kPa-sec (0.68 psi-sec). If the blast

wave loading is idealized as a triangular pulse with zero rise time, then
the duration is

$$T = \frac{2i}{P} = \frac{4.7}{4,137} = 2.2 \text{ ms}$$

Quasi-Static Loading

In addition to the blast wave, the wall must resist a quasi-static
pressure which is produced in the enclosure. Although the doors will open
during the explosion, they will retard the venting and lengthen the venting time. Because the response time of the wall will be much less than the
venting time, it will be only slightly conservative to assume that the volume is completely closed. In this case, the quasi-static pressure increases to a constant value as given by Figure 3-15 and is assumed to remain at
this value. To enter the figure, the total room volume is needed. It is
approximately

$$V = 15 \text{ ft x } 59 \text{ ft x } 75 \text{ ft} - 65,250 \text{ ft}^3 \ (1847.7 \text{ m}^3)$$

Now

$$\frac{W}{V} = \frac{147.4 \text{ kg}}{1847.7 \text{ m}^3} = 0.0798 \text{ kg/m}^3$$

and, from Figure 3-15, the quasi-static pressure is

$$P_{qs} = 330 \text{ kPa } (48 \text{ psi})$$

For a high explosive, the rise time is very rapid and will be assumed to
correspond approximately to the decay time of the blast wave. This is
shown schematically in Figure 9-3.

638

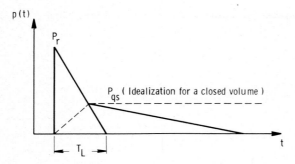

Figure 9-3. Pressure-Time History on Wall "D"

Preliminary Sizing

Preliminary estimates of the wall dimensions can be made using the transformation factors and response curves of Appendices A and B. The wall will be treated as fixed at the roof and floor because good continuity is provided by the lacing pattern. For fixed slabs (Table A-7), the ratio a/b is not less than 0.50; therefore, the transformation factors for fixed beams (Table A-2) will be used. Because large hinge rotations are permitted, plastic behavior will predominate, and the loading on the wall is approximately uniform. For this case, Table A-2 gives K_{LM} = 0.66, R_M = $8/\ell \left(M_{P_s} + M_{P_m} \right)$ = $16/\ell\ M_P$ (for equal plastic moments at the supports and mid-span), and K_E = 307 EI/ℓ^3. A preliminary value of R_M, and thus, M_P, can be estimated from Figures B-1 and B-3. For a plastic hinge rotation of 6 degrees, we will guess that $X_m/X_e \approx$ 10. The duration of the blast wave loading is very short, as is the rise time of the quasi-static pressure. Thus, in Figure B-1 for T_L/T_N = 0.1 (the smallest ratio in the figure), we estimate

$$\frac{R_M}{F} \leq 0.1$$

For the blast wave

$$F = P_r \, b \, \ell = (600 \text{ psi}) (12 \text{ in.}) (180 \text{ in.}) =$$

$$1.296 \times 10^6 \text{ lb } (5,765 \text{ kN})$$

which gives

$$R_M = \frac{16 \, M_p}{\ell} = 0.1 \, F$$

$$M_p = \frac{0.1 \, F \, \ell}{16} = \frac{(0.1) \, (1.296 \times 10^6) \, (180)}{16} =$$

$$1.458 \times 10^6 \text{ in-lb } (164.72 \text{ kN-m})$$

From Figure B-3 for $T_r/T_N = 0.1$ (again, the smallest ratio provided), we choose

$$\frac{R_M}{F} \simeq 1.05$$

For this case

$$F = P_{qs} \, b \, \ell = 48 \, (12) \, (180) = 103,630 \text{ lb } (461.2 \text{ kN})$$

and

$$R_M = 1.05 \, F = \frac{16 \, M_p}{\ell}$$

$$M_p = \frac{1.05 \, F \, \ell}{16} = \frac{1.05 \, (103,680) \, (180)}{16} =$$

$$1.225 \times 10^6 \text{ in-lb } (138.4 \text{ kN-m})$$

The blast wave loading and the quasi-static loading occur very close to-gether, but still, the required resistance is not the sum of the two com-puted separately. M_p should be bounded by

$$1.46 \times 10^6 \text{ in-lb (165 kN-m)} \leq M_p \leq 2.7 \times 10^6 \text{ in-lb (303 kN-m)}$$

Choose $M_p = 2 \times 10^6$ in-lb (226 kN-m). We can now estimate the required reinforcement.

A cross section through the wall is given in Figure 9-4. Because other walls in the building that are closer to the charge will be thick, we choose a 30 in (762 mm) thickness for Wall "D" and the following pro-perties:

Rebar spacing:	b = 12 in (305 mm)
Rebar cover (from rebar center):	d' = 3 in (50.8 mm)
Dynamic rebar strength:	σ_s = 72,000 psi (496 MPa)
Rebar elastic modulus:	29×10^6 psi (20 MPa)
Concrete strength:	σ_c = 5,000 psi (34.5 MPa)
Concrete elastic modulus:	4×10^6 psi (2.76×10^{10} Pa)

For joint rotations of over two degrees and high overpressures, <u>Structures to Resist the Effects of Accidental Explosions</u> (1969) classifies the con-crete section as Type III. This section is designed under the assumptions

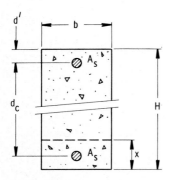

Figure 9-4. Section Through Wall "D"

that the concrete cover will spall on both wall faces and cannot be counted on to resist the blast loads. Consequently, the plastic moment for this
section is based on the rebar alone and is given by:

$$M_p = \sigma_s A_s d_c$$

This moment corresponds to a section whose width is equal to the rebar
spacing, b, or 12 in (305 mm) in this case.

Thus, noting that $d_c = H - 2d'$, A_s is simply

$$A_s = \frac{M_p}{\sigma_s d_c} = \frac{2 \times 10^6}{(62,000) \ [30 - (2) \ (3)]} = 1.16 \ in^2$$

Therefore, choose a #9 rebar with $A_s = 1.00 \ in^2$ (0.000645 m^2). Subsequent
calculations will determine whether or not a #9 rebar is adequate. Using
this value for A_s, the plastic moment and resistance become

$$M_p = 72,000 \ (1) \ (30 - 6) = 1.728 \times 10^6 \ in-lb \ (195.2 \ kN-m)$$

$$R_M = \frac{16 \ M_p}{\ell} = \frac{16 \ (1.728 \times 10^6)}{180} = 153,600 \ lb \ (683 \ kN)$$

To account properly for the combined loading and elastic-plastic behavior,
a numerical integration is required as explained in Chapter 5 under "Equivalent One-Degree-Of-Freedom Systems." To do this, we must determine the
spring constants for the elastic and elastic-plastic ranges and the yield
load at the end of the elastic range. Again, using Table A-2, we have:

Elastic Range:

$$K_{LM} = 0.77$$

$$k = \frac{384 \ EI}{\ell^3}$$

$$R_M = \frac{12 \ M_{P_s}}{\ell}$$

Elastic-Plastic Range:

$$K_{LM} = 0.78$$

$$k = \frac{384 \ EI}{5\ell^3}$$

$$R_M = \frac{16 \ M_p}{\ell} \ (as \ before)$$

Plastic Range:

$$K_{LM} = 0.66$$

$$k = 0$$

$$R_M = \frac{16 \ M_p}{\ell} \ (as \ before)$$

Only EI of the beam is needed, in addition to the information already calculated to evaluate these quantities. The neutral axis is determined by
balancing EA for the beam. Using Figure 9-4, we write the equation

$$E_s A_s \ (H - d' - x) - E_c \ b \ x \left(\frac{x}{2}\right) = 0$$

Solving for x gives:

$$x = -\frac{A_s E_s}{bE_c} \pm \sqrt{\left(\frac{A_s E_s}{bE_c}\right)^2 + \frac{2A_s E_s (H - d')}{bE_c}}$$

Substituting known values, x is found to be:

$$x = -\frac{(1) (29 \times 10^6)}{(12) (4 \times 10^6)} \pm$$

$$\sqrt{\left[\frac{(1) (29 \times 10^6)}{(12) (4 \times 10^6)}\right]^2 + \frac{2 (1) (29 \times 10^6) (30.3)}{(12) (4 \times 10^6)}} =$$

$$-0.6042 \pm 5.744 = 5.14 \text{ in } (130 \text{ mm})$$

EI can now be calculated about an axis at x from the geometry of Figure 9-4, and assuming cracked concrete on the tension side.

$$EI_x = E_s A_s (H - d' - x)^2 + E_c \frac{1}{3} b (x)^3 =$$

$$(29 \times 10^6) (1) (30 - 3 - 5.140)^2 + 4 \times 10^6 \left(\frac{1}{3}\right) (12) (5.140)^3 =$$

$$1.603 \times 10^{10} \text{ lb-in}^2 (46,000 \text{ kN-m}^2)$$

We now have all of the information required to evaluate the spring constants and resistances for the wall section in the elastic, elastic-plas-

tic, and plastic ranges. These are:

Elastic:

$$k = \frac{384 \ EI}{\ell^3} = \frac{384 \ (1.603 \times 10^{10})}{(180)^3} =$$

$$1.055 \times 10^6 \ \text{in-lb} \ (184,800 \ \text{kN-m})$$

$$R_M = \frac{12 \ M_{P_s}}{\ell} = \frac{12 \ (1.728 \times 10^6)}{180} = 115,200 \ \text{lb} \ (512.4 \ \text{kN})$$

Elastic-Plastic:

$$k = \frac{384 \ EI}{5\ell^3} = \frac{1.055 \times 10^6}{5} = 211,000 \ \text{in-lb} \ (50,600 \ \text{kN-m})$$

$$R_M = \frac{16 \ M_P}{\ell} = 153,600 \ \text{lb} \ (683 \ \text{kN}) \ \text{[as previously calculated]}$$

Plastic:

$$k = 0$$

$$R_M = \frac{16 \ M_P}{\ell} = 153,600 \ \text{lb} \ (683 \ \text{kN})$$

The mass of the beam, which will be multiplied by the load-mass factors in the integration process, is

$$m = \rho \ b \ H \ \ell = (150 \ \text{lb/ft}^3) \ (1) \left(\frac{30}{12}\right) (15) = 5,625 \ \text{lb} \ (25 \ \text{kN})$$

Numerical integration was performed according to the procedures in Chapter
5 and gave the following results:

- Maximum center displacement = 1.76 in (44.7 mm)

- Ductility ratio, μ, = 16

- Maximum hinge rotation at the support = 1.12 degrees

These requirements are well within the design criteria, and so the wall is
adequate and perhaps slightly overdesigned.

Finite Element Calculations

For comparison with this approximate analysis, calculations were also made with a finite element computer program. The program is described
in Chapter 5 and is the program which was used to calculate the results giv-
en in Tables 5-8 and 5-9.

A simple model was constructed for one-half of the beam height as
shown in Figure 9-5. The same wall cross-sectional properties already calculated were used in the analysis. Results for this calculation are:

- Maximum center displacement = 2.19 in (55.6 mm)

- Maximum hinge rotation at the support = 1.47 degrees

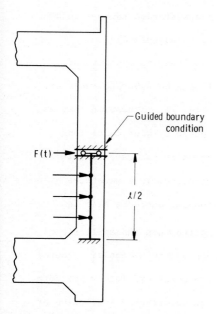

Figure 9-5. Finite Element Model of Strip From Wall "D"

These results are 25 to 31 percent greater than those calculated using a one-degree-of-freedom equivalent system. This provides quantitative and qualitative agreement with the simplified approach and confirms that the wall design is adequate. We cannot be sure that the values calculated by the finite element program are superior to those calculated by the simplified method. Beck, et al. (1981) have compared single-degree-of-freedom methods, finite element methods, and experimental results for a variety of structural elements such as thick and thin slabs, beams, and buried structures, and have found that the single-degree-of-freedom methods are as reliable as finite element methods for calculating structural response to blast loading.

Notice in Figure 9-2 that lacing and haunches at the floor and ceiling are used in Wall "D." The reader is referred to <u>Structures to Resist the Effects of Accidental Explosions</u> (1969) for these details.

HAZARD ASSESSMENT FOR A LARGE INDUSTRIAL COMPLEX

This example problem considers the hazards which can occur due to the rupture of a 2.2×10^7 kg (48,000 tons) propane storage tank at an NGL plant. For this example, the following scenario is postulated.

A propane storage tank has ruptured and spilled its contents of 2.2×10^7 kg of propane. The cause of the spill can be sabotage, weather, material failure, collision by light aircraft, etc. It is assumed that as the spill occurs, gas vapors form and are influenced by wind conditions which direct the vapor toward the refinery plant which has many structures and facilities, of which the largest enclosed structure is a warehouse of volume equal to 1.2×10^9 m^3. A plan of the proposed accident site is shown in Figure 9-6. The vapor cloud may be ignited when the spill first occurs or the vapor cloud may drift some distance before an ignition source is contacted. The amount of liquid propane which boils off into vapor form depends upon the amount of time between spill and catastrophe (explosion or

647

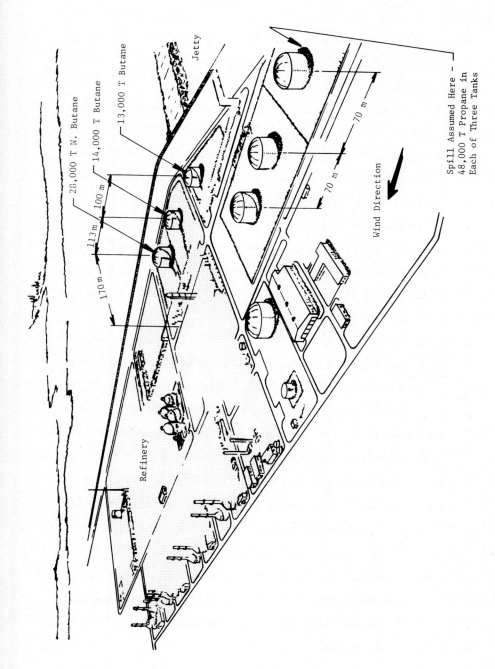

Figure 9-6. Assumed Plant Layout for Example Problem 2

28,000 T N. Butane
14,000 T Butane
13,000 T Butane
100 m
113 m
170 m

Jetty

Refinery

Wind Direction

70 m
70 m

Spill Assumed Here –
48,000 T Propane in
Each of Three Tanks

fire) at which time the remaining liquid vapor will experience extreme heat and will eventually become involved in additional burning. At the time of the event, it is assumed that 1.2×10^5 kg of propane has evaporated.

Included for consideration in this example problem are the hazards due to an explosion and blast, and hazards due to the resulting fireball.

Explosion Hazard

In order to predict the explosion hazard, two factors must be considered: (1) the size of the unconfined vapor cloud, and (2) possible sources of confinement. Confinement is more conducive for an explosive event, but is not required. Two criteria are used to fix the size of the vapor cloud intimately involved in the explosion event:

- For an unconfined vapor cloud, 10 percent of the total vapor available is considered to be involved in an explosion. This is based on a number of post-accident studies from which experience shows that the percent of vapor involvement is typically less than 10 percent.

- For a confined vapor cloud, 100 percent of the confined volume is assumed to be stoichiometrically mixed with air and involved in the explosion. This is because confinement is more efficient in including reactants in an explosion, and it is possible that all available fuel in the enclosure can be involved in the explosion. This is limited to gases in the combustible limits and also available oxygen. The worst-case condition is a stoichiometric mixture because maximum use of oxygen is made. This assumption is conservative because it would be a rare event to have a large volume at stoichiometric conditions.

Using these two criteria plus a vapor cloud size of 1.2×10^5 kg at the time of the assumed explosion, and the largest enclosed structure volume of 1.2×10^5 m^3, two explosion yields can be postulated:

- 1.2×10^4 kg propane.

- 9.4 x 10^3 kg propane (calculated using a volume = 1.2 x 10^5 m^3, a stoichiometric mix of 4 percent propane in air by volume, 0.0224 m^3/mole, MW = 0.044 kg/mole).

Using a heat of combustion of 4.6 x 10^7 J/kg, the two explosive yields are:

- 5.5 x 10^{11} J
- 4.3 x 10^{11} J

These two values are close in magnitude for the large amounts of energy calculated and would show little difference in the predicted pressure field. Only the larger will be considered from this point on, which obviously defines the greater hazard.

Before continuing the example, one needs to determine the type of explosion to be considered, i.e., whether the event is a detonation or a deflagration.

From Zabetakis (1965), detonation velocities for light hydrocarbon combustible mixtures in air are about 1500 m/s, or about Mach 4. But, calculations by Strehlow, et al. (1979) show that the scaled overpressures in blast waves generated by deflagrations traveling as slowly as Mach 1 are essentially equivalent to those from detonations, and blast wave impulses are preserved for even slower deflagrations, i.e., down to about Mach 0.25. These results are shown in Figures 2-25 and 2-26. Figure 2-25 also shows that unaccelerated flames, indicated by the dashed line labeled M_{su} = 0.01, produce only weak sound waves in the surrounding atmosphere. In these two figures, $\bar{P}_s = P_s/p_o$ is the peak overpressure in the shock wave, and $\bar{R} = R$ $p_o^{1/3}/E^{1/3}$ is dimensionless distance from the center of a spherical vapor cloud, with R being distance, p_o being ambient atmospheric pressure, and E being total energy of the cloud which is assumed to be homogeneously mixed with air and within the flammable limits. The scaled impulse is defined as $\bar{i} = i \, a_o/p_o^{2/3} \, E^{1/3}$, where i is impulse and a_o is sound velocity in the surrounding air. The vertical dashed line in Figure 2-25 indicates the edge of the expanded spherical vapor cloud.

From these calculations, and from investigations of other damaging vapor cloud explosions, it is quite clear that detonations are _not_ required for blast wave generation and transmission through the surrounding air. But, Figures 2-25 and 2-26 show that pressures within the vapor cloud are significantly lower for deflagrations than for detonations, and are strongly dependent on the wave speed M_w. Since one cannot anticipate what the wave speed will be for these scenarios, the conservative upper limit of a detonation process is considered. Using the explosion size of 5.5×10^{11} J, the scaled detonation curves can be unscaled and plotted as pressure or specific impulse versus distance. Results are given later. For this example, structural damage and human damage (ear and lung damage) are considered.

Structural Damage

Predictions of structural damage are calculated using two methods. First, two degrees of damage (minor and major) are considered using Figure 8-1. Second, several "standard" structural elements are chosen for analysis using Figure 4-26. These elements include a typical residential stud wall and roof joist, and a roof joist for a stronger-built structure such as a church.

The two degrees of structural damage assumed for this safety study are minor and major structural damage. Minor structural damage is defined as:

- Sheetrock damage
- Glass windows have been broken
- Joints are wrenched
- Partitions are out of some fittings
- Some broken joists, rafters and studs
- Repairable

Major structural damage is defined as:

- Roof partially or totally destroyed
- At least one external wall is heavily damaged

- Some load-bearing members or partitions have been destroyed

- Unrepairable

Figure 8-1 presents contours of "constant relative damage," plotted as functions of side-on specific impulse (pascal-seconds) and peak side-on overpressure (pascals). A third degree of damage termed "partial demolition" was not considered due to the implied severity at that damage level. Using the explosive energy and Figures 2-25, 2-26, and 8-1, one can determine distances for the various damage levels for a particular energy release.

The basis for these curves is British data from enemy bombing in World War II plus records of explosions dating from 1871. Although this relationship was developed for the average British dwelling house, it also works well for factories, main offices and main engineering workshops. One should expect some variation about these limits because of differences in structures, such as comparing damage of a well-built office building or church to an old inexpensive residential structure. However, variations should not be great compared to the total distance predicted. Results are discussed later.

Standard Structural Elements

Because many newer residential structures may not be as well-built as older British dwelling houses, calculations were included of loads required to damage "typical" residential stud wall and joist roof, and a "typical" roof joist for a structure such as a church. The dimensions used for these calculations are listed below and are obtained from measurements of existing structures.

- Stud Wall (0.038 m x 0.089 m stud 0.4 m on center with insulation board and asbestos shingle siding)

    b = 0.4 m

    H = 0.089 m

    ℓ = 2.3 m

$$\left.\begin{array}{l} \alpha_p = 8 \\ \alpha_i = 1.461 \end{array}\right\} \text{ (see Figure 4-26)}$$

$E = 1.2 \times 10^{10}$ Pa

$\sigma_{ult} = 5.9 \times 10^7$ Pa (degraded for knots)

$I = 2.23 \times 10^{-6}$ m^4

- Residential Roof Joist (0.038 m x 0.089 m joist 0.61 m on center with 0.013 m decking)

  $b = 0.61$ m

  $H = 0.089$ m

  $\ell = 4.0$ m

$$\left.\begin{array}{l} \alpha_p = 8 \\ \alpha_i = 1.461 \end{array}\right\} \text{ (see Figure 4-26)}$$

  $E = 1.2 \times 10^{12}$ Pa

  $\sigma_{ult} = 5.5 \times 10^7$ Pa (degraded for knots)

  $I = 2.23 \times 10^{-6}$ m^4

- Church Roof Joist (0.038 m x 0.14 m joist 0.61 m on center with 0.02 m decking)

  $b = 0.61$ m

  $H = 0.143$ m

  $\ell = 4.115$ m

$$\left.\begin{array}{l} \alpha_p = 8 \\ \alpha_i = 1.461 \end{array}\right\} \text{ (see Figure 4-26)}$$

  $E = 1.2 \times 10^{10}$ Pa

  $\sigma_{ult} = 5.9 \times 10^7$ Pa (degraded for knots)

  $I = 0.224$ m^4

Using the explosive energy and Figures 2-25, 2-26, and 4-26, loadings and distances can be determined at which these members would fail. Results are given later.

Human Damage

Two degrees of blast-induced body injury were assessed: threshold
for eardrum rupture and threshold for casualty (death) due to lung damage.
Each accident scenario has its own unique environment with trees, hills,
buildings, and various other topographical conditions which may dissipate
the energy of the blast wave or reflect it and amplify its effect on an
individual. To eliminate these complicating factors, the human body will
be assumed to be standing in the free blast field on flat and level ground
when contacted by the blast wave. The ear is the most sensitive part of
the human body to blast waves. Internal body damage will be most likely
to occur in regions of greatest density variations of adjacent tissues.
The air-filled sacs of the lungs are, therefore, more susceptible to primary blast damage than any other internal organ. The threshold to eardrum
rupture is, therefore, an estimate of the limit at which human beings are
injured by blast waves; and the threshold to lung damage presents an estimate of the limit at which deaths can be expected due to primary blast damage.

The method for assessing ear damage is as follows:

- From Figure 8-11, determine the value of P_s which corresponds to
  a given damage level. Threshold of eardrum rupture was used
  for this sample problem.

- Use the explosive energy to descale Figures 2-25 and 2-26 and
  determine distances at which the values of P_s and i_s correspond
  to that obtained in the previous step.

Results of ear damage are given later.

As presented in Figure 8-8, the severity of lung damage can be shown
to depend upon a scaled peak incident overpressure $\bar{P}_s = P_s/p_o$ and a scaled
specific impulse $\bar{i}_s$ where

$$\bar{i}_s = \frac{i_s}{P_o^{1/2} m^{1/3}}$$

The equation uses m, which is the mass of the individual exposed to the blast. It is suggested that 5 kg be used for a baby, 55 kg for a woman, and 70 kg for a man. The method for assessing lung damage is as follows:

- From Figure 8-8, determine values of P_s and i_s which correspond to a given damage level. Threshold of lung damage was used for this sample problem. Also, the mass of a woman was used (55 kg).

- Use the explosive energy to descale Figures 2-25 and 2-26 and determine distances at which the values of P_s and i_s correspond to that obtained in the previous step.

Results of lung damage are given, along with ear damage and structural damage, for an explosion involving 5.5×10^{11} joules of energy.

Explosion Hazard Results

Using the calculated energy of 5.5×10^{11} joules and Figures 2-25 and 2-26, unscaled curves of peak overpressures and specific impulse versus distance were determined and are shown in Figures 9-7 and 9-8. Using these unscaled curves and Figures 8-1, 4-26, 8-11, and 8-8, it was determined that all damage levels considered were in a pressure loading realm. Table 9-5 includes the overpressure loads required to reach the various damage

Table 9-5. Summary of Explosion Hazard Calculations

Type of Damage	Side-On Pressure Level (pascals)	Standoff (meters)
Threshold of Fatality from Lung Damage	70,000	105
Threshold of Eardrum Rupture	35,000	190
Major Structural Damage	18,000	350
Minor Structural Damage	4,700	1,600
Roof Joist Breakage (house)	3,250	1,830
Roof Joist Breakage (church)	3,450	1,800
Stud Breakage (house)	6,900	980

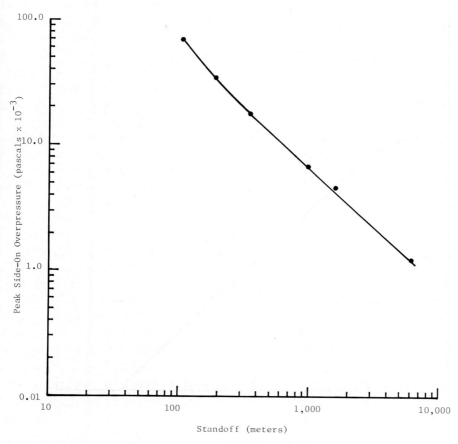

Figure 9-7.   Peak Side-On Overpressure Versus Standoff

levels. Also, Table 9-5 includes the standoff at which the various levels
of damage are predicted to occur for this energy release.

Fireball Hazard

Predictions can be made for fireball diameter and duration using
Figures 7-2 and 7-3. A second check of fireball radius can be obtained
from Figure 7-4, as well as fireball height. The amount of spill mass to
be used for these curves is difficult to predict for a spill such as this.
The data collected to define the curves in Figures 7-2 and 7-3 were from
fireballs of spills that were intimately mixed, i.e., a violent reaction
and mixing originated the spill; also the data include propellants such as

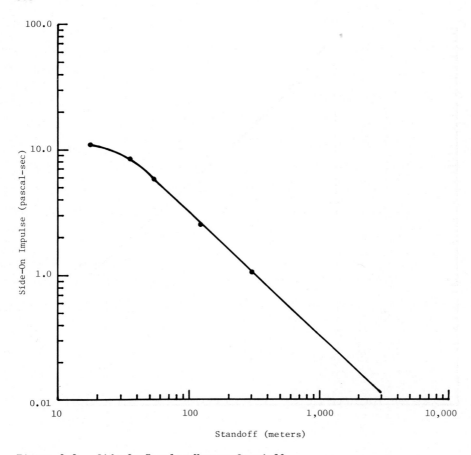

Figure 9-8. Side-On Impulse Versus Standoff

LH$_2$ and RP-1 which rapidly react in the fireball. For the postulated scenario, 2.2 x 10^7 kg of propane was spilled, 1.2 x 10^5 kg had boiled at the
time of the explosion, and 1.2 x 10^4 kg had been involved in the explosion.

It is felt that using the 1.2 x 10^4 kg value for fireball size would under-
estimate the size; also, the 2.2 x 10^7 kg value would probably overestimate
the fireball size. The most likely event is that all gas phase butane (1.2
x 10^5 kg) would partake in the fireball, with possibly some liquid propane
boiling and adding to the fireball; however, most of the liquid propane
would be a part of a pool fire after the fireball event. Therefore, 1.2 x
10^5 kg shall be used in determining fireball size and duration.

Figure 7-2 shows the relation $D = 3.86 W^{0.32}$ for W = kg and D = meters, and Figure 7-3 shows the relation $t = 0.299 W^{0.32}$ for W = kg and t = seconds. Using these two equations and the mass of propane of 1.2×10^5 kg, the following fireball size and duration were obtained:

- Diameter = 163 meters

- Duration = 12.6 seconds

Figure 7-4 gives a method for predicting fireball height and radius. Using the value of duration determined above (12.6 seconds), and a mass of 1.2×10^5 kg and a fireball temperature of 2200°K, the following values were determined:

$$\frac{[t] \, [\theta]^{10/3}}{[M]^{1/3} \times 10^{10}} = \frac{[12.6] \, [2200]^{10/3}}{\left[1.2 \times 10^5\right]^{1/3} \times 10^{10}} = 3.54$$

and from Figure 7-2:

$$\frac{[\theta]^{1/3} \, [H]}{[M]^{1/3}} = \frac{[2200]^{1/3} \, [H]}{\left[1.2 \times 10^5\right]^{1/3}} = 55$$

and

$$\frac{[\theta]^{1/3} \, [R]}{[M]^{1/3}} = \frac{[2200]^{1/3} \, [R]}{\left[1/2 \times 10^5\right]^{1/3}} = 40$$

Solving for H and R:

- H = 208 meters,

- R = 147 meters, or

- Diameter = 294 meters

The two methods give different values of fireball diameter; the average of these two can be used for the estimate:

$$\frac{294 + 163}{2} = 230 \text{ meters} = \text{Average Fireball Diameter}$$

Figure 7-11 presents different degrees of burn injury (first, second, and third degree burn) as a function of exposure duration and total energy of flash (Q). Using the calculated fireball duration for exposure duration (12.6 seconds), values of Q can be determined from Figure 7-11 for the different burn levels. These are:

- 1st Degree Burn Q = 3.5 cal/cm^2
- 2nd Degree Burn Q = 6.5 cal/cm^2
- 3rd Degree Burn Q = 9.5 cal/cm^3

From Chapter 7, Equation (7-22) relates Q to fireball size, total fuel energy, and standoff as follows:

$$\frac{Q}{(BG) \ E^{1/3} \ \theta^{2/3}} = \frac{D^2/R^2}{(F + D^2/R^2)} \qquad \begin{array}{c}(7\text{-}22)\\(\text{repeated})\end{array}$$

where Q is the radiant energy per unit area (J/m^2), D is the fireball diameter which equals 230 meters, R is the standoff (to be solved for in meters), E is the total energy of fuel = 5.5 x 10^{11} J = 1.2 x 10^4 kg (use mass value), θ is the fireball temperature which equals 2200°K, G and F are constant coefficients where G equals 2.26 x 10^{-6} J/m^2 sec^6 K^4 and F equals 161.7 (unitless), and B is equal to $\theta^{10/3}$ t/E$^{1/3}$ where t equals 12.6 sec.

Using these values and solving for values of K, the following are obtained:

Burn Level	Distance (meters)
1st Degree Burn	1200
2nd Degree Burn	900
3rd Degree Burn	740

Compared to the explosive hazard, the distances above for fireball reveal hazards to humans at much greater standoffs than the blast effects. However, the structural damage due to the blast is found to occur at farther distances than for the fireball. The fact that the fireball is predicted to be hazardous to humans at farther distances than the blast may be due to the choice of the amount of fuel assumed to be involved in the fireball compared to that assumed to be involved in the blast.

CHAPTER 9

LIST OF SYMBOLS

A	beam cross-section area
A_s	rebar area
b	beam width
C_w, C_v	nondimensional coefficients
c_τ, c_u, c_ϕ	forces per unit volume
d'	concrete cover measured to rebar center
d_c	rebar spacing (through slab thickness direction)
E, E_c, E_s	elastic modulus
g	acceleration of gravity
H	column height; concrete section thickness
h	beam depth; plate thickness
I	total impulse; second moment of area
i_r	reflected specific impulse
J	second moment of area
K_{LM}	load-mass factor
k, k_1, k_2	spring constants
L	height of center of gravity
ℓ	beam length
M_m	mass moment of inertia

M_p	plastic bending moment
M_1, M_2	masses
m	mass
P	applied pressure
P_{qs}	quasi-static pressure
P_r	peak reflected overpressure
R	standoff distance from explosive charge
R_M	maximum resistance
T	response period
T_L	load duration
T_N	fundamental vibration period
V	volume
V_1, V_2	velocities
W	charge weight
X, Y	half-spans of rectangular plates or membranes
x, z	Cartesian coordinates; distance to neutral axis
x_1, x_2	displacements
Y_1, Y_2, Z_1, Z_2	displacement functions
Z	plastic section modulus
α	length-to-width ratio
α_i, α_p	nondimensional coefficients
μ	ductility ratio
ν	Poisson's ratio
ρ	material density
σ, σ_y, σ_{max}, σ_1, σ_2, σ_s, σ_c	stresses
ϕ_p, ϕ_i	nondimensional coefficients
ϕ	angle
ψ_p, ψ_i, ψ_e	dimensionless coefficients
ω_1, ω_2	circular frequencies

APPENDIX A

TRANSFORMATION FACTORS FOR STRUCTURAL ELEMENTS

Table A-1. Transformation Factors for Beams and One-Way Slabs
[U. S. Army Corps of Engineering Manual (1975)]

Simply Supported

Loading Diagram	Strain Range	Load Factor K_L	Mass Factor K_M		Load-Mass Factor K_{LM}		Maximum Resistance R_m	Spring Constant k	Dynamic Reaction V
			Concentrated Mass*	Uniform Mass	Concentrated Mass*	Uniform Mass			
$P = pL$	Elastic	0.64	....	0.50	....	0.78	$\dfrac{8M_p}{L}$	$\dfrac{384.0\ EI}{5L^3}$	$0.39\ R\ +\ 0.11\ P$
	Plastic	0.50	....	0.33	....	0.66	$\dfrac{8M_p}{L}$	0	$0.38\ R_m\ +\ 0.12\ P$
P (concentrated at $\tfrac{L}{2},\tfrac{L}{2}$)	Elastic	1.00	1.00	0.49	1.00	0.49	$\dfrac{4M_p}{L}$	$\dfrac{48.0\ EI}{L^3}$	$0.78\ R\ -\ 0.28\ P$
	Plastic	1.00	1.00	0.33	1.00	0.33	$\dfrac{4M_p}{L}$	0	$0.75\ R_m\ -\ 0.25\ P$
$\tfrac{P}{2},\tfrac{P}{2}$ (at $\tfrac{L}{3},\tfrac{L}{3},\tfrac{L}{3}$)	Elastic	0.87	0.76	0.52	0.87	0.60	$\dfrac{6M_p}{L}$	$\dfrac{56.4\ EI}{L^3}$	$0.62\ R\ -\ 0.12\ P$
	Plastic	1.00	1.00	0.56	1.00	0.56	$\dfrac{6M_p}{L}$	0	$0.75\ R_m\ -\ 0.25\ P$

*Equal parts of the concentrated mass are lumped at each concentrated load.

Table A-2. Transformation Factors for Beams and One-Way Slabs [U. S. Army Corps of Engineering Manual (1975)]

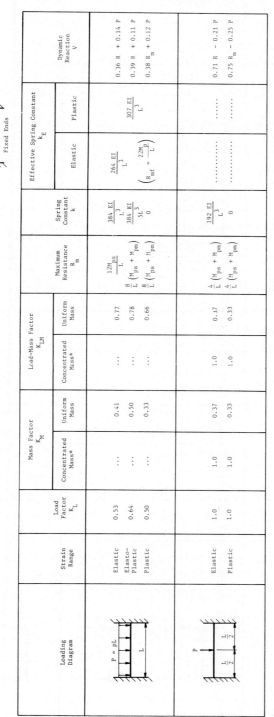

Loading Diagram	Strain Range	Load Factor K_L	Mass Factor K_M		Load-Mass Factor K_{LM}		Maximum Resistance R_m	Spring Constant k	Effective Spring Constant k_E		Dynamic Reaction V
			Concentrated Mass*	Uniform Mass	Concentrated Mass*	Uniform Mass			Elastic	Plastic	
$P = pL$; L (Fixed Ends)	Elastic	0.53	...	0.41	...	0.77	$\dfrac{12 M_{ps}}{L}$	$\dfrac{384\,EI}{L^3}$	$\dfrac{264\,EI}{L^3}$ $\left(R_{mf}=\dfrac{22M}{L}-\dfrac{P}{L}\right)$	$\dfrac{307\,EI}{L^3}$	$0.36\,R + 0.14\,P$
	Elasto-Plastic	0.64	...	0.50	...	0.78	$\dfrac{8}{L}\left(M_{ps}+M_{pm}\right)$	$\dfrac{384\,EI}{5L^3}$			$0.39\,R + 0.11\,P$
	Plastic	0.50	...	0.33	...	0.66	$\dfrac{8}{L}\left(M_{ps}+M_{pm}\right)$	0			$0.38\,R_m + 0.12\,P$
P ; $\frac{L}{2}$, $\frac{L}{2}$	Elastic	1.0	1.0	0.37	1.0	0.37	$\dfrac{4}{L}\left(M_{ps}+M_{pm}\right)$	$\dfrac{192\,EI}{L^3}$	....	......	$0.71\,R - 0.21\,P$
	Plastic	1.0	1.0	0.33	1.0	0.33	$\dfrac{4}{L}\left(M_{ps}+M_{pm}\right)$	0	....	......	$0.75\,R_m - 0.25\,P$

*Concentrated mass is lumped at the concentrated load.

Table A-3. Transformation Factors for Beams and One-Way Slabs
[U. S. Army Corps of Engineering Manual (1975)]

Simply Supported And Fixed

Loading Diagram	Strain Range	Load Factor K_L	Mass Factor K_M		Load-Mass Factor K_{LM}		Maximum Resistance R_m	Spring Constant k	Effective Spring Constant k_E		Dynamic Reaction V
			Concentrated Mass*	Uniform Mass	Concentrated Mass*	Uniform Mass			Elastic	Plastic	
	Elastic	0.58	.....	0.45	.....	0.78	$\dfrac{8M_{ps}}{L}$	$\dfrac{185\,EI}{L^3}$	$\dfrac{153.0\,EI}{L^3}$	$\dfrac{160\,EI}{L^3}$	$V_1 = 0.26\,R + 0.12\,P$ $V_2 = 0.43\,R + 0.19\,P$
	Elasto-Plastic	0.64	.....	0.50	.....	0.78	$\dfrac{4}{L}\left(M_{ps} + 2M_{pm}\right)$	$\dfrac{384\,EI}{5L^3}$	$\left(R_{mf} = \dfrac{14.60\,M}{L}\right)$		$V_1 = V_2 = 0.39\,R + 0.11\,P$
	Plastic	0.50	.....	0.33	.....	0.66	$\dfrac{4}{L}\left(M_{ps} + 2M_{pm}\right)$	0			$V_1 = V_2 = 0.38\,R_m + 0.12\,P$
	Elastic	1.00	1.00	0.43	1.00	0.43	$\dfrac{14M_{ps}}{3L}$	$\dfrac{107\,EI}{L^3}$	$\dfrac{104.0\,EI}{L^3}$	$\dfrac{106\,EI}{L^3}$	$V_1 = 0.54\,R + 0.14\,P$ $V_2 = 0.25\,R + 0.07\,P$
	Elasto-Plastic	1.00	1.00	0.49	1.00	0.49	$\dfrac{2}{L}\left(M_{ps} + 2M_{pm}\right)$	$\dfrac{48\,EI}{L^3}$	$\left(R_{mf} = \dfrac{6.63\,M}{L}\right)$		$V_1 = V_2 = 0.78\,R - 0.28\,P$
	Plastic	1.00	1.00	0.33	1.00	0.33	$\dfrac{2}{L}\left(M_{ps} + 2M_{pm}\right)$	0			$V_1 = V_2 = 0.75\,R_m - 0.25\,P$
	Elastic	0.81	0.67	0.45	0.83	0.55	$\dfrac{6M_{ps}}{L}$	$\dfrac{132\,EI}{L^3}$	$\dfrac{117.5\,EI}{L^3}$	$\dfrac{122\,EI}{L^3}$	$V_1 = 0.17\,R + 0.17\,P$ $V_2 = 0.33\,R + 0.33\,P$
	Elasto-Plastic	0.87	0.76	0.52	0.87	0.60	$\dfrac{2}{L}\left(M_{ps} + 3M_{pm}\right)$	$\dfrac{56\,EI}{L^3}$	$\left(R_{mf} = \dfrac{9.52\,M_p}{L}\right)$		$V_1 = V_2 = 0.62\,R - 0.12\,P$
	Plastic	1.00	1.00	0.56	1.00	0.56	$\dfrac{2}{L}\left(M_{ps} + 3M_{pm}\right)$	......			$V_1 = 0.56\,R_m - 0.25\,P$ $V_2 = 0.56\,R_m + 0.13\,P$

*Equal parts of the concentrated mass are lumped at each concentrated load.

Table A-4. Transformation Factors for Two-Way Slabs: Simple Supports – Four Sides, Uniform Load; For Poisson's Ratio = 0.3 [U. S. Army Corps of Engineering Manual (1975)]

Simple Support

Strain Range	a/b	Load Factor K_L	Mass Factor K_M	Load-Mass Factor K_{LM}	Maximum Resistance	Spring Constant k	Dynamic Reactions	
							V_A	V_B
Elastic	1.0	0.45	0.31	0.68	$\frac{12}{a}\left(M_{pfa} + M_{pfb}\right)$	$252\ EI/a^2$	$0.07\ F + 0.18\ R$	$0.07\ F + 0.18\ R$
	0.9	0.47	0.33	0.77	$\frac{1}{a}\left(12.0\ M_{pfa} + 11.0\ M_{pfb}\right)$	$230\ EI/a^2$	$0.06\ F + 0.16\ R$	$0.08\ F + 0.20\ R$
	0.8	0.49	0.35	0.71	$\frac{1}{a}\left(12.0\ M_{pfa} + 10.3\ M_{pfb}\right)$	$212\ EI/a^2$	$0.06\ F + 0.14\ R$	$0.08\ F + 0.22\ R$
	0.7	0.51	0.37	0.73	$\frac{1}{a}\left(12.0\ M_{pfa} + 9.8\ M_{pfb}\right)$	$201\ EI/a^2$	$0.05\ F + 0.13\ R$	$0.08\ F + 0.24\ R$
	0.6	0.53	0.39	0.74	$\frac{1}{a}\left(12.0\ M_{pfa} + 9.3\ M_{pfb}\right)$	$197\ EI/a^2$	$0.04\ F + 0.11\ R$	$0.09\ F + 0.26\ R$
	0.5	0.55	0.41	0.75	$\frac{1}{a}\left(12.0\ M_{pfa} + 9.0\ M_{pfb}\right)$	$201\ EI/a^2$	$0.04\ F + 0.09\ R$	$0.09\ F + 0.28\ R$
Plastic	1.0	0.33	0.17	0.51	$\frac{12}{a}\left(M_{pfa} + M_{pfb}\right)$	0	$0.09\ F + 0.16\ R_m$	$0.09\ F + 0.16\ R_m$
	0.9	0.35	0.18	0.51	$\frac{1}{a}\left(12.0\ M_{pfa} + 11.0\ M_{pfb}\right)$	0	$0.08\ F + 0.15\ R_m$	$0.09\ F + 0.18\ R_m$
	0.8	0.37	0.20	0.54	$\frac{1}{a}\left(12.0\ M_{pfa} + 10.3\ M_{pfb}\right)$	0	$0.07\ F + 0.13\ R_m$	$0.10\ F + 0.20\ R_m$
	0.7	0.38	0.22	0.58	$\frac{1}{a}\left(12.0\ M_{pfa} + 9.8\ M_{pfb}\right)$	0	$0.06\ F + 0.12\ R_m$	$0.10\ F + 0.22\ R_m$
	0.6	0.40	0.23	0.58	$\frac{1}{a}\left(12.0\ M_{pfa} + 9.3\ M_{pfb}\right)$	0	$0.05\ F + 0.10\ R_m$	$0.10\ F + 0.25\ R_m$
	0.5	0.42	0.25	0.59	$\frac{1}{a}\left(12.0\ M_{pfa} + 9.0\ M_{pfb}\right)$	0	$0.04\ F + 0.08\ R_m$	$0.11\ F + 0.27\ R_m$

Table A-5. Transformation Factors for Two-Way Slabs:
Short Edges Fixed - Long Edges Simply Supported;
for Poisson's Ratio = 0.3
[U. S. Army Corps of Engineering Manual (1975)]

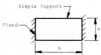

Strain Range	a/b	Load Factor K_L	Mass Factor K_M	Load-Mass Factor K_{LM}	Maximum Resistance	Spring Constant k	Dynamic Reactions	
							V_A	V_B
Elastic	1.0	0.39	0.26	0.67	$20.4\ M^\circ_{psa}$	$575\ EI_a/a^2$	$0.09\ F + 0.16\ R$	$0.07\ F + 0.18\ R$
	0.9	0.41	0.28	0.68	$10.2\ M^\circ_{psa} + \frac{11.0}{a}\ M_{pfb}$	$476\ EI_a/a^2$	$0.08\ F + 0.14\ R$	$0.08\ F + 0.20\ R$
	0.8	0.44	0.30	0.68	$10.2\ M^\circ_{psa} + \frac{10.3}{a}\ M_{pfb}$	$396\ EI_a/a^2$	$0.08\ F + 0.12\ R$	$0.08\ F + 0.22\ R$
	0.7	0.46	0.33	0.72	$9.3\ M^\circ_{psa} + \frac{9.7}{a}\ M_{pfb}$	$328\ EI_a/a^2$	$0.07\ F + 0.11\ R$	$0.08\ F + 0.24\ R$
	0.6	0.48	0.35	0.73	$8.5\ M^\circ_{psa} + \frac{9.3}{a}\ M_{pfb}$	$283\ EI_a/a^2$	$0.06\ F + 0.09\ R$	$0.09\ F + 0.26\ R$
	0.5	0.51	0.37	0.73	$7.4\ M^\circ_{psa} + \frac{9.0}{a}\ M_{pfb}$	$243\ EI_a/a^2$	$0.05\ F + 0.08\ R$	$0.09\ F + 0.28\ R$
Elasto-Plastic	1.0	0.46	0.31	0.67	$\frac{1}{a}\left[12.0\left(M_{pfa} + M_{psa}\right) + 12.0\ M_{pfb}\right]$	$271\ EI_a/a^2$	$0.07\ F + 0.18\ R$	$0.07\ F + 0.18\ R$
	0.9	0.47	0.33	0.70	$\frac{1}{a}\left[12.0\left(M_{pfa} + M_{psa}\right) + 11.0\ M_{pfb}\right]$	$248\ EI_a/a^2$	$0.06\ F + 0.16\ R$	$0.08\ F + 0.20\ R$
	0.8	0.49	0.35	0.71	$\frac{1}{a}\left[12.0\left(M_{pfa} + M_{psa}\right) + 10.3\ M_{pfb}\right]$	$228\ EI_a/a^2$	$0.06\ F + 0.14\ R$	$0.08\ F + 0.22\ R$
	0.7	0.51	0.37	0.72	$\frac{1}{a}\left[12.0\left(M_{pfa} + M_{psa}\right) + 9.7\ M_{pfb}\right]$	$216\ EI_a/a^2$	$0.05\ F + 0.13\ R$	$0.08\ F + 0.24\ R$
	0.6	0.53	0.37	0.70	$\frac{1}{a}\left[12.0\left(M_{pfa} + M_{psa}\right) + 9.3\ M_{pfb}\right]$	$212\ EI_a/a^2$	$0.04\ F + 0.11\ R$	$0.09\ F + 0.26\ R$
	0.5	0.55	0.41	0.74	$\frac{1}{a}\left[12.0\left(M_{pfa} + M_{psa}\right) + 9.0\ M_{pfb}\right]$	$216\ EI_a/a^2$	$0.04\ F + 0.09\ R$	$0.09\ F + 0.28\ R$
Plastic	1.0	0.33	0.17	0.51	$\frac{1}{a}\left[12.0\left(M_{pfa} + M_{psa}\right) + 12.0\ M_{pfb}\right]$	0	$0.09\ F + 0.16\ R_m$	$0.09\ F + 0.16\ R_m$
	0.9	0.35	0.18	0.51	$\frac{1}{a}\left[12.0\left(M_{pfa} + M_{psa}\right) + 11.0\ M_{pfb}\right]$	0	$0.08\ F + 0.15\ R_m$	$0.09\ F + 0.18\ R_m$
	0.8	0.37	0.20	0.54	$\frac{1}{a}\left[12.0\left(M_{pfa} + M_{psa}\right) + 10.3\ M_{pfb}\right]$	0	$0.07\ F + 0.13\ R_m$	$0.10\ F + 0.20\ R_m$
	0.7	0.38	0.22	0.58	$\frac{1}{a}\left[12.0\left(M_{pfa} + M_{psa}\right) + 9.7\ M_{pfb}\right]$	0	$0.06\ F + 0.12\ R_m$	$0.10\ F + 0.22\ R_m$
	0.6	0.40	0.23	0.58	$\frac{1}{a}\left[12.0\left(M_{pfa} + M_{psa}\right) + 9.3\ M_{pfb}\right]$	0	$0.05\ F + 0.10\ R_m$	$0.10\ F + 0.25\ R_m$
	0.5	0.42	0.25	0.59	$\frac{1}{a}\left[12.0\left(M_{pfa} + M_{psa}\right) + 9.0\ M_{pfb}\right]$	0	$0.04\ F + 0.08\ R_m$	$0.11\ F + 0.27\ R_m$

Table A-6. Transformation Factors for Two-Way Slabs:
Short Sides Simply Supported - Long Sides Fixed;
for Poisson's Ratio = 0.3
[U. S. Army Corps of Engineering Manual (1975)]

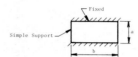

Strain Range	a/b	Load Factor K_L	Mass Factor K_M	Load-Mass Factor K_{LM}	Maximum Resistance	Spring Constant k	Dynamic Reactions	
							V_A	V_B
Elastic	1.0	0.39	0.26	0.67	$20.4\ M^\circ_{psb}$	$575\ EI_a/a^2$	0.07 F + 0.18 R	0.09 F + 0.16 R
	0.9	0.40	0.28	0.70	$19.5\ M^\circ_{psb}$	$600\ EI_a/a^2$	0.06 F + 0.16 R	0.10 F + 0.18 R
	0.8	0.42	0.29	0.69	$19.5\ M^\circ_{psb}$	$610\ EI_a/a^2$	0.06 F + 0.14 R	0.11 F + 0.19 R
	0.7	0.43	0.31	0.71	$20.2\ M^\circ_{psb}$	$662\ EI_a/a^2$	0.05 F + 0.13 R	0.11 F + 0.21 R
	0.6	0.45	0.33	0.73	$21.2\ M^\circ_{psb}$	$731\ EI_a/a^2$	0.04 F + 0.11 R	0.12 F + 0.23 R
	0.5	0.45	0.34	0.72	$22.2\ M^\circ_{psb}$	$850\ EI_a/a^2$	0.04 F + 0.09 R	0.12 F + 0.25 R
Elasto-Plastic	1.0	0.46	0.31	0.67	$\frac{1}{a}\left[12.0\ M_{pfa} + 12.0\left(M_{psb} + M_{pfb}\right)\right]$	$271\ EI_a/a^2$	0.07 F + 0.18 R	0.07 F + 0.18 R
	0.9	0.47	0.33	0.70	$\frac{1}{a}\left[12.0\ M_{pfa} + 11.0\left(M_{psb} + M_{pfb}\right)\right]$	$248\ EI_a/a^2$	0.06 F + 0.16 R	0.08 F + 0.20 R
	0.8	0.49	0.35	0.71	$\frac{1}{a}\left[12.0\ M_{pfa} + 10.3\left(M_{psb} + M_{pfb}\right)\right]$	$228\ EI_a/a^2$	0.06 F + 0.14 R	0.08 F + 0.22 R
	0.7	0.51	0.37	0.73	$\frac{1}{a}\left[12.0\ M_{pfa} + 9.8\left(M_{psb} + M_{pfb}\right)\right]$	$216\ EI_a/a^2$	0.06 F + 0.13 R	0.08 F + 0.24 R
	0.6	0.53	0.39	0.74	$\frac{1}{a}\left[12.0\ M_{pfa} + 9.3\left(M_{psb} + M_{pfb}\right)\right]$	$212\ EI_a/a^2$	0.04 F + 0.11 R	0.09 F + 0.26 R
	0.5	0.55	0.41	0.74	$\frac{1}{a}\left[12.0\ M_{pfa} + 9.0\left(M_{psb} + M_{pfb}\right)\right]$	$216\ EI_a/a^2$	0.04 F + 0.09 R	0.09 F + 0.28 R
Plastic	1.0	0.33	0.17	0.51	$\frac{1}{a}\left[12.0\ M_{pfa} + 12.0\left(M_{psb} + M_{pfb}\right)\right]$	0	0.09 F + 0.16 R_m	0.09 F + 0.16 R_m
	0.9	0.35	0.18	0.51	$\frac{1}{a}\left[12.0\ M_{pfa} + 11.0\left(M_{psb} + M_{pfb}\right)\right]$	0	0.08 F + 0.15 R_m	0.09 F + 0.18 R_m
	0.8	0.37	0.20	0.54	$\frac{1}{a}\left[12.0\ M_{pfa} + 10.3\left(M_{psb} + M_{pfb}\right)\right]$	0	0.07 F + 0.13 R_m	0.10 F + 0.20 R_m
	0.7	0.38	0.22	0.58	$\frac{1}{a}\left[12.0\ M_{pfa} + 9.8\left(M_{psb} + M_{pfb}\right)\right]$	0	0.06 F + 0.12 R_m	0.10 F + 0.22 R_m
	0.6	0.40	0.23	0.58	$\frac{1}{a}\left[12.0\ M_{pfa} + 9.3\left(M_{psb} + M_{pfb}\right)\right]$	0	0.05 F + 0.10 R_m	0.10 F + 0.25 R_m
	0.5	0.42	0.25	0.59	$\frac{1}{a}\left[12.0\ M_{pfa} + 9.0\left(M_{psb} + M_{pfb}\right)\right]$	0	0.04 F + 0.08 R_m	0.11 F + 0.27 R_m

668

Table A-7. Transformation Factors for Two-Way Slabs: Fixed Supports – Uniform Load; for Poisson's Ratio = 0.3
[U. S. Army Corps of Engineering Manual (1975)]

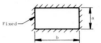

Strain Range	a/b	Load Factor K_L	Mass Factor K_M	Load-Mass Factor K_{LM}	Maximum Resistance	Spring Constant k	V_A	V_B
Elastic	1.0	0.33	0.21	0.63	$29.2\,M_{psb}^m$	$810\,EI_a/a^2$	0.10 F + 0.15 R	0.10 F + 0.15 R
	0.9	0.34	0.23	0.68	$27.4\,M_{psb}^m$	$742\,EI_a/a^2$	0.09 F + 0.14 R	0.10 F + 0.17 R
	0.8	0.36	0.25	0.69	$26.4\,M_{psb}^m$	$705\,EI_a/a^2$	0.08 F + 0.12 R	0.11 F + 0.19 R
	0.7	0.38	0.27	0.71	$26.2\,M_{psb}^m$	$692\,EI_a/a^2$	0.07 F + 0.11 R	0.11 F + 0.21 R
	0.6	0.41	0.29	0.71	$27.3\,M_{psb}^m$	$724\,EI_a/a^2$	0.06 F + 0.09 R	0.12 F + 0.23 R
	0.5	0.43	0.31	0.72	$30.2\,M_{psb}^m$	$806\,EI_a/a^2$	0.05 F + 0.08 R	0.12 F + 0.25 R
Elasto-Plastic	1.0	0.46	0.31	0.67	$\frac{1}{a}\left[12.0\left(M_{pfa}+M_{psa}\right)+12.0\left(M_{pfb}+M_{psb}\right)\right]$	$252\,EI_a/a^2$	0.07 F + 0.18 R	0.07 F + 0.18 R
	0.9	0.47	0.33	0.70	$\frac{1}{a}\left[12.0\left(M_{pfa}+M_{psa}\right)+11.0\left(M_{pfb}+M_{psb}\right)\right]$	$230\,EI_a/a^2$	0.06 F + 0.16 R	0.08 F + 0.20 R
	0.8	0.49	0.35	0.71	$\frac{1}{a}\left[12.0\left(M_{pfa}+M_{psa}\right)+10.3\left(M_{pfb}+M_{psb}\right)\right]$	$212\,EI_a/a^2$	0.06 F + 0.14 R	0.08 F + 0.22 R
	0.7	0.51	0.37	0.73	$\frac{1}{a}\left[12.0\left(M_{pfa}+M_{psa}\right)+9.8\left(M_{pfb}+M_{psb}\right)\right]$	$201\,EI_a/a^2$	0.05 F + 0.13 R	0.08 F + 0.24 R
	0.6	0.53	0.39	0.74	$\frac{1}{a}\left[12.0\left(M_{pfa}+M_{psa}\right)+9.3\left(M_{pfb}+M_{psb}\right)\right]$	$197\,EI_a/a^2$	0.04 F + 0.11 R	0.09 F + 0.26 R
	0.5	0.55	0.41	0.75	$\frac{1}{a}\left[12.0\left(M_{pfa}+M_{psa}\right)+9.0\left(M_{pfb}+M_{psb}\right)\right]$	$201\,EI_a/a^2$	0.04 F + 0.09 R	0.09 F + 0.28 R
Plastic	1.0	0.33	0.17	0.51	$\frac{1}{a}\left[12.0\left(M_{pfa}+M_{psa}\right)+12.0\left(M_{pfb}+M_{psb}\right)\right]$	0	0.09 F + 0.16 R_m	0.09 F + 0.16 R_m
	0.9	0.35	0.18	0.51	$\frac{1}{a}\left[12.0\left(M_{pfa}+M_{psa}\right)+11.0\left(M_{pfb}+M_{psb}\right)\right]$	0	0.08 F + 0.15 R_m	0.09 F + 0.18 R_m
	0.8	0.37	0.20	0.54	$\frac{1}{a}\left[12.0\left(M_{pfa}+M_{psa}\right)+10.3\left(M_{pfb}+M_{psb}\right)\right]$	0	0.07 F + 0.13 R_m	0.10 F + 0.20 R_m
	0.7	0.38	0.22	0.58	$\frac{1}{a}\left[12.0\left(M_{pfa}+M_{psa}\right)+9.8\left(M_{pfb}+M_{psb}\right)\right]$	0	0.06 F + 0.12 R_m	0.10 F + 0.22 R_m
	0.6	0.40	0.23	0.58	$\frac{1}{a}\left[12.0\left(M_{pfa}+M_{psa}\right)+9.3\left(M_{pfb}+M_{psb}\right)\right]$	0	0.05 F + 0.10 R_m	0.10 F + 0.25 R_m
	0.5	0.42	0.25	0.59	$\frac{1}{a}\left[12.0\left(M_{pfa}+M_{psa}\right)+9.0\left(M_{pfb}+M_{psb}\right)\right]$	0	0.04 F + 0.08 R_m	0.11 F + 0.27 R_m

Fixed

Table A-8. Transformation Factors for Circular Slabs;
for Poisson's Ratio = 0.3
[U. S. Army Corps of Engineering Manual (1975)]

Fixed Edges

Simple Supports

Edge Condition	Strain Range	Load Factor K_L	Mass Factor K_M	Load-Mass Factor K_{LM}	Maximum Resistance	Spring Constant	Dynamic Reaction
Simple Supports	Elastic	0.46	0.30	0.65	18.8 M_{pc}	216 EI/a^2	0.28 F + 0.72 R
	Plastic	0.33	0.17	0.52	18.8 M_{pc}	0	0.36 F + 0.64 R_m
Fixed Supports	Elastic	0.33	0.20	0.61	25.1 M_{ps}	880 EI/a^2	0.40 F + 0.60 R
	Elasto-Plastic	0.46	0.30	0.65	$18.8\left(M_{pc} + M_{ps}\right)$	216 EI/a^2	0.28 F + 0.72 R
	Plastic	0.33	0.17	0.52	$18.8\left(M_{pc} + M_{ps}\right)$	0	0.36 F + 0.64 R_m

Table A-9. Dynamic Design Factors for Flat Slabs:
Square Interior, Uniform Load
[U. S. Army Corps of Engineering Manual (1975)]

Strain Phase	d/a	Load Factor K_L	Mass Factor K_M	Load-Mass Factor K_{LM}	Spring Constant k kips/ft	Maximum Resistance R_m kips	Dynamic Column Load V_c kips
	0.05	8/15	0.34	0.64	$1.45 \ EI_a/a^2$	$4.2 \ \Sigma \ M_p$	
	0.10	8/15	0.34	0.64	$1.60 \ EI_a/a^2$	$4.4 \ \Sigma \ M_p$	
Elastic	0.15	8/15	0.34	0.64	$1.75 \ EI_a/a^2$	$4.6 \ \Sigma \ M_p$	0.16 P + 0.84 R + Load on Capital
	0.20	8/15	0.34	0.64	$1.92 \ EI_a/a^2$	$4.8 \ \Sigma \ M_p$	
	0.25	8/15	0.34	0.64	$2.10 \ EI_a/a^2$	$5.0 \ \Sigma \ M_p$	
	0.05	1/2	7/24	7/12	0	$4.2 \ \Sigma \ M_p$	
	0.10	1/2	7/24	7/12	0	$4.4 \ \Sigma \ M_p$	
Plastic	0.15	1/2	7/24	7/12	0	$4.6 \ \Sigma \ M_p$	0.14 P + 0.86 R_m + Load on Capital
	0.20	1/2	7/24	7/12	0	$4.8 \ \Sigma \ M_p$	
	0.25	1/2	7/24	7/12	0	$5.0 \ \Sigma \ M_p$	

d   = width of column capital

a   = column spacing, ft

E   = compressive modulus of elasticity of concrete, ksi

I_a = average of gross and transformed moments of inertia per unit width, equal in both directions, in.^{4}/ft

P   = total load on one slab panel, excluding capitals

R   = total resistance of one slab panel, excluding capitals

$\Sigma \ M_p = M_{pmp} + M_{pmm} + M_{pcp} + M_{pcm}$

APPENDIX B

RESPONSE CHARTS FOR A SINGLE-DEGREE-OF-FREEDOM OSCILLATOR

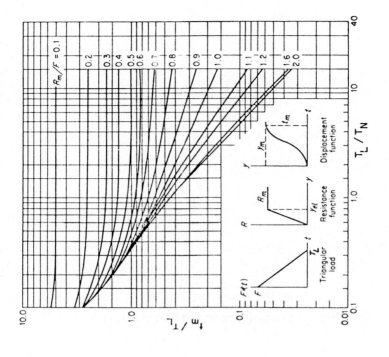

(a) Maximum Deflection

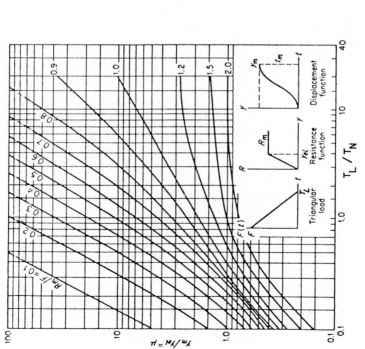

(b) Time of Maximum Deflection

Figure B-1. Maximum Response of an Undamped One-Degree-Of-Freedom
Elastic-Plastic System for a Triangular Load
[U. S. Army Corps of Engineers Manual (1975)]

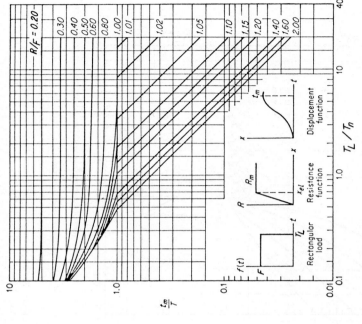

(a) Maximum Deflection

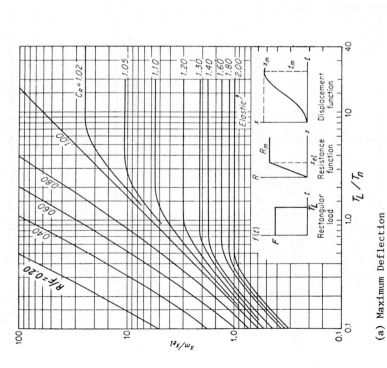

(b) Time of Maximum Deflection

Figure B-2. Maximum Response of an Undamped One-Degree-Of-Freedom
Elastic-Plastic System for a Step Load
[U. S. Army Corps of Engineers Manual (1975)]

674

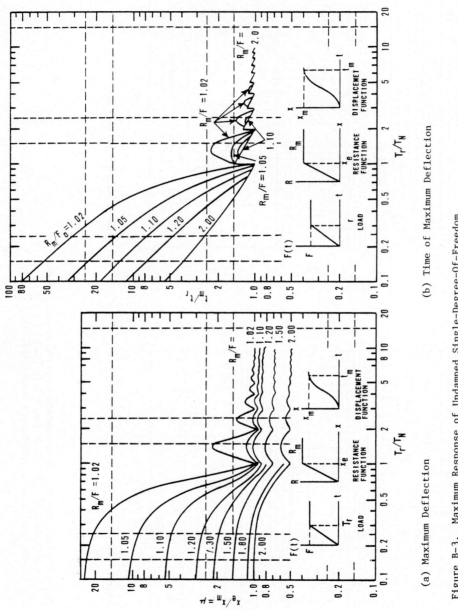

(a) Maximum Deflection

(b) Time of Maximum Deflection

Figure B-3. Maximum Response of Undamped Single-Degree-Of-Freedom
Elastic-Plastic System to Step Pulse with Finite Rise Time
[U. S. Army Corps of Engineers Manual (1975)]

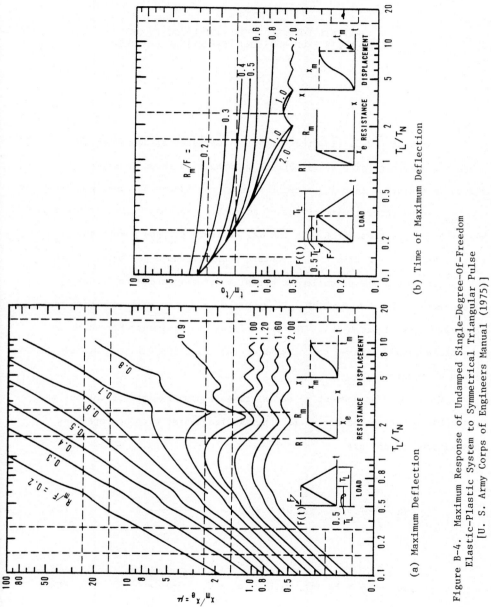

(a) Maximum Deflection

(b) Time of Maximum Deflection

Figure B-4. Maximum Response of Undamped Single-Degree-Of-Freedom
Elastic-Plastic System to Symmetrical Triangular Pulse
[U. S. Army Corps of Engineers Manual (1975)]

APPENDIX C

ANALYSIS FOR PREDICTING THE VELOCITY OF
FRAGMENTS FROM BURSTING PRESSURIZED SPHERES

A schematic depicting the essential characteristics of the Taylor

and Price solution for a sphere bursting in half is shown in Figure 6-1.

Before accelerating into an exterior vacuum, the sphere has internal vol-

ume V_{oo} and contains a perfect gas of adiabatic exponent (ratio of specific

heats) γ and gas constant R with initial pressure p_{oo} and temperature θ_{oo}

(Figure 6-1a). At a time t = 0, rupture occurs along a perimeter Π, and

the two fragments are propelled in opposite directions due to forces ap-

plied against the area F which is perpendicular to the axis of motion of

the fragments (Figure 6-1b). The mass of the two fragments M_1 and M_2 is

considered large relative to the mass of the remaining gas at elevated

pressure (Figure 6-1c).

With the displacement x_1 and x_2 taken along the axis of motion, the

applicable equations of motion for the two fragments, with accompanying

initial conditions, are

$$M_1 \frac{d^2 x_1}{dt^2} = Fp_1(t), \text{ with } x_1(o) = 0, \frac{dx_1(o)}{dt} = 0 \qquad \text{(C-1)}$$

$$M_2 \frac{d^2 x_2}{dt^2} = Fp_2(t), \text{ with } x_2(o) = 0, \frac{dx_2(o)}{dt} = 0 \qquad \text{(C-2)}$$

The equation of state for the unaccelerated gas remaining within the confinement of the container fragments is

$$p_o(t) \ V_o(t) \ = \ C(t) \ R \ \theta_o(t) \qquad \qquad (C-3)$$

where subscript "o" denotes reservoir conditions immediately after failure,
R is the gas constant, p is absolute pressure, V is volume, θ is temperature and C(t) is the mass of gas confined at high pressure as a function of
time. The rate of change of the confined mass is

$$\frac{dC(t)}{dt} \ = \ -k \ \Pi \ x \ \rho_* a_* \qquad \qquad (C-4)$$

where

$$x \ = \ x_1 \ + \ x_2 \qquad \qquad (C-5)$$

k is the coefficient of discharge of the area between the fragments, ρ_*
is the gas density at critical gas velocity a_* and Π is the perimeter of
the opening. Gas density ρ_* and gas velocity a_* are standard expressions
and are

$$\rho_* \ = \ \rho_o(t) \left(\frac{2}{\gamma + 1} \right)^{1/(\gamma \ - \ 1)} \qquad \qquad (C-6)$$

$$a_* \ = \ a_o(t) \left(\frac{2}{\gamma + 1} \right)^{1/2} \qquad \qquad (C-7)$$

where γ is the adiabatic exponent (ratio of specific heats) for an ideal
gas, ρ_o is the gas density and a_o is the gas velocity at time t. The vol-

ume is assumed to be variable and can be described by

$$v_o(t) = V_{oo} + Fx \qquad\qquad (C-8)$$

Nearly all of the gas is assumed to be accelerated with the fragments, with gas immediately adjacent to the fragments being accelerated to the velocity of the fragments. From simple one-dimensional flow relationships,

$$P_1(t) = P_o(t) \left[1 - \left(\frac{\gamma - 1}{2 \, [a_o(t)]^2} \right) \left(\frac{dx_1(t)}{dt} \right)^2 \right]^{\left(\frac{\gamma}{\gamma - 1} \right)} \qquad (C-9)$$

$$P_2(t) = P_o(t) \left[1 - \left(\frac{\gamma - 1}{2 \, [a_o(t)]^2} \right) \left(\frac{dx_2(t)}{dt} \right)^2 \right]^{\left(\frac{\gamma}{\gamma - 1} \right)} \qquad (C-10)$$

To generalize the solution, one can use the following nondimensional forms of the variables:

Displacement: $X(t) = \chi g(\xi)$, $x_1(t) = \chi g_1(\xi)$, $x_2(t) = \chi g_2(\xi)$

$$\text{Time:} \quad t = \tau \xi \qquad\qquad (C-11)$$

Pressure: $p_o(t) = P_{oo} \, p_*(\xi)$

From appropriate substitutions and initial conditions:

$$\frac{dx_1(t)}{dt} = \frac{\chi}{\tau} g_1', \quad \frac{dx_2(t)}{dt} = \frac{\chi}{\tau} g_2'$$

$$\frac{d^2 x_1(t)}{dt^2} = \frac{\chi}{\tau^2} g_1'', \quad \frac{d^2 x_2(t)}{dt^2} = \frac{\chi}{\tau^2} g_2''$$

$$\frac{dP_o(t)}{dt} = \frac{P_{oo}}{\tau} P_*' \tag{C-12}$$

$$x_1(o) = x_2(o) = \frac{dx_1(o)}{dt} = \frac{dx_2(o)}{dt} = g_1(o) = g_2(o) = g_1'(o) = g_2'(o) = 0$$

$$P_*(o) = 1$$

where primes denote differentiation with respect to ξ. The pair of characteristic values for dimension χ and time τ chosen by Taylor and Price are:

$$\chi = \frac{M_t a_{oo}^2}{F P_{oo}} \left(\frac{2}{\gamma - 1} \right) \tag{C-13}$$

$$\tau = \frac{M_t a_{oo}}{F P_{oo}} \left(\frac{2}{\gamma - 1} \right)^{1/2} \tag{C-14}$$

where M_t is the total mass of the reservoir. For the adiabatic case,

$$\frac{P_o(t)}{P_{oo}} = \left[\frac{\rho_o(t)}{\rho_{oo}} \right]^{\gamma} = \left[\frac{\theta_o(t)}{\theta_{oo}} \right]^{\frac{\gamma}{\gamma - 1}} = \left[\frac{a_o(t)}{a_{oo}} \right]^{\frac{2\gamma}{\gamma - 1}} \tag{C-15}$$

Substitution of Equations (C-9) through (C-15) into Equations (C-1) and

(C-2) and letting $M_1 = M_2 = M_t/2$, one has

$$g'' = 4 \ p_* \left[1 - \frac{g'^2}{4(p_*)^{\left(\frac{\gamma - 1}{\gamma}\right)}} \right]^{\frac{\gamma}{\gamma - 1}} \tag{C-16}$$

If we define

$$\alpha = \frac{P_{oo} \ V_{oo}}{M_t \ a_{oo}^2} \tag{C-17}$$

and

$$\beta = k \left(\frac{2}{\gamma + 1} \right)^{\frac{\gamma + 1}{2 \ (\gamma - 1)}} \left(\frac{2}{\gamma - 1} \right)^{1/2} \frac{\Pi \ V_{oo}}{F^2} \tag{C-18}$$

Then differentiation of Equation (C-3) and substitution of Equations (C-4) through (C-8) and Equations (C-11) and (C-12) yields

$$p_*' = \frac{\frac{-\beta\gamma}{\alpha} \ g \ p_*^{\left[\frac{3\gamma - 1}{2\gamma} \right]} - \gamma g' \ p_*}{\left(\frac{\gamma - 1}{2} \right) \alpha + g} \tag{C-19}$$

For initial conditions, $g(o) = o$, $g'(o) = 0$, and $p_*(o) = 1$, nondimensional values of distance, velocity, and acceleration and pressure as a function of time can be calculated by solving Equations (C-16) and (C-19) simultaneously using the Runge-Kutta method of numerical iteration. Dimensional values can then be calculated from

$$t = \tau\xi, \ x_1(t) = \frac{xg(\xi)}{2}, \ x_1'(t) = \frac{x}{2\theta} g'(\xi), \ x_1''(t) = \tag{C-20}$$

$$\frac{x}{2\theta^2} g''(\xi), \ \gamma_o(t) = P_{oo} \ p_*(\xi)$$

APPENDIX D

ANALYSIS FOR PREDICTING THE VELOCITY OF
CONSTRAINED OBJECTS EXPOSED TO AIR BLAST

The second half of Westine's analysis for predicting the velocity
of constrained fragments exposed to air blast consisted of the development
of a method to determine the amount of energy consumed in freeing the appurtenance from its moorings. The strain energy U consumed in fracturing
a cantilever beam was estimated by assuming a deformed shape and substituting the appropriate mechanics relationships for different modes of response in both ductile and brittle beams. A number of different solutions
resulted which had sufficient similarities to permit generalizations after
the strain energies were developed.

Using this method, one determines the strain energy at fracture by
assuming a deformed shape given by

$$y = w_o \left[1 - \cos \frac{\pi x}{2\ell} \right] \qquad\qquad (D-1)$$

where w_o is the tip deflection, ℓ is the total length of the beam, x is
the position along the beam, and y is the deformation at any position x.
This deformed shape has no deformation at the root of the beam, a maximum deformation at the tip of the beam, no slope at the root of the beam,
a maximum slope at the tip of the beam, a maximum elastic moment or curvature at the root of the beam, and no elastic moment or curvature at the
tip of the beam. If one performs a bending analysis for a rigid, perfect-

ly plastic beam, the strain energy stored in the beam is given by the moment-curvature relationship:

$$U = \int_{o}^{\ell} M_y \frac{d^2 y}{dx^2} \, dx \qquad (D-2)$$

Substituting the yield stress times the plastic section modulus $\sigma_y Z$ for the yield moment M_y, and differentiating Equation (D-1) twice before substituting gives

$$U = \frac{\pi^2 \, w_o \, \sigma_y Z}{4 \, \ell^2} \int_{o}^{\ell} \cos \frac{\pi x}{2 \ell} \, dx \qquad (D-3)$$

Completing the integration gives

$$U = \frac{\pi \, \sigma_y Z \, w_o}{2 \, \ell} \qquad (D-4)$$

But, the maximum strain ε_{max} is

$$\varepsilon_{max} = \frac{h}{2} \left(\frac{d^2 y}{dx^2} \right)_{max} \qquad (D-5)$$

which occurs at the root of the cantilever where $d^2 y/dx^2$ equals $\pi^2 \, w_o/4\ell^2$, and

$$\varepsilon_{max} = \frac{\pi^2 \, h \, w_o}{8 \, \ell^2} \qquad (D-6)$$

Substituting Equation (D-6) for w_o in Equation (D-4) and rearranging terms gives

$$U = \frac{4 \, Z \, \ell \, \sigma_y \, \varepsilon_{max}}{\pi \, h} \qquad (D-7)$$

The quantity $\sigma_y \, \varepsilon_{max}$ in a rigid, perfectly plastic, ductile material as in this analysis is the area under the stress-strain curve, which is called toughness T. Expressing U in terms of toughness and dividing by toughness times beam volume $A\ell$ gives the desired result.

$$\frac{U}{TA\ell} = \frac{4}{\pi} \left(\frac{Z}{Ah} \right) \qquad \text{(bending solution ductile material)} \qquad (D-8)$$

In a similar manner, Westine (1977) also developed an elastic bending solution for the strain energy in brittle beams, a solution for the strain energy due to plastic shear in a cantilever, and a solution for strain energy from shear in an elastic, brittle material. The results of these analyses are shown in Table D-1 where S is the elastic section modulus and μ is Poisson's ratio.

The major point which should be made from these solutions for strain energy for the four different modes of failure is that, no matter what mode of failure is hypothesized, the strain energy at failure equals $(TA\ell)$ times a constant. For some modes of failure, the constant may be a weak function of the cross-sectional shape of the fragment (a function of S/Ah or Z/Ah), but this constant varies very little. Table D-1 demonstrates the limited variation in these constants.

The second major conclusion is that toughness T appears to be the only mechanical property of importance. All four solutions give the result that this area under the stress-strain curve times the volume of the

Table D-1. Variations in Strain Energy Coefficients $\frac{U}{TA\ell}$

Type of Failure — Shape of Beam Cross Section	Ductile Bending	Brittle Bending	Ductile Shear	Brittle Shear
General Solution	$\frac{4}{\pi}\left(\frac{Z}{Ah}\right)$	$\left(\frac{S}{Ah}\right)$	$\frac{1}{2}$	$\frac{1}{2(1+\nu)}$
Circular Solid	0.270	0.125	0.500	0.385
Rectangular Solid	0.318	0.167	0.500	0.385
I-Beam	≈0.637	≈0.500	0.500	0.385

specimen times a constant equals the strain energy U expended in fracturing the specimen.

For analysis purposes these conclusions indicate that the mode of failure does not have to be determined. The solution can proceed by assuming that strain energy is given by Equation (D-9) and that the constant C can be obtained from experimental test results.

$$U = C(TA\ell) \qquad\qquad (D-9)$$

The use of a different deformed shape will not change the conclusions that U is directly proportional to (TAℓ); however, a different shape will result in a slightly different numerical proportionality constant C. Because C is determined experimentally, the qualitative conclusions still can be applied in the development of a solution.

All four modes of failure were developed for failure in a cantilever beam. Other boundary conditions such as clamped-clamped, simply supported, etc., will give similar qualitative results; however, the proportionality coefficient C is a function of support conditions.

Using the conservation of momentum and allowing the structural constraint to reduce the imparted impulse by an amount I_{st}, one has

$$I - I_{st} = mV \qquad \text{(D-10)}$$

where I is the total impulse acquired by the target. Substituting $\sqrt{2mE}$ for I_{st} and rearranging terms yields the conservation of momentum relationship

$$\frac{I}{\sqrt{mE}} = \sqrt{2} + \frac{\sqrt{m}\ V}{\sqrt{E}} \qquad \text{(D-11)}$$

Total impulse I equals $ib\ell$ where b is the loaded width, ℓ is the loaded length, and total mass m of the fragment equals $\rho_s A\ell$ where ρ_s is the density of the appurtenance and A is its cross-sectional area in the plane perpendicular to the long axis of the target. Substituting these relationships and the strain energy U as given by Equation (D-9) into Equation (D-11) one has

$$\frac{i\ b\ \ell}{\sqrt{(\rho_s A\ell)\ (CTA\ell)}} = \sqrt{2} + \frac{\sqrt{\rho_s A\ell}\ V}{\sqrt{CTA\ell}} \qquad \text{(D-12)}$$

Or, after reduction:

$$\frac{i\ b}{\sqrt{\rho_s T\ A}} = \sqrt{2C} + \frac{\sqrt{\rho_s}\ V}{\sqrt{T}} \qquad \text{(D-13)}$$

Equation (D-13) is a two-parameter space of nondimensional energy ratios. If the term $\sqrt{\rho_s}\, V/\sqrt{T}$ is squared, this group is the ratio of fragment kinetic energy per unit volume to strain energy expended per unit volume. The square of the term $i\, b/A\, \sqrt{\rho_s T}$ represents the energy put into the fragment per unit length divided by the strain energy expended per unit length of the fragment. This solution infers that the constrained secondary fragment velocity is independent of beam length ℓ. Test results show that this conclusion is not quite accurate.

After curve fitting to experimental data, Westine (1977) concluded that the velocity of the constrained fragment could be described by

$$\frac{\sqrt{\rho_s}\, V}{\sqrt{T}} = -\,0.2369 + 0.3931 \left(\frac{i\, b}{\sqrt{\rho_s T}\, A}\right) \left(\frac{\ell}{b/2}\right)^{0.3} \qquad (D-14)$$

$$\text{for} \quad \left(\frac{i\, b}{\sqrt{\rho_s T}\, A}\right) \left(\frac{\ell}{b/2}\right)^{0.3} \geq 0.602$$

$$V = 0 \quad \text{for} \quad \left(\frac{i\, b}{\sqrt{\rho_s T}\, A}\right) \left(\frac{\ell}{b/2}\right)^{0.3} \leq 0.602$$

where V is the fragment velocity, ρ_s is the fragment mass density, T is the toughness of fragment material, b is the loaded width of the beam, ℓ is the length of the target, A is the cross-sectional area, and i is the specific impulse.

This pair of equations works for cantilever beams of any materials and any cross-sectional area. To estimate the velocity, the specific impulse i imparted to the beam is an estimate from the standoff distance and line charge geometry using the equation in Figure 6-9. Substituting this impulse, beam properties, and beam geometry into Equation (D-14) gives the fragment velocity (see Figure 6-10). If the quantity

$$\left(\frac{i \, b}{\sqrt{\rho_s T \, A}}\right)\left(\frac{\ell}{b/2}\right)^{0.3}$$

is less than 0.602, the fragment will not break free; hence, its velocity
is zero.

An equation similar in format to Equation (D-14) but with different
coefficients for slope and intercept can also be used for beams with other
boundary or support conditions. Although Westine did not have a large
quantity of data available to demonstrate this observation, enough data existed on clamped-clamped beams to show that the coefficients -0.6498 instead
of -0.2369 and 0.4358 instead of 0.3931 work better for this boundary condition.

APPENDIX E

ANALYSIS FOR CALCULATING THE LIFT-TO-DRAG RATIO
FOR FLYING PLATES HAVING SQUARE OR CIRCULAR SHAPES

Most of the fragments from explosions are usually "chunky" in shape and have a lift coefficient (C_L) of 0.0. However, in some cases, where one predicts a breakup pattern which involves a large number of plate-like fragments, lift on fragments can become an important consideration. The discussion which follows gives a technique for calculating the normal force coefficients and lift and drag force coefficients of <u>plates</u> having square or circular shapes.

Consider Figure E-1 which shows a square plate moving with velocity v from left to right at an angle of attack α_1. The lift area for the largest surface is A_1 and the normal force coefficient for this surface is C_{N_1}. This normal force coefficient is divided into a lift component C_{L_1} in the vertical direction and a drag component C_{D_1} in the horizontal direction. Likewise, examining the smallest surface, it has an angle of attack α_2 which is $\alpha_1 + 90°$ and an area A_2. The normal force coefficient is C_{N_2} and

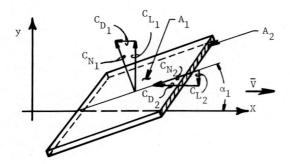

Figure E-1. Square Plate in Flight

is divided into a lift component C_{L_2} and a drag component C_{D_2}. Note that this surface has a negative lift component, but because A_1 is much larger than A_2, the fragment will experience a net positive lift force. Thus, the drag and lift forces, F_D and F_L respectively, can be expressed by:

$$F_D = (1/2)\ \rho v^2 \left(C_{D_1}\ A_1 + C_{D_2}\ A_2 \right) \qquad \text{(E-1)}$$

$$F_L = (1/2)\ \rho v^2 \left(C_{L_1}\ A_1 + C_{L_2}\ A_2 \right) \qquad \text{(E-2)}$$

From Figure E-1, one can readily obtain the following relationships for the lift and drag coefficients:

$$C_{L_1} = C_{N_1}\ \cos \alpha_1 \qquad \text{(E-3)}$$

$$C_{D_1} = C_{N_1}\ \sin \alpha_1 \qquad \text{(E-4)}$$

$$C_{L_2} = C_{N_2}\ \cos \alpha_2 \qquad \text{(E-5)}$$

$$C_{D_2} = C_{N_2}\ \sin \alpha_2 \qquad \text{(E-6)}$$

When α_1 equals 0° and 180°, $\cos \alpha_1$ equals 1 and C_{L_1} and C_{N_1} must both equal 0. Also, when α_2 equals 0° and 180°, $\cos \alpha_2$ equals 1 and C_{L_2} and C_{N_2} must both equal 0. When α_1 equals 90°, area A_1 is traveling face-on and (from Table 3-2) C_{D_1} equals 1.17 and [from Equation (C-2)] C_{N_1} also equals 1.17. Likewise, when α_2 equals 90°, area A_2 is traveling face-on and (from Table 3-2) C_{D_2} equals 2.05 and [from Equation (E-6)] C_{N_2} also equals 2.05. Intermediate values for C_{N_1} and C_{N_2} can be derived from Hoerner (1958). The results are shown graphically in Figure E-2 and are tabulated in Table E-1. Table E-1 also contains calculated values for the drag and lift coefficients.

In order to use Figure 6-12 for lifting fragments, it is necessary to determine the ratio $C_{L}A_{L}/C_{D}A_{D}$ or, more accurately, $C_{L_1}A_1 + C_{L_2}A_2/$

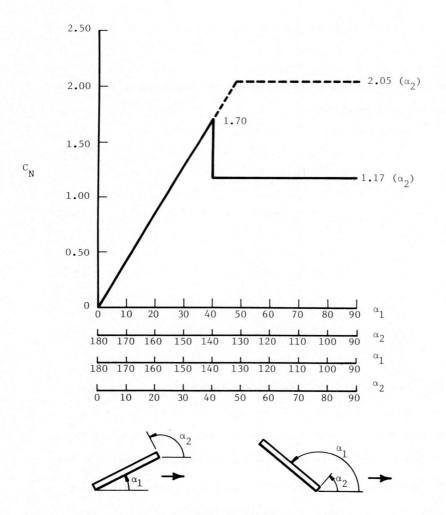

Figure E-2.   Determination of the Normal Force Coefficient as a Function of Angle of Attack

Table E-1. Tabulation of Normal Force, Lift and Drag Coefficients
as a Function of Angle of Attack

Angle of Attack (α_1)	(α_2)	C_{N_1}	C_{N_2}	C_{L_1} $\left(C_{N_1} \cos \alpha_1 \right)$	C_{D_1} $\left(C_{N_1} \sin \alpha_1 \right)$	C_{L_2} $\left(C_{N_2} \cos \alpha_2 \right)$	C_{D_2} $\left(C_{N_2} \sin \alpha_2 \right)$
0°	90°	0.00	2.05	0.00	0.00	0.00	2.05
10°	100°	0.43	2.05	0.42	0.075	−0.36	2.02
20°	110°	0.85	2.05	0.80	0.29	−0.70	1.93
30°	120°	1.28	2.05	1.11	0.64	−1.03	1.78
40°	130°	1.70	2.05	1.30	1.09	−1.32	1.57
50°	140°	1.17	1.70	0.75	0.90	−1.30	1.09
60°	150°	1.17	1.28	0.59	1.01	−1.11	0.64
70°	160°	1.17	0.85	0.40	1.10	−0.80	0.29
80°	170°	1.17	0.43	0.20	1.15	−0.42	0.075
90°	0°	1.17	0.00	0.00	1.17	0.00	0.00
100°	10°	1.17	0.43	−0.20	1.15	0.42	0.075
110°	20°	1.17	0.85	−0.40	1.10	0.80	0.29
120°	30°	1.17	1.28	−0.59	1.01	1.11	0.64
130°	40°	1.17	1.70	−0.75	0.90	1.30	1.09
140°	50°	1.70	2.05	−1.30	1.09	1.32	1.57
150°	60°	1.28	2.05	−1.11	0.64	1.03	1.78
160°	70°	0.85	2.05	−0.80	0.29	0.70	1.93
170°	80°	0.43	2.05	−0.42	0.075	0.36	2.02

$C_{D_1} A_1 + C_{D_2} A_2$. If $A_2 \ll A_1$, then one would normally use the simpler expression. As an example calculation of the lift-to-drag ratio, consider a square plate of dimensions 1.0 m x 1.0 m x 0.01 m (see Table E-2). That is, A_1 equals 1.0 m^2 (1.0 m x 1.0 m) and A_2 equals 0.01 m^2 (0.01 m x 1.0 m). Using the lift and drag coefficients in Table E-1 and the appropriate values for A_1 and A_2, one can readily calculate the lift-to-drag ratio. The complete and approximate forms of this expression are contained in Table E-2 for various angles of attack α_1. For this particular example, one can readily see that the approximate values of the lift-to-drag ratio do not differ greatly from the complete values.

Table E-2. Example Calculation for Determining the
Lift-to-Drag Ratio

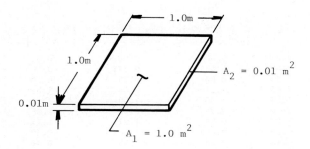

α_1	$\left(C_{L_1}A_1 + C_{L_2}A_2\right)$	$\left(C_{D_1}A_1 + C_{D_2}A_2\right)$	$\dfrac{\left(C_{L_1}A_1 + C_{L_2}A_2\right)}{\left(C_{D_1}A_1 + C_{D_2}A_2\right)}$	$\dfrac{C_{L_1}A_1}{C_{D_1}A_1}$
(degrees)	(square meters)	(square meters)	(-)	(-)
0	0.00	0.02	0.00	0.00
10	0.42	0.10	4.20	5.60
20	0.79	0.31	2.55	2.76
30	1.10	0.66	1.67	1.73
40	1.29	1.11	1.16	1.19
50	0.74	0.91	0.81	0.83
60	0.58	1.02	0.57	0.58
70	0.39	1.10	0.35	0.36
80	0.20	1.15	0.17	0.17
90	0.00	1.17	0.00	0.00
100	-0.20	1.15	-0.17	-0.17
110	-0.39	1.10	-0.35	-0.36
120	-0.58	1.02	-0.57	-0.58
130	-0.74	0.91	-0.81	-0.83
140	-1.29	1.11	-1.16	-1.19
150	-1.10	0.66	-1.67	-1.73
160	-0.79	0.31	-2.55	-2.76
170	-0.42	0.10	-4.20	-5.60

APPENDIX F

FLOW CHARTS FOR DYNAMIC STRUCTURAL DESIGN PROCEDURES

The flow charts in Figures F-1 and F-2 outline procedures for the blast-resistant design of a building or structure. Figure F-1 applies to buildings that are subjected to external loads. Figure F-2 applies to buildings that are subjected to internal loads. Information in these flow charts is intended to give the AE guidance in the design of blast-resistant structures, and not to supplant other design manuals.

As noted earlier, external loads that can be produced from an accidental explosion are:

- overpressures from the blast wave

- ground shock

- impact from soil if cratering occurs

- impacts from fragments

When the explosion occurs internally, additional loads are produced by an increase in the ambient pressure, but ground shock and cratering are not usually significant loads. All of these loads may act independently or they may occur in combination, depending upon the nature of the accident and the position of the building relative to the explosion. Treatment of multiple loads acting on the structure is covered in the procedure.

Nomenclature used in Figures F-1 and F-2 is equally applicable to steel frame and reinforced concrete structures. Dashed lines are used to suggest optional feedback loops or information exchange that should occur. Phantom lines enclose blocks that are not active tasks, but that contain comments, instructions, or conclusions.

Each block in the flow charts is identified and specific comments on each block are given in the following sections. When it is suggested

694

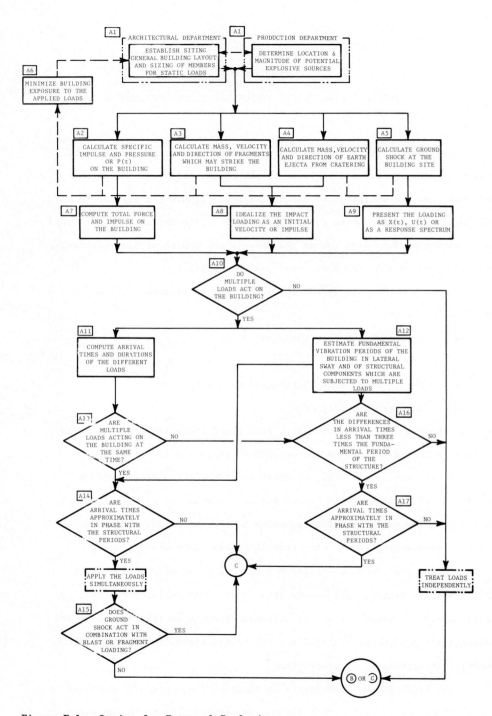

Figure F-1. Design for External Explosions

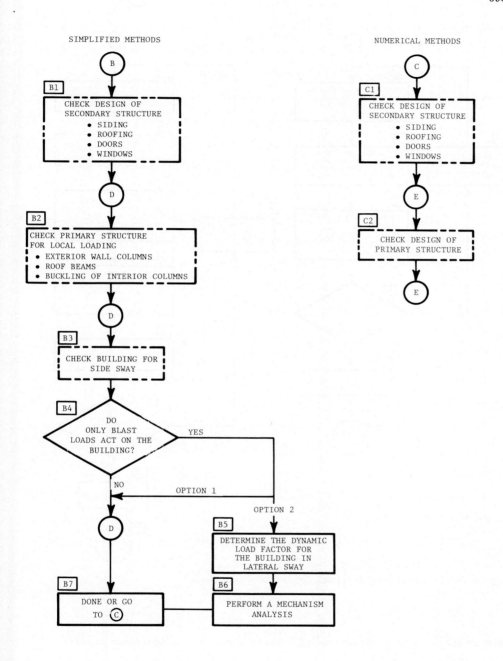

Figure F-1. Design for External Explosions (Continued)

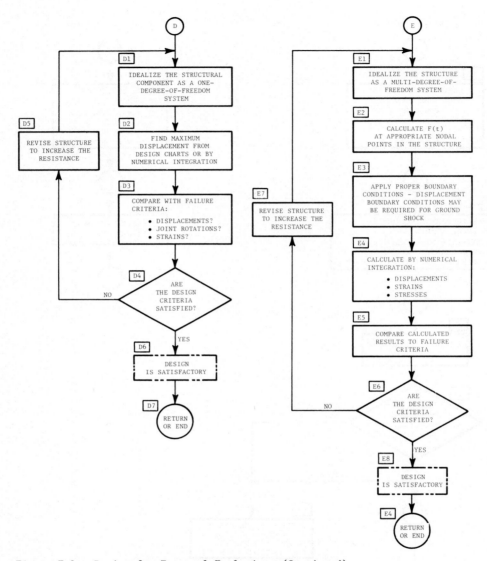

Figure F-1. Design for External Explosions (Continued)

that an analysis be performed using numerical methods, a nonlinear transient type of calculation is intended. Simplified methods can be nonlinear
transients or other types of approximations.

DESIGN FOR EXTERNAL LOADS (FIGURE F-1)

[A1] These two blocks are considered as basic input to the design-
er/analyst who will be responsible for assuring that the building meets the
requirements for blast resistance that have been placed upon it. Although
the source of these input data will vary within each organization, for purposes of the flow chart, it has been assumed that it is provided by the architectural and production departments in consultation with the safety engineer. Within some companies, a single design department may perform the
entire job, but the architectural engineering work and the prediction of
probable accidents are not covered in this book. Data provided to the analyst should include:

- position and orientation of the building relative to other
  facilities
- building plan and elevation drawings that give exterior dimensions
- detail drawings giving the primary and secondary structural
  members that have been sized to meet the static design loads
- location and magnitude of the energy release in postulated
  accidents which the building must be designed to withstand
- level of protection and post-accident service that the building must provide

[A2] Overpressures and specific impulse on the building can be calculated from the data in Chapters 2 and 3 by knowing the energy release in
the accident, whether detonation or deflagration occurs, and the distance
from the accident to the loaded face of the building. Ground reflection at
the accident location and local blast wave reflection on the building must
be considered in the calculations. Note that each surface of the building
may have a different p(t) and that both the reflected shock wave and the
"wind" behind the shock will impart loads to the building.

A3 Procedures for calculating the distribution and energy in fragments that may occur in an accidental explosion are covered in Chapter 6 and by Baker, Westine, et al. (1980). To avoid penetration, the "worst-case" fragment that can strike the building must be determined. To account for the loading on the building, i.e., the force-time or impulse that the fragments can impart to the building, the total energy in the fragments (mass, velocity, and number) that strike the building must be known. Usually, the energy imparted to a building by fragments is small relative to the energy imparted to the building by overpressures; however, for individual building components, it may be a significant part of the applied load.

A4 A nearby explosion at or below ground level can cause earth ejecta to be thrown against the building. This loading can be a significant part of the total load on the building, and perhaps, the major load for some building components. In this case, individual fragments are usually not important, but the total mass and velocity of the soil impacting the building must be known. Means of estimating the loading produced by earth ejecta from cratering are provided by Baker, Westine, et al. (1980).

A5 Ground shock occurs in every explosion, whether it is above or below the ground surface, but usually the effect of the shock is small compared to the effect of the air blast; however, for some explosions, this will not be the case, particularly for subsurface ones. Also, some buildings may be underground and only subjected to the ground shock loads. Methods of predicting ground shock can be found in the work by Baker, Westine, et al. (1980), Nicholls, et al. (1971), and Biggs (1964).

A6 Siting and orientation influence the loads on the building, and if possible, these two factors should be selected to reduce the loads on the building to some minimum value. Feedback between load estimates and the building placement and orientation is necessary to accomplish this objective.

A7 From the overpressure and specific impulse, the total forces and impulses on the different faces of the building and for various structural components can be calculated. For lateral sway of the building, it is customary to neglect the pressures on the side of the building opposite the explosion, but these pressures can easily be included when using numerical methods. Depending upon the frequency of the response of the building and the arrival times of the loads, the "back side" pressure can increase, but it usually decreases the response.

A8 Loading from both the fragment impacts and the earth ejecta thrown against the building can be idealized as an impulse or as an initial velocity imparted to the building. These estimates are based upon change in momentum, and when penetration occurs, residual fragment velocities can be taken as the final velocity if the fragment does not strike another part of the building. It is conservative and customary to neglect the residual velocity.

A9 It is often difficult to determine the displacement, $d(t)$, or acceleration, $a(t)$, produced at the building by ground shock. When these values are not well known, a good approximation of a shock spectrum can still be obtained as explained by Biggs (1964). A shock spectrum is adequate for calculating the elastic response of the building using mode superposition; however, for a nonlinear dynamic analysis, the ground displacement must be applied.

A10 A single accident will very likely produce air blast, fragments, and ground shock, all of which can excite the structure. It is necessary to consider multiple loads from either a single or multiple explosion and to decide how they must be applied to the structure for design purposes.

A11 Arrival times and durations of the different loads that act on the building must be determined. Data are readily available to make this calculation for air blast, but only estimates are available for ground shock and fragments. Because fragment loads are treated as an initial impulse, their duration is zero.

[A12] To determine the effect of multiple loads on the building, the fundamental period of vibration must be estimated for the building in lateral sway and for building components that are subjected to multiple loads. [See the section in Chapter 5 entitled "One-Degree-Of-Freedom Equivalent Systems," or any vibration handbook, such as Blevins (1979), which contains formulas for computing fundamental periods of vibration for structural components.]

[A13] If multiple loads act on the building at the same time, then simplified methods may not be appropriate. Because impact loads are represented as an impulse, they have no duration. If the arrival times of two separate loads, one of which is an impact load, are within one period of the fundamental mode of vibration of the structure, then the loads should be treated as acting at the same time.

[A14] If multiple loads are acting at the same time and the arrival times are approximately in phase with the fundamental periods of vibration, then the loading can be idealized as acting simultaneously for purposes of performing a simplified analysis (see Chapter 4). If they are not approximately in phase, then numerical methods, as described in Chapter 5, must be applied to compute the structural response of the building.

[A15] Even if the loads act together and the arrival times are approximately in phase with the vibration period of the structure, there is no simple method of combining ground shock and fragment or air blast loading. For this combination of loads, numerical methods are again recommended.

[A16] Even if the loads are not acting at the same time, the loads should be treated together if loads after the first are applied while the structure is still vibrating. A separation time of three times the fundamental vibration period has been chosen as a criterion for deciding wheth-

er or not to treat the loads together or separately. After three periods
of vibration have passed, damping has reduced the amplitude substantially,
particularly if plastic straining has occurred. In this case, treating
the loads separately will permit a simplified analysis to be performed and
should give a good estimate of the maximum response.

$\boxed{\text{A17}}$ At this point, it has been determined that the loads do not
act together; yet loads after the first are applied within a time equal to
three times the fundamental vibration period of the structure. If the arrival times are approximately in phase with the period of the structure,
then the loads should be applied together. For this case, numerical methods are recommended so that proper phasing of the loads can be accounted
for. If the arrival times are not in phase with the structural period,
then it is usually safe to treat the loads independently and use simplified methods of analysis.

$\boxed{\text{B}}$ Simplified Methods

When using simplified methods, individual components of the
building, and the building itself, are idealized as a one- (or possibly
two-) degree-of-freedom system. The response of each one-degree-of-freedom
system is then found for the assumption that coupling between the various
parts of the structure does not occur, i.e., the support for each component
is treated as rigid and no alteration of the loading occurs as it passes
through the structure. This approach usually yields conservative results.

Because the procedure of idealizing the parts of the structure
as a one-degree-of-freedom is repeated many times for different parts of
the building, it has been treated in a separate flow diagram and referred
to as $\boxed{\text{D}}$. This part of the flow chart functions as a subroutine that is
entered and used, and it then returns control back to the location from
which it was entered. Contents are described under Subroutine $\boxed{\text{D}}$.

$\boxed{\text{B1}}$ As the first step in the simplified analysis of a building,
the secondary structure is analyzed and sized for blast and fragment loads.

Design of secondary structure is performed first because its mass must be included in frequency calculations for the primary structure. Only significant changes, such as changes in frame spacing which may occur subsequently during the analysis of the primary structure, will require redesign of the secondary structure. When the secondary structure is analyzed first, it is also possible to compute boundary reactions and apply them as loads to the primary structure. This approach is not recommended for simplified analysis because the boundary reactions cannot be determined accurately.

Secondary structure includes wall panels, roofing, windows, and doors. Wall panels are usually constructed with siding, girts, and purlins. Each component must be analyzed for the applied loads. Roofing includes both the decking and roof joists, and each of these components must be checked also. Doors and windows that are blast resistant require strong supports so the frames also must be checked for adequate strength.

$\boxed{B2}$ After the secondary structure has been designed, the primary structure should be checked for local loads. This includes:

- exterior columns
- interior columns
- roof beams
- wall panels that are part of the primary structure

If the building is first designed for adequate strength in lateral sway, then any increase in the above members for local loading will result in some overdesign of the building. However, if the above members are first sized for local loads, then any additional strength that may be required for these loads will also contribute to the strength of the building in lateral sway.

$\boxed{B3}$ At this point in the analysis, the building is checked for side sway. If one axis of the building is oriented along the direction from the building to the explosion, then side sway in only one direction may need to be checked. If this is not the case, or if the building is irregular in shape, then side sway must be checked in two perpendicular directions. When biaxial bending of the columns occur, the bending capacity in each direction must be reduced accordingly. Stea, et al. (1977) give formulas for the reduction of bending capacity under these conditions. Columns can also be subjected to combined axial and lateral loads; however, as shown in Chapter 5, the effect produced by axial loads is usually small and can be neglected. Neal (1977) gives formulas for reducing the fully plastic bending moment for combined bending and axial loads.

[B4] If only air blast loads are acting on the building, side sway can be checked using either a mechanism analysis or by one-degree-of-freedom methods. If a mechanism analysis is used with impact loads, then the kinetic energy imparted to the structure by the impact is equated to the energy absorbed by the rotation of plastic hinges in the mechanism. This kinetic energy can also be added to kinetic energy from impulsive blast loading or other impact loads.

[B5],[B6] Option 1 is to use one-degree-of-freedom methods and proceed to [D]; Option 2 is to perform a mechanism analysis. A mechanism analysis is also an iterative process, but this is inherent in the method and so no looping to increase the resistance is shown. As the first step in the analysis, dynamic load factors must be determined for the roof loads and for lateral loads. These loads are a function of the structural frequencies and the nature of the loading. Stea, et al. (1977) give guidance for determining the dynamic load factors and for performing the mechanism analysis.

[B7] If only simplified methods are being used, then the analysis is complete. If numerical methods will also be used to confirm or refine the design, then the results obtained from the amplified methods provide input for the more exact analyses. Preliminary sizing by simplified methods is always recommended before undertaking a more complex and expensive numerical analysis.

C Numerical Methods

The flow chart for numerical methods includes a section (identified as E) that is utilized more than once, and so it has been separated from the main flow diagram. Section E outlines, in general terms,
the process of performing a multi-degree-of-freedom analysis of a structure
or structural component. This section is referred to twice because it is
suggested in the flow diagram that the primary and secondary structures of
the building be analyzed separately. The reason for performing separate
analyses is to minimize the degrees-of-freedom that must be included in the
model. If one attempts to analyze an entire structure with one model, the
model size (degrees-of-freedom, number of elements, etc.) would be large
and the analysis might not be economically feasible. The primary structure
alone can be represented by a model of reasonable size. Models of individual components would be used to analyze the secondary structure. Alternately, some combination of numerical and simplified methods can be used. For
example, the primary structure can be analyzed with numerical methods and
secondary structure with simplified methods. Regardless of which approach
is used, it is recommended that, before numerical methods are used to analyze a structure or structural component, preliminary sizing of the structure be made using simplified methods.

C1 As for simplified methods, analysis and sizing of secondary
structure have been placed ahead of the analysis of the primary structure.
Secondary structure is analyzed first so that the mass of these components
will be known, and so that the boundary reactions can be used as the applied loads for the primary structure. This approach has merit when numerical methods are used because the boundary reactions are more accurate than
those obtained with simplified methods; however, even with numerical methods, the reactions are still approximate because flexibility of the primary
structure is neglected in the analysis. Thus, this approach is optional
and the approach taken with simplified methods can be used, whereby flexibility of the secondary structure is ignored when loads for the primary

structure are computed. Refer to paragraph ⬚B1⬚ under <u>Simplified Methods</u> for additional comments. The steps in setting up a multi-degree-of-freedom model and computing the transient response of a structure are covered in Subsection ⬚E⬚ .

⬚C2⬚ Primary structure of the building can be analyzed in total using multi-degree-of-freedom methods or it can be analyzed in parts. How the structure is analyzed will depend upon the structural symmetry and the symmetry of the loading. For example, a multi-frame structure with a loading direction parallel to the plane of the framework can be analyzed by treating individual frames. This assumes that the frames are identical except for the outside ones, or have small variations. For nonsymmetric structures with loading from arbitrary directions, a three-dimensional analysis is required. Loading on the primary structure is often transferred through secondary structure. The exceptions are concrete slab or cylindrical structures where the walls provide the strength and the covering. When the loading is transferred through secondary structure, the analyst has three choices:

- analyze the coupled problem (include primary and secondary structure in the model),
- use support reactions from separate analyses of the secondary structure, or
- assume that the loading is unaltered by the secondary structure.

The last choice is the simplest, and it most often results in conservative results as noted in Chapter 5. In the most general case, the loading on the structure will not be uniform or simultaneous. Variations in the magnitude, arrival time, and duration of the loads can be readily treated with numerical methods, but computing the loading at multiple points on the structure is tedious. Note that when the third approach is used, the secondary structure is represented as an added mass on the primary structure. When the second approach is chosen, the mass of the secondary structure is ignored in the model.

D One-Degree-Of-Freedom Analysis

This part of the flow chart functions like a subroutine in a FORTRAN program in that it is entered from different parts of the flow chart and returns control to the position from which it was entered. This part of the flow chart is entered from   B   , Simplified Methods (Figure F-1) and from Figure F-2.

D1 , D2   Appendix A contains transformation factors for deriving one-degree-of-freedom approximations for distributed systems. Factors are given for beams and plates with different boundary conditions and different loads. Solutions for elastic-plastic behavior are contained in the design charts of Appendix B, or they can be obtained by numerical integration procedures as described in Chapter 5.

Chapter 4 gives a similar, but slightly different, approach for finding a one-degree-of-freedom solution. This chapter gives solutions for the asymptotes of the response of quasi-static and impulsive loads and also gives procedures for approximating the intermediate (dynamic) region of response. The result is a P-i (pressure-impulse) diagram for the one-degree-of-freedom system. If the deformation pattern (or failure mode) assumed is the same for the two methods, very similar results are obtained. Asymptotes of the response are expressed in equation form so that graphical or numerical methods are not required to obtain a solution; however, as a consequence, no information on the transient nature of the response is obtained. Solutions for the simultaneous application of a long duration pressure pulse and an impulse are given by Cox, et al. (1978) and in Shelter Design and Analysis, Volume 4 (1972).

D3   One-degree-of-freedom equivalent systems were developed to give displacements that are the same as the maximum displacement in the distributed system. Thus, the most reliable information that can be obtained is the displacement. This displacement can be compared to the displacement at which yielding occurs in the structure and a value for the ductility ratio, μ, which is the ratio of the maximum deflection to the deflection at yield, can be computed. This is one suitable criterion for design, and recommended design values are given in Chapter 8. From the assumed deformation pattern and the maximum displacement, additional information can be obtained, such as the maximum strain and joint rotations. These, too, can be used as design criteria, but the values obtained from the one-degree-of-freedom approximation are less reliable than the displacements.

Accurate values for the maximum reactions at supports are difficult to obtain from these simple methods; however, as explained in Chapter 5, estimates are given in the tables of Appendix A for the maximum reaction at the support as a function of the applied load and the system resistance. These estimates are good for loads of long duration relative to the fundamental period of the structure, but can be nonconservative for loads of very short duration and high intensity.

D4 If the design criteria are not satisfied at this stage, then design modification of the structure is necessary. A logical step is to increase the resistance of the structure. Alternatives to increasing the resistance would be to reduce the loading or relax the criteria, but it is assumed at this point in the design that these values are set.

D5 Increasing the resistance of the structure to satisfy the design criteria is an iterative process; however, the previous calculation and response charts (if they are being used) provide guidance for selecting a new resistance. Because the frequency changes with changes to the structure, estimates of the proper resistance to provide a given ductility ratio are only approximate. From the resistance chosen, values of the yield moment, etc., and thus, the size of the structural member, can be obtained.

D6 , D7 When the design criteria are met, the design is satisfactory and, if the analyses of all components have been completed, the job is ended; otherwise, the analyst returns to the position in the flow diagram from which the subroutine was entered.

E | Multi-Degree-Of-Freedom Analysis

This part of the flow chart is entered repeatedly from different points in Figures F-1 and F-2. Return is to the point in the flow chart from which it was entered. It outlines the steps in setting up and performing a multi-degree-of-freedom analysis.

E1 | To perform a multi-degree-of-freedom numerical analysis, the structure or component to be analyzed must first be represented by an analytical model, just as for simplified methods. In this case, the model can be much more detailed and represent the structure more accurately. Every model will be a compromise between accuracy and cost. When designing structures to resist accidental explosions, plasticity is usually permitted (so that a more efficient structure is obtained), and, because the loading is dynamic, a nonlinear transient solution is required. This type of analysis can be very costly in terms of computer time, and so it is important that large models be avoided. Also, a detailed (fine mesh) model should be avoided for this type of analysis because flow rules for metal plasticity are not exact, and usually there is some uncertainty in the loads which are applied to the structure.

Modeling structures for multi-degree-of-freedom analyses must be learned by experience. To keep the model small, the analyst can take advantage of symmetry in the structure and in the loading. If one plane of symmetry exists in the structure and if the loading is also symmetric about this plane, then only one-half of the structure need be included in the model. If this same condition exists for two planes of symmetry, then only one-fourth of the structure need be modeled. This condition can easily occur for structures loaded internally, but will seldom occur for external loads.

Because of the complexity and expense associated with a non-linear dynamic analysis, three-dimensional problems are often treated in only two dimensions. Guidance for the modeling of two-dimensional frame type structures is given in Chapter 5 and by Stea, et al. (1977). As previously described (see Chapter 5), the program DYNFA was written specifically for two-dimensional frameworks, and programs such as MARC [MARC Analysis Corporation (1975)] and ANSYS [DeSalvo and Swanson (1979)] can be used for modeling more general configurations and for three-dimensional problems. Example problems have been solved using these programs, and the problem manuals should be consulted before attempting to prepare a model for either code. The user is again cautioned to keep models as small as practical and also to solve a very small sample problem, using the features of the program that will be required in the analysis, before attempting to solve the actual problem.

$\boxed{\text{E2}}$ The applied forces, which will be a function of time, must be calculated and applied at selected nodal points on the model. If the blast wave travels perpendicular to the wall, normal reflection will occur and reflected values of pressure and/or impulse must be used when computing the forces. For surfaces loaded by normal blast waves, the loading is applied simultaneously to all nodes and will have the same duration. For surfaces loaded by oblique blast waves or side-on blast waves that sweep across a surface (such as the roof), the loads arrive at different times at different nodes (see the section in Chapter 5 entitled "Multiple-Degree-Of-Freedom Systems"). These loads will also have slightly different magnitudes and durations. An adequate approximation is to assume a linear variation of arrival times, durations, and pressure magnitudes over the building surfaces. Chapter 3 provides the analyst with sufficient information to calculate side-on and reflected values of pressure and impulse, arrival times, durations, and the drag phase of the loading.

Impulsive loads associated with fragment or soil impact should be applied as an initial velocity or as an impulse. An alternative way of applying an impulse is with a very short duration, but high intensity load. The load duration should be shorter than about one-fourth of the shortest period in the model. Shock loads will be applied as a displacement-time history at the base of the building for a nonlinear analysis.

E3 Normal boundary conditions are specified by setting to zero those displacements on the model that are fixed (do not move) in the building or structural component. These will usually be points where the building is attached to the foundation. If the foundation is included in the model, then some points on the foundation must either be fixed or it must be attached to points in the soil which are, in turn, attached at some point in the model. As a minimum, sufficient displacements must be set to zero to prevent rigid body motions of the model in translation and rotation.

When planes of symmetry in the structure (refer to E1) are used to reduce the size of the model, then the symmetry boundary conditions must be applied at these model boundaries. As an example, if the plane of symmetry is the X-Y plane, then the boundary conditions require that no translations occur in the Z-direction and no rotations occur about the X- and Y-axes.

Ground shock is applied to the model as a displacement-time history at points corresponding to the building foundation. These will be vertical or lateral displacements, or a combination of both. Without ground shock, these points would normally be fixed points in the structure.

E4 When analyzing structures with numerical methods, a computer program is used that solves the analytical model based upon the input data provided. With such a code, detailed procedures are followed in the preparation and coding of input data for the multi-degree-of-freedom model developed in E1 . Instructions are provided in the user's manual for the program.

Once the input data have been coded, the load case is run to obtain displacements, strains, and stresses. To obtain the solution for a nonlinear transient problem, a numerical integration in time is performed with a specified time step, and, when yielding occurs, iterations to obtain convergence are sometimes required within the time step. These elaborate integration and convergence procedures require substantial computation time. Specification of the integration time step is the choice of the user, but guidance is provided in the user's manual. Again, some compromise between accuracy and cost is often sought. If the integration time step is too large, excessive numerical damping or instability can result. If the integration time step is too small, costs can be excessive. A few trial runs with different time steps may be required before the correct choice can be made. Because each program uses a somewhat different approach to numerical integration, recommendations in the user's manual should be followed.

From the numerical results, displacements, strain, and stresses are obtained at points within elements or a nodal point in the model. Many programs offer graphical output of data that are very useful in evaluating the results, which are often voluminous.

E5 Results from the numerical solution are compared to preset criteria to determine whether or not the design is satisfactory. For elastic-plastic structures, these will usually be maximum allowable strains, translations, rotations, or boundary reactions at points of attachment.

E6 If the design criteria are not satisfied, then an iteration is required with a modified design.

E7 If the design criteria are not satisfied, it is necessary to increase the resistance of the structure, alter the loading or change the design criteria. It is assumed that the latter two possibilities have been properly treated earlier in the solution and are not considered here.

Increasing the strength or resistance of a structure also changes its stiffness and frequency. For elastic structures, increasing frequency will sometimes offset an increase in strength because the load amplification is increased by the increase in frequency. For elastic-plastic behavior, this factor is not so important because of the substantial damping produced by structural yielding. Little guidance can be offered the analyst in selecting ways to increase the resistance. It depends upon the type of structure and the type of "failure." The analyst must depend upon the results of the previous analysis in order to select the best alteration to increase the structural resistance. In some situations (with simple models), it may be helpful to consider the additional energy that must be absorbed within a specified strain or displacement in order to determine how much the resistance should be increased.

E8 , E9 When the design criteria have been satisfied, then the design is satisfactory and control is returned to the main flow diagram at the point from which E was entered. If the primary structure has been analyzed, then the design for blast resistance is complete.

DESIGN FOR INTERNAL EXPLOSIONS (FIGURE F-2)

The flow diagram of Figure F-2 pertains primarily to containment type structures. It references parts of the flow chart in Figure F-1. In this procedure, it is assumed that the structure is designed to contain fragments that may be generated by the accident; however, this may not always be the case for structures with blow-out walls. It is also assumed that the foundation will be designed to contain the explosion and prevent cratering. No reference is made to secondary and primary structures in the design for internal explosions. All structure in the containment boundary is considered to be primary structure.

F1 , F2 It is assumed that general building layout and design specifications have been provided to the designer/analyst. If a factor of safety is to be included in the design of the building, it will be included in the Design Criteria.

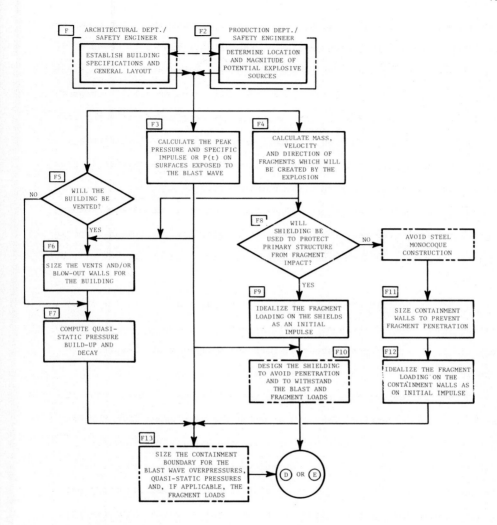

Figure 2. Design for Internal Explosions

F3   If the explosion involves high explosives, then data are readily available (see Chapters 2 and 3) for determining the peak pressures,
specific impulse, and arrival times, both at standard atmospheric conditions and for reduced pressures. It is only necessary to know the energy
release, the distance from the explosion, the angle of incidence of the

surface exposed to the blast wave, and the ambient pressure. If the explosion is not produced by a high explosive, the HE data can still be used once an equivalent amount of HE is defined for the accident. For vapors or dusts, different conditions exist inside the dust or vapor cloud. Chapter 3 explains how the overpressures and impulses are determined for vapor and dust explosions. If detonation does not occur, then only a low intensity shock (or none at all) will be produced. In this case, a quasi-static pressure buildup can still occur from the burning process.

F4 Fragments may be created by the accident if machines or other structures are located near the explosion. Chapter 6 gives guidance and information necessary to determine the mass, velocity, and directions of these fragments.

F5 If venting occurs in the structure, it will affect the decay time and peak value of the quasi-static pressure produced in the enclosure. Predictions for vented gas pressures, with and without vent closures, are given in Chapter 3.

F6 If venting does occur, then the vent areas and blow-out walls must be designed properly. The rate at which the vent opens (for blow-out walls) and the total effective vent area are the important parameters. [Peak pressure and vent times for covered vents are given by Baker, Westine, et al. (1980).] Venting is most effective for deflagration processes such as with burning propellant. For these relatively slow processes, both the peak pressure produced in the enclosure and the decay time can be significantly reduced. For high explosives, the chemical reactions are very fast, and very large vent areas, without covers, are required to attenuate the peak pressure. The decay time can be shortened, however, and this can be important in the design of some types of structures. In general, initial shock and peak quasi-static pressure loads on the structure cannot be significantly reduced by venting when the explosion is produced by a high explosive charge.

F7 The quasi-static pressure in a confined volume is produced by the heating of the air and the mass addition from the explosion. It is affected by venting and by the availability of sufficient oxygen for complete combustion. Graphical solutions are provided in Chapter 3 for the maximum pressure and decay time for HE explosions in enclosures with small vents. For other explosives, the maximum pressure is estimated from the products of the chemical reaction. Graphical solutions are given by Baker, Westine, et al. (1980) for vent times with and without blow-out walls. These data were generated by a computer program that is described and documented.

F8 Sometimes it is more economical to provide fragment shields in some locations than to design a large containment boundary to avoid fragment penetration. Also, some types of containment structures should not be used to both contain the explosive and stop the fragments. This applies primarily to single skin steel structures where stress risers produced by fragment impact can cause fracture of the stressed steel skin at lower than normal design stresses. Reinforced concrete, layered steel, or frame and panel construction is recommended when the primary containment boundary must resist both the blast loads and fragment impacts.

F9 Loading on the shields from fragment impacts can be idealized as an initial impulse for design purposes. When the fragments are stopped by the shield, their momentum is converted to an impulse on the structure. This approach assumes that inelastic impact occurs and that the fragment is fully arrested by the shield. If the fragment imbeds in the shield, then its mass should be included when computing the initial shield velocity. If fragment rebound occurs, then an initial shield velocity can be estimated by considering elastic impact with a low coefficient of restitution.

F10 The shields must be designed, not only to avoid penetration, but also to remain in place when the explosion occurs. (They must not become fragments themselves.) Thus, the penetration resistance of the shields must be designed for the worst-case fragment, and the shield sup-

port must be designed to resist the blast and fragment loads. Either simplified methods or numerical methods can be used for this design analysis.

F11 If shielding is not used and if the fragments are to be contained, then it is necessary to size the walls and ceiling of the structure to prevent fragment penetration. The thickness required to prevent
penetration can be calculated for various materials from the equations in
Chapter 6. The thickness set by penetration requirements may be larger
or smaller than that required to resist the blast and fragment loading.

F12 As for the shields, the fragment loads on the walls and ceiling are idealized as an initial impulse for the conditions of inelastic
impact and no residual fragment velocity.

F13 Once the fragment penetration criteria have been satisfied,
the primary containment boundary is designed to withstand the blast loading (pressure and impulses from the blast wave and the quasi-static pressure) and the fragment loads. For internal explosions, multiple reflections of the blast wave are usually assumed as explained in Chapter 3.
For HE charges, the quasi-static pressure reaches its peak very quickly,
and can be assumed to occur simultaneously with the blast wave. This is
particularly convenient for simplified analysis. For numerical analyses,
this assumption is not necessary. Dead loads are usually small relative
to the blast loading and can be omitted in the analysis.

Arrival times of the blast wave and fragments can be calculated to establish their phasing; however, it is acceptable for simplified
analyses to apply them simultaneously. It is often found that the fragment loads are small relative to the blast and quasi-static loads and can
be neglected.

Either simplified or numerical methods can be used to analyze
the structure. Symmetric shell and concrete slab structures can often be
analyzed conveniently with one-degree-of-freedom models. The analysis of
penetrations through the boundary of shell structures can be difficult,

and off-the-shelf door designs should be used when possible (if it has been established that they can withstand the applied loads). Penetrations through the boundary are more straightforward for reinforced concrete or frame and panel construction.

APPENDIX G

BRIEF ANNOTATED BIBLIOGRAPHY

This relatively short list highlights some general references, and gives our evaluation of these references. In most instances, references in this list will be quite useful to readers who wish to consider various aspects of explosion hazards evaluation more thoroughly than they can using only this book. But, in a few cases, we have included references which we feel should be avoided or which may be misleading, and we tell you why.

Annals of the New York Academy of Sciences (1968), Volume 152, Article 1, *"Prevention of and Protection Against Accidental Explosion of Munitions, Fuels and Other Hazardous Mixtures," October 28, 1968, 913pp.*

This is a very unlikely source for information on accidental explosions, but this one volume of the Annals contains a wealth of useful information in papers contributed by many researchers and safety engineers from the U. S. and Europe. A number of individual papers are cited in our bibliography, and considerable material has been incorporated in this book. It is a very good general reference, if you can get a copy.

Baker, W. E. (1973), Explosions in Air, University of Texas Press, Austin, Texas, 268pp.

We cannot give an unbiased review of this book, because one of our coauthors wrote it. So, we will defer to Joseph Petes, recently retired from U. S. Naval Surface Weapons Center ...

"Although this book will not satisfy all the requirements of either the casual, neophyte, or experienced investigator, it provides by far the most comprehensive treatment of the subject available in a single volume,

and as such it offers something worthwhile to all. A good balance between
theoretical and experimental approaches is maintained throughout the book
with adequate mention of theoretical-experimental relationships and their
importance to the understanding of the blast phenomena or solving practical
explosion problems."

"Baker does a good job of describing many types of instrumentation
in current use for laboratory or field applications. Mechanical, electro-
mechanical, and piezoelectric gages are discussed. Mechanical, CRO, mag-
netic tape, and photographic systems and techniques for recording transducer output or blast phenomena directly are given adequate treatment."

"In light of the rapid changes in instrumentation available for researchers' use, it would have been well to devote a page or two to the requirements of a blast-measuring system. A curve showing frequency response
requirements versus charge weight (or energy, the term Baker is prone to
use) would have been quite helpful."

"The relatively large bibliography is a fine feature; the serious
investigator is given guidance to sources in greater depth."

"All in all, Baker's Explosions in Air is a welcomed and notewor-
thy addition to a sparsely documented field."

Baker, W. E., Kulesz, J. J., Ricker, R. E., Bessey, R. L., Westine, P. S.,
Parr, V. B., and Oldham, G. A. (1975), "Workbook for Predicting Pressure
Wave and Fragment Effects of Exploding Propellant Tanks and Gas Storage
Vessels," NASA CR-134906, NASA Lewis Research Center, November 1975, (re-
printed September 1977), 549pp.

This is the first of two workbooks prepared for NASA, to help safety
engineers predict blast and fragment effects for accidental explosions of
liquid-fueled rockets or compressed gas vessels in or near launch facilities. The material and presentations are often quite different from that
appearing in high explosive safety manuals. We have drawn freely from this
reference in preparing this book.

Baker, W. E., Kulesz, J. J., Ricker, R. E., Westine, P. S., Parr, V. B., Vargas, L. M., and Moseley, P. K. (1978), "Workbook for Estimating the Effects of Accidental Explosions in Propellant Handling Systems," NASA Contractor Report 3023, Contract NAS3-20497, NASA Lewis Research Center, August 1978, 269pp.

This is the second of two workbooks prepared for NASA, for safety engineers to use in assessing explosion hazards from liquid propellant explosions and gas pressure vessel bursts in ground storage and transport related to space vehicles. It also includes the earlier workbook [Baker, et al. (1975)] as a microfiche copy. Extensive experimental work on blast from bursting pressure spheres is included, as is a rather detailed treatment of fragmentation of such vessels. Again, we have drawn freely on this reference in writing this book.

Baker, W. E., Westine, P. S., Kulesz, J. J., Wilbeck, J. S., and Cox, P. A. (1980), "A Manual for the Prediction of Blast and Fragment Loading on Structures," DOE/TIC-11268, U. S. Department of Energy, Amarillo, Texas, November 1980, 768pp.

This voluminous work is directed toward structural engineers who design blast-resistant structures for accidental explosions involving high explosives. Many graphs and equations, and many example problems are included. Topics covered are air blast from single and multiple high explosive sources, both bare and encased; blast loading of structures for internal and external explosions; cratering and ground shock for explosions in earth-covered structures; and a very detailed treatment of fragmentation and impact effects of fragments. General information is given on dynamic properties of construction materials and on methods for dynamic structural design. An extensive bibliography is also included.

Bartknecht, W. (1978a), Explosionen: Ablauf and Schutzmassnahmen, Spring-er-Verlag, Berlin, 264pp.

The title of this book should really be "Gas and Dust Explosions." It is the most comprehensive reference on the pressures which can be developed by contained and partially vented explosions of combustible gas mixtures and dusts, and on practical means for controlling and mitigating these effects. The book is very well illustrated with color and black-and-white pictures giving experimental results and showing test and explosion control equipment.

Biggs, J. M. (1964), Introduction to Structural Dynamics, McGraw-Hill Book Company, New York, New York, 341pp.

This is an excellent introductory text for any engineer engaged in dynamic structural design. Biggs is one of the authors of the earlier Army Corps of Engineers manual, and Norris, et al. (1959). This book draws heavily on the earlier work, but adds considerable material. Presentation is very clear and understandable.

Bodurtha, F. T. (1980), Industrial Explosion Prevention and Protection, McGraw-Hill Book Company, New York, New York, 167pp.

This book is neither a textbook nor a research monograph. Instead, it is a compilation of the many different types of explosion hazards that are present in an industrial environment and the techniques that have been devised to evaluate the hazards presented by particular materials and to prevent the occurrence of explosions or reduce their incidence. Because of the author's holistic approach to the industrial explosion hazard problem, the book contains a tremendous range of technical subject matter. A brief mention of the contents of the individual chapters will emphasize this range: 1) Introduction: Risk analysis and the responsibility of management, engineers, and regulatory agencies; 2) Flammability limits: Including the flash point mixing rules and fuels mixed with oxidizers other than air; 3) Ignition sources: Including ignition by compression, friction, electrical sparks and autoignition; 4) Explosion pressure: Including maximum pressure and rate of pressure rise, transition-to-detonation and blast wave effects; 5) Explosion protection: Including flame arrestors, purging, pressure relief valves and unconfined vapor cloud explosions; 6) Hazardous compounds: Including hazardous reactions and hazardous operations; and 7) Dust explosions: Including explosibility limits, ignition and explosion protection.

In our opinion, the text is much too short for the material that the author attempts to cover. It is true that he makes extensive use of references and that his reference lists are by and large complete. However, the treatment is quite often uncritical, in places it is spotty, and many times is too terse to benefit the unexperienced reader.

However, even with its shortcomings, this volume does fill a gap in the current literature because this is the first time that most of the information pertinent to all aspects of industrial explosion safety has been gathered in one place in the open literature.

Bureau for Industrial Safety TNO (1980a and 1980b), "Methods for the Calculation of the Physical Effects of the Escape of Dangerous Material (Liquids and Gases), Parts I and II," Directorate-General of Labour, Voorburg, the Netherlands, March 1980, 222pp. and 230pp.

This two volume manual is nicknamed "yellow book" for the color of its looseleaf binders, and is intended for use in safety studies involving accidental release of dangerous liquids and gases. Chapter 6 on heat radiation, Chapter 7 on vapor cloud explosions, and Chapter 9 on vessel rupture are of most interest in explosion hazards evaluation. Methods throughout are kept simple, and a number of example problems are given. We can probably expect periodic modifications and updating of the manual. Tables of properties of a number of potentially dangerous gases and liquids are included.

Design of Structures to Resist the Effects of Atomic Weapons (1965), TN 5-856, 1 Through 9, Department of the Army, U. S. Army AG Publications Center, 1655 Woodson Road, St. Louis, Missouri, 63114, March 1965, (This is an up-dated version of the U. S. Army COE Manual EM1110-345-415, March 1957).

This multi-volume design manual was one of the first to present simplified design procedures for plastically deforming structures. It was prepared by staff at MIT. The methods reported in this manual reappear in many later manuals, with no or minimal change.

Field, P. (1982), "Dust Explosions," Handbook of Powder Technology, Volume 4, Elsevier Scientific Publishing Company, Amsterdam.

This new book gives very readable coverage of the title problem, and accurately reflects the state of current research and development efforts in Europe and the United States to understand dust explosions and to control or mitigate their effects. The author states that he intended the book to be a valuable reference for engineers and plant managers experienced in the hazards of dust explosions, as well as an introductory text for a neophyte. He has accomplished that aim very well. The only criticism we can level is his retention of the older explosion relief vent area prediction methods which are physically incorrect, as well as the proven cube root law.

Henrych, J. (1979), "The Dynamics of Explosion and Its Use," Amsterdam, Elsevier Scientific Publishing , 558pp.

This book is a very useful reference work, but flawed. In some respects, it is encyclopedic, with coverage of many aspects of explosions of chemical high explosives and the effects of such explosions in air, water, and earth. Some treatment of nuclear explosions is also included. But, like an encyclopedia, coverage of some topics is shallow and does not reflect the depth of material available on these topics.

Topics which are well covered include the stress wave theory, detonations and close-in effects of explosions in high explosives, explosions in soils, underground blasting and cratering, and response of elastic and elastoplastic structural elements to blast loading. Coverage of elastic vibrations of structures is particularly exhaustive. Topics covered in a more superficial manner include explosions in air, explosions in water, use of explosives in demolition and seismic effects of explosions. Scaling laws for explosions in air or water, which are essential to these topics, are barely mentioned.

As is probably natural for a book written in Czechoslovakia, references to work in eastern Europe and Russia are extensive, and the inclusion of such references is very valuable for western readers. But, many readily available references from the United States and other western world sources are lacking, and many of the references which are listed are now dated and superceded by later work. In general, there are too few references to work more recent than 1969.

The writing is clear and the exposition easy to follow, but in some instances, too much mathematical detail is included. We would have preferred to see less mathematical development and more experimental data verifying some of the theory.

The choice of a system of units in the book is unfortunate in an era when we are attempting conversion to System International (SI). The unit used for pressure or stress is kp/cm^2 rather than the SI unit of $Pa = N/m^2$. This forces use of strange units for other physical parameters.

In summary, this is a voluminous book containing much useful information on detonative explosions of chemical explosives, effects of these explosions, and a number of peripheral topics. But the reader is cautioned

that the coverage is quite incomplete on some topics, and perhaps too detailed on other topics. References to good recent works in this field are
also omitted.

*Kinney, G. F. (1962), Explosive Shocks in Air, MacMillan, New York, New
York, 198pp.*

For some years, this was one of the few readily available references
on air blast waves. It evolved from a course taught by its author of the
U. S. Naval Postgraduate School. But, the material is relatively superficial and does not at all reflect the breadth of experimental and analytical
work which had been done before it was published. Literature citations are
very limited. It is now out of print, and has been supplanted by later and
more comprehensive works.

*Lees, F. P. (1980a and 1980b), Loss Prevention in the Process Industries.
Hazard Identification, Assessment and Control, Volumes I and II, Butterworths, London, 641pp. and 675pp.*

This two volume work is encyclopedic, and should be required reading for safety engineers in industrial (particularly chemical) plants. It
covers much more than explosion hazards, having 28 chapters on almost all
conceivable aspects of loss prevention. The text is very readable, and
references are extensive. Of particular value are the case histories of
serious accidents. The "Flixborough" Explosion is treated in depth in a
separate appendix. Chapter 17 on explosions provides a good overview of
this topic, but is necessarily much less comprehensive than the treatment
in this more specialized book.

*Norris, C. H., Hansen, R. J., Holley, M. J., Biggs, J. M., Namyet, S., and
Minami, J. K. (1959), Structural Design for Dynamic Loads, McGraw-Hill Book
Company, New York, New York, 453pp.*

This book first appeared as a set of course notes for a short course taught by MIT staff. Procedures carry over directly from an earlier U. S. Army Corps of Engineers manual. As with the Army manual, these methods reappear in many later manuals. Although it is a good introductory text on dynamic structural design, it is unfortunately difficult to obtain because it is out of print.

Palmer, K. N. (1973), Dust Explosions and Fires, Chapman and Hall, Ltd., London, 396pp.

This reference is required reading for anyone involved in studies of dust explosions. The coverage is thorough, and the writing very clear and readable. It should, however, be supplemented by Bartknecht's later book [Bartknecht, W. (]978a)].

Stull, Daniel R. (1977), Fundamentals of Fire and Explosion, AIChE Monograph Series, 73 (10), 124 pp.

We must give this monograph a mixed review. It contains much useful information on combustion and contained or vented gaseous explosions, and also discusses shock waves and their effects. The reference list is extensive and excellent, but the treatment of blast and its effects is superficial compared to that found in other books and monographs.

Structures to Resist the Effects of Accidental Explosions (1969), Department of the Army Technical Manual TM 5-1300, Department of the Navy Publication NAVFAC P-397, Department of the Air Force Manual AFM 88-22, Department of the Army, the Navy, and the Air Force, June 1969, 203pp.

This design manual is the "Bible" for most structural engineers involved in blast-resistant design in reinforced concrete. Its strengths are in presentation of detailed procedures for estimating blast loading for internal explosions, failure modes for reinforced concrete, structural elements, and design of reinforcing. Some of the blast loading and fragment impact data in this manual are, however, now outdated and should be supplanted by later information. Basic structural design procedures are identical to those presented earlier by MIT authors.

Suppressive Shields Structural Design and Analysis Handbook (1977), HNDM-1110-1-2, U. S. Army Corps of Engineers, Huntsville Division, November 1977, *510pp.*

This manual contains much useful information on internal blast loads and vented gas pressures for high explosive accidents. It also includes methods for predicting penetration of high-speed fragments through multi-layered steel panels. Chapter 5 in this manual is directed primarily to simplified methods for dynamic elastic and elastic-plastic design of steel structures subjected to internal blast loading. The primary methods are those previously developed by the MIT staff and by Newmark, but special attention is paid to response to loading pulses for initial shock loads and gas venting pressures for the structures with small or no venting which typify suppressive shields. More sophisticated dynamic design methods are identified, but not used.

White, M. T. (Editor) (1946), *Effects of Impacts and Explosions*, Summary Technical Report of Division 2, National Defense Research Council, Volume 1, Washington, D.C., AD 221-586, *512pp.*

Although this reference is rather old, it is a thorough review of U. S. research studies during World War II on explosions in air, water, and earth; ballistic impact effects on steel, concrete, and soil; and other topics which are not of particular interest in explosion hazards studies. Some of the Weapon Data Sheets included in the compendium, particularly those on penetration of projectiles into various media, are useful.

The discussions and descriptions in this work are very clear and readable, and some of the compiled and scaled test data are still the most definitive available. The volume was originally classified, but is now readily available through the U. S. National Technical Information Service (NTIS).

Zabetakis, M. G. (1965), "Flammability Characteristics of Combustible Gases and Vapors," Bulletin 627, Bureau of Mines, U. S. Department of the Interior, 121pp.

This report compiles much of the U. S. Bureau of Mines work on flammability and detonability limits for most combustible gases and vapors. It also includes an excellent reference list and tables of properties of hydrocarbons and other fuels.

APPENDIX H

BIBLIOGRAPHY

ABLOW, C. M. and Woolfolk, R. W. (1972), "Blast Effects from Non-Ideal Explosions," SRI Final Report, Contract No. 0017-71-C-4421, Stanford
Research Institute, Menlo Park, California (December 1972).

ABRAHAMSON, G. R. and Lindberg, H. E. (1976), "Peak Load-Impulse Characterization of Critical Pulse Loads in Structural Dynamics," Nuclear Engineering and Design, 37, pp. 35-46.

ABRAHAMSSON, E. (1967), "Dome Action in Slabs with Special Reference to
Blast Loaded Concrete Slabs," Stockholm, Sweden.

ABRAHAMSSON, E. (1978), "Blast Loaded Windows," Fortifikations Förvaltningen
Forskningsbyrån, C-rapport nr 169, Stockholm, Sweden.

ADAMCZYK, A. A. (1976), "An Investigation of Blast Waves Generated from Non-
Ideal Energy Sources," Technical Report AAE 76-6, UILU-Eng 76 0506,
University of Illinois, 147 pp.

ADAMCZYK, A. A. and Strehlow, R. A. (1977), "Terminal Energy Distribution of
Blast Waves from Bursting Spheres," NASA CR 2903, 26 pp. (September
1977).

AFFENS, W. A. (1966), "Flammability Properties of Hydrocarbon Fuels, Inter-
relations of Flammability Properties of n-Alkanes in Air," J. Chem.
and Eng. Data, 11, pp. 197-202.

AFFENS, W. A., Carhart, H. W., and McLaren, G. W. (1977a), "Determination of
Flammability Index of Hydrocarbon Fuels by Means of a Hydrogen Flame
Ionization Detector," J. Fire and Flammability, 8, pp, 141-151.

AFFENS, W. A., Carhart, H. W., and McLaren, G. W. (1977b), "Variations of Flammability Index with Temperature and the Relationship to Flash Point of Liquid Hydrocarbons," J. Fire and Flammability, 8, pp. 152-159.

AFFENS, W. A. and McLaren, G. W. (1972), "Flammability Properties of Hydrocarbon Solution in Air," J. Chem. and Eng. Data, 17, pp. 482-488.

AFZAL, S. M. M. and Chiccarine (1979), "SCOTCH, Safe Containment of Total Chemical Hazards. A Portable Alternative to a Fixed Cell Design," Paper Presented at the American Society of Chemical Engineers Meeting, San Francisco, California (November 1979).

AHLERS, E. B. (1969), "Fragment Hazard Study," Minutes of the Eleventh Explosives Safety Seminar, Memphis, Tennessee (September 1969).

ALDIS, D. F. and Lai, F. S. (1979), "Review of Literature Related to Engineering Aspects of Grain Dust Explosions," U. S. Department of Agriculture Miscellaneous Publication No. 1375 (August 1979).

ALLEN, D. J. and Rao, M. S. M. (1980), "New Algorithms for the Synthesis and Analysis of Fault Trees," Ind. and Eng. Chem. Fundamentals, 19, No. 1, pp. 79-85.

ALROTH, F. D., Briesch, E. M., and Schram, P. J. (1976), "An Investigation of Additional Flammable Gases or Vapors with Respect to Explosion-Proof Electrical Equipment," Underwriters' Laboratories, Inc. Bulletin of Research No. 58A, Supplements Bulletin of Research No. 58.

AMES, S. A. (1973), "Gas Explosions in Buildings, Part 2: The Measurement of Gas Explosion Pressures," Fire Research Note No. 985, Fire Research Station (December 1973).

AMSDEN, A. A. (1973), "YAQUI: An Arbitrary Lagrangian-Eulerian Computer Program for Fluid Flow at all Speeds," LA-5100 (March 1973).

ANDERSON, R. P. and Armstrong, D. R. (1974), "Comparison Between Vapor Explosion Models and Recent Experimental Results," AIChE Symposium Series 138, Volume 70, Heat Transfer-Research and Design, pp. 31-47.

ANDERSON, W. H. and Louie, N. A. (1975), "Effect of Energy Release Rate on the Blast Produced by Fuel-Air Explosions," SH-TR-75-01, Shock Hydrodynamics Division, Whitakker Corporation, North Hollywood, California (January 1975).

ANDREWS, G. E. and Bradley, D. (1972), "Determination of Burning Velocities: A Critical Review," Combustion and Flame, 18, pp. 1-33.

ANDREWS, G. E., Bradley, D., and Lwakavamba, S. G. (1975), "Turbulence and Turbulent Flame Propagation: A Critical Appraisal," Combustion and Flame, 24, pp. 285.

ANGIULLO, F. J. (1975), "Explosion of a Chloronitrotoluene Distillation Column," Loss Prevention Journal, 8, Paper No. 90a, Presented at the AIChE Symposium on Loss Prevention in the Chemical Industry, Houston, Texas (March 18-20, 1975).

ANNALS OF THE NEW YORK ACADEMY OF SCIENCES (1968), "Prevention of and Protection Against Accidental Explosion of Munitions, Fuels, and Other Hazardous Mixtures," Volume 152, Article 1 (October 28, 1968).

ANONYMOUS (1970), "Spacecraft Incident Investigation, Panel I," Volumes I, II, and III, NASA TMX-66922, 66921, and 66934 (June, July, and September 1970).

ANONYMOUS (1976), "Overpressure Effects on Structures," HNDTR-75-23-ED-SR, U. S. Army Corps of Engineers, Huntsville Division, Huntsville, Alabama (February 1976).

ANTHONY, E. J. (1977a), "Some Aspects of Unconfined Gas and Vapor Cloud Explosions," J. Hazardous Materials, 1, pp. 289-301.

ANTHONY, E. J. (1977b), "The Use of Venting Formulae in the Design and Protection of Building and Industrial Plant from Damage by Gas or Vapor Explosions," J. Hazardous Materials, 2, pp. 23-49.

ASHRAE HANDBOOK OF FUNDAMENTALS (1972), American Society of Heating, Refrigerating, and Air Conditioning Engineers, Inc., New York, New York.

ASSHETON, R. (1930), "History of Explosions on Which the American Table of
    Distances was Based, Including Other Explosions of Large Quantities
    of Explosives," The Institute of Makers of Explosives, AD 493246.

ASTBURY, N. F. and Vaughan, G. N. (1972), "Motion of a Brickwork Structure
    Under Certain Assumed Conditions," The British Ceramic Research Asso-
    ciation, Technical Note No. 191 (September 1972).

ASTBURY, N. F., West, H. W. H., and Hodgkinson, H. R. (1972), "Experimental
    Gas Explosions: Report of Further Work at Potters Marston," The
    British Ceramic Research Association, Special Publication No. 74.

ASTBURY, N. F., West, H. W. H., Hodgkinson, H. R., Cubbage, P. A., and
    Clare, R. (1970), "Gas Explosions in Load-Bearing Brick Structures,"
    The British Ceramic Research Association, Special Publication No. 68.

ASTM (1970), D 2155-69 ASTM Standard 17, pp. 724-727 (November 1970).

AXELSSON, H. and Berglund, S. (1978), "Cloud Development and Blast Wave Mea-
    surements from Detonation Fuel-Air Explosive Charges," FOA Rapport C
    20225-D4, Försvarets Forskningsanstalt, Huvudavdelning 2, Stockholm,
    Sweden (March 1978).

BACH, G. G., Knystautas, R., and Lee, J. H. S. (1971), "Initiation Criteria
    for Diverging Gaseous Detonations," Thirteenth Symposium (Interna-
    tional) on Combustion, The Combustion Institute, pp. 1097-1100.

BACH, G. G. and Lee, J. H. S. (1970), "An Analytic Solution for Blast
    Waves," AIAA Journal, 8, pp. 271-275.

BACIGALUPI, C. M. (1980), "Design of a Maze Structure to Attenuate Blast
    Waves," UCRL-52921, Lawrence Livermore Laboratory, Livermore, Cali-
    fornia (March 1980).

BADER, B. E., Donaldson, A. B., and Hardee, H. C. (1971), "Liquid-Propel-
    lant Rocket Abort Fire Model," J. Spacecraft, 8, 12, pp. 1216-1219
    (December 1971).

BAKER, W. E. (1958), "Scale Model Tests for Evaluating Outer Containment
    Structures for Nuclear Reactors," Proceedings of the Second Inter-

national Conference on the Peaceful Uses of Atomic Energy, United Nations, Geneva, Volume II, pp. 79-84.

BAKER, W. E. (1960), "The Elastic-Plastic Response of Thin Spherical Shells to Internal Blast Loading," J. of Appl. Mech., 27, Series E, 1, pp. 139-144 (March 1960).

BAKER, W. E. (1967), "Prediction and Scaling of Reflected Impulse From Strong Blast Waves," Int. J. Mech. Sci., 9, pp. 45-51.

BAKER, W. E. (1973), Explosions in Air, University of Texas Press, Austin, Texas.

BAKER, W. E., Esparza, E. D., Hokanson, J. C., Funnell, J. E., Moseley, P. K., and Deffenbaugh, D. M. (1978), "Initial Feasibility Study of Water Vessels for Arresting Lava Flow," AMSAA FEAT Interim Note No. F-13, U. S. Army Material Systems Analysis Activity, Aberdeen Proving Ground, Maryland (June 1978).

BAKER, W. E., Esparza, E. D., and Kulesz, J. J. (1977), "Venting of Chemical Explosions and Reactions," Proceedings of the Second International Symposium on Loss Prevention and Safety Promotion in the Process Industries, Heidelberg, Germany (September 1977).

BAKER, W. E., Ewing, W. O., Jr., and Hanna, J. W. (1958), "Laws for Large Elastic and Permanent Deformation of Model Structures Subjected to Blast Loading," BRL Report No. 1060, Aberdeen Proving Ground, Maryland (December 1958).

BAKER, W. E., Hokanson, J. C., and Cervantes, R. A. (1976), "Model Tests of Industrial Missiles," Final Report, SwRI Project No. 02-9153-001, Southwest Research Institute, San Antonio, Texas (May 1976).

BAKER, W. E., Hokanson, J. C., and Kulesz, J. J. (1980), "A Model Analysis for Vented Dust Explosions," Third International Symposium on Loss Prevention in the Process Industries, Basel, Switzerland, pp. 1339-1347 (September 1980).

734

BAKER, W. E., Hu, W. C. L., and Jackson, T. R. (1966), "Elastic Response of Thin Spherical Shells to Axisymmetric Blast Loading," _J. of Appl. Mech._, 33, Series E, 4, pp. 800-806 (December 1966).

BAKER, W. E., Kulesz, J. J., Ricker, R. E., Bessey, R. L., Westine, P. S., Parr, V. B., and Oldham, G. A. (1975), "Workbook for Predicting Pressure Wave and Fragment Effects of Exploding Propellant Tanks and Gas Storage Vessels," NASA CR-134906, NASA Lewis Research Center (November 1975) (reprinted September 1977).

BAKER, W. E., Kulesz, J. J., Ricker, R. E., Westine, P. S., Parr, V. B., Vargas, L. M., and Moseley, P. K. (1978), "Workbook for Estimating the Effects of Accidental Explosions in Propellant Handling Systems," NASA Contractor Report 3023, Contract NAS3-20497, NASA Lewis Research Center (August 1978).

BAKER, W. E. and Oldham, G. A. (1975), "Estimates of Blowdown of Quasi-Static Pressures in Vented Chambers," Edgewood Arsenal, Contractor Report EM-CR-76029, Report No. 2 (November 1975).

BAKER, W. E., Parr, V. B., Bessey, R. L., and Cox, P. A. (1974a), "Assembly and Analysis of Fragmentation Data for Liquid Propellant Vessels," _Minutes of the Fifteenth Explosives Safety Seminar_, Volume II, Department of Defense Explosives Safety Board, Washington, D.C., pp. 1171-1203 (September 1973).

BAKER, W. E., Parr, V. B., Bessey, R. L., and Cox, P. A. (1974b), "Assembly and Analysis of Fragmentation Data for Liquid Propellant Vessels," NASA Contractor Report 134538, NASA Lewis Research Center, Cleveland, Ohio (January 1974).

BAKER, W. E., Silverman, S., Cox, P. A., and Young, D. (1969), "Methods of Computing Structural Response of Helicopters to Weapons' Muzzle and Breech Blast," _The Shock and Vibration Bulletin_, Bulletin 40, Part 2, pp. 227-241 (December 1969).

BAKER, W. E. and Westine, P. S. (1974), "Methods of Predicting Loading and Blast Field Outside Suppressive Structures," Minutes of the Sixteenth Annual Explosives Safety Seminar, Department of Defense Safety Board.

BAKER, W. E. and Westine, P. S. (1975), "Methods of Predicting Blast Loads Inside and Outside Suppressive Structures," Edgewood Arsenal, Contractor Report No. EM-CR-76026, Report No. 5 (November 1975).

BAKER, W. E., Westine, P. S., and Bessey, R. L. (1971), "Blast Fields About Rockets and Recoilless Rifles," Final Technical Report, Contract No. DAAD05-70-C-0170, Southwest Research Institute, San Antonio, Texas (May 1971).

BAKER, W. E., Westine, P. S., and Cox, P. A. (1977), "Methods for Prediction of Damage to Structures from Accidental Explosions," Proceedings of the Second International Symposium on Loss Prevention and Safety Promotion in the Process Industries, Heidelberg, Germany (September 1977).

BAKER, W. E., Westine, P. S., and Dodge, F. T. (1973), Similarity Methods in Engineering Dynamics: Theory and Practice of Scale Modeling, Spartan Books, Rochelle Park, New Jersey.

BAKER, W. E., Westine, P. S., Kulesz, J. J., Wilbeck, J. S., and Cox, P. A. (1980), "A Manual for the Prediction of Blast and Fragment Loading on Structures," DOE/TIC-11268, U. S. Department of Energy, Amarillo, Texas (November 1980).

BALEMANS, A. W. M. and van de Putte, T. (1977), "Guideline for Explosion-Resistant Control Buildings in the Chemical Process Industry," Second International Symposium on Loss Prevention and Safety Promotion in the Process Industries, Heidelberg, Germany, pp. 215-222 (September 1977).

BARBER, R. B. (1973), "Steel Rod/Concrete Slab Impact Test (Experimental Simulation)," Technical Development Program, Final Report, Job No. 90142, Scientific Development, Bechtel Corporation (October 1973).

BARKAN, D. D. (1962), Dynamics of Bases and Foundations, McGraw-Hill Book Company, New York, New York.

BARTHEL, H. O. (1974), Physics in Fluids, 17, pp. 1547-1554.

BARTHEL, H. O. and Strehlow, R. A. (1979), "Direct Detonation Initiation by Localized Enhanced Reactivity," AIAA Paper No. 79-0286, New Orleans, Louisiana (January 1979).

BARTKNECHT, W. (1977), "Explosion Pressure Relief," CEP 73 (9), pp. 45-53 (September 1977).

BARTKNECHT, W. (1978a), Explosionen: Ablauf und Schutzmassnahmen, Springer-Verlag, Berlin, Germany.

BARTKNECHT, W. (1978b), "Gas, Vapor and Dust Explosions. Fundamentals, Prevention, Control," International Symposium on Grain Elevator Explosions, U. S. National Research Council (July 1978).

BATEMAN, T. L., Small, F. H., and Snyder, G. E. (1974), "Dinitrotoluene Pipeline Explosion," Loss Prevention, 8, pp. 117-122.

BATHE, K. J. (1976), "ADINA, A Finite Element Program for Automatic Dynamic Incremental Nonlinear Analysis," Report 82448-1, Acoustics and Vibration Lab, Medical Engineering Department, MIT.

BECK, J. E., Beaver, B. M. LeVine, H. S., and Richardson, E. Q. (1981), "Single-Degree-Of-Freedom Evaluation," AFWL-TR-80-99 (March 1981).

BENDLER, A. J., Roros, J. K., and Wagner, N. H. (1958), "Fast Transient Heating and Explosion of Metals Under Stagnant Liquids," AECU-3623, Contract AT(30-3)-187, Task II, Columbia University, Department of Chemical Engineering (February 1958).

BENEDICK, W. B. (1979), "High Explosive Initiation of Methane-Air Detonations," Combustion and Flame, 35, pp. 89-94.

BENZLEY, S. E., Bertholf, L. D., and Clark, G. E. (1969), "TOODY II-A -- A Computer Program for Two-Dimensional Wave Propagation -- CDC 6600 Version," Sandia Report SC-DR-69-516, NTIS PB 187809 (November 1969).

BERNSTEIN, B., Hall, D. A., and Trent, H. M. (1958), "On the Dynamics of a Bull Whip," J. of the Acoust. Soc. of Am., 30, pp. 1112-1115.

BESSEY, R. L. (1974), "Fragment Velocities from Exploding Liquid Propellant Tanks," The Shock and Vibration Bulletin, Bulletin 44 (August 1974).

BESSEY, R. L. and Kulesz, J. J. (1976), "Fragment Velocities from Bursting Cylindrical and Spherical Pressure Vessels," The Shock and Vibration Bulletin, Bulletin 46 (August 1976).

BETHE, H. A., Fuchs, K., Hirschfelder, H. O., Magee, J. L., Peierls, R. E., and vonNeumann, J. (1947), "Blast Wave," LASL 2000, Los Alamos Scientific Laboratory (August 1947) (distributed March 27, 1958).

BETHE, H. A., Fuchs, K., vonNeumann, J., Peierls, R., and Penney, W. G. (1944), "Shock Hydrodynamics and Blast Waves," AECD 2860 (October 1944).

BIGGS, J. M. (1964), Introduction to Structural Dynamics, McGraw-Hill Book Company, New York, New York.

BLANDER, M. and Katz, J. L. (1975), "Bubble Nucleation in Liquids," AIChE Journal, 21, 5, pp. 833-848 (September 1975).

BLEVINS, R. D. (1979), Formulas for Natural Frequency and Mode Shape, Van Nostrand Reinhold Company, New York, New York.

BOARD, S. J., Farmer, C. L., and Poole, D. H. (1974), "Fragmentation in Thermal Explosions," Int. J. Heat Mass Transfer, 17, pp. 331-339.

BOARD, S. J., Hall, R. W., and Hall, R. S. (1975), "Detonation of Fuel-Coolant Explosions," Nature, 254, pp. 319-321.

BODURTHA, F. T. (1980), Industrial Explosion Prevention and Protection, McGraw-Hill Book Company, New York, New York.

BOGER, R. C. and Waldman, G. D. (1973), "Blast Wave Interactions from Multiple Explosions," Proceedings of the Conference on Mechanisms of Explosion and Blast Waves, Paper No. XII, J. Alstor, Editor, Sponsored by the Joint Technical Coordinating Group for Air Launched Non-Nuclear Ordnance Working Party for Explosives (November 1973).

BOWEN, I. G., Fletcher, E. R., and Richmond, D. R. (1968), "Estimate of Man's Tolerance to the Direct Effects of Air Blast," Technical Re-

738

port to Defense Atomic Support Agency, DASA 2113, Lovelace Foundation for Medical Education and Research, AD 693105 (October 1968).

BOWEN, J. A. (1972), "Hazard Considerations Relating to Fuel-Air Explosive
Weapons," Minutes of the Fourteenth Explosives Safety Seminar, New
Orleans, Louisiana, pp. 27-35 (November 1972).

BOWLEY, W. and Prince, J. F. (1971), "Finite Element Analysis of General
Fluid Flow Problems," AIAA Fourth Fluid and Plasma Dynamics Conference, AIAA Paper No. 71-602 (June 21-23, 1971).

BOYER, D. W., Brode, H. L., Glass, I. I., and Hall, J. G. (1958), "Blast
From a Pressurized Sphere," UTIA Report No. 48, Institute of Aero-
physics, University of Toronto.

BRADLEY, D. and Mitcheson, A. (1978a), "The Venting of Gaseous Explosions
in Spherical Vessels: I - Theory," Combustion and Flame, 32, pp.
221-236.

BRADLEY, D. and Mitcheson, A. (1978b), "The Venting of Gaseous Explosions
in Spherical Vessels: II - Theory and Experiment," Combustion and
Flame, 32, pp. 237-255.

BRASIE, W. C. and Simpson, D. W. (1968), "Guidelines for Estimating Damage
from Chemical Explosions," Preprint 21A, Presented at the Symposium
on Loss Prevention in the Process Industries, Sixty-Third National
Meeting, St. Louis, Missouri (February 18-21, 1968).

BRINKLEY, S. R. (1969), "Determination of Explosion Yields," AIChE Loss
Prevention, 3, pp. 79-82.

BRINKLEY, S. R. (1970), "Shock Waves in Air Generated by Deflagration Explosions," Paper Presented at Disaster Hazards Meeting of CSSCI,
Houston, Texas (April 1970).

BRINKLEY, S. R. and Kirkwood, J. G. (1947), "Theory of the Propagation of
Shock Waves," Phys. Rev., 71, pp. 606.

BRINN, L. G. (1974), "A Select Bibliography on Explosions in Industrial
Plant with Particular Reference to Design Aspects of Explosion Con-

tainment and Relief," SM/BIB/851, British Steel Corporation, PB-236
352 (August 1974).

BRISCO, F. and Shaw, P. (1980), "Spread and Evaporation of Liquid," _Prog.
in Energy and Comb. Sci._, _6_, No. 2, pp. 127-140.

BRODE, H. L. (1955), "Numerical Solutions of Spherical Blast Waves," _J.
App. Phys._, _26_, pp. 766-775.

BRODE, H. L. (1959), "Blast Wave From a Spherical Charge," _Physics of Flu-
ids_, _2_, pp. 217.

BRODE, H. L. (1977), "Quick Estimates of Peak Overpressure from Two Simultaneous Blast Waves," Defense Nuclear Agency Report No. DNA4503T
(December 1977).

BROWN, J. A. (1973), "A Study of the Growing Danger of Detonation in Unconfined Gas Cloud Explosions," John Brown Associates, Inc., Berkeley
Heights, New Jersey (December 1973).

BROWN, T. (1943), "Minimizing Explosions in Compressed-Air Systems," _The
Oil Weekly_, pp. 12-14 (May 17, 1943).

BUETTNER, K. (1950), "Effects of Extreme Heat on Man," _Journal of American
Medical Association_, _144_, pp. 732-738 (October 1950).

BUREAU FOR INDUSTRIAL SAFETY TNO (1980a), "Methods for the Calculation of
the Physical Effects of the Escape of Dangerous Material (Liquids
and Gases), Part I," Directorate-General of Labour, Voorburg, the
Netherlands (March 1980).

BUREAU FOR INDUSTRIAL SAFETY TNO (1980b), "Methods for the Calculation of
the Physical Effects of the Escape of Dangerous Material (Liquids
and Gases), Part II," Directorate-General of Labour, Voorburg, the
Netherlands (March 1980).

BURENIN, P. I. (1974), "Effect of Shock Waves," Final Report on Contract
NASA-2485, Techtran Corporation (March 1974).

BURGESS, D. S., Murphy, J. N., Hanna, N. E., and Van Dolah, R. W. (1968),
"Large Scale Studies of Gas Detonations," Report of Investigations

7196, U. S. Department of the Interior, Bureau of Mines, Washington, D.C. (November 1968).

BURGESS, D. S., Murphy, J. N., and Zabetakis, M. G. (1970a), "Hazards of LNG Spillage in Marine Transportation," SRC Report No. S-4105, Final Report, MIPR No. Z-70099-9-92317, Project 714152, U. S. Coast Guard, Washington, D.C. (February 1970).

BURGESS, D. S., Murphy, J. N., and Zabetakis, M. G. (1970b), "Hazards Associated with Spillage of Liquefied Natural Gas on Water," U. S. Bureau of Mines, RI 7448, 27 pp. (November 1970).

BURGESS, D. S., Murphy, J. N., Zabetakis, M. G., and Perlee, H. E. (1975), "Volume of Flammable Mixtures Resulting from the Atmospheric Dispersion of a Leak or Spill," Fifteenth International Symposium on Combustion, The Combustion Institute, Pittsburgh, Pennsylvania, Paper No. 29.

BURGESS, D. S. and Zabetakis, M. G. (1962), "Fire and Explosion Hazards Associated with Liquefied Natural Gas," U. S. Department of the Interior, RI 6099.

BURGESS, D. S. and Zabetakis, M. G. (1973), "Detonation of a Flammable Cloud Following a Propane Pipeline Break: The December 9, 1970, Explosion in Port Hudson, Missouri," Report of Investigation 7752, U. S. Department of the Interior, Bureau of Mines, Washington, D.C.

BURGOYNE, J. H. (1978), "The Testing and Assessment of Materials Liable to Dust Explosion or Fire," Chemistry and Industry, pp. 81-87 (February 4, 1978).

BURGOYNE, J. H. and Craven, A. D. (1973), "Fire and Explosion Hazards in Compressed Air Systems," Loss Prevention, 7, pp. 79-87.

BUTLIN, R. N. (1975), "A Review of Information on Experiments Concerning the Venting of Gas Explosions in Buildings," Fire Research Note No. 1026, Fire Research Station, Borehamwood, Hertfordshire, England (February 1975).

BUTLIN, R. N. (1976), "Estimation of Maximum Explosion Pressure from Damage to Surrounding Buildings. Explosion at Mersey House, Bootle -- 28 August 1975," Fire Research Note No. 1054, Fire Research Station, Borehamwood, Hertfordshire, England (July 1976).

BUTLIN, R. N. and Tonkin, P. S. (1974), "Pressures Produced by Gas Explosions in a Vented Compartment," Fire Research Note No. 1019, Fire Research Station, Borehamwood, Hertfordshire, England (September 1974).

CARDILLO, P. and Anthony, E. J. (1979), "Dust Explosions and Fires, Guide to Literature (1957-1977)," Stazione Sperimentale per I Combustibili, San Donato Milanese (March 1979).

CARMICHAEL, J. B., JR., and von Gierke, H. E. (1973), "Biodynamic Applications Regarding Isolation of Humans from Shock and Vibration," Aerospace Medical Research Laboratory, Wright-Patterson Air Force Base, Ohio, AD 770316 (September 1973).

CARROLL, J. R. (1979), Physical and Technical Aspects of Fire and Arson Investigation, C. C. Thomas, New York.

CAVE, L. (1980), "Risk Assessment for Vapor Cloud Explosions," Prog. in Energy and Comb. Sci., 6, No. 2, pp. 167-176.

CHAMPION, K. S. W., O'Sullivan, W. J., Jr., and Jeweles, S. (1962), U. S. Standard Atmosphere, U. S. Government Printing Office, Washington, D. C. (December 1962).

CHAPMAN, D. L. (1899), Phil Mag., 213, Series 5, No. 47, pp. 90.

CHAR, W. T. (1978), "Simplified Approach for Design of Buildings Containing Accidental Explosions," Minutes of the Eighteenth Explosives Safety Seminar, San Antonio, Texas.

CHARNEY, M. (1967), "Explosive Venting versus Explosion Venting," Presented at AIChE Petrochemical and Refining Exposition, Houston, Texas (February 2, 1967).

CHARNEY, M. (1969), "Flame Inhibition of Vapor Air Mixture," Presented to AIChE Petrochemical and Refining Exposition, New Orleans, Louisiana (March 18, 1969).

CHASE, J. D. (1966), "Ignition Caused by the Adiabatic Compression of Air in Contact with a Lubricating-Oil-Wetted Surface," Proc. Instn. Mech. Engrs., 181, Part 1, No. 11, pp. 243-258.

CHEMICAL DICTIONARY, THE CONDENSED (1971), Eighth Edition, Revised by Gressner G. Hawley, Editor, Van Nostrand Reinhold Company, New York, New York.

CHIU, K., Lee, J., and Knystautas, R. (1976), "The Blast Waves from Asymmetrical Explosions," Department of Mechanical Engineering, McGill University, Montreal, Canada, Internal Report.

CHOROMOKOS, J. (1972), "Detonable Gas Explosions -- SLEDGE," Proceedings of the Third International Symposium on Military Applications of Blast Simulation, Schwetzingen, Germany, pp. B4-1 through B4-10 (September 1972).

CHOU, P. C., Karpp, R. R., and Huang, S. L. (1967), "Numerical Calculation of Blast Waves by the Method of Characteristics," AIAA Journal, 5, pp. 618-723.

CLEMEDSON, C. J., Hellstrom, G., and Lingren, S. (1968), "The Relative Tolerance of the Head, Thorax, and Abdomen to Blunt Trauma," Annals of the New York Academy of Sciences, 152, Article 1, pp. 187+ (October 1968).

COEVERT, K., Groothuizen, T. M., Pasman, H. J., and Trense, R. W. (1974), "Explosions of Unconfined Vapor Clouds," Loss Prevention Symposium, The Hague, Netherlands (May 1974).

COLE, R. H. (1965), Underwater Explosions, Dover Publications, Inc.

CONSTANCE, J. D. (1971), "Pressure Ventilate for Explosion Protection," Chemical Engineering (September 20, 1971).

COOK, R. D. (1974), <u>Concepts and Applications of Finite Element Analysis</u>, John Wiley and Sons, Inc., New York, New York.

CORBEN, H. C. (1958), "Power Bursts in Nuclear Reactors," RWC 22-127, Contract AT (04-3)-165 with U. S. AEC, The Ramo-Wooldridge Corporation, Los Angeles, California (September 1958).

COWARD, H. F. and Jones, G. W. (1952), "Limits of Flammability of Gases and Vapors," Bureau of Mines Bulletin 503.

COX, P. A. and Esparza, E. D. (1974), "Design of a Suppressive Structure for a Melt Loading Operation," <u>Minutes of the Sixteenth Annual Explosive Safety Seminar</u>, Department of Defense Explosives Safety Board.

COX, P. A., Westine, P. S., Kulesz, J. J., and Esparza, E. D. (1978), "Analysis and Evaluation of Suppressive Shields," Edgewood Arsenal Contractor Report, ARCFL-CR-77028, Report No. 10, Contract No. DAAA15-75-C-0083, Edgewood Arsenal, Aberdeen Proving Ground, Maryland (January 1978).

COX, R. A. (1980), "Methods of Predicting the Atmospheric Dispersion of Massive Releases of Flammable Vapor," <u>Prog. in Energy and Comb. Sci.</u>, <u>6</u>, No. 2, pp. 141-150.

CRANZ, C. (1926), <u>Lehrbuch der Ballistik</u>, Springer-Verlag, Berlin, Germany.

CRAVEN, A. D. and Grieg, T. R. (1968), "The Development of Detonation Overpressures in Pipelines," <u>I. Chem. E.</u>, Series No. 25, Institution of Chemical Engineers, London, England.

CROCKER, M. J. and Hudson, R. R. (1969), "Structural Response to Sonic Booms," <u>J. Sound and Vibration</u>, <u>9</u>, pp. 454-468.

CROUCH, W. W. and Hillyer, J. C. (1972), "What Happens When LNG Spills?" <u>Chemtech</u>, pp. 210-215 (April 1972).

CUBBAGE, P. A. (1959), "Flame Traps for Use with Town Gas/Air Mixtures," The Gas Council, Research Communication GC63, 1, Grosvenor Place, London, England, S.W.1.

CUBBAGE, P. A. and Marshall, M. R. (   ), "Chemical Process Hazards with Special Reference to Plant Design," _V. I. Chem. E. Symposium_, Series No. 39, Institution of Chemical Engineers, London, England, pp. 1-15.

CUBBAGE, P. A. and Marshall, M. R. (1972), "Pressures Generated in Combustion Chambers by the Ignition of Air-Gas Mixtures," _I. Chem. E. Symposium_, Series No. 33, Institution of Chemical Engineers, London, England, pp. 24-31.

CUSTARD, G. H. and Thayer, J. R. (1970), "Evaluation of Explosive Storage Safety Criteria," Falcon Research and Development Company, Contract DAHC 04-69-C-0095 (March 1970) [also, "Target Response to Explosive Blast" (September 1970)].

CYBULSKI, . . (1975), "Coal Dust Explosions and Their Suppression," U. S. Department of Commerce NTIS, Springfield, Virginia, Translation TT73-54001.

DABORA, E. K., Ragland, K. W., and Nicholls, J. A. (1966), "A Study of Heterogeneous Detonations," _Astronautica Acta_, 12, pp. 9-16.

DALY, B. J., Harlow, F. H., and Welch, J. E. (1964), "Numerical Fluid Dynamics Using the Particle-and-Force Method," Los Alamos Scientific Laboratories, LA-3144 (September 1964).

DAMON, E. G., Henderson, E. A., and Jones, R. K. (1973), "The Effects of Intermittent Positive Pressure Respiration on Occurrence of Air Embolism and Mortality Following Primary Blast Injury," Technical Report to Defense Nuclear Agency, DNA 2989F, Lovelace Foundation for Medical Education and Research, AD 754448 (January 1973).

DAMON, E. G., Richmond, D. R., Fletcher, E. R., and Jones, R. K. (1974), "The Tolerance of Birds to Airblast," Final Report to Defense Nuclear Agency, DNA 3314F, Lovelace Foundation for Medical Education and Research, AD 785259 (July 1974).

DAMON, E. G., Yelverton, J. T., Luft, U. C., and Jones, R. K. (1970), "Recovery of the Respiratory System Following Blast Injury," Technical Progress Report to Defense Atomic Support Agency, DASA 2580, Lovelace Foundation for Medical Education and Research, AD 618369 (October 1970).

DAMON, E. G., Yelverton, J. T., Luft, U. C., Mitchell, K., Jr., and Jones, R. K. (1970), "The Acute Effects of Air Blast on Pulmonary Function in Dogs and Sheep," Technical Progress Report to Defense Atomic Support Agency, DASA 2461, Lovelace Foundation for Medical Education and Research, AD 709972 (March 1970).

DARTNELL, R. C., JR. and Ventrone, T. A. (1971), "Explosion in a Para-Nitro-Meta-Cresol Unit," Loss Prevention, 5, pp. 53-56.

DAVENPORT, J. A. (1977a), "A Survey of Vapor Cloud Incidents," CEP 73 (9), pp. 54-63 (September 1977).

DAVENPORT, J. A. (1977b), "A Study of Vapor Cloud Incidents," AIChE Loss Prevention Symposium, Houston, Texas (March 1977).

DAVIS, J. D. and Reynolds, D. A. (1929), "Spontaneous Heating of Coal," Technical Paper 409, U. S. Bureau of Mines, Department of the Interior.

DEISENBERG, N. A., Lynch, C. J., and Breeding, R. J. (1975), "Vulnerability Model: A Simulation System for Assessing Damage Resulting from Marine Spills," Report No. CG-D-136-75, AD-A015 245 (June 1975).

DENBIGH, K. G. and Turner, J. C. R. (1971), Chemical Reactor Theory, An Introduction, Cambridge, The University Press, Second Edition.

DEN HARTOG, J. P. (1949), Strength of Materials, Dover Publications, New York, New York.

DEN HARTOG, J. P. (1956), Mechanical Vibrations, McGraw-Hill Book Company, Inc., New York, New York.

DESALVO, G. J. and Swanson, J. A. (1979), ANSYS Engineering Analysis Systems, User's Manual, Swanson Analysis Systems, Inc. Houston, Pennsylvania.

746

DESIGN OF STRUCTURES TO RESIST THE EFFECTS OF ATOMIC WEAPONS (1965), TM 5-856, 1 Through 9, Department of the Army, U. S. Army AG Publications Center, 1655 Woodson Road, St. Louis, Missouri, 63114 (March 1965) [This is an updated version of the U. S. Army COE Manual EM110-345-415 (March 1957)].

DEWEY, J. M., Johnson, O. T., and Patterson, J. D., II (1962), "Mechanical Impulse Measurements Close to Explosive Charges," BRL Report No. 1182, Aberdeen Proving Ground, Maryland (November 1962).

DEWEY, J. M. and Sperrazza, J. (1950), "The Effect of Atmospheric Pressure and Temperature on Air Shock," BRL Report 721, Aberdeen Proving Ground, Maryland.

DICK, R. A. (1968), "Factors in Selecting and Applying Commercial Explosives and Blasting Agents," Bureau of Mines Information Circular 8405.

DIETRICH, J. R. (1954), "Experimental Investigation of the Self-Limitation of Power During Reactivity Transients in a Subcooled, Water-Moderated Reactor (Borax-I Experiments)," AECD-3668, Argonne National Laboratories, Lemont, Illinois.

DIN, F. (EDITOR) (1962), Thermodynamic Functions of Gases. Volume 1 - Ammonia, Carbon Monoxide and Carbon Dioxide. Volume 2 - Air, Acetylene, Ethylene, Propane, and Argon, Butterworths, London, England.

DISS, E., Karam, H., and Jones, C. (1961), "Practical Way to Size Safety Disks," Chemical Engineering, pp. 187-190 (September 18, 1961).

DOBRATZ, B. M. (1981), "LLNL Explosives Handbook, Properties of Chemical Explosives and Explosive Simulants," UCRL-52997, Lawrence Livermore National Laboratory (March 16, 1981).

DOERING, W. and Burkhardt, G. (1949), "Contributions to the Theory of Detonation," Translation from the German as Technical Report No. F-TS-1227-IA (GDAM A9-T-4G), Headquarters, Air Materiel Command, Wright-Patterson Air Force Base, Ohio, AD 77863 (May 1949).

DOMALSKI, E. S. (1977), "A Second Appraisal of Methods for Estimating Self-Reaction Hazards," NBSIR 76/1149, Report No. DOT/MTB/OHMO-76/6, Washington, D.C.

DÖRGE, K. J. and Wagner, H. G. (1975), "Acceleration of Spherical Flames," Deuxieme Symposium Europeen sur la Combustion, Orleans, France, pp. 253-258.

DORSETT, H. G., JR., Jacobson, M., Nagy, J., and Williams, R. P. (1960), "Laboratory Equipment and Test Procedures for Evaluating Explosibility of Dusts," RI 5624, Bureau of Mines.

DORSETT, H. G., JR. and Nagy, J. (1968), "Dust Explosibility of Chemicals, Drugs, Dyes, and Pesticides," U. S. Bureau of Mines, RI 7132.

DOW CHEMICAL COMPANY (1973), "Fire and Explosion, Dow's Safety and Loss Prevention Guide. Hazard Classification and Protection," Prepared by the Editor of Chem. Eng. Progress, American Institute of Chemical Engineering, New York.

DOYLE, W. H. (1969), "Industrial Explosions and Insurance," Loss Prevention, 3, pp. 11-17.

DOYLE, W. H. (1970), "Estimating Losses," Paper Presented at CSSCI Meeting in Houston, Texas (April 1970).

DUBIN, M., Hull, A. R., and Champion, K. S. W. (1976), U. S. Standard Atmosphere, NOAA-S/T 76-1562, U. S. Government Printing Office.

DUFOUR, R. E. and Westerberg, . . (1970), "An Investigation of Fifteen Flammable Gases or Vapors with Respect to Explosion-Proof Electrical Equipment," UL Bulletin of Research No. 58 (April 2, 1970).

DUXBURY, H. A. (1976), "Gas Vent Sizing Methods," Loss Prevention, 10, pp. 147-149.

ECKER, H. W., James, B. A., and Toensing, R. H. (1974), "Electrical Safety: Designing Purged Enclosures," Chemical Engineering (May 13, 1974).

ECKHOFF, R. K. (1975), "Towards Absolute Minimum Ignition Energies for Dust Clouds," Combustion and Flame, 24, pp. 53-64.

ECKHOFF, R. K. (1976), "A Study of Selected Problems Related to the Assessment of Ignitability and Explosibility of Dust Clouds," Chr. Michelsens Institutt for Videnskap og Andsfrihet Beretninger XXXCIII, 2, Bergen, A. S. John Griegs Boktrykkeri.

ECKHOFF, R. K. (1977), "The Use of the Hartmann Bomb for Determining K_{St} Values of Explosible Dust Clouds," Staub-Reinhalt, 37, pp. 110-112.

ECKHOFF, R. K. and Enstad, G. (1976), "Why Are 'Long' Electrical Sparks More Effective Dust Explosion Initiators Than 'Short' Ones?" Combustion and Flame, 27, pp. 129-131.

ECKHOFF, R. K., Fuhre, K., Krest, C. M., Guirao, C. M., and Lee, J. H. S. (1980), "Some Recent Large Scale Gas Explosion Experiments in Norway," The Chr. Michelsen Institute, Report CMI No. 790750-1 (January 1980).

EICHLER, T. V. and Napadensky, H. S. (1977), "Accidental Vapor Phase Explosions on Transportation Routes Near Nuclear Plants," IITRI Final Report J6405, Argonne National Laboratory (April 1977).

ELLIS, O. C. DE C. (1928), Fuel and Science in Practice, 7, pp. 502.

ENGER, T. (1972), "Explosive Boiling of Liquefied Gases on Water," Presented at the National Research Council Conference on LNG Importation and Terminal Safety, Boston, Massachusetts, National Academy of Sciences, Washington, D.C. (June 13, 1972).

ENGER, T. and Hartman, D. E. (1971), "LNG Spillage on Water. I - Exploratory Research on Rapid Phase Transformations," Shell Pipe Line Corporation, Research and Development Laboratory, Technical Progress Report No. 1-71 (February 1971).

ENGER, T. and Hartman, D. E. (1972), "LNG Spillage on Water. II - Final Report on Rapid Phase Transformations," Shell Pipe Line Corporation, Research and Development Laboratory, Technical Progress Report No. 1-72 (February 1972).

ENGINEERING DESIGN HANDBOOK (1972), "Principles of Explosive Behavior,"
    AMCP 706-180, Headquarters, U. S. Army Material Command, Washington,
    D.C. (April 1972).

ESPARZA, E. D. (1975), "Estimating External Blast Loads from Suppressive
    Structures," Edgewood Arsenal Contractor Report EM-CR-76030, Report
    No. 3 (November 1975).

ESPARZA, E. D. and Baker, W. E. (1977a), "Measurement of Blast Waves From
    Bursting Pressurized Frangible Spheres," NASA CR-2843, Southwest Re-
    search Institute, San Antonio, Texas (May 1977).

ESPARZA, E. D. and Baker, W. E. (1977b), "Measurements of Blast Waves From
    Bursting Frangible Spheres Pressurized with Flash-Evaporating Vapor
    or Liquid," NASA Contractor Report 2811, Contract NSG-3008, National
    Aeronautics and Space Administration (November 1977).

ESPARZA, E. D., Baker, W. E., and Oldham, G. A. (1975), "Blast Pressures
    Inside and Outside Suppressive Structures," Edgewood Arsenal Con-
    tractor Report EM-CR-76042, Report No. 8 (December 1975).

FAUSKE, H. K. (1973), "On the Mechanisms of Uranium Dioxide-Sodium Explo-
    sive Interactions," Nuclear Science and Engineering, 51, pp. 95-101.

FAUSKE, H. K. (1974), "Some Aspects of Liquid-Liquid Heat Transfer and Ex-
    plosive Boiling," Proceedings of the Fast Reactor Safety Meeting,
    Beverly Hills, California, pp. 992-1005 (April 2-4, 1974).

FAY, J. A. (1973), "Unusual Fire Hazard of LNG Tanker Spills," Combustion
    Science and Technology, 7, pp. 47-49.

FAY, J. A. and Lewis, D. H., Jr. (1977), "Unsteady Burning of Unconfined
    Fuel Vapor Clouds," Sixteenth Symposium (International) on Combus-
    tion, The Combustion Institute, Pittsburgh, Pennsylvania, pp. 1397-
    1405.

FAY, J. A. and MacKenzie, J. J. (1972), "Cold Cargo," Environment, 14, No.
    9, pp. 21-29 (November 1972).

FEEHERY, J. (1977), "Disaster at Texas City," Amoco Torch, pp. 5-17 (Novem-
    ber/December 1977).

FEINSTEIN, D. I. (1973), "Fragmentation Hazard Evaluations and Experimental Verification," Minutes of the Fourteenth Explosives Safety Seminar, New Orleans, Louisiana, Department of Defense Explosives Safety Board, pp. 1099-1116 (November 8-10, 1972).

FENVES, S. J., Perrone, N., Robinson, A. R., and Schnobrich, W. C. (Editors) (1973), Numerical and Computer Methods in Structural Mechanics, Academic Press, New York, New York.

FIQUET, G. and Dacquet, S. (1977), "Study of the Perforation of Reinforced Concrete by Rigid Missiles - Experimental Study, Part II," Nuclear Engineering and Design, No. 41, pp. 103-120.

FLEMING, K. N., Houghton, W. J., and Scaletta, F. P. (   ), "A Methodology for Risk Assessment of Major Fires and Its Application to an HTGR Plant," General Atomics Corporation, GA-A154902, UC-77.

FLETCHER, E. R. (1971), "A Model to Simulate Thoracic Responses to Air Blast and to Impact," Aerospace Medical Research Laboratory, Paper No. 1, Wright-Patterson Air Force Base, Ohio, AD 740438 (December 1971).

FLETCHER, E. R., Richmond, D. R., and Jones, R. K. (1973), "Airblast Effects on Windows in Buildings and Automobiles on the Eskimo II Event," Minutes of the Fifteenth Explosives Safety Seminar, 1 (September 1973).

FLETCHER, E. R., Richmond, D. R., and Jones, R. K. (1976), "Velocities, Masses, and Spatial Distributions of Glass Fragments from Windows Broken by Air Blast," Lovelace Foundation Final Report for Defense Nuclear Agency (September 1976).

FLETCHER, R. F. (1968a), "Characteristics of Liquid Propellant Explosions," Annals of the New York Academy of Science, 152, I, pp. 432-440.

FLETCHER, R. F. (1968b), "Liquid-Propellant Explosions," Journal of Spacecraft and Rockets, 5, 10, pp. 1227-1229 (October 1968).

FLORENCE, A. L. and Firth, R. D. (1965), "Rigid Plastic Beams Under Uniformly Distributed Impulses," Journal of Applied Mechanics, 32, Series E, 1, pp. 7-10 (March 1965).

FLORY, K., Paoli, R., and Mesler, R. (1969), "Molten Metal-Water Explosions," _Chemistry Engineering Progress_, 65, 12, pp. 50-54 (December 1969).

FONTEIN, R. J. (1970), "Disastrous Fire at the Shell Oil Refinery, Rotterdam," _Inst. Fire Engr. Quarterly_, pp. 408-417.

FORBES, D. J. (1976), "Design of Blast-Resistant Buildings in Petroleum and Chemical Plants," Paper Presented at API 1976 Autumn Meeting, Subcommittee on Facilities and Maintenance (October 1976).

FOWLE, T. I. (1965), "Compressed Air: The Explosion Risk," _Engineering_, pp. 523 (April 23, 1965).

FREEMAN, R. H. and McCready, M. P. (1971a), "Butadiene Explosion at Texas City - 2," _Chemical Engineering Progress_, 67, pp. 45-50 (June 1971).

FREEMAN, R. H. and McCready, M. P. (1971b), "Butadiene Explosion at Texas City - 2," _Loss Prevention_, 5, pp. 61-66.

FREESE, R. W. (1973), "Solvent Recovery from Waste Chemical Sludge -- An Explosion Case History," _Loss Prevention_, 7, pp. 108-112.

FREYTAG, H. H. (EDITOR) (1965), _Handbuch der Raumexplosionen_, Verlag Bhemie, GMBH, Weinheim/Bergstr.

FRY, M. A., Durrett, R. E., Ganong, G. P., Matuska, D. A., Stucker, M. D., Chambers, B. S., Needham, C. E., and Westmoreland, C. D. (1976), "The HULL Hydrodynamics Computer Code," AFWL-TR-76-183, Air Force Weapons Laboratory, Kirtland Air Force Base, New Mexico (September 1976).

FRY, R. S. and Nicholls, J. A. (1974), "Blast Wave Initiation of Gaseous and Heterogeneous Cylindrical Detonation Waves," _Fifteenth Symposium (International) on Combustion_, The Combustion Institute, Pittsburgh, Pennsylvania, pp. 43-51.

FUGELSO, L. E., Weiner, L. M., and Schiffman, T. H. (1974), "A Computation Aid for Estimating Blast Damage from Accidental Detonation of Stored Munitions," _Minutes of the Fourteenth Explosives Safety Seminar_, New

Orleans, Louisiana, Department of Defense Explosives Safety Board, pp. 1139-1166 (November 8-10, 1973).

GALLAGHER, R. H. (1975), Finite Element Analysis: Fundamentals, Prentice-Hall, Englewood Cliffs, New Jersey.

GAYLE, J. B. and Bransford, J. W. (1975), "Size and Duration of Fireballs from Propellant Explosions," NASA TM X-53314, George C. Marshall Space Flight Center, Huntsville, Alabama (August 1975).

GENTRY, R. A., Martin, R. E., and Daly, B. J. (1966), "An Eulerian Differencing Method for Unsteady Compressible Flow Problems," Journal of Computational Physics, 1, pp. 87-118.

GERBERICH, W. W. and Baker, G. S. (1968), "Toughness of Two-Phase 6A ℓ-4V Titanium Microstructures," Applications Related Phenomena In Titanium Alloys, ASTM Special Publication No. 432.

GERSTEIN, M. and Stine, W. B. (1973), "Anomalies in Flash Points of Liquid Mixtures," I&EC Product Research and Development, 12, pp. 253.

GIBSON, N. and Harris, G. F. P. (1976), "The Calculation of Dust Explosion Vents," Chemical Engineering Progress, pp. 62-67 (November 1976).

GLASSTONE, S. (1962), The Effects of Nuclear Weapons, U. S. Government Printing Office, Revised Edition (April 1962).

GLASSTONE, S. and Dolan, P. J. (1977), The Effects of Nuclear Weapons, U. S. Department of Defense and U. S. Department of Energy, Third Edition.

GLAZ, H. M. (1979), "Development of Random Choice Numerical Methods for Blast Wave Problems," Naval Surface Weapons Center, Report NSWC/WOL TR 78-211, AD A071156 (March 1979).

GODBERT, A. L. (1952), "A Standard Apparatus for Determining the Inflammability of Coal Dusts and Mine Dusts," Safety in Mines Research Establishment (British), Research Report 58.

GODBERT, A. L. and Greenwald, H. P. (1936), "Laboratory Studies of the Inflammability of Coal Dusts," Bureau of Mines Bulletin 389.

GOLDSMITH, W. (1960), Impact, Edward Arnold, Ltd., London, England.

GOLDSTEIN, S. and Berriaud, C. (1977), "Study of the Perforation of Reinforced Concrete Slabs by Rigid Missiles -- Experimental Study, Part
III," Nuclear Engineering and Design, No. 41, pp. 121-128.

GOODMAN, H. J. (1960), "Compiled Free Air Blast Data on Bare Spherical Pentolite," BRL Report 1092, Aberdeen Proving Ground, Maryland.

GOODMAN, H. J. and Giglio-Tos, L. (1978), "Equivalent Weight Factors for
Four Plastic Bonded Explosives: PBX-108, PBX-109, AFX-103, and AFX-
702," Technical Report ARBRL-TR-02057, U. S. Army Ballistic Research
Laboratory, Aberdeen Proving Ground, Maryland (April 1978).

GOODWIN, R. D. (1974), "The Thermophysical Properties of Methane from 90 to
500 k at Pressures to 700 bar," NBS Technical Note 653, U. S. Department of Commerce, National Bureau of Standards (April 1974).

GOODWIN, R. D., Roder, H. M., and Straty, G. C. (1976), "Thermophysical Properties of Ethane, from 90 to 600 k at Pressures to 700 bar," NBS
Technical Note 684, National Bureau of Standards (August 1976).

GRANT, R. L., Murphy, J. N., and Bowser, M. L. (1967), "Effect of Weather
on Sound Transmission from Explosive Shots," Bureau of Mines Report
of Investigation 6921.

GREENFIELD, S. H. (1969), "Hail Resistance of Roofing Products," U. S. Department of Commerce, National Bureau of Standards, Building Science
Series 23 (August 1969).

GREENSPON, J. E. (1976), "Energy Approaches to Structural Vulnerability
with Application of the New Bell Stress-Strain Laws," BRL Contract
Report No. 291, Aberdeen Proving Ground, Maryland (March 1976).

GREGORY, F. H. (1976), "Analysis of the Loading and Response of a Suppressive Shield When Subjected to an Internal Explosion," Minutes of the
Seventeenth Explosive Safety Seminar, Denver, Colorado (September
1976).

GRIBBIN, J. (1978), This Shaking Earth, Putnam, New York.

754

GRIFFITHS, R. F. (1981), Dealing with Risk, Halstead Press, Wiley, New York.

GRODZOVSKII, G. L. and Kukanov, F. A. (1965), "Motion of Fragments of a Vessel Bursting in a Vacuum," Inzhenemyi Zhumal, 5, No. 2, pp. 352-355.

GUERAUD, R., Sokolovsky, A., Kavyrchine, M., and Astruc, M. (1977), "Study of the Perforation of Reinforced Concrete Slabs by Rigid Missiles -- General Introduction and Experimental Study, Part I," Nuclear Engineering and Design, No. 41, pp. 97-102.

GUGAN, K. (1978), Unconfined Vapor Cloud Explosions, The Institute of Chemical Engineering, Rigley Wark, England.

HÄGGLUND, B. (1977), "The Heat Radiation from Petroleum Fires," FoU-brand, Published by Swedish Fire Protection Association (January 1977).

HAISLER, W. E. (1977), "AGGIE I - A Finite Element Program for Nonlinear Structural Analysis," Report TEES-3275-77-1, Aerospace Engineering Department, Texas A&M University (June 1977).

HAISLER, W. E. (1978), "Status Report of AGGIE I Computer Program," Technical Report No. 3275-78-2, Aerospace Engineering Department, Texas A&M University (April 1978).

HALE, D. (1972), "LNG Continuous Spectacular Growth," Pipeline and Gas Journal, pp. 41-46 (June 1972) (and following articles).

HALL, R. W. and Board, S. J. (1979), "The Propagation of Large Scale Thermal Explosions," Int. J. Heat and Mass Trans., 22, pp. 1083-1093.

HALLQUIST, J. O. (1979), "NIKE2D: An Implicit, Finite-Deformation, Finite-Element Code for Analyzing the Static and Dynamic Response of Two-Dimensional Solids," Lawrence Livermore Laboratory, Report No. UCRL-52678 (March 1979).

HALVERSON, LCDR F. H. (1975), "A Review of Some Recent Accidents in the Marine Transportation Mode," Paper No. 52e, Presented at the AIChE Symposium on Loss Prevention in the Chemical Industry, Houston, Texas, Loss Prevention Journal, 9, pp. 76-81 (March 18-20, 1975).

HAMMER, W. (1980), Product Safety Management and Engineering, Prentice-Hall, New York, New York.

HARDEE, H. C., Lee, D. O., and Bendick, W. B. (1978), "Thermal Hazards from LNG Fireballs," Combustion Science and Technology, 17, pp. 189-197.

HARDESTY, D. R. and Weinberg, F. J. (1974), "Burners Producing Large Excess Enthalpies," Combustion Science and Technology, 8, pp. 201-214.

HARLOW, F. H. (1957), "Hydrodynamic Problems Involving Large Fluid Distortion," J. Assn. Comp. Machinery, 4, No. 2.

HARLOW, F. H. and Amsden, A. A. (1970), "Fluid Dynamics - An Introductory Text," LA-4100, Los Alamos Scientific Laboratory, University of California, Los Alamos, New Mexico (February 1970).

HARLOW, F. H., Dickman, D. O., Harris, D. E., and Martin, R. E. (1959), "Two-Dimensional Hydrodynamic Calculation," LA-2301, Los Alamos Scientific Laboratory, University of California, Los Alamos, New Mexico.

HASEGAWA, K. and Sato, K. (1978), "Experimental Investigation of the Unconfined Vapour-Cloud Explosions of Hydrocarbons," Technical Memorandum of Fire Research Institute, No. 12, Fire Research Institute, Fire Defence Agency, Japan.

HEALEY, J., Ammar, A., Vellozzi, J., Pecone, G., Weissman, S., Dobbs, N., and Price, P. (1975), "Design of Steel Structures to Resist the Effects of HE Explosions," Technical Report 4837, Picatinny Arsenal, Dover, New Jersey (August 1975).

HEALTH AND SAFETY EXECUTIVE, HM FACTORY INSPECTORATE (1977), "The Explosion at the Dow Chemical Factory, King's Lynn, 27 June 1976," Her Majesty's Stationery Office (March 1977).

HEINRICH, H. J. (1974), "Zum Ablauf von Gasexplosionen in mit Rohrleitungen verbudnenen Behältern," BAM Berichte No. 28, Berlin, Germany (August 1974).

HENRY, G. A. (1945), "Blast Injuries of the Ear," Laryngoscope, 55, pp. 663-672.

HENRY, R. E., Gabor, J. D., Winsch, I. O., Spleha, E. A., Quinn, D. J., Erickson, E. G., Heiberger, J. J., and Goldfuss, G. T. (1974), "Large Scale Vapor Explosions," Proceedings of the Fast Reactor Safety Meeting, Beverly Hills, California, pp. 922-934 (April 2-4, 1974).

HENRYCH, J. (1979), The Dynamics of Explosion and Its Use, Amsterdam, Elsevier Scientific Publishing.

HESS, P. D. and Brondyke, K. J. (1969), "Causes of Molten Aluminum-Water Explosions and Their Prevention," Metal Progress, pp, 93-100 (April 1969).

HIGH, R. W. (1968), "The Saturn Fireball," Annals of the New York Academy of Science, 152, I, pp. 441-451.

HIKITA, T. (1978), "On the Investigations of Investigations of Accidents at Mizushima Complex," The Safety Information Center, The Institution for Safety of High Pressure Gas Engineering, Journal of the Institution for Safety of High Pressure Gas Engineering, No. 80, pp. 43-51.

HILADO, C. J. and Clark, S. W. (1972), "Autoignition Temperatures of Organic Chemicals," Chemical Engineering, pp. 75-80 (September 4, 1972).

HIRSCH, A. E. (1968), "The Tolerance of Man to Impact," Annals of the New York Academy of Sciences, 152, Article 1, pp. 168+ (October 1968).

HIRSCH, F. G. (1968), "Effects of Overpressure on the Ear -- A Review," Annals of the New York Academy of Sciences, 152, Article 1, pp. 147+ (October 1968).

HOERNER, S. F. (1958), Fluid-Dynamic Drag, Published by the Author, Midland Park, New Jersey.

HOOKE, R. and Rawlings, B. (1969), "An Experimental Investigation of the Behavior of Clamped, Rectangular, Mild Steel Plates Subjected to Uniform Transverse Pressure," Institute of Civil Engineers, Proceedings, 42, pp. 75-103.

HOPKINSON, B. (1915), British Ordnance Board Minutes 13565.

HOWARD, W. B. (1972), "Interpretation of a Building Explosion Accident," Loss Prevention, 6, pp. 68-73.

HOWARD, W. B. (1976), "Tests of Orifices and Flame Arresters to Prevent Flashback of Hydrogen Flames," Loss Prevention Journal, 8.

HOWARD, W. B. and Russell, W. W. (1972), "A Procedure for Designing Gas Combustion Venting Systems," Loss Prevention, 6.

HUANG, S. L. and Chou, P. C. (1968), "Calculations of Expanding Shock Waves and Late-State Equivalence," Final Report, Contract No. DA-18-001-AMC-876 (X), Report 125-12, Drexel Institute of Technology, Philadelphia, Pennsylvania (April 1968).

HUMPHREYS, J. R., JR. (1958), "Sodium-Air Reactions as They Pertain to Reactor Safety and Containment," Proceedings of the Second International Conference on the Peaceful Uses of Atomic Energy, United Nations, Geneva, Volume II, pp. 177-185.

INGARD, U. (1953), "A Review of the Influence of Meteorological Conditions on Sound Propagation," J. Acous. Soc. Am., 25, pp. 405-411.

INSTITUTE OF CHEMICAL ENGINEERING (1977), "Chemical Process Hazards," Symposium Series 58.

IOTTI, R. C., Krotiuk, W. J., and DeBoisblanc, D. R. (1974), "Hazards to Nuclear Plants from On (Or Near) Site Gaseous Explosions," Ebasco Services, Inc., 2 Reactor Street, New York, New York.

ISA (1972), "Electrical Safety Practices," Monograph 113, Instrument Society of America, Philadelphia, Pennsylvania.

JACK, W. H., JR. (1963), "Measurements of Normally Reflected Shock Waves from Explosive Charges," BRL Memorandum Report No. 1499, Aberdeen Proving Ground, Maryland, AD 422886 (July 1963).

JACK, W. H., JR. and Armendt, B. F., Jr. (L965), "Measurements of Normally Reflected Shock Parameters Under Simulated High Altitude Conditions," BRL Report No. 1280, Aberdeen Proving Ground, Maryland, AD 469 015 (April 1965).

JACOBSON, M., Cooper, A. R., and Nagy, J. (1964), "Explosibility of Metal Powders," Bureau of Mines Report RI 6516.

JACOBSON, M., Nagy, J., and Cooper, A. R. (1962), "Explosivility of Dusts in the Plastics Industry," Bureau of Mines Report RI 5971.

JACOBSON, M., Nagy, J., Cooper, A. R., and Ball, F. J. (1961), "Explosibility of Agricultural Dusts," Bureau of Mines Report RI 5753.

JANKOV, Z. D., Shanahan, J. A., and White, M. P. (1976), "Missile Tests of Quarter Scale Reinforced Concrete Barriers," Proceedings of the Symposium on Tornadoes, Assessment of Knowledge and Implications for Man, Texas Tech University, pp. 608-622 (June 22-24, 1976).

JANSSEN, E., Cook, W. H., and Hikido, K. (1958), "Metal-Water Reactions: I. A Method for Analyzing a Nuclear Excursion in a Water Cooled and Moderated Reactor," GEAP-3073, General Electric Atomic Power Equipment Department, San Jose, California (October 1958).

JARRETT, D. E. (1968), "Derivation of British Explosive Safety Distances," Annals of the New York Academy of Sciences, 152, Article 1, pp. 18-35 (October 1968).

JARVIS, H. C. (1971a), "Butadiene Explosions at Texas City - 1," Chemical Engineering Progress, 67, 6, pp. 41-44 (June 1971).

JARVIS, H. C. (1971b), "Butadiene Explosions at Texas City - 1," Loss Prevention, 5, pp. 57-60.

JOHNSON, G. R. (1977), "EPIC-3, A Computer Program for Elastic-Plastic Impact Calculations in Three-Dimensions," Ballistic Research Laboratory, Contract Report No. 343 (July 1977).

JOHNSON, G. R. (1978), "EPIC-2, A Computer Program for Elastic-Plastic Impact Computations in Two-Dimensions Plus Spin," Ballistic Research Laboratory, Contract Report ARBRL-CR-00373 (June 1978).

JOHNSON, O. T. (1967), "A Blast-Damage Relationship," BRL Report No. 1389, U. S. Army Ballistic Research Laboratory, Aberdeen Proving Ground, Maryland, AD 388 909 (September 1967).

JOHNSON, O. T., Patterson, J. D., II, and Olson, W. C. (1957), "A Simple Mechanical Method for Measuring the Reflected Impulse of Air Blast Waves," BRL Report No. 1088, Aberdeen Proving Ground, Maryland (July 1957).

JOHNSON, W. E. (1971), "A Two-Dimensional, Two Material Continuous Eulerian Hydrodynamic Code," ATWL-TR-70-244 (April 1971).

JOUGUET, E. (1905), J. Pure Appl. Math., 70, Series 6, 1, pp. 347 [also (1906), 2, 1].

KALEPS, Ints, and von Gierke, H. E. (1971), "A Five-Degree-Of-Freedom Mathematical Model of the Body," Aerospace Medical Research Laboratory, Paper No. 8, Wright-Patterson Air Force Base, Ohio, AD 740445.

KANGAS, P. (1950), "Hailstone Impact Tests on Aircraft Structural Components," Civil Aeronautics Administration Technical Development and Evaluation Center, Technical Development Report No. 124, Indianapolis, Indiana (September 1950).

KATZ, D. L. and Sliepcevich, C. M. (1971), "LNG/Water Explosions: Cause and Effect," Hydrocarbon Processing, 50, No. 11, pp. 240-244 (November 1971).

KAUFFMAN, C. W., Walanski, P, Ural, E., Nicholls, J. A., and Van Dyke, R. (1979), "Shock Wave Initiated Combustion of Grain Dust," Proceedings of the International Symposium on Grain Dusts, Manhattan, Kansas, Kansas State University (October 2-4, 1979).

KEENAN, J. H., Keyes, F. G., Hill, P. G., and Moore, J. G. (1969), Steam Tables, John Wiley and Sons, Inc., New York, New York.

KEENAN, W. A. and Tancreto, J. E. (1973), "Effects of Venting and Frangibility of Blast Environment from Explosions in Cubicles," Minutes of the Fourteenth Explosives Safety Seminar, New Orleans, Louisiana, Department of Defense Explosives Safety Board, pp. 125-162 (November 8-10, 1972).

KEENAN, W. A. and Tancreto, J. E. (1974), "Blast Environment from Fully and Partially Vented Explosions in Cubicles," Technical Report No. 51-

027, Civil Engineering Laboratory, Naval Construction Battalion Center, Port Hueneme, California (February 1974).

KEISTER, R. G., Pesetsky, B. I., and Clark, S. W. (1971), "Butadiene Explosion at Texas City - 3," Loss Prevention, 5, pp. 67-75.

KENNEDY, W. D. (1946), "Explosions and Explosives in Air," Effects of Impact and Explosion, M. T. White (Editor), Summary Technical Report of Division 2, NDRC, Volume I, Washington, D.C., AD 221 586.

KINEKE, J. H., JR. (1976), "Secondary Fragment Speed with Unconfined Explosives: Model and Validation," Minutes of the Seventeenth Explosives Safety Seminar, Denver, Colorado, pp. 1225-1246 (September 1976).

KINGERY, C. N. (1966), "Air Blast Parameters Versus Distance for Hemispherical TNT Surface Bursts," BRL Report No. 1344, Aberdeen Proving Ground, Maryland (September 1966).

KINGERY, C. N. (1968), "Parametric Analysis of Sub-Kiloton Nuclear and High Explosive Air Blast," BRL Report No. 1393, Aberdeen Proving Ground, Maryland (February 1968).

KINGERY, C. N., Schumacher, R. N., and Ewing, W. O., Jr. (1975), "Internal Pressures from Explosions in Suppressive Structures," BRL Interim Memorandum Report No. 403, Aberdeen Proving Ground, Maryland (June 1975).

KINNERSLY, P. (1975), "What Really Happened at Flixborough?" New Scientist, 65 (938), pp. 520-522 (February 27, 1975).

KINNEY, G. F. (1962), Explosive Shocks in Air, MacMillian, New York, New York.

KINNEY, G. F. (1968), "Engineering Elements of Explosions," NWC TP4654, Naval Weapons Center, China Lake, California, AD 844 917 (November 1968).

KINNEY, G. F. and Sewell, R. G. S. (1974), "Venting of Explosions," NWC Technical Memorandum Report No. 2448, Naval Weapons Center, China Lake, California (July 1974).

KIWAN, A. R. (1970a), "Self Similar Flow Outside an Expanding Sphere," BRL Report No. 1495, Aberdeen Proving Ground, Maryland (September 1970).

KIWAN, A. R. (1970b), "Gas Flow During and After the Deflagration of a Spherical Cloud of Fuel-Air Mixture," BRL Report No. 1511, Aberdeen Proving Ground, Maryland (November 1970).

KIWAN, A. R. (1971), "FAE Flow Computations Using AFAMF Code," BRL Report No. 1547, Aberdeen Proving Ground, Maryland (September 1971).

KIWAN, A. R. and Arbuckle, A. L. (1975), "Fuel Air Explosions in Reduced Atmospheres," BRL Memorandum Report No. 2506, Aberdeen Proving Ground, Maryland (July 1975).

KIWAN, A. R., Arbuckle, A. L., and Giglio-Tos, L. (1975), "Experimental Study of Fuel Air Detonations at High Altitudes," BRL Memorandum Report No. 2554, Aberdeen Proving Ground, Maryland (October 1975).

KLETZ, T. A. (1975), "Lessons to be Learned from Flixborough," Paper No. 67f, Presented at the AIChE Symposium on Loss Prevention in the Chemical Industry, Houston, Texas, Loss Prevention Journal, 8 (March 18-20, 1975).

KNYSTAUTAS, R. and Lee, J. H. S. (1976), "On the Effective Energy for Direct Initiation of Gaseous Detonation," Combustion and Flame, 27, pp. 221.

KNYSTAUTAS, R., Lee, J. H. S., Moen, I., and Wagner, H. G. (1979), "Direct Initiation of Spherical Detonation by a Hot Turbulent Gas Jet," Seventeenth Symposium (International) on Combustion, The Combustion Institute, Pittsburgh, Pennsylvania, pp. 1235-1245.

KOCH, C. and Bökemeien, V. (1977), "Phenomenology of Explosions of Hydrocarbon Gas-Air Mixtures in the Atmosphere," Nuclear Engineering and Design, 41, pp. 69-74.

KOGARKO, Adushkin, and Lyamin (1965), "Investigation of Spherical Detonation of Gas Mixtures," Combustion, Explosion and Shock Waves, 1, No. 2, pp. 22-34.

KOGLER, . . (1971), "Vinyl Tank Car Incident," Loss Prevention, 5, pp. 26-28.

KOHNKE, P. C. (1977), "ANSYS Engineering Analysis System Theoretical Manual," Swanson Analysis Systems, Inc., Elizabeth, Pennsylvania (November 1977).

KOKINAKIS, W. (1974), "A New Methodology for Wounding and Safety Criteria," Proceedings of the Sixteenth Explosives Safety Seminar, pp. 1209-1226 (September 1974).

KOLB, J. and Ross, S. S. (1980), Product Safety and Liability, McGraw-Hill Book Company, New York, New York.

KOMAMIYA, K. (1969), "The Quenching Ability of Flame Arrestors for N-Hexane," Research Report of the Research Institute of Industrial Safety, RR-17-4, pp. 1-5 (March 1969).

KOROBEINIKOV, V. P., Mil'nikova, N. S., and Ryazanov, Y. V. (1962), The Theory of Point Explosion, Fizmatgiz, Moscow, 1961, English Translation, U. S. Department of Commerce, JPRS: 14,334, CSO: 69-61-N, Washington, D.C.

KOT, C. A. and Turula, P. (1976), "Air Blast Effect on Concrete Walls," Technical Memorandum ANL-CT-76-50, Argonne National Laboratory (July 1976).

KOT, C. A., Valentin, R. A., McLennan, D. A., and Turula, P. (1978), "Effects of Air Blast on Power Plant Structures and Components," NUREG/CR-0442, ANL-CT-78-41, Argonne National Laboratory (October 1978).

KRAZUISKI, J. L., Buckius, R. O., and Krier, H. (1979), "Coal Dust Flames: A Review and Development of a Model for Flame Propagation," Prog. Energy Comb. Sci., 5, pp. 31-71.

KRIER, H. and Foo, C. L. (1973), "A Review and Detailed Derivation of Basic Relations Describing the Burning of Droplets," Oxidation and Combustion Reviews, 6, Tipper, Editor, Elsevier, Amsterdam, pp. 111-144.

KUHL, A. L., Kamel, M. M., and Oppenheim, A. K. (1973), "On Flame Generated Self-Similar Blast Waves," Fourteenth Symposium (International) on Combustion, The Combustion Institute, pp. 1201-1214.

LASSEIGNE, A. H. (1973), "Static and Blast Pressure Investigation for the Chemical Agent Munition Demilitarization System: Sub-Scale," Report EA-FR-4C04, Edgewood Arsenal Resident Office, Bay Saint Louis, Mississippi (November 1973).

LATHROP, J. K. (1978), "Fifty-Four Killed in Two Grain Elevator Explosions," Fire Journal, 72 (5), pp. 29-35 (September 1978).

LAW, C. K. and Chung, S. H. (1980), "An Ignition Criterion for Droplets in Sprays," Combustion Science and Technology, 22, pp. 17-26.

LAWRENCE, W. W. (1976), Of Acceptable Risk, William Kaufmann, Inc., Los Altos, California.

LAWRENCE, W. W. and Johnson, E. E. (1974), "Design for Limiting Explosion Damage," Chemical Engineering (January 7, 1974).

LE CHATELIER, H. and Boudouard, O. (1898), Comptes Rendu, 126, pp. 1344-1347.

LEE, J. F. and Sears, F. W. (1959), Thermodynamics, Addison-Wesley Publishing Company, Inc., Reading, Massachusetts (August 1959).

LEE, J. H. S. (1977), "Initiation of Gaseous Detonation," Annual Review of Physical Chemistry, 28, Annual Reviews, Inc., Palo Alto, California, pp. 75-104.

LEE, J. H. S., Guirao, C., Chiu, K., and Bach, G. (1977), "Blast Effects from Vapor Cloud Explosions," 1977 AIChE Loss Prevention Symposium, Houston, Texas (March 1977).

LEE, J. H. S., Knystautas, R., and Bach, G. G. (1969), "Theory of Explosions," Department of Mechanical Engineering, McGill University, AFOSR Scientific Report 69-3090 TR.

LEE, J. H. S., Knystautas, R., and Yoshikawa, A. (1979), _Acta Astronautica_, <u>5</u>, pp. 972-982.

LEE, J. H. S. and Matusi, H. (1977), "A Comparison of the Critical Energies for Direct Initiation of Spherical Detonations in Acetylene-Oxygen Mixtures," _Combustion and Flame_, <u>28</u>, p. 61.

LEE, J. H. S. and Moen, I. O. (1980), "The Mechanism of Transition From Deflagration to Detonation in Vapor Cloud Explosions," _Prog. in Energy and Comb. Sci._, <u>6</u>, No. 4, pp. 359-389.

LEE, J. H. S. and Ramamurthi, K. (1976), "On the Concept of the Critical Size of a Detonation Kernel," _Combustion and Flame_, <u>27</u>, pp. 331.

LEES, F. P. (1980a), _Loss Prevention in the Process Industries. Hazard Identification, Assessment and Control, Volume 1_, Butterworths, London, England.

LEES, F. P. (1980b), _Loss Prevention in the Process Industries. Hazard Identification, Assessment and Control, Volume 2_, Butterworths, London, England.

LEHTO, D. L. and Larson, R. A. (1969), "Long Range Propagation of Spherical Shock Waves from Explosions in Air," NOLTR 69-88, Naval Ordnance Laboratory, White Oak, Maryland.

LEIGH, B. R. (1974), "Lifetime Concept of Plaster Panels Subjected to Sonic Boom," UTIAS Technical Note 91, Institute for Space Studies, University of Toronto (July 1974).

LEVENTUEV, V. P. and Nemchinov, I. V. (1975), "Shock Propagation from an Expanding Hot Volume," _Fizika Goveniyer i Vzryva_, <u>11</u>, pp. 776-781 [Translated by Plenum Press (1976) <u>11</u>, pp. 663-666].

LE VINE, R. Y. (1972a), "Electrical Equipment and Installation for Hazardous Areas A-4," _Chemical Engineering_ (May 1, 1972).

LE VINE, R. Y. (1972b), "Electrical Safety in Process Plants ... Classes and Limits of Hazardous Areas," _Chemical Engineering_ (May 1, 1972).

LEWIS, B. and von Elbe, G. (1961), _Combustion, Flames and Explosions of Gases_, Academic Press, New York, New York.

LEWIS, D. J. (1980a), "Unconfined Vapor Cloud Explosions: Definition of Source of Fuel," Progress in Energy and Combustion Science, 6, No. 2, pp. 121-126.

LEWIS, D. J. (1980b), "Unconfined Vapor Cloud Explosions: Historical Perspective and Prediction Method Based on Incident Records," Progress in Energy and Combustion Science, 6, No. 2, pp. 151-166.

LEYER, J. C., Guerrand, C., and Manson, N. (1974), "Flame Propagation in Small Spheres of Unconfined and Slightly Confined Flammable Mixtures," Fifteenth Symposium on Combustion, pp. 645-653.

LIEBMAN, I. and Richmond, J. K. (1974), "Suppression of Coal Dust Explosions by Passive Water Barriers in a Single Entry Mine," U. S. Bureau of Mines RI 7815.

LIEPMANN, H. W. and Roshko, A. (1967), Elements of Gasdynamics, John Wiley and Sons, Inc., New York, New York.

LIND, C. D. (1975a), "Unconfined Vapor Cloud Explosion Studies," Paper No. 67e, Presented at the AIChE Symposium on Loss Prevention in the Chemical Industry, Houston, Texas, Loss Prevention Journal, 8 (March 18-20, 1975).

LIND, C. D. (1975b), "What Causes Vapor Cloud Explosions?" Loss Prevention, 9, pp. 101-105.

LIND, C. D. and Whitson, J. (1977), "Explosion Hazards Associated with Spills of Large Quantities of Hazardous Materials," Phase III Report No. CG-D-85-77, Department of Transportation, U. S. Coast Guard Final Report, ADA 047585.

LINDBERG, H. E., Anderson, D. L., Firth, R. D., and Parker, L. V. (1965), "Response of Reentry Vehicle-Type Shells to Blast Loads," Final Report, SRI Project FGD-5228, P. O. 24-14517 under Contract No. AF 04(694)-655, Stanford Research Institute, Menlo Park, California (September 1965).

LIPSETT, S. G. (1966), "Explosions from Molten Materials and Water," _Fire Technology_, pp. 118-126 (May 1966).

LITCHFIELD, E. L., Kubala, T. A., Schellinger, T., Perzak, F. J., and Burgess, D. (1980), "Practical Ignition Problems Related to Industrial Safety in Mine Equipment," U. S. Bureau of Mines RI 8464.

LIVESLEY, R. K. (1964), _Matrix Methods of Structural Analysis_, Pergamon and MacMillan.

LOSS PREVENTION HANDBOOK (1967), Second Edition - Properties of Flammable Liquids, Gases and Solids, Factory Mutual.

LUCKRITZ, R. T. (1977), "An Investigation of Blast Waves Generated by Constant Velocity Flames," Technical Report AAE 77-2, UILU-Eng 77 0502, University of Illinois.

LUNN, G. A. and Phillips, H. (1973), "A Summary of Experimental Data on the Maximum Experimental Safe Gap," SMRE Report R2, Department of Trade and Industry, Sheffield, England.

LUTZSKY, M. and Lehto, D. (1968), "Shock Propagation in Spherically Symmetric Exponential Atmospheres," _Physics of Fluids_, $\underline{11}$, pp. 1466.

MACEK, A. (1962), "Sensitivity of Explosives," _Chem. Rev._, $\underline{62}$, pp. 41-62.

MACH, E. and Sommer, J. (1877), "Uber die Fortpflanzunggeshwindigkeit von Explosionsschallwellen," Akademie der Wissenschaften, Sitzangberichte der Wiener, $\underline{74}$.

MAINSTONE, R. J. (1971), "The Breakage of Glass Windows by Gas Explosions," Building Research Establishment Current Paper CP 26/71, Building Research Establishment (September 1971).

MAINSTONE, R. J. (1973), "The Hazard of Internal Blast in Buildings," Building Research Establishment Current Paper CP 11/73, Building Research Establishment (April 1973).

MAINSTONE, R. J. (1974), "The Hazard of Explosion, Impact and Other Random Loadings on Tall Buildings," Building Research Establishment Current Paper CP 64/74, Building Research Establishment (June 1974).

MAINSTONE, R. J. (1976), "The Response of Buildings to Accidental Explo-
     sions," Building Research Establishment Current Paper CP 24/76,
     Building Research Establishment, Garston, Watford, England (April
     1976).

MAJID, K. I. (1972), Non-Linear Structures, Wiley-Interscience, A Division
     of John Wiley and Sons, Inc., New York, New York.

MAKINO, R. (1951), "The Kirkwood-Brinkley Theory of Propagation of Spheri-
     cal Shock Waves and Its Comparison with Experiment," BRL Report No.
     750 (April 1951).

MANUAL OF STEEL CONSTRUCTION (1980), American Institute of Steel Construc-
     tion, Eighth Edition, New York, New York.

MARC ANALYSIS CORPORATION and Control Data Corporation (1975), "Non-Linear
     Finite Element Analysis Program, Volume I: User's Information Man-
     ual."

MARKSTEIN, G. H. (1958), "Combustion and Propulsion," Third AGARD Colloqui-
     um, Pergamon, pp. 153.

MASTROMONICO, C. R. (1974), "Blast Hazards of CO/N_2O Mixtures," Minutes of
     the Fifteenth Explosives Safety Seminar, Department of Defense Ex-
     plosives Safety Board, San Francisco, California, pp. 1305-1357 (Sep-
     tember 18-20, 1973).

MATUSI, H. and Lee, J. H. S. (1976), "Influence of Electrode Geometry and
     Spacing on the Critical Energy for Direct Initiation of Spherical
     Gaseous Detonation," Combustion and Flame, 27, pp. 217.

MAUER, B., Schneider, H., Hess, K., and Leuckel, W. (1975), "Modellversuche
     zur Flash-Entspannung, atmosphärischen Vermischung und Deflagration
     von Flüssiggasen nach deren Freisetzung bei Behälterzerknall," In-
     ternational Seminar on Extreme Load Conditions and Limit Analysis
     Procedures for Structural Reactor Safeguards and Containment Struc-
     tures, Berlin, Germany (September 1975).

768

MCCARTHY, W. J., JR., Nicholson, R. B., O'Krent, D., and Jankus, V. Z. (1958), "Studies of Nuclear Accidents in Fast Power Reactors," Paper P/2165, Second United Nations International Conference on the Peaceful Uses of Atomic Energy (June 1958).

MCCORMICK, C. W. (EDITOR) (1976), "MSC/NASTRAN User's Manual," MSR-39, The MacNeal-Schwendler Corporation, 7442 North Figueroa Street, Los Angeles, California (May 1976).

MCCRACKEN, D. J. (1970), "Hydrocarbon Combustion and Physical Properties," BRL Report No. 1496, AD 714674 (September 1970).

MCCRACKEN, G. M. (1973), "Investigation of Explosions Produced by Dropping Liquid Metals into Aqueous Solutions," Safety Research Bulletin of the Safety and Reliability Directorate, Issue No. 11, pp. 20.

MCNAMARA, J. F. (1975), "Solution Scheme for Problems of Non-Linear Structural Dynamics," Paper No. 74-PVP. 30, Journal of Pressure Vessel Technology, American Society of Mechanical Engineers.

MCNAUGHTAN, I. I. and Chisman, S. W. (1969), "A Study of Hail Impact at High Speed on Light Alloy Plates," Proceedings of the Ninth Annual National Conference on Environmental Effects on Aircraft and Propulsion Systems, Naval Air Propulsion Test Center, pp. 16-1+ (October 7-9, 1969).

MERCK INDEX OF CHEMICALS AND DRUGS (1960), Seventh Edition, Merck and Company, Inc.

MEYER, J. W., Urtiew, P. A., and Oppenheim, A. K. (1970), "On the Inadequacy of Gasdynamic Processes for Triggering to Detonation," Combustion and Flame, 14, pp. 13-20.

MINISTRY OF LABOUR (1965), "Guide to the Use of Flame Arresters and Explosion Reliefs," Safety, Health and Welfare Series No. 34, Her Majesty's Stationery Office, London, England.

MINISTRY OF SOCIAL AFFAIRS AND PUBLIC HEALTH (1968), "Report Concerning an Inquiry into the Cause of the Explosion on 20th January 1968, at

the Premises of Shell Nederland Raffinaderij N.V. in Pernis," State
Publishing House, The Hague, Netherlands.

MOEN, I. O., Lee, J. H. S., Hjertager, B. H., Fuhre, K., and Eckhoff, R. K.
(1982), "Pressure Development Due to Turbulent Flame Propagation in
Large Scale Methane -Air Explosions," Combustion and Flame (in press).

MOORE, C. V. (1967), "The Design of Barricades for Hazardous Pressure Systems," Nuclear Engineering and Design, 5, pp. 81-97.

MOORE, T. D. (EDITOR) (1975), Structural Alloys Handbook, Mechanical Properties Data Center, Traverse City, Michigan.

MULLINS, B. P. and Penner, S. S. (1959), Explosions, Detonations, Flammability and Ignition, Agardograph 31.

MUNCK, J. (1965), "How to Avoid Fires and Explosions in Compressed Air Systems," Power, pp. 82-83 (April 1965).

MUNDAY, G. (1976), "Unconfined Vapour-Cloud Explosions," The Chemical Engineer, pp. 278-281 (April 1976).

NABERT, K. (1967a), "The Significance of a Standard Apparatus for Testing
Safe Gaps for the Safety of Electrical Equipment in Respect of Explosion Hazard," IEC/SC 31 A/WG 3, Physikalisch-Technische Bundesanstalt, Braunschweig, Germany.

NABERT, K. (1967b), "The Significance of a Standard Apparatus for Testing
Safe Gap for the Safety of Electrical Equipment in Respect of Explosion Hazard, Chapter 8 - The M.E.S.G. as a Function of the Bound-
ing Volumes and the Obstacles in These Volumes," Appendix to the
Paper IEC/SC 31 A/WGe, Mr. Nabert-PTB, Germany.

NABERT, K. and Schön, G. (1963), "Sicherheitstechnische Kennzahlen brenn-
barer Gase und Dämpfe," Deutscher Eichverlag GmbH, Braunschweig,
Germany 2, erweiterte Auflage.

NABERT, K. and Schön, G. (1970), "Sicherheitstechnische Kennzahlen brenn-
barer Gase und Dämpfe," 4, Nachtrag 2, erweiterte Auflage [Stand
1.11 (1970)].

770

NACA (1959), "Basic Considerations in the Combustion of Hydrocarbon Fuels with Air," NACA Report 1300, Lewis Flight Propulsion Laboratory.

NAGY, J., Cooper, A. R., and Stupar, J. M. (1964), "Pressure Development in Laboratory Dust Explosions," RI 6561, U. S. Bureau of Mines.

NAGY, J., Dorsett, H. G., Jr., and Cooper, A. R. (1965), "Explosibility of Carbonaceous Dusts," RI 6597, U. S. Bureau of Mines.

NAGY, J., Dorsett, H. G., Jr., and Jacobson, M. (1964), "Preventing Ignition of Dust Dispersions by Inerting," RI 6543, U. S. Bureau of Mines.

NAGY, J., et al. (1968), "Explosibility of Miscellaneous Dusts," RI 7208, U. S. Bureau of Mines.

NAKANISHI, E. and Reid, R. C. (1971), "Liquid Natural Gas-Water Reactions," Chemical Engineering Progress, 67, No. 12, pp. 36-41 (December 1971).

NAPADENSKY, H. S., Swatosh, J. J., Jr., and Morita, D. R. (1973), "TNT Equivalency Studies," Minutes of the Fourteenth Explosives Safety Seminar, New Orleans, Louisiana, Department of Defense Explosives Safety Board, pp. 289-312 (November 8-10, 1972).

NATIONAL ACADEMY OF SCIENCES (1973), "Fire Hazard Classification of Chemical Vapors Relative to Explosion-Proof Electrical Equipment, Report III," Supplemental Report Prepared by the Electrical Hazards Panel of the Committee on Hazardous Materials, Division of Chemistry and Chemical Technology, National Research Council, Under Contract No. GC-11, 775-A DOT-OS-00035, Task Order 13, for the U. S. Coast Guard (May 1973).

NATIONAL ACADEMY OF SCIENCES (1975a), "Fire Hazard Classification of Chemical Vapors Relative to Explosion-Proof Electrical Equipment, Report IV," Supplementary Report Prepared by the Electrical Hazards Panel of the Committee on Hazardous Materials, Assembly of Mathematical and Physical Sciences, National Research Council.

NATIONAL ACADEMY OF SCIENCES (1975b), "Matrix of Electrical and Fire Hazard Properties and Classification of Chemicals," A Report Prepared for

771

the Committee on Hazardous Materials, Assembly of Mathematical and
Physical Sciences, National Research Council, Washington, D.C.

NATIONAL ACADEMY OF SCIENCES (1976), "Analysis of Risk in the Water Transportation of Hazardous Materials," A Report of the Risk Analysis and
Hazard Evaluation Panel of the Committee on Hazardous Materials, Assembly of Mathematical and Physical Sciences, National Research Council, Washington, D.C.

NATIONAL ACADEMY OF SCIENCES (1978), "Flammability, Smoke, Toxicity, and
Corrosive Gases of Electric Cable Materials," Publication NMAB-342.

NATIONAL ACADEMY OF SCIENCES (1979a), "Matrix of Combustion-Relevant Properties and Classifications of Gases, Vapors, and Selected Solids,"
Publication NMAB-353-1.

NATIONAL ACADEMY OF SCIENCES (1979b), "Test Equipment for Use in Determining Classifications of Combustible Dusts," Publication NMAB-353-2.

NATIONAL FIRE CODES (1980), 16 Volumes plus 2 Addenda Volumes, National Fire
Protection Association, Boston, Massachusetts.

NEAL, B. G. (1977), The Plastic Methods of Structural Analysis, Chapman and
Hall, London, England.

NELSON, C. W. (1973), "RIPPLE - A Two-Dimensional Unsteady Eulerian Hydrodynamic Code," Ballistic Research Laboratory, Report No. BRL R 1632,
AD 758157 (February 1973).

NELSON, W. (1973), "A New Theory to Explain Physical Explosions," Combustion, pp. 31-36 (May 1973).

NEWMARK, N. M. (1950), "Methods of Analysis for Structures Subjected to Dynamic Loading," University of Illinois, Urbana, Illinois (December
1950).

NEWMARK, N. M. (1953), "An Engineering Approach to Blast-Resistant Design,"
Proceedings of ASCE, 79, Separate No. 309 (October 1953) [Also,
Trans. ASCE, 121, pp. 45-64 (1956) and Selected Papers of N. M. New-
mark, ASCE, New York, pp. 495-515 (1976)].

NFPA (1972), "Halogenated Extinguishing Agent Systems," Halon 1301, NFPA 12A, Halon 1211, NFPA 12B.

NFPA (1974), "Purged and Pressurized Enclosures for Electrical Equipment," NFPA 496.

NFPA (1975), "Hazardous Chemicals Data," NFPA 49.

NFPA (1976), "Standard on Basic Classifications of Flammable and Combustible Liquids," NFPA 321, National Fire Codes, 3.

NFPA (1977a), "Flammable and Combustible Liquids," NFPA 30.

NFPA (1977b), "Fire Hazard Properties of Flammable Liquids, Gases, Volatiles, and Solids," NFPA 325 M.

NFPA (1978a), "Explosion Prevention Systems," NFPA 69.

NFPA (1978b), "Guide for Explosion Venting," NFPA 68.

NFPA (1978c), "Standard for Intrinsically Safe Apparatus for Use in Class 1 Hazardous Locations and Its Associated Apparatus," NFPA 493.

NFPA (1981), "Hazardous (Classified) Locations," National Electric Code Article 500, NFPA 70.

NICHOLLS, J. A., Sichel, M., Dry, T., and Glass, D. R. (1974), "Theoretical and Experimental Study of Cylindrical Shock and Heterogeneous Detonation Waves," Acta Astronautica, 1, pp. 385-404.

NICHOLLS, R. W., Johnson, C. F., and Duvall, W. I. (1971), "Blast Vibrations and Their Effects on Structures," Bureau of Mines Bulletin 656.

NICKERSON, J. I. (1976), "Dryer Explosion," Loss Prevention Journal, 8.

NORRIE, H. and de Vries, G. (1973), The Finite Element Method: Fundamentals and Applications, Academic Press, New York, New York.

NORRIS, C. H., Hansen, R. J., Holley, M. J., Biggs, J. M., Namyet, S., and Minami, J. V. (1959), Structural Design for Dynamic Loads, McGraw-Hill Book Company, New York, New York.

NTSB (1971), "Highway Accident Report, Liquefied Oxygen Tank Explosion Followed by Fires in Brooklyn, New York, May 30, 1970," NTSB-HAR-71-6, Washington, D.C.

NTSB (1972a), "Pipeline Accident Report, Phillips Pipe Line Company Propane Gas Explosion, Franklin County, Missouri, December 9, 1970," Report NTSB-PAR-72-1, Washington, D.C.

NTSB (1972b), "Railroad Accident Report, Derailment of Toledo, Peoria and Western Railroad Company's Train No. 20 with Resultant Fire and Tank Car Ruptures, Crescent City, Illinois, June 21, 1970," NTSB-RAR-72-2, Washington, D.C.

NTSB (1972c), "Highway Accident Report, Automobile-Truck Collision Followed by Fire and Explosion of Dynamite Cargo on U. S. Highway 78, Near Waco, Georgia, on June 4, 1971," NTSB-HAR-72-5, Washington, D.C.

NTSB (1972d), "Railroad Accident Report, Derailment of Missouri Pacific Railroad Company's Train 94 at Houston, Texas, October 19, 1971," NTSB-RAR-72-6, Washington, D.C.

NTSB (1973a), "Railroad Accident Report, Hazardous Materials Railroad Accident in the Alton and Southern Gateway Yard in East St. Louis, Illinois, January 22, 1972," NTSB-RAR-73-1, Washington, D.C.

NTSB (1973b), "Highway Accident Report, Propane Tractor-Semitrailer Overturn and Fire, U. S. Route 501, Lynchburg, Virginia, March 9, 1972," NTSB-HAR-73-3, Washington, D.C.

NTSB (1973c), "Highway Accident Report, Multiple-Vehicle Collision Followed by Propylene Cargo-Tank Explosion, New Jersey Turnpike, Exit 8, September 21, 1972," NTSB-HAR-73-4, Washington, D.C.

NTSB (1973d), "Pipeline Accident Report, Phillips Pipeline Company, Natural Gas Liquids Fire, Austin, Texas, February 22, 1973," NTSB-PAR-73-4, Washington, D.C.

NTSB (1974a), "Railroad Accident Report, Derailment and Subsequent Burning of Delaware and Hudson Railway Freight Train at Oneonta, New York, February 12, 1974," NTSB-RAR-74-4, Washington, D.C.

NTSB (1974b), "Hazardous Materials Accident at the Southern Pacific Transportation Company's Englewood Yard, Houston, Texas, September 21, 1974," NTSB-RAR-75-7, Washington, D.C. (May 1975).

NTSB (1975), "Highway Accident Report, Surtigas, S. A., Tank-Semitrailer Overturn, Explosion, and Fire Near Eagle Pass, Texas, April 29, 1975," NTSB-HAR-76-4, Washington, D.C. (May 1976).

NTSB (1976), "Pipeline Accident Report, Consolidated Edison Company Explosion at 305 East 45th Street, New York, New York, April 22, 1974," NTSB-PAR-76-2, Washington, D.C.

OCHIAI, M. and Bankoff, S. G. (1976), "Liquid-Liquid Contact in Vapor Explosions," Ind. Cont. on Fast Reactor Safety, American Nuclear Society, Chicago, Illinois.

ODEN, J. T., Zienkiewicz, O. C., Gallagher, R. H., and Taylor, C. (1974), Finite Element Methods in Flow Problems, University of Alabama Press, Huntsville, Alabama.

OFFICE OF PIPELINE SAFETY (1974), "Griffith Indiana Butane Leak," Report No. OPS-PFR-74-1.

OIL INSURANCE ASSOCIATION (1974), "Hydrorefining Process Units, Loss Causes and Guidelines for Loss Prevention," Pamphlet No. 101.

OPPENHEIM, A. K. (1972), "Introduction to Gas Dynamics of Explosions," Course Lecture Notes No. 48, CISM, Udine, Italy, Springer-Verlag, New York, New York, pp. 43.

OPPENHEIM, A. K. (1973), "Elementary Blast Wave Theory and Computations," Paper No. I, Proceedings of the Conference on Mechanisms of Explosion and Blast Waves, J. Alstor, Editor, Sponsored by the Joint Technical Coordinating Group for Air Launched Non-Nuclear Ordnance Working Party for Explosives (November 1973).

OPPENHEIM, A. K., Kuhl, A. L., and Kamel, M. M. (1972), "On Self-Similar Blast Waves Headed by the Chapman Jouguet Detonation," J. Fluid Mech., 55, Part 2, pp. 257-270.

OPPENHEIM, A. K., Kuhl, A. L., Lundstrom, E. A., and Kamel, M. M. (1972), "A Parametric Study of Self-Similar Blast Waves," J. Fluid Mech., 52, Part 4, pp. 657-682.

OPPENHEIM, A. K., Lundstrom, E. A., Kuhl, A. L., and Kamel, M. M. (1971),
"A Systematic Exposition of the Conservation Equations for Blast
Waves," Journal of Applied Mechanics, pp. 783-794 (December 1971).

OPSCHOOR, Ir. G. (1975), "Investigation Into the Spreading and Evaporation
of LNG Spilled on Water," Subproject 4, Report 2, Central Technical
Institute T.N.O., Rijswijk (January 1975).

ORAN, E., Young, T., and Boris, J. (1978), "Application of Time-Dependent
Numerical Methods to the Description of Reactive Shocks," Seventeenth International Symposium on Combustion, The Combustion Institute, pp. 43-54 (August 1978).

ORDIN, P. M. (1974), "Review of Hydrogen Accidents and Incidents in NASA
Operations," NASA Technical Memorandum X-71565.

OSTROOT, G., JR. (1972), "Explosions in Gas and Oil Fired Furnaces," Loss
Prevention, 6, pp. 112-117.

OWCZAREK, J. A. (1964), Fundamentals of Gas Dynamics, Int. Textbook Company,
Scranton, Pennsylvania.

OWEN, T. E. (1979), "Laboratory Scale Model Studies of Electrostatics in
Ship Tanks," Paper Presented at the Conference on Electrostatic Haz-
ards in the Storage and Handling of Powders and Liquids, Chicago,
Illinois (October 16-17, 1979).

OWENS, J. I. (1959), "Metal-Water Reactions: II. An Evaluation of Severe
Nuclear Excursions in Light Water Reactors," GEAP-3178, General Electric Atomic Power Equipment Department, San Jose, California (June
1959).

PALMER, H. B., Sibulkin, M., Strehlow, R. A., and Yang, C. H. (1976), "An
Appraisal of Possible Combustion Hazards Associated with a High-
Temperature Gas Cooled Reactor," BNL-NUREG-50764.

PALMER, K. N. (1973), Dust Explosions and Fires, Chapman and Hall, Ltd.,
London, England.

PALMER, K. N. (1974), "Loss Prevention: Relief Venting of Dust Explosions,"
Chemical Engineering Progress, 70 (4), pp. 57-61 (April 1974).

PARKER, R. J., Pope, J. A., Davidson, J. F., and Simpson, W. J. (1975), "The Flixborough Disaster. Report of the Court of Inquiry," Her Majesty's Stationery Office, London, England (April 1975).

PASMAN, H. J., Groothuizen, T. M., and de Gooijer, H. (1974), "Design of Pressure Relief Vents," Proceedings of the First Symposium on Loss Prevention and Safety Promotion in the Process Industries, Elsevier.

PENNINGER, J. M. L. and Okazaki, J. K. (1980), "Designing a High-Pressure Laboratory," Chemical Engineering Progress, pp. 65-71 (June 1980).

PENNY, W. G., Samuels, D. E. J., and Scorgie, G. C. (1970), "The Nuclear Explosive Yields at Hiroshima and Nagasaki," Phil. Trans. Roy. Soc., 266, pp. 357.

PERKINS, B., JR. and Jackson, W. F. (1964), "Handbook for Prediction of Air Blast Focusing," BRL Report 1240, Ballistics Research Laboratory, Aberdeen Proving Ground, Maryland.

PERKINS, B., JR., Lorrain, P. H., and Townsend, W. H. (1960), "Forecasting the Focus of Air Blasts due to Meteorological Conditions in the Lower Atmosphere," BRL Report 1118, Aberdeen Proving Ground, Maryland.

PERRONE, N. and Pilkey, W. (Editors) (1977), Structural Mechanics Software Series, Volume I, University Press of Virginia, Charlottesville, Virginia.

PERRONE, N. and Pilkey, W. (Editors) (1978), Structural Mechanics Software Series, Volume II, University Press of Virginia, Charlottesville, Virginia.

PETERSON, P. and Cutler, H. R. (1973), "Explosion Protection for Centrifuges," Chemical Engineering Progress, 69, (4), pp. 42-44 (April 1973).

PFÖRTNER, H. (1977), "Gas Cloud Explosions and Resulting Blast Effects," Nuclear Engineering and Design, 41, pp. 59-67.

PHILLIPS, E. A. (1975), "RPI-AAR Tank Car Safety Research and Test Project -- Status Report," Paper No. 52c, Presented at the AIChE Symposium on

777

Loss Prevention in the Chemical Industry, Houston, Texas, Loss Prevention Journal, 8 (March 18-20, 1975).

PHILLIPS, H. (1971), "The Mechanism of Flameproof Protection," Department of Trade and Industry, Safety in Mines Research Establishment Research Report 275, Sheffield, England.

PHILLIPS, H. (1972a), "A Nondimensional Parameter Characterizing Mixing Processes in a Model of Thermal Gas Ignition," Combustion and Flame, 19, pp. 181-186.

PHILLIPS, H. (1972b), "Ignition in a Transient Turbulent Jet of Hot Inert Gas," Combustion and Flame, 19, pp. 187-195.

PHILLIPS, H. (1973), "Use of a Thermal Model of Ignition to Explain Aspects of Flameproof Enclosure," Combustion and Flame, 20, pp. 121-126.

PHILLIPS, W. (CHAIRMAN) (1977), Proceedings of the International Symposium on Grain Dust Explosions, Grain Elevator and Processing Society, Minneapolis, Minnesota (October 1977).

PILKEY, W., Saczalski, K., and Schaeffer, H. (Editors) (1974), Structural Mechanics Computer Program, University Press of Virginia, Charlottesville, Virginia.

PIROTIN, S. D., Berg, B. A., and Witmer, E. A. (1976), "PETROS 4: New Developments and Program Manual for the Finite-Difference Calculation of Large Elastic-Plastic, and/or Viscoelastic Transient Deformations of Multilayer Variable-Thickness (1) Hard-Bonded, (2) Moderately-Thick Hard-Bonded, or (3) Thin Soft-Bonded Shells," BRL Contract Report No. 316, MIT (September 1976).

PITTMAN, J. F. (1972a), "Blast and Fragment Hazards from Bursting High Pressure Tanks," NOLTR 72-102 (May 1972).

PITTMAN, J. F. (1972b), "Pressures, Fragments, and Damage from Bursting Pressure Tanks," Minutes of the Fourteenth Explosives Safety Seminar, New Orleans, Louisiana, Department of Defense Explosives Safety Board, pp. 1117-1138 (November 8-10, 1972).

PITTMAN, J. F. (1976), "Blast and Fragments from Superpressure Vessel Rupture," NSWC/WOL/TR 75-87, Naval Surface Weapons Center, White Oak, Maryland (February 1976).

PLASTIC DESIGN IN STEEL, A GUIDE AND COMMENTARY (1971), American Society of Civil Engineers, Manuals and Reports on Engineering Practice, No. 41.

PORZEL, F. B. (1972), "Introduction to a Unified Theory of Explosions," (UTE), NOLTR 72-209, Naval Ordnance Laboratory, White Oak, Silver Spring, Maryland, AD 758 000 (September 14, 1972).

POTTER, E. A., JR. (1960), "Flame Quenching," Progress in Combustion Science and Technology, 1, (editors) Ducarme, Gerstein, and Lefebvre, Pergamon Press, New York, New York.

PRICE. D. J. and Brown, H. H. (1922), "Dust Explosions: Theory and Nature of Phenomena, Causes and Methods of Prevention," NFPA, Boston, Massachusetts.

PROCTOR, J. F. and Filler, W. S. (1972), "A Computerized Technique for Blast Loads from Confined Explosions," Minutes of the Fourteenth Annual Explosives Safety Seminar, New Orleans, Louisiana, pp. 99-124 (November 8-10, 1972).

PRUGH, R. W. (1980), "Applications of Fault Tree Analysis," Chemical Engineering Progress, 76, No. 7, pp. 59-67.

PRZEMIENIECKI, J. S. (1968), Theory of Matrix Structural Analysis, McGraw-Hill, Inc., New York, New York.

PTB (1967), "Correlation of Maximum Experimental Safe Gaps (M.E.S.G.) with Minimum Igniting Current (M.E.C.)," Comments on IEC-Document 31, United Kingdom, 22, Physikalisch-Technische Bundesanstalt (PTB), Braunschweig (December 31, 1967).

PUTNAM, A. A. (1971), Combustion Driven Oscillations in Industry, Elsevier, New York, New York.

RABIE, R. L., Fowles, G. R., and Fickett, W. (1979), "The Polymorphic Detonation," Physics of Fluids, 22, pp. 222-235.

RAE, D. (1973), "Initiation of Weak Coal-Dust Explosions in Long Galleries and the Importance of the Time Dependence of the Explosion Pressure," <u>Fourteenth Symposium (International) on Combustion</u>, The Combustion Institute, Pittsburgh, Pennsylvania, pp. 1225-1234.

RAFTERY, M. M. (1975), "Explosibility Tests for Industrial Dusts," Fire Research Technical Paper No. 21, Her Majesty's Stationery Office, London, England.

RAGLAND, K. W., Dabora, E. K., and Nicholls, J. A. (1968), "Observed Structure of Spray Detonations," <u>Physics of Fluids</u>, <u>11</u>, pp. 2377-2388.

RAGLAND, K. W. and Nicholls, J. A. (1969), <u>AIAA Journal</u>, <u>7</u>, 5, pp. 859-863 (May 1969).

RAJ, P. P. K. and Emmons, H. W. (1975), "On the Burning of a Large Flammable Vapor Cloud," Paper Presented at the Joint Meeting of the Western and Central States Sections of the Combustion Institute, San Antonio, Texas (April 21-22, 1975).

RAJU, M. S. (1981), "The Blast Wave from Axisymmetric Unconfined Vapor Cloud Explosions," Ph.D. Thesis, University of Illinois at Urbana, Champaign, Illinois (in press).

RAKACZKY, J. A. (1975), "The Suppression of Thermal Hazards from Explosions of Munitions: A Literature Survey," BRL Interim Memorandum Report No. 377, Aberdeen Proving Ground, Maryland (May 1975).

RAMSHAW, J. D. and Dukowicz, J. K. (1979), "APACHE: A Generalized Mesh Eulerian Computer Code for Multicomponent Chemically Reactive Fluid Flow," Los Alamos Scientific Laboratory, Report No. LA-7427 (January 1979).

RAO, C. S., Sichel, M., and Nicholls, J. A. (1972), <u>Combustion Science and Technology</u>, <u>4</u>, pp. 209-220.

RASBASH, D. J. (1970), "Fire Protection Engineering with Particular Reference to Chemical Engineering," <u>The Chemical Engineer</u>, <u>243</u> (November 1970).

RASBASH, D. J. (1973), "Present and Future Design Philosophy for Fire and Explosion Hazards in the Chemical Industry," Fire Prevention Science and Technology, 8 [Text of Lecture Given by Professor Rasbash in June 1973 at the Imperial College/FPA Summer School on "Aspects of Design Relating to Fire and Explosion Hazards in the Chemical Industry"].

RASBASH, D. J. and Rogowski, . . (1961), "Explosion Relief," Institution of Chemical Engineering, 1, pp. 58-68.

RASBASH, D. J. and Rogowski, . . (1964), "Explosion Relief," Institution of Chemical Engineering, pp. 21-28.

RAYLEIGH, J. W. S. (1878a), The Theory of Sound, Volume II, Dover Publications, pp. 109-114.

RAYLEIGH, J. W. S. (1978b), Nature, 18, pp. 319-321 (July 18, 1878).

RECHT, R. F. (1970), "Containing Ballistic Fragments," Engineering Solids Under Pressure, H. Pugh, Editor, Papers Presented at the Third International Conference on High Pressure, Aviemore, Scotland, pp. 51-60.

REED, J. W. (1968), "Evaluation of Window Pane Damage Intensity in San Antonio Resulting from Medina Facility Explosion on November 13, 1963," Annals of the New York Academy of Science, 152, I, pp. 565-584.

REED, J. W. (1973), "Distant Blast Predictions for Explosions," Minutes of the Fifteenth Explosives Safety Seminar, Department of Defense Explosives Safety Board, Washington, D.C., Volume II, pp. 1403-1424.

REID, R. C. (1976), "Superheated Liquids," American Scientist, 64, pp. 146-156 (March-April 1976).

REID, R. C. (1977), "Superheated Liquids. A Laboratory Curiosity and, Possibly, an Industrial Curse. Part 1: Laboratory Studies and Theory," Chemical Engineering Education, pp. 60-87 (Spring 1978).

REISLER, R. C. (1973), "Explosive Yield Criteria," Minutes of the Fourteenth Explosives Safety Seminar, New Orleans, Louisiana, Department of Defense Explosives Safety Board, pp. 271-288 (November 8-10, 1972).

REMPEL, J. R. and Wiehle, C. K. (1978), "Collateral Air Blast Damage," DNA 4609Z, Interim Report for Period May 1977-March 1978, SRI International, Prepared for Defense Nuclear Agency, Washington, D.C. (April 1978).

RICHMOND, D. R., Damon, E. G., Fletcher, E. R., Bowen, I. G., and White, C. S. (1968), "The Relationship Between Selected Blast Wave Parameters and the Response of Mammals Exposed to Air Blast," Annals of the New York Academy of Sciences, 152, Article 1, pp. 103-121 (October 1968).

RICHMOND, J. K. and Liebman, I. (1975), "A Physical Description of Coal Mine Explosions," Fifteenth Symposium (International) on Combustion, The Combustion Institute, Pittsburgh, Pennsylvania, pp. 115-126.

RICHMOND, J. K., Liebman, I., Bruszak, A. E., and Miller, L. F. (1979), "A Physical Description of Coal Mine Explosions. Part II," Seventeenth Symposium (International) on Combustion, The Combustion Institute, Pittsburgh, Pennsylvania, pp. 1257-1268.

RICKER, R. E. (1975), "Blast Waves from Bursting Pressurized Spheres," Department of Aeronautical and Astronautical Engineering, Master of Science Thesis, University of Illinois at Urbana, Champaign, Illinois (May 1975).

RINEHART, J. S. (1975), Stress Transients in Solids, Hyper-Dynamics, Santa Fe, New Mexico.

RISK ANALYSIS (1981), An International Journal, Society of Risk Analysis, R. B. Cummings, Editor, Oak Ridge National Laboratory, Tennessee, Plenum Press.

ROACHE, P. J. (1972), Computational Fluid Dynamics, Hermosa Publishers, Albuquerque, New Mexico.

ROBINSON, C. A., JR. (1973), "Special Report: Fuel Air Explosives. Services Ready Joint Development Plan," Aviation Week and Space Technology, pp. 42-46 (February 19, 1973).

ROBINSON, C. S. (1944), Explosions, Their Anatomy and Destructiveness, McGraw-Hill Book Company, New York, New York.

782

ROSS, R., et al. (1967), "Criteria for Assessing Hearing Damage Risk from Impulse-Noise Exposure," Human Engineering Laboratory, Aberdeen Proving Ground, Maryland, AD 666206 (August 1967).

ROTZ, J. V. (1975), "Results of Missile Impact Tests on Reinforced Concrete Panels," Second ASCE Specialty Conference on Structural Design of Nuclear Plant Facilities, 1-A, New Orleans, Louisiana, pp. 720-738 (December 1975).

RUDINGER, G. (1955), Wave Diagrams for Nonsteady Flow in Ducts, D. van Nostrand.

RUNES, E. (1972), "Explosion Venting," Loss Prevention, 6, pp. 63-67.

RUSSELL, W. W. (1976), "Special Hazards with Hydrocarbons," Symposium on Prevention of Explosions, Fires, and Injuries in the Gas and Petroleum Industries, Institute of Gas Technology and Illinois Institute of Technology, Chicago, Illinois (July 19 - August 5, 1976).

SACHS, R. G. (1944), "The Dependence of Blast on Ambient Pressure and Temperature," BRL Report 466, Aberdeen Proving Ground, Maryland.

SADEE, C., Samuels, D. E., and O'Brien, T. P. (1977), "The Characteristics of the Explosion of Cyclohexane at the Nypro (UK) Flixborough Plant on July 1, 1974," Journal of Occupational Accidents, 1, pp. 203-235.

SAKURAI, A. (1965), "Blast Wave Theory," Basic Developments in Fluid Mechanics, 1, Morris Holt, Editor, Academic Press, New York, New York, pp. 309-375.

SAPKO, M. J., Furno, A. L., and Kuchta, J. M. (1976), "Flame and Pressure Development of Large Scale CH_4-Air-N_2 Explosions," Bureau of Mines Report of Investigation 8176, U. S. Department of the Interior.

SAX, N. I. (1965), Dangerous Properties of Industrial Materials, Rheinhold Publishing Company, New York, New York.

SCHMITT, J. A. (1979), "A New Internal Energy Calculation for the HELP Code and Its Implications to Conical Shaped Charge Simulations," Ballistic Research Laboratory, Technical Report ARBRL-TR-02168 (June 1979).

SCHNEIDER, A. L., Lind, C. D., and Parnarouskis, M. C. (1979), "U. S. Coast Guard Liquefied Natural Gas Research at China Lake," Draft Copy, Office of Merchant Marine Safety, U. S. Coast Guard Headquarters, Washington, D.C.

SCHNEIDMAN, D. and Strobel, L. (1974), "Indiana Gas Leak Capped; 1700 Return," Chicago Daily Tribune (Sunday, September 15, 1974).

SCHURING, D. J. (1977), Scale Models in Engineering, Fundamentals and Applications, Pergamon Press, Oxford, England.

SCHWARTZ, R. and Keller, M. (1977), "Environmental Factors Versus Flare Application," Chemical Engineering Progress, 73, No. 9, pp. 41 (September 1977).

SEERY, D. J. and Bowman, C. T. (1970), "An Experimental and Analytical Study of Methane Oxidation Behind Shock Waves," Combustion and Flame, 14, pp. 37-47.

SENIOR, M. (1974), "Gas Explosions in Buildings, Part IV, Strain Measurements on the Gas Explosion Chamber," Fire Research Note No. 987, Fire Research Station (March 1974).

SETCHKIN, N. (1954), "Self-Ignition Temperatures of Combustible Liquids," J. Res. of National Bureau of Standards, 53, pp. 49-66.

SETTLES, J. E. (1968), "Deficiencies in the Testing and Classification of Dangerous Materials," Annals of the New York Academy of Sciences, 152, Article 1, pp. 199-205 (October 1968).

SEWELL, R. G. S. and Kinney, G. F. (1968), "Response of Structures to Blast: A New Criterion," Annals of the New York Academy of Sciences, 152, I, pp. 532-547.

SEWELL, R. G. S. and Kinney, G. F. (1974), "Internal Explosions in Vented and Unvented Chambers," Minutes of the Fourteenth Explosives Safety Seminar, New Orleans, Louisiana, Department of Defense Explosives Safety Board, pp. 87-98 (November 8-10, 1972).

SHEAR, R. E. (1964), "Incident and Reflected Blast Pressures for Pentolite,"
BRL Report No. 1262, Aberdeen Proving Ground, Maryland (September
1964).

SHEAR, R. E. and Arbuckle, A. L. (1971), "Calculated Explosive and Blast
Properties for Selected Explosives," Annex of the Explosive Steering
Committee, NOL 61-JTCG/ME-70-10.

SHEAR, R. E. and Wright, E. (1962), "Calculated Peak Pressure -- Distance
Curves for Pentolite and TNT," BRL Memorandum Report No. 1423, Aberdeen Proving Ground, Maryland (August 1962).

"SHELTER DESIGN AND ANALYSIS, VOLUME 4, PROTECTIVE CONSTRUCTION FOR SHEL-
TERS," (1972), Defense Civil Preparedness Agency, TR-20, Reviewer's
Draft (July 1972).

SHIMPI, S. A. (1978), "The Blast Waves Produced by Bursting Spheres with
Simultaneous or Delayed Explosion or Implosion of the Contents,"
Ph.D. Thesis, University of Illinois at Urbana-Champaign.

SCHUMACHER, R. N., Kingery, C. N., and Ewing, W. O., Jr. (1976), "Airblast
and Structural Response Testing of a 1/4 Scale Category 1 Suppressive Shield," BRL Memorandum Report No. 2623, Aberdeen Proving
Ground, Maryland (May 1976).

SICHEL, M. and Foster, J. C. (1979), "The Ground Impulse Generated by a
Plane Fuel-Air Explosion with Side Relief," Acta Astronautica, 6,
pp. 243-256.

SICHEL, M. and Hu, C. (1973), "The Impulse Generated by Blast Waves Propa-
gating Through Combustible Mixtures," Paper VIII, Proceedings of the
Conference on Mechanisms of Explosions and Blast Waves, J. Alstor,
Editor, Sponsored by the Joint Technical Coordinating Group for Air-
Launched Non-Nuclear Ordnance Working Party for Explosives (Novem-
ber 1973).

SICHEL, M., Rao, C. S., and Nicholls, J. A. (1971), Thirteenth Symposium
(International) on Combustion, The Combustion Institute, Pittsburgh,
Pennsylvania, pp. 1141-1149.

SIEWERT, R. D. (1972), "Evacuation Areas for Transportation Accidents In-
    volving Propellant Tank Pressure Bursts," NASA Technical Memorandum
    X-68277.

SIMMONDS, W. A. and Cubbage, P. A. (1960), "Explosion Relief," First Sym-
    posium on Chemical Process Hazards, I. Chem. E., pp. 69-77.

SIMON, H. and Thomson, S. J. (   ), "Relief System Optimization," Loss
    Prevention.

SINGH, J. (1978), "Identification of Needs and Sizing of Vents for Pressure
    Relief," Insurance Technical Bureau, Internal Report No. R 037,
    Petty, France, London, England (February 1978).

SINGH, J. (1979), "The Sizing of Relief for Exothermic Reactions," Paper
    Presented at I. Chem. E. Symposium, Chester, England (March 1979).

SISKIND, D. E. (1973), "Ground and Air Vibrations from Blasting," Subsec-
    tion 11.8 SME and Mining Engineering Handbook, A. B. Cummings and I.
    A. Given, Editors, Soc. of Min. Eng. of the Am. Inst. of Min. Metal-
    lur. and Pet. Eng. Inc., New York, VI, pp. 11-99 to 11-112.

SISKIND, D. E. and Summers, C. R. (1974), "Blast Noise Standards and In-
    strumentation," Bureau of Mines Environmental Research Program,
    Technical Progress Report 78, U. S. Department of the Interior (May
    1974).

SLOVIC, P., Fischhoff, B., and Lichtenstein, S. (1979), "Rating the Risks,"
    Chem. Tech., 9, pp. 738-744.

SMITH, T. L. (1959), "Explosion in Wind Tunnel Air Line," BRL Memorandum
    Report 1235, Ballistic Research Laboratory, Aberdeen Proving Ground,
    Maryland (September 1959).

SMOOT, L. D., Hecker, W. C., and Williams, G. A. (1976), "Prediction of
    Propagating Methane Air Flames," Combustion and Flame, 26, pp. 323-
    342.

SOLBERG, D. M., Pappas, J. A., and Skramstad, E. (1981), "Observation of
    Flame Instabilities in Large Scale Vented Gas Explosions," Eigh-

teenth International Symposium on Combustion, The Combustion Institute, Pittsburgh, Pennsylvania, pp. 1607-1614.

SOLBERG, D. M., Skramstad, E., and Pappas, J. A. (1979), "Experimental Investigations on Partly Confined Gas Explosions. Analysis of Pressure Loads, Part I," Det norske Veritas, Research Division, Report No. 79-0483, NTMF Project 1830 6500 (October 1975).

SORENSON, S. C., Savage, L. D., and Strehlow, R. A. (1975), "Flammability Limits - A New Technique," Combustion and Flame, 24, pp. 347-355.

SPERRAZZA, J. (1951), "Dependence of External Blast Damage to A25 Aircraft on Peak Pressure and Impulse," BRL Memorandum Report 575, AD 378275 (September 1951).

SPERRAZZA, J. (1963), "Modeling of Air Blast," Use of Modeling and Scaling in Shock and Vibration, W. E. Baker, Editor, ASME, New York, pp. 65-78 (November 1963).

SPERRAZZA, J. and Kokinakis, W. (1967), "Ballistic Limits of Tissue and Clothing," Technical Note No. 1645, U. S. Army Ballistic Research Laboratory, RDT&E Project No. 1P025601A027 (January 1967).

STARLING, K. E. (1973), "Fluid Thermodynamic Properties for Light Petroleum Systems," Gulf Publishing Company, Houston, Texas.

STEA, W., Dobbs, N., and Sock, F. E. (1981), "Shear Response of One- and Two-Way Elements Subjected to Blast Loads," Civil Engineering Laboratory, Report No. CR 31-010 (March 1981).

STEA, W., Tseng, G., Kossover, D., Price, P., and Caltagirone, J. (1977), "Nonlinear Analysis of Frame Structures Subjected to Blast Overpressures," Contractor Report ARLCD-CR-77008, U. S. Army Armament Research and Development Command, Large Caliber Weapon Systems Laboratory, Dover, New Jersey (May 1977).

STEIN, L. R., Gentry, R. A., and Hirt, C. W. (1977), "Computational Simulation of Transient Blast Loading on Three-Dimensional Structures," Computer Methods in Applied Mechanics and Engineering, 11, pp. 57-74.

STEPHENSON, A. E. (1976), "Full-Scale Tornado-Missile Impact Tests," EPRI-NP-148, Project 399, Interim Report, Electric Power Research Institute, Palo Alto, California (April 1976).

STEPHENSON, A. E., Sliter, G., and Burdette, E. (1975), "Full-Scale Tornado-Missile Impact Tests Using A Rocket Launcher," Second ASCE Specialty Conference on Structural Design of Nuclear Plant Facilities, 1-A, New Orleans, Louisiana, pp. 611-636 (December 1975).

STOKES, G. G. (1849), "On Some Points on the Received Theory of Sound," Phil. Mag., 34 (3), p. 52.

STRATTON, W. R., Colvin, T. H., and Lazarus, R. B. (1958), "Analysis of Prompt Excursions in Simple Systems and Idealized Fast Reactors," Paper P/431, Second United Nations International Conference on the Peaceful Uses of Atomic Energy (June 1958).

STREHLOW, R. A. (1968a), Fundamentals of Combustion, International Textbook Company, Scranton, Pennsylvania [Reprinted by Robert E. Krieger Publishing Company, Box 542, Huntington, New York, 11743].

STREHLOW, R. A. (1968b), "Gas Phase Detonation - Recent Developments," Combustion and Flame, 12, pp. 81-101.

STREHLOW, R. A. (1973a), "Equivalent Explosive Yield of the Explosion in the Alton and Southern Gateway, East St. Louis, Illinois, January 22, 1972," AAE TR 73-3, Department of Aeronautical and Astronautical Engineering, University of Illinois, Urbana, Illinois (June 1973).

STREHLOW, R. A. (1973b), "Unconfined Vapor-Cloud Explosions -- An Overview," Fourteenth Symposium (International) on Combustion, The Combustion Institute, Pittsburgh, Pennsylvania, pp. 1189-1200.

STREHLOW, R. A. (1975), "Blast Waves Generated by Constant Velocity Flames -- A Simplified Approach," Combustion and Flame, 24, pp. 257-261.

STREHLOW, R. A. (1976), "Blast Waves from Non-Ideal Sources," Seventeenth Department of Defense Explosives Safety Seminar, Denver, Colorado.

STREHLOW, R. A. (1980a), "The Blast Wave From Deflagrative Explosions, an Acoustic Approach," Paper Presented at the Fourteenth Loss Prevention Symposium, AIChE, Philadelphia, Pennsylvania (June 8-12, 1980).

STREHLOW, R. A. (1980b), "Accidental Explosions," American Scientist, pp. 420-428 (July-August 1980).

STREHLOW, R. A. (1981), Loss Prevention (AIChE), 14, pp. 145-153.

STREHLOW, R. A. (1982), private communication.

STREHLOW, R. A. and Baker, W. E. (1976), "The Characterization and Evaluation of Accidental Explosions," Progress in Energy and Combustion Science, 2, 1, pp. 27-60 [Also NASA CR 134779 (1975)].

STREHLOW, R. A., Crooker, A. J., and Cusey, R. E. (1967), Combustion and Flame, 11, p. 339.

STREHLOW, R. A., Luckritz, R. T., Adamczyk, A. A., and Shimpi, S. A. (1979), "The Blast Wave Generated by Spherical Flames," Combustion and Flame, 35, pp. 297-310.

STREHLOW, R. A., Nicholls, J. S., Schram, P., and Magison, E. (1979), "An Investigation of the Maximum Experimental Safe Gap Anomaly," Journal of Hazardous Materials, 3, pp. 1-15.

STREHLOW, R. A. and Ricker, R. E. (1976), "The Blast Wave from a Bursting Sphere," AIChE, 10, pp. 115-121.

STREHLOW, R. A. and Shimpi, S. A. (1978), "The Blast Waves Produced by Bursting Spheres with Simultaneous or Delayed Explosion or Implosion of the Contents," Final Report, Part A, Prepared for Der Bundesminister des Innern Die Bundesrepublic Deutschland, Contract SR-69, University of Illinois at Urbana-Champaign (November 1978).

STRICKLIN, J. A. and Haisler, W. E. (1974), "Survey of Solution Procedures for Nonlinear Static and Dynamic Analyses," Proceedings of SAE International Conference on Vehicle Structural Mechanics, Detroit, Michigan (March 26-28, 1974).

STRUCTURES TO RESIST THE EFFECTS OF ACCIDENTAL EXPLOSIONS (1969), Department of the Army Technical Manual TM 5-1300, Department of the Navy Publication NAVFAC P-397, Department of the Air Force Manual AFM

88-22, Department of the Army, the Navy, and the Air Force (June 1969).

STULL, D. R. (1977), "Fundamentals of Fire and Explosion," AIChE Monograph Series, 73, (10).

STULL, D. R. and Prophet, H. (1972), JANNAF Thermochemical Tables, Second Edition, NSRDS-NBS 37, National Bureau of Standards (June 1971) [They may be purchased from the Superintendent of Documents, U. S. Government Printing Office, Washington, D.C. 20402 (Catalog No. C 13.48:37). Supplement tables containing correction tables and new species tables appear at irregular intervals in the Journal of Physical and Chemical Reference Data, published by the American Chemical Society and the American Institute of Physics for the National Bureau of Standards].

STULL, D. R., Westrum, E. F., and Sinke, . . (1969), The Chemical Thermodynamics of Organic Compounds, John Wiley and Sons, Inc., New York, New York.

SUPPRESSIVE SHIELDS STRUCTURAL DESIGN AND ANALYSIS HANDBOOK (1977), HNDM-1110-1-2, U. S. Army Corps of Engineers, Huntsville Division, Huntsville, Alabama (November 1977).

SUTHERLAND, L. C. (1974), "A Simplified Method for Estimating the Approximate TNT Equivalent From Liquid Propellant Explosions," Minutes of the Fifteenth Explosives Safety Seminar, Department of Defense Explosives Safety Board, Washington, D.C., Volume II, pp. 1273-1277.

SUTHERLAND, L. C. (1978), "Scaling Law for Estimating Liquid Propellant Explosive Yields," Journal of Spacecraft, 15, No. 2, pp. 124-125.

SUTHERLAND, M. E. and Wegert, H. W. (1973), "An Acetylene Decomposition Incident," Loss Prevention, 7, pp. 99-103.

SWEGLE, J. W. (1978), "TOODY IV - A Computer Program for Two-Dimensional Wave Propagation," Sandia Report SAND-78-0552 (September 1978).

SWISDAK, M. M., JR. (1975), "Explosion Effects and Properties: Part I - Explosion Effects in Air," NSWC/WOL/TR 75-116, Naval Surface Weapons Center, White Oak, Silver Spring, Maryland (October 1975).

TAYLOR, D. B. and Price, C. F. (1971), "Velocities of Fragment From Bursting Gas Reservoirs," ASME Transactions, Journal of Engineering for Industry, 93B, pp. 981-985.

TAYLOR, G. I. (1946), "The Air Wave Surrounding an Expanding Sphere," Proceedings Roy. Soc., A186, pp. 273-292.

TAYLOR, G. I. (1950), "The Formation of a Blast Wave by a Very Intense Explosion," Proceedings Roy. Soc., A201, p. 159.

TAYLOR, N. and Alexander, S. J. (1974), "Structural Damage in Buildings Caused by Gaseous Explosions and Other Accidental Loadings," Building Research Establishment Current Paper CP 45/74, Building Research Establishment (March 1974).

THOMAS, A. and Williams, G. T. (1966), "Flame Noise: Sound Emission from Spark Ignited Bubble of Combustible Gas," Proceedings Roy. Soc., A294, pp. 449-466.

THOMPSON, S. L. (1975), "CSQ - A Two-Dimensional Hydrodynamic Program with Energy Flow and Material Strength," Sandia National Laboratories Report SAND 74-0122 (August 1975).

THORNHILL, C. K. (1960), "Explosions in Air," ARDE Memo (B) 57/60, Armament Research and Development Establishment, England.

TILLERSON, J. R. (1975), The Shock and Vibration Digest, 7, No. 4 (April 1975).

TIMOSHENKO, S. (1928), Vibration Problems in Engineering, D. Van Nostrand Company, Inc., New York, New York.

TING, R. M. L. (1975), "Non-Composite and Composite Steel Panels for Tornado Missile Barrier Walls," Second ASCE Specialty Conference on Structural Design of Nuclear Plant Facilities, 1-A, New Orleans, Louisiana, pp. 663-687 (December 1975).

TONKIN, P. S. and Berlemont, C. F. J. (1974), "Gas Explosions in Buildings, Part I, Experimental Explosion Chamber," Fire Research Note No. 984, Fire Research Station (February 1974).

TOONG, T. Y. (1982), _Combustion and Flame_ (in press).

TOWNSEND, D. I. (1977), "Hazard Evaluation of Self-Accelerating Reactions," _Chemical Engineering Progress_, pp. 80-81 (September 1977).

TSANG, W. and Domalski, E. S. (1974), "An Appraisal of Methods for Estimating Self Reaction Hazards," NBSIR 74-551, DOT Report No. TES-20-74-8 (June 1974).

TSATSARONIS, G. (1978), "Prediction of Propagating Laminar Flames in Methane, Oxygen, Nitrogen Mixtures," _Combustion and Flame_, 33, pp. 217-240.

TSENG, G., Weissman, S., Dobbs, N., and Price, P. (1975), "Design Charts for Cold-Formed Steel Panels and Wide-Flange Beams Subjected to Blast Loading," Technical Report 4838, Picatinny Arsenal, Dover, New Jersey (August 1975).

TUCK, C. A., JR. (EDITOR) (1976), "NFPA Inspection Manual," Fourth Edition, National Fire Protection Association, Boston, Massachusetts.

TUCKER, D. M. (1975), "The Explosion and Fire at Nypro (UK), Ltd., Flixborough, on 1 June 1974," Building Research Establishment, Fire Research Station, Borehamwood, Hertfordshire, England (July 1975).

TUCKER, D. M. (1976), "The Explosion and Fire at Nypro (UK), Ltd., Flixborough, on 1 June 1974," Department of the Environment, Building Research Establishment Note N 60/76, Borehamwood, Hertfordshire, England.

U. S. ARMY CORPS OF ENGINEERS MANUAL EM 1110-345-415 (1975), "Design of Structures to Resist the Effects of Atomic Weapons."

U. S. ARMY MATERIEL COMMAND (1972), _Engineering Design Handbook: Principles of Explosive Behavior_," AMC Pamphlet AMCP 706-180.

U. S. ATOMIC ENERGY COMMISSION (1966), "The Study of Missiles Resulting from Accidental Explosions," Safety and Fire Protection Bulletin 10, Division of Operational Safety, Washington, D.C. (March 1966).

U. S. COAST GUARD (1974), "S.S. V. A. Fogg: Sinking in the Gulf of Mexico on 1 February 1972, with Loss of Life," USCG/NTSB-MAR-74-8 (November 21, 1974).

U. S. COAST GUARD (1977), "Marine Casualty Report, S.S. Sansinena (Liberian): Explosion and Fire in Los Angeles Harbor, California, on 17 December 1976, with Loss of Life," Report No. USCG 16732/71895.

UTRIEW, P. A., Laderman, A. J., and Oppenheim, A. K. (1965), "Dynamics of the Generation of Pressure Waves by Accelerating Flames," Tenth Symposium on Combustion, pp. 797-804.

UTRIEW, P. A. and Oppenheim, A. K. (1967), "Detonation Initiation by Shock Merging," Eleventh Combustion Symposium, pp. 665-676.

VADALA, A. J. (1930), "Effects of Gun Explosions on the Ear and Hearing Mechanism," Milit. Surg., 66, pp. 710-822.

VAN BUIJTENEN, C. J. P. (1980), "Calculation of the Amount of Gas in the Explosive Region of a Vapour Cloud Released in the Atmosphere," J. Hazardous Materials, 3, pp. 201-220.

VAN DOLAH, R. W. and Burgess, D. S. (1970), "Explosion Problems in the Chemical Industry," The American Chemical Society.

VASSALLO, F. A. (1975), "Missile Impact Testing of Reinforced Concrete Panels," Calspan Report No. HC-5609-D-1, Final Report, Buffalo, New York (January 1975).

VINCENT, G. C. (1971), "Rupture of a Nitroaniline Reactor," Loss Prevention, 5, pp. 46-52.

VON GIERKE, H. E. (1967), "Mechanical Behavior of Biological Systems," Aerospace Medical Research Laboratory, Wright-Patterson Air Force Base, Ohio, AD 758963 (September 1967).

VON GIERKE, H. E. (1971a), "Biodynamic Models and Their Applications," Aerospace Medical Research Laboratory, Wright-Patterson Air Force Base, Ohio, AD 736985.

VON GIERKE, H. E. (1971b), "On the Dynamics of Some Head Injury Mechanisms," Aerospace Medical Research Laboratory, Wright-Patterson Air Force Base, Ohio, AD 728885.

VON GIERKE, H. E. (1971c), "Man to Model, or Model to Man," Aerospace Medical Research Laboratory, Wright-Patterson Air Force Base, Ohio, AD 771670.

VON GIERKE, H. E. (1973), "Dynamic Characteristics of the Human Body," Aerospace Medical Research Laboratory, Wright-Patterson Air Force Base, Ohio, AD 769022.

VON NEUMANN, J. and Goldstine, H. (1955), "Blast Wave Calculation," Communication on Pure and Applied Mathematics, 8, pp. 327-353 [Reprinted in John von Neumann Collected Works, A. H. Taub, Editor, Volume VI, Pergamon Press, New York, New York, pp. 386-412].

WALLS, W. L. (1963), "LP-Gas Tank Truck Accident and Fire, Berlin, New York," National Fire Protection Association Quarterly, 57, pp. 3-8 (July 1963).

WALLS, W. L. (1978), "Just What is a BLEVE?" Fire Journal, 72, (6), pp. 46 (November 1978).

WARREN, A. (1958), "Blast Pressures from Distant Explosions," ARDE Memo 18/58, AD 305 732.

WEAST, R. C. (1979), Handbook of Chemistry and Physics, 60th Edition, CRC Press, West Palm Beach, Florida.

WEATHERFORD, W. D., JR. (1975), "U. S. Army Helicopter Modified Fuel Development Program - Review of Emulsified and Gelled Fuel Studies," Final Report AFLRL No. 69 (June 1975).

WEATHERFORD, W. D., JR. and Schaekel, F. W. (1971), "Emulsified Fuels and Aircraft Safety," AGARD/NATO 37th Propulsion and Energetics Panel Meeting: Aircraft Fuels, Lubricants, and Fire Safety, The Hague, Netherlands (May 10-14, 1971).

WEATHERFORD, W. D., JR. and Wright, B. R. (1975), "Status of Research on Antimist Aircraft Turbine Engine Fuels in the United States," AGARD/NATO 45th Meeting, Propulsion and Energetics Panel: Aircraft Fire Safety, Rome, Italy (April 7-11, 1975).

WEIBULL, H. R. W. (1968), "Pressures Recorded in Partially Closed Chambers at Explosion of TNT Charges," Annals of the New York Academy of Sciences, 152, Article 1, pp. 356-361 (October 1968).

WELLS, C. H. and Cook, T. S. (1976), "EPRI RP 502, Reliability of Steam Turbine Rotors, Task I. Lifetime Prediction Analysis System," SwRI Project No. 02-4448-001, Prepared for Electric Power Research Institute, 3412 Hillview Avenue, Palo Alto, California, 94303 (July 1, 1976).

WELLS, G. L. (1979), "Safety in Plant Design," The Institution of Chemical Engineers, London, England.

WENZEL, A. B. and Bessey, R. L. (1969), "Barricaded and Unbarricaded Blast Measurements," Contract No. DAHC04-69-C-0028, Subcontract 1-OU-431, Southwest Research Institute, San Antonio, Texas.

WENZEL, A. B. and Esparza, E. D. (1972), "Measurements of Pressures and Impulses at Close Distances from Explosive Charges Buried and in Air," Final Report on Contract No. DAAK02-71-C-0393 with U. S. Army MERDC, Ft. Belvoir, Virginia (August 1972).

WESTINE, P. S. (1972), "R-W Plane Analysis for Vulnerability of Targets to Air Blast," The Shock and Vibration Bulletin, Bulletin 42, Part 5, pp. 173-183 (January 1972).

WESTINE, P. S. (1977), "Constrained Secondary Fragment Modeling," SwRI Final Report for U. S. Army Ballistic Research Laboratory (June 1977).

WESTINE, P. S. and Baker, W. E. (1974), "Energy Solutions for Predicting Deformations in Blast Loaded Structures," Minutes of the Sixteenth Annual Explosives Safety Seminar, Department of Defense Safety Board.

WESTINE, P. S. and Baker, W. E. (1975), "Energy Solutions for Predicting Deformations in Blast-Loaded Structures," Edgewood Arsenal Contractor Report EM-CR-76027, Report No. 6 (November 1975).

WESTINE, P. S. and Cox, P. A. (1975), "Additional Energy Solutions for Predicting Structural Deformations," Edgewood Arsenal Contractor Report EM-CR-76031, Report No. 4 (November 1975).

WESTINE, P. S. and Vargas, L. M. (1978), "Design Guide for Armoring Criti-
    cal Aircraft Components to Protect from High Explosive Projectiles,"
    Final Report, Contract No. F33615-77-C-3006, U. S. Air Force Flight
    Dynamics Laboratory, Wright-Patterson Air Force Base, Ohio (August
    1978).

WHEATON, E. L. (1948), Texas City Remembers, Naylor Company, San Antonio,
    Texas.

WHITE, C. S. (1968), "The Scope of Blast and Shock Biology and Problem
    Areas in Relating Physical and Biological Parameters," Annals of the
    New York Academy of Sciences, 152, Article 1, pp. 89-102 (October
    1968).

WHITE, C. S. (1971), "The Nature of the Problems Involved in Estimating the
    Immediate Casualties from Nuclear Explosions," CEX-71.1, Civil Ef-
    fects Study, U. S. Atomic Energy Commission, DR-1886 (July 1971).

WHITE, C. S., Gowen, I. G., Richmond, D. R., and Corsbie, R. L. (1961), "Com-
    parative Nuclear Effect of Biomedical Interest," CEX-58.8, Civil Ef-
    fects Study, U. S. Atomic Energy Commission (January 1961).

WHITE, C. S., Jones, R. K., Damon, E. G., Fletcher, E. R., and Richmond, D.
    R. (1971), "The Biodynamics of Airblast," Technical Report to Defense
    Nuclear Agency, DNA 2738T, Lovelace Foundation for Medical Education
    and Research, AD 734208 (July 1971).

WHITE, M. T. (EDITOR) (1946), "Effects of Impacts and Explosions," Summary
    Technical Report of Division 2, National Defense Research Council,
    Volume I, Washington, D.C., AD 221586.

WHITMAN, G. B. (1950), "The Propagation of Spherical Blast," Proceedings
    Roy. Soc., A203, pp. 571-581.

WIEKEMA, B. J. (1980), "Vapor Cloud Explosion Model," J. Hazardous Materi-
    als, 3, pp. 221-232.

WILKINS, M. L. (1969), "Calculation of Elastic-Plastic Flow," University of
    California, Lawrence Radiation Laboratory, Report UCRL-7322, Revision
    1 (January 1969).

WILLIAMS, A. (1968), "The Mechanism of Combustion of Droplets and Sprays of Liquid Fuels," Oxidation and Combustion Reviews, 3, Tipper Ed., Elsevier, Amsterdam, pp. 1-45.

WILLIAMS, F. A. (1974), "Qualitative Theory of Non-Ideal Explosions," in Phase I. Final Report entitled, Explosion Hazards Associated with Spills of Large Quantities of Hazardous Materials, by D. C. Lind, Naval Weapons Center, China Lake, California, U. S. Coast Guard, Washington, D.C. (August 5, 1974).

WILLOUGHBY, A. B., Wilton, C., and Mansfield, J. (1968a), "Liquid Propellant Explosion Hazards. Final Report - December 1968. Volume I - Technical Documentary Report," AFRPL-TR-68-92, URS-652-35, URS Research Company, Burlingame, California.

WILLOUGHBY, A. B., Wilton, C., and Mansfield, J. (1968b), "Liquid Propellant Explosion Hazards. Final Report - December 1968. Volume II - Test Data," AFRPL-TR-68-92, URS-652-35, URS Research Company, Burlingame, California.

WILLOUGHBY, A. B., Wilton, C., and Mansfield, J. (1968c), "Liquid Propellant Explosion Hazards. Final Report - December 1968. Volume III - Prediction Methods," AFRPL-TR-68-92, URS-652-35, URS Research Company, Burlingame, California.

WILSE, T. (1974), "Fire and Explosions Onboard Ships," Veritas, 80, pp. 12-16 (September 1974).

WILTON, C. and Gabrielson, B. (1972), "House Damage Assessment," Minutes of the Fourteenth Explosives Safety Seminar, New Orleans, Louisiana (November 8-10, 1972).

WILMER, W. W., Wright, B. R., and Weatherford, W. D., Jr. (1974), "Ignition and Flammability Properties of Fire-Safe Fuels," Interim Report, AFLRL No. 39, AD 784281.

WISOTSKI, J. and Snyer, W. H. (1965), "Characteristics of Blast Waves Obtained from Cylindrical High Explosive Charges," University of Denver, Denver Research Institute (November 1965).

WITTE, L. C., Cox, J. E., and Bouvier, J. E. (1970), "The Vapor Explosion,"
J. Metals, 22, 2, pp. 39-44 (February 1970).

WITTE, L. C., Vyas, T. J., and Gelabert, A. A. (1973), "Heat Transfer and
Fragmentation During Molten-Metal/Water Interactions," Journal of
Heat Transfer, Transactions of the ASME, pp. 521-527 (November 1973).

WOOLFOLK, R. W. (1971), "Correlation of Rate of Explosion with Blast Effects for Non-Ideal Explosions," SRI Final Report for Contract No.
0017-69-C-4432, Stanford Research Institute, Menlo Park, California
(January 25, 1971).

WOOLFOLK, R. W. and Ablow, C. M. (1973), "Blast Waves from Non-Ideal Explo-
sions," Paper No. IV, Proceedings of the Conference on Mechanisms of
Explosion and Blast Waves, J. Alstor, Editor, Sponsored by the Joint
Technical Coordinating Group for Air Launched Non-Nuclear Ordnance
Working Party for Explosives (November 1973).

WU, R. W-H. and Witmer, E. A. (1972), "Finite Element Analysis of Large
Transient Elastic-Plastic Deformations of Simple Structures, with
Application to the Engine Rotor Fragment Containment/Deflection Program," NASA CR-120886, ASRL TR 154-4, Aeroelastic and Structures Research Laboratory, MIT, Cambridge, Massachusetts (January 1972).

YAMADA, Y., Kawai, T., and Yoshimura, N. (1968), Proceedings of the Second
Conference on Matrix Methods in Structural Mechanics, AFFDL-TR-68-
150.

YAO, C. (1974), "Explosion Venting of Low-Strength Equipment and Structures," Loss Prevention, 8, AIChE, New York.

YAO, C. and Friedman, R. (1977), "Technical and Cost Proposal Explosion
Venting Test Program for Initially Pressurized Vessels," FMRC Serial
No. 22576P, RC77-P-21 (May 1977).

ZABETAKIS, M. G. (1960), "Explosion of Dephlegmator at Cities Service Oil
Company Refinery, Ponca City, Oklahoma, 1959," Bureau of Mines Report of Investigation 5645.

ZABETAKIS, M. G. (1965), "Flammability Characteristics of Combustible Gases and Vapors," Bulletin 627, Bureau of Mines, U. S. Department of the Interior.

ZAKER, T. A. (1975a), "Computer Program for Predicting Casualties and Damage from Accidental Explosions," Technical Paper No. 11, Department of Defense Explosives Safety Board, AD A012847 (May 1975).

ZAKER, T. A. (1975b), "Fragment and Debris Hazards," Technical Paper No. 12, Department of Defense Explosives Safety Board, AD A013 634 (July 1975).

ZALOSH, R. G. (1980), "Gas Explosion Tests in Room-Size Vented Enclosures," Loss Prevention, 13, pp. 98-110.

ZENKIEWICZ, O. C. (1971), The Finite Element Method in Engineering Science, McGraw-Hill Book Company, London, England.

ZILLIACUS, S., Phyillaier, W. E., and Shorrow, P. K. (1974), "The Response of Clamped Circular Plates to Confined Explosive Loadings," Naval Ship R&D Center Report 3987, NSRDC, Bethesda, Maryland (February 1974).

SUBJECT INDEX

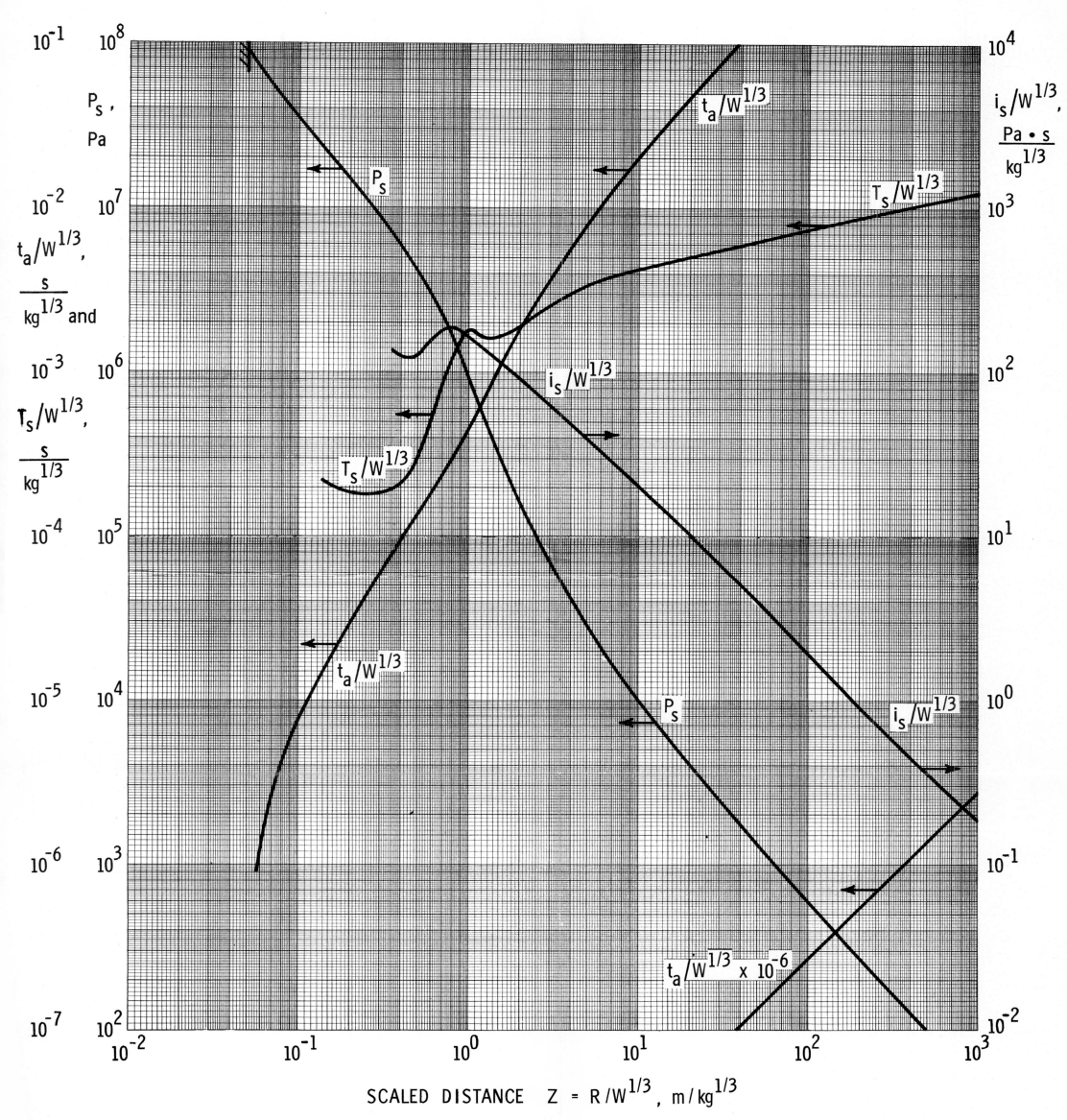

Figure 2-45. Side-On Blast Parameters for TNT

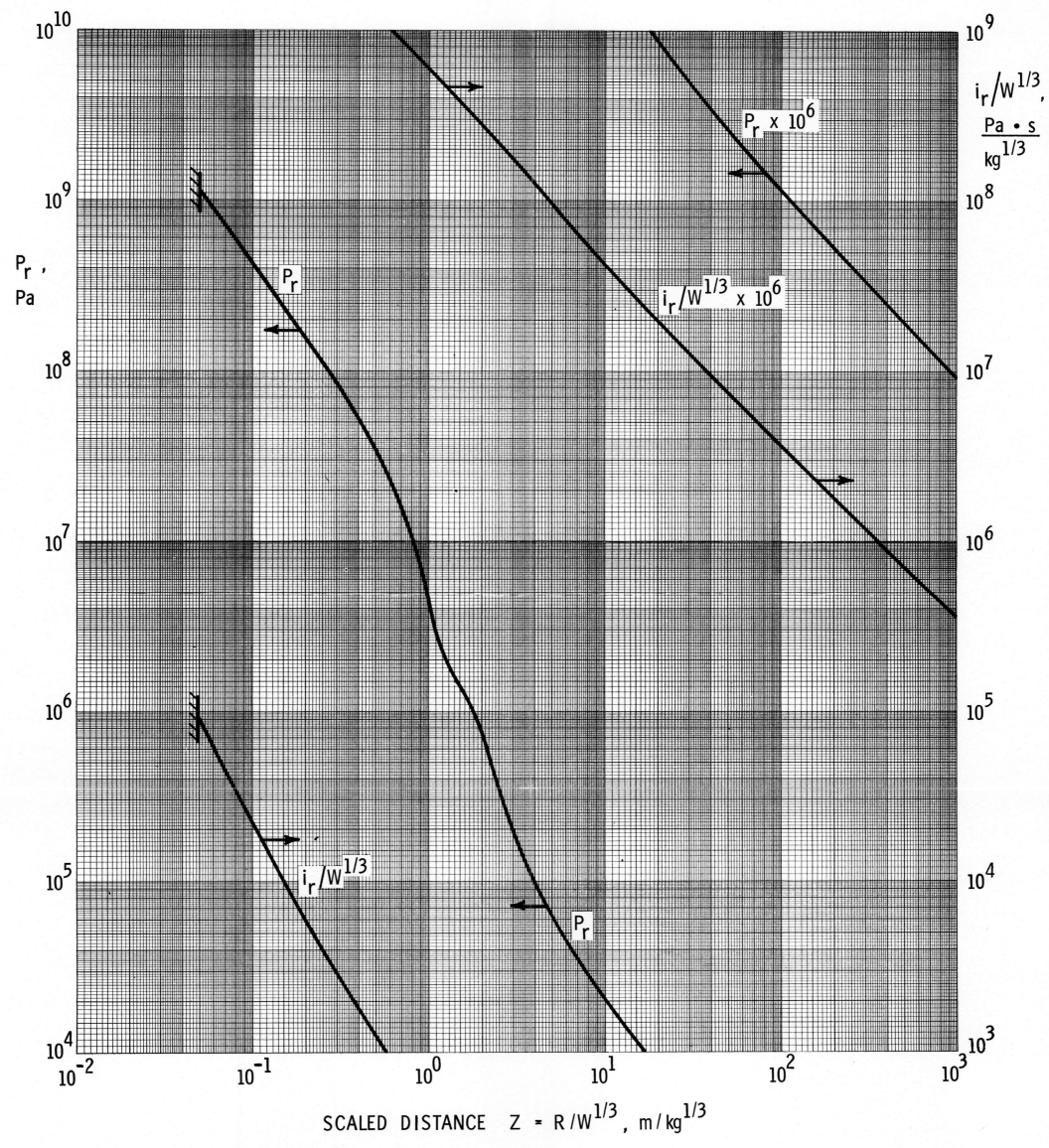

$$i_r/W^{1/3},$$
$$\frac{Pa \cdot s}{kg^{1/3}}$$

$P_r \times 10^6$

P_r,
Pa

P_r

$i_r/W^{1/3} \times 10^6$

$i_r/W^{1/3}$

P_r

SCALED DISTANCE $Z = R/W^{1/3}$, $m/kg^{1/3}$

Figure 2-46. Normally Reflected Blast Parameters for TNT

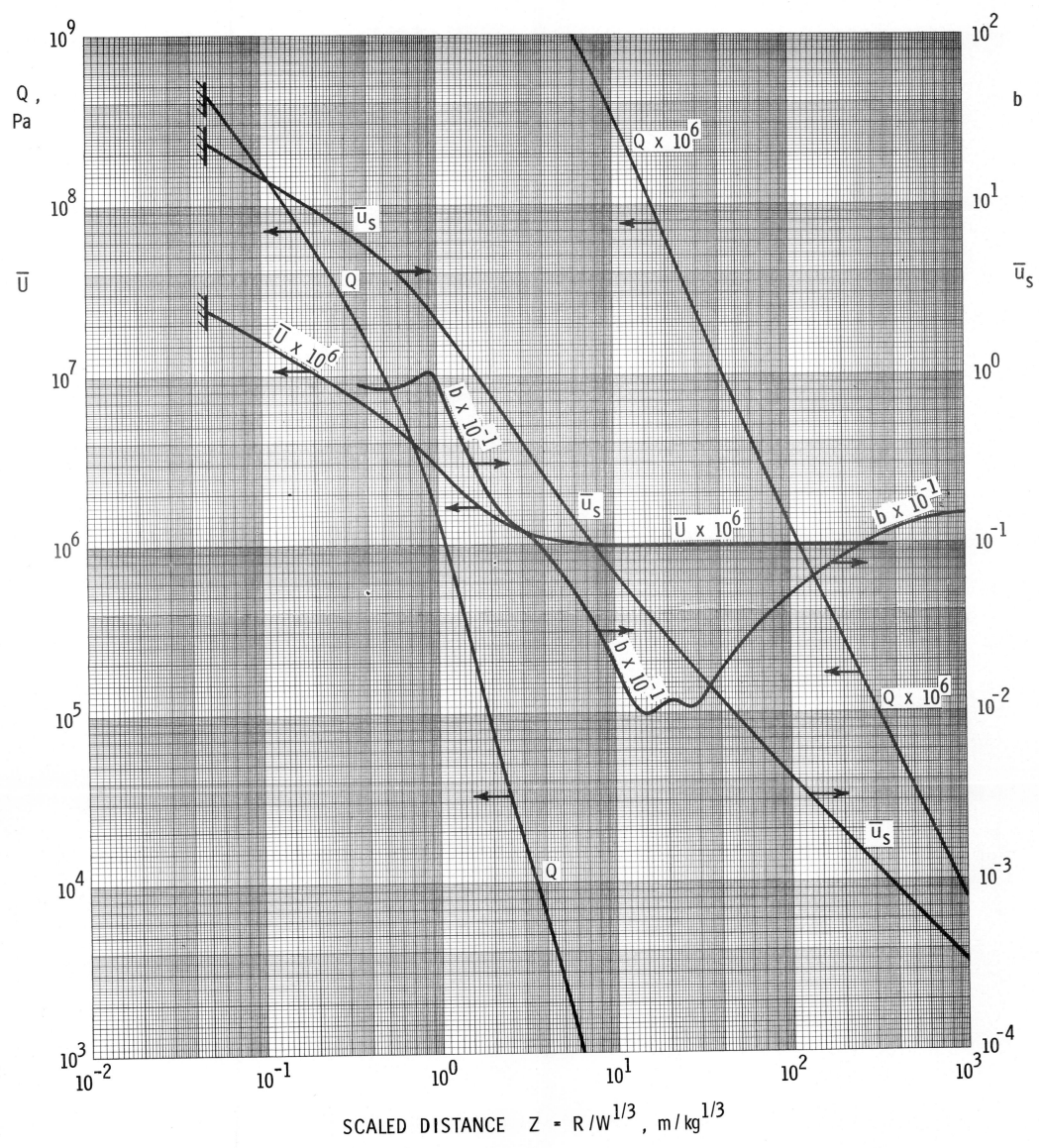

SCALED DISTANCE $Z = R/W^{1/3}$, $m/kg^{1/3}$

Figure 2-47. Additional Side-On Blast Parameters for TNT